TI-82 GRAPHICS CALCULATOR PROCEDURE INDEX

MW00723304

Calculate Poisson probabilities (p. 258)

Generate a random number (p. 287)
- MATH , ▶ , ▶ , ▶ , 1 , ENTER .

Calculate sample statistics (1-Var Stats) (p. 294)

Interpret 1-Var Stats for matched-pair differences (p. 417)

Use 1-Var Stats for a Mann-Whitney confidence interval (p. 423)

Calculate a chi-square test statistic (p. 452)

Calculate the Spearman rank-correlation coefficient (pp. 910–912)
- Enter data in two lists, STAT , ▶ , 5 , ENTER .

Calculate variation among the samples in an ANOVA test (pp. 510–511)

Calculate variation within the samples in an ANOVA test (pp. 511–512)

Calculate the Pearson coefficient of correlation (pp. 530–531)
- Enter data in two lists, STAT , ▶ , 5 , ENTER .

Construct a scatter plot (pp. 530–532)
- Enter the data in two lists, 2nd , Y= , 1 ; use arrow and ENTER keys to activate plot, choose appropriate icon, specify Xlist, Ylist, and choose Mark; press WINDOW to set limits for the display, GRAPH .

Quickly set limits for a graphical display (p. 532)
- ZOOM , 9 .

Calculate the equation of the line of best fit (pp. 547–548)
- Follow procedure to determine the Pearson coefficient of correlation.

Foundations of Statistics

WARREN HAWLEY

The Latin School of Chicago
DePaul University

SAUNDERS COLLEGE PUBLISHING

Harcourt Brace College Publishers

Fort Worth   Philadelphia   San Diego   New York   Orlando
San Antonio   Toronto   Montreal   London   Sydney   Tokyo

This book is dedicated to Kate and Alex—two scientists of the future.

Text Typeface: Galliard
Compositor: Monotype Composition
Executive Editor: Jay Ricci
Developmental Editor: Marc Sherman
Project Editor: Linda Boyle
Copy Editor: Mary Patton
Art Director: Joan Wendt
Cover Designer: Kathryn Needle
Text Designer: Rebecca Lemna
Text Artwork: Monotype Composition
Production Manager: Joanne Cassetti

Printed in the United States of America

FOUNDATIONS OF STATISTICS

ISBN 0-03-098253-7

Library of Congress Catalog Card Number: 95-069129

567890123 039 10 987654321

Preface

In a world increasingly dominated by technology and flooded with statistics, Mark Twain's words are more important than ever. Our future as a society may depend in large measure on our ability to evaluate information, determine truth, and make correct decisions.

This text was written for the college student who is taking a first course in statistics, and whose mathematical expertise does not go beyond a year of college algebra. This is often the group that has become most frustrated with mathematics in general, and one that chooses not to continue its further study. However, I believe the study of statistics is ideal for demonstrating the relevance and importance of mathematics, and has the potential for reaching many students.

The aim of this text is to engage these new statisticians. As a result, this book has many features that set it apart from others. Some of these follow.

CONTENT FEATURES

- The text gives equal coverage to data analysis and inferential statistics. The extra attention given to data analysis provides a practical foundation for using statistics.
- The Technology Tools sections give equal attention to Minitab and the graphics calculator, allowing flexibility in the use of technology.
- The initial treatment of probability is spread over two chapters, one required and one optional. This allows for a more flexible treatment of a topic that nearly all students find difficult.
- The concept of ranking is developed over the first two chapters and is used with graphic methods of examining data well before measures of central tendency are examined. In place of the more traditional approach of beginning by computing measures of central tendency, the central concept is ranking scores, with associated graphic approaches such as box plots with inner and outer fences to identify outliers. These ideas are then developed further and used to compare several sets of data. The student is thus required at the very start to focus on the whole picture—on the set's distribution rather than its mean—and this approach can pay off later in the text when inferential statistics is studied.
- The concept of a confidence interval is first introduced and used in the context of overlapping box plots (Chapter 2), free from the potentially confusing notion of Z-scores. This allows the student to concentrate on the concept of a confidence interval and digest it before it is applied in a more sophisticated manner in the inferential statistics unit that begins with Chapter 8.

- The chi-square test is used in gradually more applications, culminating in goodness of fit for both normal and binomial distributions.
- The formulas for the determination of the line of best fit are simplified and easy to work with.

The text is divided into three main units:

Descriptive Statistics and Data Analysis (Chapters 1–4)
Foundations of Inference (Chapters 5A–7)
Inferential Statistics (Chapters 8–14)

Obviously, an important goal of this book is to lead the student through carefully developed steps in understanding the material. A special effort has been made in the first third of the book to provide a good grounding in the fundamental thought process that underlie the later work in the text, and as a result the chance of initial success for the student is high in the first unit.

PEDAGOGICAL FEATURES

- The **spiral approach** of revisiting topics in successively more complex ways is used extensively throughout the text.
- There is a practical **case study** at the end of each chapter. The case study summarizes and enriches the material in the chapter and provides the student with the opportunity of using the library to obtain more information about that particular topic. The case study questions are varied, and all of them ask the student to write about statistical observations.
- Many historical, biographical, and statistical references are placed throughout the margins of the text as **Stat Facts.** This interesting feature helps to reinforce the notion of Statistics as a practical and "real" application of mathematics.
- Concise **Chapter Summaries** aid efficient review, and include page references to important concepts, terms, and formulas.
- Each chapter is cross-referenced with the popular video series *Against All Odds: Inside Statistics*.
- A Cumulative Overview, Procedure Index and Review Exercises are presented at four places in the text. These features cut across chapter boundaries and may be used to integrate different parts of the text.
- A **removable formula card** has been bound into the book for possible use during examinations.
- To further help students see the importance and application of statistics in their daily lives an **Applications Index** is included in addition to a regular index of topics.

EXERCISES AND EXAMPLES

Much time and effort, along with many years of experience has gone into the selection of the exercises for this text. A priority has been placed on obtaining **real data** to reinforce the practical application and value of statistics. In many instances sources of the data used have been cited to remind students of this value. The exercise sets have these characteristics:

- A special effort has been made to draw data from the broadest range of topics of interest to both men and women.
- Many exercises are composed of multiple parts. This allows the student to examine the same data from several different levels.
- Some of the exercises are labeled "GROUP," representing good choices for collaborative work and presentation back to the entire class. However, these same exercises can be used quite effectively by individual students as well.
- Some exercises marked with an asterisk will provide a good challenge for the more able student.

All examples are very carefully worked and illustrated. In all parts of the text I have recognized that while the mathematics may not be beyond college algebra, it can look very intimidating. To alleviate student anxiety in this regard, much care has been taken with the examples. All important steps have been shown in the technology illustrations as well.

The answers to the odd-numbered exercises may be found in the back of the text, and a student supplement is available that details the solutions to these exercises.

USE OF TECHNOLOGY TOOLS

The **Technology Tools** sections show how technology can be used to reinforce and extend statistical concepts introduced in the chapters. The Technology Tools sections are placed at the point in the chapters where the student can appreciate the advantage of using a computer, or graphics calculator, to work with large sets of data, or many repeated operations.

Minitab®, Release 8, is most often discussed as the software of choice, using both its interactive and macro-driven modes; illustrations in the text feature the Macintosh format. However, directions are general enough so that users of the DOS version of Minitab, Minitab for Windows, or the newest dual platform release of Minitab, should have no little or no difficulty in applying the text information to their own systems. Appendix A-9, a User's Guide to the DOS Version of Minitab, is also provided. Spreadsheets are illustrated and used in several places of the text as well.

The **TI-82 graphics calculator** is an important technology tool that is demonstrated often throughout the text. Although directions are specific for the TI-82, this text also lends itself to the use of other brands of graphics calculators, and their use presents a viable alternative to using computer software. Appendix A-10 is provided for users of the **TI-85 graphics calculator.**

The instructor has several options for treating the **Technology Tools** sections in this text:

1. Ignore them completely.
2. Use a little class time to discuss the illustrations, encouraging students to experiment on their own, using as much technology with which they feel comfortable.

3. Require some exercises to be done with computer software.
4. Require the use of a graphics calculator throughout the course, with or without additional computer software.

POSSIBLE CONTENT COVERAGE

This book is written so that the instructor has flexibility in the topics that are covered. The following flow chart may be used to suggest some possibilities:

Core Course

- Most of Chapters 1 through 4
  Descriptive stats and data analysis
- Most of Chapters 5A, 6, and 7
  Foundations of inference
- Chapters 8 and 9
  Basic inferential statistics

Estimating the mode, median, and control charts in Chap. 4 may be omitted or postponed

Chap. 5B Further Topics in Probability is optional

The Poisson random variable from Chap. 6 may be omitted or postponed

Highly Recommended

- Chapter 10
  Inferences about the difference of population means

Optional

- Chapter 11
  The chi-square distribution
- Chapter 12
  Nonparametric tests
- Chapter 13
  Analysis of variances
- Chapter 14
  Correlation and the line of best fit

These chapters may be treated in any order after the core material.

In my own teaching, I became aware that the need for this book was real. This book is an effort to present statistics for that broad range of people who need this information, beyond the level of rudimentary number crunching but short of the demands of high-powered mathematical analysis.

SUPPLEMENTS AND OTHER SUPPORT

For the Student

- A **Student Solutions Manual,** prepared by the author, contains the work and complete explanations of all odd-numbered exercises.
- A **Minitab Lab Manual,** by Lloyd Jaisingh and Robert Lindahl of Morehead State University, contains labs using the Windows version of Minitab®. Each lab begins with a precise Purpose and Background Information, followed by Procedures and Examples.
- A **Graphics Calculator Manual** by Stephen Kokoska of Bloomsburg University is a practical and readable graphics calculator supplement based on the TI-82, but useful for any graphics calculator. This supplement focuses on problem solving, and closely parallels the organization of the text.

For the Instructor

- An **Instructor's Solutions Manual and Resource Guide,** prepared by the author, has the complete solutions to all exercises and the case studies. It also contains Chapter Objectives, Teaching Suggestions, and alternate case studies for all chapters. Another convenient feature is the full reprint of all journal articles cited in the text's case studies.
- **Test Bank** and **Computerized Test Banks.** Tests written specifically for the text are available in printed form and on disk (Mac, DOS, or Windows).
- **Data Disk** containing all the large data sets (usually 10 or more elements) in the text are available free upon request. Data sets are also available online; contact Saunders for more information.
- The set of *Against All Odds: Inside Statistics* videotapes is a valuable educational tool. The tapes comprise 26 half-hour programs and were produced by the Consortium for Mathematics and Its Applications (COMAP). More information may be obtained about them by calling 1-800-LEARNER.

ACKNOWLEDGMENTS

I am pleased to thank the following reviewers for their valuable comments and suggestions in writing *Foundations of Statistics:*

Tom Carnevale, *Shawnee State University*
Carol Freeman, *Community College of the Finger Lakes*
Sheldon Gordon, *Suffolk Community College*
William Fox, *Moberly Area Community College*
John Hayes, *Worcester State College*
Robert Horvath, *El Camino College*
Syed Kirmani, *University of Northern Iowa*
Leonard Lipkin, *University of North Florida*
Lynn Mack, *Piedmont Technical College*
Donald Mason, *Elmhurst College*

Christine Mclaren, *Morehead State University*
William McClure, *Golden West College*
Allen Newhart, *West Virginia University at Parkersburg*
John Reeder, *American River College*
Donal B. Staake, Jr., *Jackson Community College*
David Stout, *University of West Florida*
Jan Vandever, *South Dakota State University*

This text has been developed carefully over several years and has been thoroughly class tested, ensuring the highest degree of accuracy. Many of the above reviewers contributed to the maintenance of a virtual error-free text, and several others specialized in checking specific parts of the text. I would like to thank:

Elizabeth Betzel, *Columbus State Community College,* for checking the odd-numbered answers and graphics calculator sections.

Sudhir Goel, *Valdosta State University,* for checking the even-numbered answers.

Lloyd Jaisingh, *Morehead State University,* for checking the Minitab sections.

Jan Vandever, *South Dakota State University,* for checking the odd-numbered answers and graphics calculator sections.

I also would like to thank Dr. Roger Jones, Professor of Mathematics at DePaul University in Chicago, who read my very first draft nearly five years ago. Our discussions helped me to see where I was going, and any stumbles I may have made are mine alone. In addition, my students at the Latin School of Chicago and DePaul University were some of my most helpful critics.

Jonathan Slater, former Headmaster, and the trustees of the Latin School of Chicago have given me extensive support. My good friends Margaret Jasinski and John Moorhead have given me the benefit of their wide experience, encouragement, and advice. Among my wife Mary Ellen's many skills, her extraordinary expertise as a reference librarian has added enormously to the quality of this text.

Finally, I am indebted to the staff of Saunders College Publishing. I especially want to thank Jay Ricci, Executive Editor, for his consistent and enthusiastic support; Marc Sherman, Developmental Editor, for his expertise, patience and skillful guidance; and Linda Boyle, Project Editor, for her knowledge, organization and efficiency in bringing this project to a successful completion.

I am sincerely interested in comments and suggestions regarding this text. I may be reached at the following e-mail address: whawley@latin.pvt.k12.16.us.

Warren J. Hawley
Chicago
December 1995

Contents

Index of Applications

The Andromeda galaxy represents a universe of objects. We can accurately describe and predict behaviors of a large set of elements using statistical methods. *(National Optical Astronomy Observatories)*

1 Data Analysis and Descriptive Statistics

Chapters 1 through 4 of this text treat the study of statistics from a more graphic and intuitive point of view than later chapters. An ability to grasp the overall picture is important for any practical application of statistics, and the first four chapters develop this ability.

Chapter 1 includes methods of ranking data from least to greatest, along with some good reasons for using this approach. The following chapters build upon ranking by using graphic methods to examine one set of data and to compare several sets of data. Confidence intervals and statistical significance, important ideas in inferential statistics (Unit 3), are introduced graphically in Chapter 2.

Chapters 3 and 4 investigate the traditional measures of center and spread of a set of data. Less traditionally, ways to estimate some of these measures for data that are summarized and presented graphically are also explored.

The case studies at the ends of the chapters are designed to reinforce and extend the ideas in the chapters as well as provide opportunities to practice basic library research.

Educational achievement
in different public
schools can be
examined by identifying
key factors and then
sorting and arranging
data efficiently.

C h a p t e r

1

Stem-and-Leaf Displays and Percentiles

Florence Nightingale pioneered the discipline of descriptive statistics. *(North Wind Picture Archives)*

I n 1992 the State of Illinois reported that some of its elementary schools spent nearly $10,000 per year to educate each student, while others spent less than $3000. Some administrators' salaries were almost $100,000, but others were less than $40,000. Before inequities such as these can be identified, data must be first sorted—that is, arranged efficiently for examination. This chapter presents several methods of ranking data and summarizing the information it contains. The **case study** at the end of the chapter examines more closely the substantial differences in per-pupil spending, teacher and administrator salaries, and standardized test scores for some fortunate and not-so-fortunate schools in Illinois.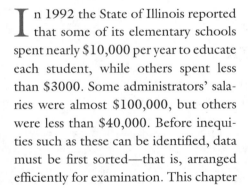

1.1 STATISTICS AND VARIABLES

Statistics is a branch of mathematics that deals with the collection, presentation, analysis, and implications of numbers that reflect measurable events. **Descriptive statistics,** covered in the first part of this text, is the branch of statistics that handles the collection and presentation of data. The latter part of this text addresses **inferential statistics,** which is concerned with properly analyzing and drawing conclusions from numerical information.

3

In the world around us we can observe many variables, which are of two basic types.

Definition ▶

> **Quantitative variables** measure size.
> **Qualitative variables** measure basic differentiating characteristics.

The number of students in your math class, the average age of your teachers, and your mother's age when you were born are all examples of quantitative variables. The sex of a newborn child, the religious preference of the president of the United States, and the political party affiliation of a voter are all examples of qualitative variables.

Definition ▶

> The **value** of a variable is an observation of that variable.

Stat Fact

> The first duty of government listed in the Constitution—mentioned even before the power to declare war or levy taxes—is to conduct a census every 10 years.

The values of a quantitative variable are numbers. The values of a qualitative variable are not numbers; they are gender distinctions, political party affiliations, or any other differentiating characteristic.

1.2 DATA

Data are the raw material with which the statistician works. Data can be found through surveys, experiments, historical records, and many other modes of research.

Definition ▶

> A set of **data** is a collection of values for a particular variable and may be qualitative or quantitative.

When one particular item in the data set is to be considered, the term **"score"** is usually used. The terms **"data value," "data item,"** and **"member of the data set"** can also be used.

Data can be presented in two ways.

Definitions ▶

> **Ungrouped data,** sometimes called raw data, are presented simply as an unorganized list of scores.
> **Grouped data** are presented in an organized manner that makes their interpretation easier.

The organization of grouped data can take the form of tables, graphs, or other displays. Some later chapters will examine methods of interpreting grouped data, but for now the exercises will concentrate on ways of organizing ungrouped data to identify key scores and characteristics.

Data analysis involves the examination of collected data and has generally broad goals, as the examples and exercises of this chapter illustrate. Two of the common applications of data analysis are identification of trends and exploration of relationships.

The method of collecting data is very important and will be covered at length in Chapter 7.

High-powered special interest groups often try to make their points by emphasizing one set of data over another. The application of "selective statistics" presents an ethical dilemma for the professional statistician.

1.3   STEM-AND-LEAF DISPLAYS

Before any data can be analyzed, they must first be arranged in a manner that makes them easy to examine and manipulate. Data are often presented in the form of an **array,** which is simply a block of numbers with no particular ordering. The following array shows the numbers of calls received by a 911 dispatcher between midnight and 7 A.M. for 36 days, chosen at random.

22	76	6	23	54	31
30	27	35	19	71	48
17	30	48	28	105	22
63	41	26	37	35	44
11	41	64	65	52	63
8	34	38	32	43	30

When the data comprise a relatively small number of scores, one of the most efficient ways to sort them for analysis is through a stem-and-leaf display.

Definition ▶ | A **stem-and-leaf-display** is a vertical[1] arrangement of data that sorts the scores from least to greatest.

Scores are copied from the array and entered on the stem-and-leaf display as shown in Figure 1-1. This figure shows that the creation of a stem-and-leaf display involves the splitting of each score into two parts. Exactly how many digits of the score go into the stem part and how many digits go into the leaf part depend on the nature of the data being examined. A stem-and-leaf display starts to look like a list as more digits of each score are placed in the stem part. However, expanding the stem part reduces the speed with which the display can be constructed. A good compromise between speed of construction and detail of display should be the goal.

---

[1] The actual orientation of a stem-and-leaf display is arbitrary. It can also be drawn horizontally or can list the scores such that the largest ones are at the bottom. (This latter construction will appear in Chapter 3 when the software program Minitab is used to construct stem-and-leaf displays.)

Commas are used to separate scores, and only stems that have leaf entries are shown. The second row from the bottom shows the scores 11, 17, and 19.

Stem	Leaf
10	5
7	1, 6
6	3, 3, 4, 5
5	2, 4
4	1, 1, 3, 4, 8, 8
3	0, 0, 0, 1, 2, 4, 5, 5, 7, 8
2	2, 2, 3, 6, 7, 8
1	1, 7, 9
0	6, 8

Figure 1-1 A stem-and-leaf display for the numbers of calls received by a 911 dispatcher.

Scores are taken from the array in any order. In this case, the tens and hundreds parts of the score are written in the stem column, and the units part is written in the leaf column. Once all the possibilities for the stem are entered, only the leaf part needs to be recorded as new scores are added. It is important to allow room in the leaf part of each row for the next score to be entered so that the entire leaf ends up ordered correctly, in ascending order.

A very important piece of information that the stem-and-leaf display can provide is the data rank of a particular score.

Definition ▶ The **data rank** of a score is its position from the bottom in an ordered list of all the scores.

For example, the score 06 (or 6) in the preceding stem-and-leaf display has data rank 1, the score 08 has data rank 2, and the score 11 has data rank 3. The score 22 has data ranks 6 and 7, since that score repeats two times. The score 105 has data rank 36.

An important characteristic of data that can be observed easily in a stem-and-leaf display is their general distribution, or dispersion. In this case, it is clear that the number of calls varies from just a few to over a hundred, while most of the time it numbers in the thirties. Several extremes are immediately apparent, especially 105. An important question that can arise from the study of this display is whether or not the score 105 can be considered a one-time-only occurrence, for which no future planning is necessary. Although there is no way of knowing the answer with absolute certainty, the next chapter will explain one way to make the decision—a method to identify unusual and rare scores.

A stem-and-leaf display is also very useful for determining the median of the data.

Definition ▶ The **median** of a set of data is the score that has equal numbers of scores greater than it and less than it.

Stat Fact

An early use of the word "median" was in the 12th U.S. Census report in 1901, which stated, "The median point is the point of intersection of the line dividing the population equally north and south with the line dividing it equally east and west."

For certain types of data, the median is the measure used to indicate the "typical" score. If the data comprise an odd number of scores, then the median is the middle-ranked score. However, if the data comprise an even number of scores, the median is half the sum of the two scores that are ranked in the middle, and it may not be an actual score.[2]

Example 1

For the 36 days' worth of calls to a 911 dispatcher, listed on page 5, find the median number of calls.

Solution Because 36 is an even number of scores, the median is half the sum of the two scores that are ranked in the middle. In Figure 1-1, the median is therefore half the sum of the scores with data ranks 18 and 19, or $\frac{35 + 35}{2} = 35$. ●

Example 2

Stat Fact

31% of U.S. homes today have personal computers, and the time between the introduction and discontinuation of a new model is about 15 months.

The following array shows unexplained computer downtimes (in minutes) for a particular company in the last 25 working days.

1.24	1.45	2.08	1.35	1.29
1.32	1.29	2.78	1.77	2.05
1.21	2.13	1.28	1.44	1.33
1.28	2.56	2.09	0.89	2.34
0.86	1.28	1.45	1.55	2.16

Determine the following:

a. The median downtime.
b. The time(s) at which most of the downtimes cluster.
c. The most extreme downtimes.

Solution A stem-and-leaf display will help to solve all parts of this example. If it is constructed so that the stem column has the unit parts of the scores, it looks like Figure 1-2.

Stem	Leaf
2	05, 08, 09, 13, 16, 34, 56, 78
1	21, 24, 28, 28, 28, 29, 29, 32, 33, 35, 44, 45, 45, 55, 77
0	86, 89

Figure 1-2

[2]Finding half the sum of two scores is also referred to as finding the **mean** of the two scores.

A much better display would be this one:

Stem	Leaf
2.7	8
2.5	6
2.3	4
2.1	3, 6
2.0	5, 8, 9
1.7	7
1.5	5
1.4	4, 5, 5
1.3	2, 3, 5
1.2	1, 4, 8, 8, 8, 9, 9
0.8	6, 9

Figure 1-3

The first stem-and-leaf display (Figure 1-2) is too rough. If computer down-time is a serious concern, it is important to analyze it in more detail by breaking the data into more than just three groups. A more useful breakdown places the units and tenths part of the scores in the stem column and the hundredths part in the leaf column. The second stem-and-leaf display (Figure 1-3) is more informative at a glance.

a. Because 25 is an odd number of scores the median is the score ranked in the middle. This is the score with data rank 13 (there are 12 scores ranked below it and 12 scores ranked above it). Thus, the median is 1.44.
b. There is a clustering of downtimes in the 1.2 stem.
c. The most extreme times (0.86 and 2.78) are readily seen. ●

E x a m p l e 3

The number of female physicians under the age of 35 in each of the 50 states, as of December 31, 1986, is shown below. This distribution is obviously related, in some degree, to the populations, educational opportunities, and other factors of the states. A researcher wishes to determine whether these factors have the same influence on males as on females.

The decision is made to find which states contribute the "typical" scores to this data by defining as typical those states that rank in the middle 20% of the data. (Exercise 12 on page 13 will determine whether these are the same states that are judged typical on the basis of numbers of male physicians.)

Determine the states that compose the middle 20% of this data. (Two-letter postal abbreviations are used to identify the states.)

Numbers of Female Physicians Under 35

AL	350	AK	38	AZ	425	AR	190
CA	3989	CO	470	CT	699	DE	73
FL	984	GA	639	HI	149	ID	36

IL	2012	IN	479	IA	264	KS	274
KY	376	LA	538	ME	106	MD	1321
MA	1778	MI	1168	MN	736	MS	187
MO	767	MT	34	NE	161	NV	41
NH	86	NJ	1097	NM	153	NY	4484
NC	814	ND	45	OH	1515	OK	308
OR	304	PA	2106	RI	215	SC	324
SD	47	TN	531	TX	2033	UT	169
VT	88	VA	916	WA	584	WV	163
WI	574	WY	24				

Solution 20% of 50 = .20 · (50) = 10. Thus, ten states make up the middle 20% of this data, which means that there are 20 states with more female physicians, and 20 states with fewer female physicians, than the typical states (Figure 1-4).

Often, however, partitions do not work out as neatly as these. For instance, if the District of Columbia were a state, 20% of 51 would be 10.2. Most statisticians would round this to 10 and perhaps arbitrarily assign the extra state to one of the other two groups. Also, there are occasions when the same score repeats many times, making it impossible to divide scores in a specific way.

If a stem-and-leaf display is used to order the data, the stem should contain the digits that are in the hundreds and thousands places of each score. The boldface scores in Figure 1-5 represent the middle 20%, or middle ten scores. These scores correspond to the states of Kansas (274), Oregon (304), Oklahoma (308), South Carolina (324), Alabama (350), Kentucky (376), Arizona (425), Colorado (470), Indiana (479), and Tennessee (531).

20 states	More physicians
10 states	"Typical" states
20 states	Fewer physicians

Figure 1-4 A distribution of states.

Stem	Leaf
44	84
39	89
21	06
20	12, 33
17	78
15	15
13	21
11	68
10	97
9	16, 84
8	14
7	36, 67
6	39, 99
5	**31**, 38, 74, 84
4	**25, 70, 79**
3	**04, 08, 24, 50, 76**
2	15, 64, **74**
1	06, 49, 53, 61, 63, 69, 87, 90
0	24, 34, 36, 38, 41, 45, 47, 73, 86, 88

Figure 1-5

Exercises 1.1–1.3

1. Can a particular score hold more than one data rank? Explain.

2. What is the purpose of a stem-and-leaf display?

3. How does an array differ from a stem-and-leaf display?

4. Under what conditions, if any, is one assured that the median will be an actual score? Are there any conditions under which the median may not be an actual score?

5. Can the median be closer in value to the maximum score than to the minimum score in a set of data? Explain.

* 6. Give some examples of quantitative variables whose values, if shown in a stem-and-leaf display, would cluster around two different rows.

7. The winning times for the women's 100-meter dash in the last 15 Summer Olympic games are as follows:

Stat Fact

Since 1928 the winning times for women have been steadily approaching those for men. In 1928 Percy Williams of Canada (a male) ran this event in 1.38 seconds less time than did Elizabeth Robinson. By 1992 the difference between men's and women's times had shrunk to 0.84 seconds.

Winner	Country	Time (seconds)	Year
Evelyn Ashford	United States	10.97	1984
Fanny Blankers	Netherlands	11.9	1948
Betty Cuthbert	Australia	11.5	1956
Gail Devers	United States	10.82	1992
Florence Griffith-Joyner	United States	10.54	1988
Marjorie Jackson	Australia	11.5	1952
Lyudmila Kondratyeva	U.S.S.R.	11.06	1980
Wilma Randolph	United States	11.0	1960
Annegret Richter	West Germany	11.08	1976
Elizabeth Robinson	United States	12.2	1928
Helen Stephens	United States	11.5	1936
Renate Stecher	East Germany	11.07	1972
Stella Walsh	Poland	11.9	1932
Wyomia Tyus	United States	11.4	1964
Wyomia Tyus	United States	11.0	1968

Source: *The 1994 Information Please Almanac,* p. 900.

a. Find the median of these 15 winning times.

b. What general conclusion seems possible from close examination of these data?

*Exercises marked with an asterisk are especially challenging.

8. The 20 top-grossing films of 1993 were the following:

Film	Box-Office Gross (millions of dollars)
A Few Good Men	78.2
Aladdin	117.9
Cliffhanger	84.0
Cool Runnings	59.4
Dave	63.3
Demolition Man	56.0
Dennis the Menace	51.3
Free Willy	77.7
Groundhog Day	70.8
Indecent Proposal	106.6
In the Line of Fire	102.3
Jurassic Park	337.8
Mrs. Doubtfire	89.2
Rising Sun	62.5
Rookie of the Year	53.1
Scent of a Woman	62.2
Sleepless in Seattle	126.5
The Crying Game	59.3
The Firm	158.3
The Fugitive	179.3

Source: *The Universal Almanac, 1995,* p. 235.

Use a stem-and-leaf display to identify the median gross revenue for these 20 films. Identify which films, if any, grossed at least twice the median. Which, if any, grossed less than half the median?

9. The 30 batting champions of the American League in the years 1964 through 1993 are as follows:

1964:	.323 (Oliva)	1979:	.333 (Lynn)
1965:	.321 (Oliva)	1980:	.390 (Brett)
1966:	.316 (F. Robinson)	1981:	.336 (Lansford)
1967:	.326 (Yastrzemski)	1982:	.332 (Wilson)
1968:	.301 (Yastrzemski)	1983:	.361 (Boggs)
1969:	.332 (Carew)	1984:	.343 (Mattingly)
1970:	.328 (Johnson)	1985:	.368 (Boggs)
1971:	.337 (Oliva)	1986:	.357 (Boggs)
1972:	.318 (Carew)	1987:	.363 (Boggs)
1973:	.350 (Carew)	1988:	.366 (Boggs)
1974:	.364 (Carew)	1989:	.339 (Puckett)
1975:	.359 (Carew)	1990:	.329 (Brett)
1976:	.333 (Brett)	1991:	.342 (Franco)
1977:	.388 (Carew)	1992:	.343 (Martinez)
1978:	.333 (Carew)	1993:	.363 (Olerud)

Find the median batting average and the player who came closest to batting the median average. Identify the players whose batting averages are in the highest 20% of this data.

10. The estimated number of injuries in the United States in 1991 that can be directly associated with certain products are as follows:

Product	Estimated Injuries	Product	Estimated Injuries
Baby strollers	14,900	Jewelry	46,027
Bathtubs and showers	139,434	Knives	448,524
Bicycles	600,649	Ladders	134,341
Bottles and jars	105,840	Lawn mowers	74,400
Bunk beds	42,969	Metal containers	96,353
Carpets	87,979	Nails and screws	241,859
Chain saws	44,019	Razors and shavers	43,604
Drinking glasses	134,840	Scissors	30,823
Drugs	113,511	Skateboards	56,435
Fences	122,111	Sleds	33,562
Footwear	83,369	Telephones	16,004
Hammers	50,036	Trampolines	38,823
Irons	17,326	Wheelchairs	51,732

What is the median number of injuries? Which product(s) have numbers of injuries that come closest to the median? Does it seem reasonable to use the ranking of these scores in a stem-and-leaf display to determine which products are more dangerous than others? Why or why not?

GROUP   11. The following table shows average hours of household TV usage, in hours (h) and minutes (m) per day.

	Yearly Average	February	July
1983–84	7 h, 8 m	7 h, 38 m	6 h, 26 m
1984–85	7 h, 7 m	7 h, 49 m	6 h, 34 m
1985–86	7 h, 10 m	7 h, 48 m	6 h, 37 m
1986–87	7 h, 5 m	7 h, 35 m	6 h, 32 m
1987–88	6 h, 55 m	7 h, 38 m	6 h, 31 m
1988–89	7 h, 2 m	7 h, 32 m	6 h, 27 m
1989–90	6 h, 55 m	7 h, 16 m	6 h, 24 m
1990–91	6 h, 56 m	7 h, 30 m	6 h, 26 m
1991–92	7 h, 4 m	7 h, 32 m	6 h, 39 m

a. Find the median yearly average, the median average for February, and the median average for July.

b. Speculate why the months of February and July were chosen for this study.

12. The following array shows the number of male physicians under the age of 35 in each state.

AL	1446	AK	127	AZ	1267	AR	788	CA	11,186
CO	1306	CT	1940	DE	199	FL	4158	GA	2273
HI	494	ID	163	IL	5359	IN	1639	IA	948
KS	909	KY	1218	LA	1960	ME	300	MD	3420
MA	4124	MI	3635	MN	2107	MS	755	MO	2264
MT	137	NE	592	NV	233	NH	322	NJ	3223
NM	445	NY	11,648	NC	2668	ND	227	OH	4573
OK	1045	OR	714	PA	6060	RI	469	SC	1268
SD	185	TN	2108	TX	6318	UT	610	VT	270
VA	2645	WA	1570	WV	640	WI	1856	WY	131

Refer to Example 3, and determine what percentage of the middle 20% of states with female physicians under the age of 35 are also in the middle 20% of states with male physicians under the age of 35. Do the results seem to indicate that outside factors affect males and females in the same way? Does the strategy used to reach this conclusion seem valid? Why or why not?

13. The following array shows the ages of the last 45 actresses to receive the Oscar for best actress from 1950 through 1993.

Actress	Age	Actress	Age	Actress	Age
Judy Holliday	28	Julie Christie	24	Sally Field	33
Vivien Leigh	38	Elizabeth Taylor	34	Sissy Spacek	30
Shirley Booth	45	Katharine Hepburn	60	Katharine Hepburn	74
Audrey Hepburn	24	Katharine Hepburn	61	Meryl Streep	33
Grace Kelly	26	Barbra Streisand	26	Shirley MacLaine	49
Anna Magnani	48	Maggie Smith	35	Sally Field	38
Ingrid Bergman	41	Glenda Jackson	34	Geraldine Page	61
Joanne Woodward	27	Jane Fonda	34	Marlee Matlin	21
Susan Hayward	40	Liza Minelli	26	Cher	41
Simone Signoret	38	Glenda Jackson	37	Jodie Foster	26
Elizabeth Taylor	28	Ellen Burstyn	42	Jessica Tandy	80
Sophia Loren	27	Louise Fletcher	41	Kathy Bates	42
Anne Bancroft	31	Faye Dunaway	35	Jodie Foster	29
Patricia Neal	37	Diane Keaton	31	Emma Thompson	33
Julie Andrews	30	Jane Fonda	41	Holly Hunter	35

Source: *The World Almanac and Book of Facts, 1995*, pp. 327–329, 357–370.

a. What is the median age, and what percentage of the actresses were within 5 or fewer years of this median age when they received their Oscars?

b. Identify the actresses who belong to the group made up of the oldest 10% of the ages and those who belong to the group made up of the youngest 10% of the ages.

* 14. Refer to the preceding exercise, and use trial and error to find the smallest value of c such that the percentage of actresses who received Oscars at age (median $\pm$ c) is 90%.

15. In 1991 the salaries of the governors of the 50 states were as follows:

AL	81,151	AK	81,648	AZ	75,000	AR	35,000	CA	114,286
CO	70,000	CT	78,000	DE	80,000	FL	103,909	GA	91,080
HI	94,780	ID	75,000	IL	97,370	IN	74,100	IA	76,700
KS	74,235	KY	79,255	LA	73,440	ME	70,000	MD	120,000
MA	75,000	MI	106,690	MN	109,053	MS	75,600	MO	88,541
MT	54,254	NE	65,000	NV	82,391	NH	79,541	NJ	85,000
NM	90,692	NY	130,000	NC	123,000	ND	67,800	OH	99,986
OK	70,000	OR	80,000	PA	105,000	RI	69,900	SC	98,000
SD	60,890	TN	85,000	TX	95,301	UT	72,800	VT	80,730
VA	105,882	WA	112,000	WV	72,000	WI	92,283	WY	70,000

a. Determine which state(s) have salaries for their governors that are out of line in relation to the rest of the states. Define "out of line" as at least one and one-half times the median salary. What factors could justify such salaries?

b. Which two states have governors' salaries farthest from the median?

1.4 PERCENTILES

A stem-and-leaf display conveys some information, but is often the starting point for further analysis and comparison of the data. When this further work is undertaken, it is very useful to identify certain key scores, such as percentiles, that give information about the entire data set.

Definition ▶ A **percentile** is a number that divides the data into two parts. Ideally, a given percentage of the data, $k\%$, will be less than this number, and $(100 - k)\%$ will be greater than this number.

As an example, the 30th percentile of a set of data, denoted by P_{30}, is a number such that, as near as possible, 30% of the data are less than P_{30} and 70% are greater than P_{30}. This dividing number, the 30th percentile may or may not be an actual score. The 50th percentile, P_{50}, is another name for the median, since ideally 50% of the data are less than P_{50} and 50% are greater than P_{50}.

To determine a desired percentile, the first task is to order, or sort, the data—that is, arrange it in a list from least to greatest. A stem-and-leaf display can be used for this purpose. A later part of this chapter will show how some computer software programs can sort data.

Since percentiles are usually used only for rough discrimination among scores, statisticians do not universally agree on the proper method of determining a given percentile. The procedure illustrated in the following flow chart is a common one, and it will be used throughout this text for manual computation of percentiles. Percentiles determined by statistical software programs can differ slightly from those computed by hand, because many programs use a more complicated method involving interpolation.

How to Find the Score That Corresponds to a Given Percentile

Consider Figure 1-6.

P_k = Given percentile.
k = The percentage which defines that percentile.
n = The total number of scores.
R = Data rank of score.

STEP 1

Arrange all the scores from the least to greatest. A stem-and-leaf display or certain computer software programs can be applied to accomplish this.

STEP 2

Compute the Data Rank, R, by using the formula

$$R = \left(\frac{k}{100}\right) \cdot n$$

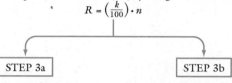

STEP 3a

If R is not a whole number, then for simplicity, always round R up to the next whole number. Using the rounded value of R, consult the stem-and-leaf display to find the score with data rank R. That score represents the given percentile P_k.

STEP 3b

If R is a whole number, then the number which corresponds to P_k will be half the sum of the scores with data ranks R and $(R + 1)$. Use the stem-and-leaf display to find these scores.

Figure 1-6 A flow chart for determining the score that corresponds to a given percentile.

E x a m p l e
4

The 1990 State Fuel Efficiency Ranking gave the following average miles traveled per gallon of gasoline and gasohol.

Stat Fact

In 1991 Nebraska's 23.9¢ excise tax on a gallon of gasoline was the greatest of any state in the union. Florida's tax of 4¢ per gallon was the least.

AK	11.23	**AL**	15.77	**AZ**	17.51	**AR**	12.86
CA	17.24	**CO**	15.64	**CT**	17.24	**DE**	16.37
FL	15.76	**GA**	16.38	**ID**	15.86	**HI**	19.74
IL	13.24	**IN**	15.82	**IA**	13.23	**KS**	14.77
KY	14.39	**LA**	16.63	**MA**	17.29	**MD**	16.90
ME	16.49	**MI**	16.80	**MN**	16.29	**MS**	15.43
MO	15.22	**MT**	14.55	**NE**	13.75	**NM**	15.74
NC	15.83	**ND**	13.32	**NH**	17.65	**NV**	13.14
NY	15.65	**NJ**	15.56	**OH**	15.22	**OK**	15.70
OR	15.69	**PA**	15.45	**RI**	16.79	**SC**	15.06
SD	14.26	**TN**	15.00	**TX**	15.80	**UT**	16.23
VA	16.89	**VT**	17.71	**WA**	16.75	**WI**	17.43
WV	14.87	**WY**	12.08				

Source: *The 1993 Information Please Environmental Almanac,* p. 84.

Find the mileages that mark the following percentiles:

a. The 50th percentile (P_{50}), or median.
b. The 33rd percentile (P_{33}).

Solution   There are $n = 50$ scores in the data. Figure 1-7 is a stem-and-leaf display with the whole-number part in the stem column and the decimal part in the leaf column. Notice that rows with many scores are broken into two parts.

a. To find P_{50}, first calculate R:

$$R = \frac{50}{100} \cdot 50 = 25.$$

Since R is a whole number, P_{50} is the mean of the scores with data ranks 25 and 26. The score with data rank 25 is 15.74, and the score with data rank 26 is 15.76. Therefore, the median is

$$P_{50} = \frac{15.74 + 15.76}{2}$$
$$= 15.75.$$

This means that 50% of the data are below 15.75. A check of the preceding display verifies that 25 out of 50, or 50%, actually are below the median (15.75).

b. To find P_{33}, let $R = \frac{33}{100} \cdot 50 = 16.5$. Since R is not a whole number, it is rounded up to 17. (The directions require one to *always* round up, never down.) Thus, P_{33} is equal to the score with data rank 17, which is 15.22. Therefore, 15.22 represents the 33rd percentile.

Interestingly, only 15 out of 50, or $\frac{15}{30} = 30\%$, of the scores are less than P_{33}. This is because of the relatively few scores in the data set and the duplication of

Stem	Leaf
19	7 4
17	24, 24, 29, 43, 51, 65, 71
16	63, 75, 79, 80, 89, 90
16	23, 29, 37, 38, 49
15	56, 64, 65, 69, 70, 74, 76, 77, 80, 82, 83, 86
15	00, 06, 22, 22, 43, 45
14	26, 39, 55, 77, 87
13	14, 23, 24, 32, 75
12	08, 86
11	23

Figure 1-7

scores at the 33rd percentile. Also as a result of these factors, 15.22 is the 32nd percentile (P_{32}) as well as the 31st percentile (P_{31}). However, as the number of scores in the data set increases and the duplications of scores decrease, these types of ambiguities become less likely.   ●

How to Find the Percentile Corresponding to a Given Score

Very few steps are needed for this type of calculation.

To find the percentile corresponding to a given score, let S = the score in question. Then form the fraction:

$$P_k = \frac{\text{Number of scores less than } S}{\text{Total number of scores in data}}.$$

Express that fraction as a percent, rounded to the nearest whole percent if necessary.

Example 5

For the data in Example 4, determine the percentile that corresponds to the gasoline mileage for Illinois.

Solution   In this case, the score is given and P_k is to be determined.

The gasoline mileage for Illinois is 13.24. By checking the stem-and-leaf display it can be seen that five states have gasoline mileages less than 13.24. Therefore,

$$P_k = \frac{5}{50}$$
$$= .10$$
$$= P_{10}, \text{ or the 10th percentile.}$$

This means that only 10% of the states have gasoline mileages less than that of Illinois.   ●

Deciles, Quintiles, and Quartiles

Deciles, quintiles, and quartiles are special, shortened names given to certain frequently used percentiles.

The prefix "dec-" implies 10, and ideally from one decile to the next are one-tenth of the scores. The first decile is P_{10}, the second decile is P_{20}, and so on. The prefix "quint-" implies five, and from one quintile to the next are one-fifth of the scores. The first quintile is P_{20}, the second quintile is P_{40}, and so on.

By similar reasoning, the first quartile is P_{25}, the second quartile is P_{50} (the median), and the third quartile is P_{75}. The letter Q followed by a subscript is a shorthand designation for a specific quartile. For example, Q_3 denotes the third quartile.

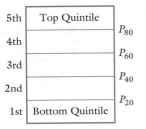

Figure 1-8

Figure 1-8 shows quintile relationships. Quintiles (and quartiles) are often used informally to indicate intervals of scores, as well. In this text, the percentiles shown above, such as P_{40}, will mark the "tops" of the groups they define. For instance, in order for a score, S, from the data to be in the middle quintile, it must be true that

$$P_{40} < S \le P_{60}.$$

The bottom quintile is not universally called the first quintile. In fact, most colleges and universities refer to the bottom quintile as the fifth quintile and the top quintile as the first quintile in determining admissions. That can be very misleading because percentiles, and hence quintiles, are determined by counting from the least, or bottom, numbers. Usually, however, the context in which these terms are used minimizes confusion.

Example 6

The GPAs of 62 members of a high-school senior class, rounded to the nearest hundredth, are as follows:

2.28	3.56	3.35	2.62	1.89	2.39	3.12	2.48	2.46	3.78
1.87	3.52	2.27	1.09	2.37	3.67	3.00	2.45	2.14	3.88
2.78	4.00	1.78	1.94	2.55	2.45	2.98	3.87	3.92	1.99
1.56	2.87	2.34	3.56	3.76	2.80	2.33	1.89	1.78	2.67
3.83	3.67	2.66	3.50	2.78	3.98	1.13	2.98	3.71	3.08
2.88	2.56	2.47	3.06	2.05	1.02	2.19	3.28	3.56	3.45
3.33									2.79

Stat Fact

For those students who took the SAT test in 1993 and had GPAs in the highest category, 44% of them were males and 56% were females.

Among other requirements, very selective schools require that applicants place in the top quintile, and selective schools require placement in one of the top two quintiles. Determine the minimum GPAs required for "very selective" and "selective" consideration.

Solution   "Placing in the top quintile" will denote a score greater than P_{80}.

Start with a stem-and-leaf display, placing the units and tenths places in the stem column. To determine the cut-off GPAs, first find P_{60} and P_{80}. P_{60} marks the score that must be exceeded in order to place in the fourth quintile, and $R = \frac{60}{100} \cdot 62 = 37.2$. Therefore, use the score in Figure 1-9 with data rank 38: $P_{60} = 2.98$. P_{80} marks the score that must be exceeded in order to place in the top, or fifth, quintile, and $R = \frac{80}{100} \cdot 62 = 49.6$. Therefore, use the score in Figure 1-9 with data rank 50: $P_{80} = 3.56$.

Stem	Leaf
4.0	0
3.9	2, 8
3.8	3, 7, 8
3.7	1, 6, 8
3.6	7, 7
3.5	0, 2, 6, **6**, 6
3.4	5
3.3	3, 5
3.2	8
3.1	2
3.0	0, 6, 8
2.9	8, **8**
2.8	0, 7, 8
2.7	8, 8, 9
2.6	2, 6, 7
2.5	5, 6
2.4	5, 5, 6, 7, 8
2.3	3, 4, 7, 9
2.2	7, 8
2.1	4, 9
2.0	5
1.9	4, 9
1.8	7, 9, 9
1.7	8, 8
1.5	6
1.1	3
1.0	2, 9

Boldface scores represent P_{60} and P_{80}.

Figure 1-9

Thus, a GPA greater than 2.98 will satisfy selective schools, and a GPA greater than 3.56 will satisfy very selective schools.   •

As shown in the previous examples, some scores repeat a number of times; in Figure 1-9, for example, 1.78 occurs twice, and so do some others. Unfortunately, this duplication of scores occurs in two awkward places: data rank 38 and data rank 50. The reason for the awkwardness is that these data ranks are used to determine P_{60} and P_{80}. For example, both data ranks 37 and 38 are equal to 2.98. As a result, P_{60} is located at the common score of 2.98, which means that there are fewer scores than desired below 2.98: $\frac{36}{62}$, or 58.06%, of the scores are less than the theoretical 60th percentile.

As already mentioned, the data are relatively few and include duplicate scores, so such errors are unavoidable. In actual practice, however, the data set could be much larger and more precise GPAs could be available.

1.5   THE RANGE AND THE INTERQUARTILE RANGE

For a given set of data, it is often useful to know how far apart the high score and low score are.

Definition ▶   The difference between the high score and the low score is called the **range** of the data.

Stem	Leaf
10	5
7	1, 6
6	3, 3, 4, 5
5	2, 4
4	1, 1, 3, 4, 8, 8
3	0, 0, 0, 1, 2, 4, 6, 6, 7, 8
2	2, 2, 3, 6, 7, 8
1	1, 7, 9
0	6, 8

Figure 1-10 A stem-and-leaf display for the numbers of calls received by a 911 dispatcher.

Stat Fact

Among people born in the United States today, the life expectancy of a woman is about 7 years greater than that of a man; that is, the range of life expectancies is about 7 years.

In a comparison of several sets of similar data, the range can be used to indicate which set has the most spread-apart scores. By itself, however, the range can sometimes be misleading.

Consider the data presented near the start of this chapter—the numbers of calls received by a 911 dispatcher as shown in Figure 1-10. The range of this set of data is $105 - 6 = 99$, but more than half of the scores are very close together—between 22 and 48—and the range does not indicate that fact. Often another measure of the spread called the interquartile range is used.

Definition ▶ The **interquartile range (IQR)** is the difference between the third and first quartiles: $Q_3 - Q_1$, or $P_{75} - P_{25}$.

The IQR is the range of the middle 50%, or middle half, of the scores and is not affected by unusual extremes. For the data compiled by the 911 dispatcher, the IQR is computed as follows:

$$\begin{aligned} IQR &= Q_3 - Q_1 \\ &= P_{75} - P_{25} \\ &= 50 - 26.5 \\ &= 23.5 \\ &\approx 24. \end{aligned}$$

The IQR will be used extensively in the next chapter to compare several sets of data.

Exercises 1.4–1.5

1. What is the difference between a percentage and a percentile?

2. Is it possible for a score to correspond to more than one percentile? Explain.

3. A score, S, is equal to P_{50}. What quartile is it in?

4. Is it possible for the percentile P_{70} to *not* be an actual score in the data set? Explain.

5. To say that "a score is in the bottom quartile" is not precise. Why? What is the probable meaning of such a statement?

6. Is the range a number or an interval?

7. Percentiles usually become more accurate as the number of scores increases. Why? When would they not?

8. Is there a greater percentage of scores between the first and third quartiles or between the first and fourth quintiles?

GROUP

9. For each of the following parts, give an example of two sets of data, A and B, each with ten scores, that satisfy the pair of statements.
   a. The range of A is greater than the range of B.
      The IQR of A is greater than the IQR of B.
   b. The range of A is less than the range of B.
      The IQR of A is greater than the IQR of B.
   c. The range of A is twice the range of B.
      The IQR of A is half the IQR of B.
   d. The range of A is half the range of B.
      The IQR of A is one-fourth the IQR of B.

10. Refer to Exercise 9 (in Exercises 1.1–1.3) on page 11, which contains batting averages for 30 American League batting champions.
   a. Find the players whose batting averages correspond to the first, second, and third quartiles.
   b. Find the range and the IQR.
   c. Find the range of percentiles for Rod Carew's seven batting championships.

11. The following array shows the numbers (in thousands) of singles, LPs, cassettes, and compact discs (CDs) shipped for distribution in the United States in the years 1983–1992.

Year	Singles	LPs/EPs	Cassettes	CDs
1983	124,800	209,600	236,800	800
1984	131,500	204,600	332,000	5800
1985	120,700	167,000	339,100	22,600
1986	93,900	125,200	344,500	53,000
1987	82,000	107,000	410,000	102,100
1988	65,600	72,400	450,100	149,700
1989	36,600	34,600	446,200	207,200
1990	27,600	11,700	442,200	286,500
1991	22,000	4800	360,100	333,300
1992	19,800	2300	366,400	407,500

   a. Find the median number shipped in each of these four categories, and then rank the four medians from least to greatest.
   b. What percentile does the median number of CDs shipped represent with respect to the number of singles shipped? With respect to LPs? With respect to cassettes?
   c. If the trend shown in this array continues, estimate the numbers of singles, LPs, cassettes, and CDs shipped in the year 2000.

12. Queen Elizabeth II has been queen of Great Britain since 1952. The following table lists the lengths of the reigns, in years, of the monarchs of England and Great Britain since the Battle of Hastings.

Monarch	Reign	Monarch	Reign
William I	21	Mary I	5
William II	13	Elizabeth I	44
Henry I	35	James I	22
Stephen	19	Charles I	24
Henry II	35	Charles II	25
Richard I	10	James II	3
John	17	William III	13
Henry III	56	Mary II	6
Edward I	35	Anne	12
Edward II	20	George I	13
Edward III	50	George II	33
Richard II	22	George III	59
Henry IV	13	George IV	10
Henry V	9	William IV	7
Henry VI	39	Victoria	63
Edward IV	22	Edward VII	9
Edward V	0	George V	25
Richard III	2	Edward VIII	1
Henry VII	24	George VI	15
Henry VIII	38	Elizabeth II (as of 1993)	41
Edward VI	6		

Name the monarchs whose reigns determine the second, third, fourth, and eighth deciles.

Hint: More than one monarch may have the length of reign that determines the required decile.

13. Refer to the preceding exercise. Name the monarchs whose reigns would *not* be included in the middle half (middle two quartiles) of all the reigns listed. (This group is sometimes called the **outer half** of a set of data.)

14. The voter turnout in a presidential election can be defined as the percentage of the voting-age population actually voting. The following table lists the voter turnout for the years 1928–1992:

Year	Voter Turnout	Year	Voter Turnout
1928	51.8%	1964	61.9%
1932	52.6	1968	60.9
1936	56.8	1972	55.2
1940	58.8	1976	53.5
1944	56.1	1980	52.6
1948	51.1	1984	53.1

Year	Voter Turnout	Year	Voter Turnout
1952	61.6	1988	50.2
1956	59.4	1992	55.2
1960	62.8		

Sources: *The Universal Almanac, 1993,* p. 91 and *Statistical Abstract of the United States,* 1994, p. 289.

a. Identify voter turnouts that mark the first, second, and third quartiles. Which years had voter turnouts less than the third quartile?

b. Suppose that in a future presidential election 119,250,000 votes are cast by a voting-age population of 225,000,000. Find the percentile of this voter turnout.

15. The following table shows the numbers of cases of AIDS in selected cities in 1992.

City	Number of Cases
Anaheim, CA	2046
Atlanta, GA	4583
Baltimore, MD	2635
Boston, MA	3438
Chicago, IL	5583
Dallas, TX	3690
Denver, CO	1771
Detroit, MI	2110
Fort Lauderdale, FL	3677
Houston, TX	6258
Jersey City, NJ	2273
Los Angeles, CA	14,567
Miami, FL	6625
Nassau-Suffolk, NY	2053
New Orleans, LA	2048
New York, NY	38,326
Newark, NJ	5523
Oakland, CA	2678
Philadelphia, PA	4521
San Diego, CA	3117
San Francisco, CA	11,912
Seattle, WA	2193
Tampa–St. Pete, FL	2188
Washington, DC	6226
West Palm Beach, FL	1937

a. Determine the median, IQR, and range of these numbers of AIDS cases.

b. Approximately how many times greater than the IQR is the range? Does this seem unusual, and what implications, if any, would seem to follow?

c. Scores in a set of data are considered **skewed** if they are not evenly spaced— that is, if they are more stretched out toward one end of the set of data than toward the other. This is often the case when the data include a few extreme scores. Do you consider the scores for the data in this exercise skewed? Why or why not?

1.6 Technology Tools

With care, a stem-and-leaf display can be an efficient and accurate means of arranging data from least to greatest for further analysis. However, it can be slow and tedious, particularly with a great number of scores. A graphics calculator or a computer with a spreadsheet program, a database program, or a statistical software program can make the task of sorting a great number of scores quite easy.[3] The rest of this chapter examines how two of these tools, a graphics calculator and a spreadsheet program, can be used to sort scores.

Spreadsheets

A spreadsheet program, or simply **spreadsheet,** is a software program that allows the user to enter a practically unlimited amount of data in a grid of boxes, or cells, of adjustable size, as partially shown in Figure 1-11. The data can be entered to form one row or column or as an array, depending upon the intended use. Besides providing the option of performing standard mathematical calculations on the data entered in each cell, most spreadsheets can easily be used to sort the data in either ascending or descending order.[4]

		Spreadsheet (SS)			
	A	B	C	D	E
1					
2					
3					
4					
5					
6					
7					
8					
9					
10					
11					

Figure 1-11 A typical spreadsheet.

[3] Many word processing programs sort as well.

[4] For illustration, this chapter uses Microsoft Works Spreadsheet, version 2.0, for the Macintosh. Many spreadsheets for PCs, such as Excel, are designed to work with Microsoft Windows and look and perform very similarly to this one.

The next example, which involves stanines, illustrates the use of a spreadsheet.

Definition ▶ | A **stanine** is a *group* of scores, each of which contains one-ninth of the data.

S t a t   F a c t

Stanines were first developed by the U.S. Air Force for scoring aptitude tests for pilot, bombadier, and navigator students.

The word "stanine" is derived from "standard nine." The nine stanines form three groups of equal size, defining lower, middle, and upper scores; each of those groups, in turn, may be divided into lower, middle, and upper (Figure 1-12). Some standardized tests in the primary grades report their scores in terms of stanines.[5] After the primary grades, however, the use of stanines is discontinued in the reporting of standardized test scores.

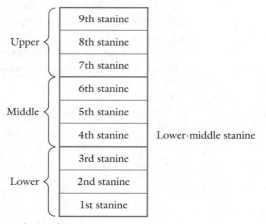

Figure 1-12 Stanine relationships.

E x a m p l e 7

A mathematics placement test is administered to 80 students entering the eighth grade of a local public school, with the following results:

13	27	29	78	19	67	9	78	87	69
80	11	67	23	87	78	75	4	32	37
66	54	76	55	48	19	29	38	85	8
44	75	81	80	71	60	82	76	72	19
59	80	71	63	52	28	33	71	40	38
18	26	71	64	82	85	48	65	7	18
59	80	33	21	39	44	59	67	21	30
78	12	56	19	23	76	56	45	80	27

[5]The ERBs (Educational Records Bureau tests) and Iowa Test of Basic Skills are two examples.

No one is expected to answer all of the questions correctly. The results are to be used to form one class of lower ability, one class of higher ability, and three classes of average ability. Use stanines and a spreadsheet to accomplish this.

Solution   Eighty students divided into 5 classes is $\frac{80}{5}$, or 16 students per class. Also, there are $\frac{80}{9}$, or about 9, students per stanine—giving not quite two stanines (18 students) per class. This would imply the following distribution of students as shown in Figure 1-13. A few "borderline" scores between the second and third stanines and between the seventh and eighth stanines will go into the average classes, depending upon the actual distribution of scores.[6]

Next, the data is entered onto a spreadsheet in one long column, in no particular order. Figure 1-14 shows how to begin. When all the scores have been entered, the entire column A is selected, and the **Sort** option (in descending order) chosen from the **Edit** menu.[7] This option takes only a few seconds and results in a perfectly sorted list, part of which is shown in Figure 1-15.

After studying the entire list of sorted scores,[8] the school principal might decide that it is more important to have a slightly smaller lower-ability class. In that case, a higher-ability class of 16 students could be taken from those with raw scores of at least 78, and a lower-ability class of 13 students taken from those with raw scores of 19 or less. Both classes are roughly two stanines.

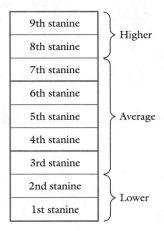

Figure 1-13

---

[6] In real life, there might be pedagogical, philosophical, or political factors governing the sizes of certain types of classes. This solution is based on a strictly mathematical approach.

[7] PC users employing a spreadsheet that does not operate under Microsoft Windows will have a screen that looks a bit different and will accomplish this sorting through appropriate keystrokes.

[8] This spreadsheet prints boldface numbers to the left of the data. Each one corresponds to a score's data rank, if that rank is determined from the highest score instead of the lowest score. In this text, these numbers will be ignored.

Spreadsheet			
	A	**B**	**C**
1	13		
2	80		
3	66		
4	44		
5	59		
6	18		
7	59		
8	78		
9	27		
10	11		
11	54		
12	75		
13	80		

Figure 1-14

Spreadsheet			
	A	**B**	**C**
1	87		
2	87		
3	85		
4	85		
5	82		
6	82		
7	81		
8	80		
9	80		
10	80		
11	80		
12	80		

Figure 1-15 ●

The Graphics Calculator

S t a t   F a c t

Today's computers and calculators make it possible for the data analyst to expend more computation on a single problem than was expended on the world's yearly total of statistical computation 70 years ago.

A graphics calculator can perform a wide range of calculations on data and can also be used to construct some types of statistical graphs.[9] In addition to these capabilities, the graphics calculator can sort data. The next example illustrates this function.

[9]The Texas Instruments TI-82 will be used for illustration purposes throughout this text, but *any* graphics calculator may be used.

Appendix A-10, "User's Guide to the TI-85 Graphics Calculator," has specific information on how to modify procedures given in the text for the TI-85 graphics calculator.

Figure 1-16 is a sketch of a TI-82 graphics calculator, with important landmarks identified.

Figure 1-16

Example
8

The Environmental Protection Agency developed an indicator called the Pollutant Standard Index (PSI) to measure air quality in urban areas. The PSI indicates, as a single number, emission levels of five major pollutants. The following chart shows the numbers of days in the years 1987–90 on which the PSI exceeded 100, in 15 major urban areas of the United States.

City	Total Number of Days PSI Exceeded 100
Atlanta	53
Boston	18
Chicago	32
Dallas	17
Denver	70
Detroit	41
Houston	116
Kansas City	12
Los Angeles	788

City	Total Number of Days PSI Exceeded 100
New York	70
Philadelphia	100
Pittsburgh	68
San Francisco	3
Seattle	27
Washington, D.C.	69

Use a graphics calculator to determine the percentile that corresponds to Denver's PSI level.

First clear any previous data. An entire data list can be cleared at any time by the keystrokes (STAT) (4), followed by the name of the list to be cleared. The lists are designated L1 through L6 and are specified as the secondary function of the 1 through 6 number keys.

Solution   The first step is to sort, or rank, the scores. To enter scores for statistical calculation, first make the following keystrokes:

• Press the (STAT) button.
• Press (1).

The following screen appears, as shown in Figure 1-17. Data may be entered in any one of the lists on the screen, and the arrow keys may be used to move up and down lists or to switch between lists. Data corrections are made by simply typing over.

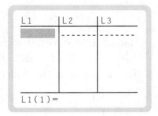

Figure 1-17   The screen of a TI-82 graphics calculator.

The rectangle indicates the active list. Type the scores one at a time, pressing the (ENTER) key after each score. When you have entered the 15th score (Washington D.C.'s 69), the screen should appear as follows:

```
L1     L2     L3
70   ------ ------
100
68
3
27
69
──────────────────
L1(16)=
```

The data in list L1 are now ready to be sorted. Do this by pressing the (STAT) key again, which brings up the following menu:

To sort this data from greatest to least, choose the decreasing sort option, denoted by "SortD(" in the menu, by typing ③. The following dialog screen then appears:

This dialog screen is asking the user to name the list that is to be sorted in descending order. Specify the data in list L1 by typing (2nd) followed by (1). When the (ENTER) key is pressed, the following confirmation appears:

All that remains is to view the sorted list. Do this by pressing (STAT) followed by (1). The up and down arrow keys may be used to scroll through the sorted list in Figure 1-18, and it can be seen that the two scores of 70 (Denver's and New York's) are greater than ten other scores. Thus, Denver's PSI level is higher than the levels of 10 of 15 cities. Using the formula to determine percentiles,

The active cell's position from the top of the list is indicated at the bottom of the screen. If the "SortA(" option had been chosen earlier instead of "SortD(", the scores would have been sorted in ascending order and the position of the active cell would be the data rank of that score.

L1	L2	L3
788	------	------
116		
100		
70		
70		
69		
68		

L1(1) = 788

Figure 1-18

$$P_k = \frac{\text{Number of scores less than } S}{\text{Total number of scores in data}}$$
$$= \frac{10}{15}$$
$$\approx .667.$$

Therefore, Denver's PSI level corresponds to approximately the 67th percentile.   •

Exercises 1.6

Some of the following exercises can be worked efficiently with a graphics calculator or with a spreadsheet or other software program that can sort. However, these exercises can also be done by hand.

1. How are stanines like deciles? How are they different from deciles?

2. What are some of the advantages and disadvantages of the use of stanines? Why does their use seem to be restricted to the reporting of scores for younger children?

3. Ideally, what percentage of the scores would be between the middle of the second stanine and the middle of the seventh stanine?

4. For the eighth-grade class of Example 7 on page 25, determine the cutoff scores for the higher- and lower-ability classes if an arbitrary decision were made to make the higher-ability class slightly smaller.

5. For the eighth-grade class of Example 7 on page 25, 30 additional students enroll, with the following placement test scores:

50	78	82	38	29	82	45	75	67	52
77	69	27	85	80	74	67	66	81	12
48	68	76	74	82	24	11	38	47	81

The decision is made to form two classes of higher ability, one class of lower ability, and four regular classes. All class sizes must be greater than 12 but less than 20. If all the classes are to be as close as possible to the same size, use stanines to find the placement test scores that would define these three groups.

6. Refer to Example 3 on page 8 and Exercise 12 (in Exercises 1.1–1.3) on page 13. Identify which states, if any, have numbers of female physicians that rank significantly different from their numbers of male physicians. Consider a data-rank difference of 5 or more as "significantly different."

   Hint: If you are using a spreadsheet, you will enter the data in four columns, started as follows:

	A	B	C	D	E
				Spreadsheet (SS)	
1	AL	350	AL	1446	
2	AK	38	AK	127	
3	AZ	425	AZ	1267	

   Once all the data is entered, select both columns A and B, choose **Sort** from the **Edit** menu, and make B the key column for sorting. This will place the correct state next to the sorted scores in column B. Repeat this process with columns C and D.

GROUP

7. The following chart shows the percentages of women in 15 selected occupations for three different years.

Occupation	1975	1985	1992
Librarian	81.1	87.0	87.6
Registered nurse	97.0	95.1	94.3
Waitress	91.1	84.0	79.8
Dental assistant	100.0	99.0	98.6
Dentist	1.8	6.5	8.5
Elementary school teacher	85.4	84.0	85.4
College-level teacher	31.1	35.2	40.9
Physician	13.0	17.2	20.4
Carpenter	0.6	1.2	1.0
Data entry worker	92.8	90.7	84.9
Computer programmer	25.6	34.3	33.0
Mail carrier	8.7	17.2	27.8
Airline pilot	0.0	2.6	2.3
Bartender	35.2	47.9	55.0
Social worker	60.8	66.7	68.9

   a. Are there any occupations that rank somewhere in the top five for all three years? Which occupations rank somewhere in the bottom five for all three years?

b. Which of these occupations have the most consistent rankings over the three years shown? Which have the most variable?

Hint: See the hint for Exercise 6. Use of the **Copy** *and* **Paste** *options from the* **Edit** *menu can reduce the amount of typing.*

GROUP

8. Some of the most popular majors among incoming freshmen for 1990 and 1993 are as follows:

Major	Percentage 1990	1993
Accounting	5.3%	5.5%
Arts	2.0	1.7
Business administration	5.5	4.1
Communications	2.3	2.0
Electrical engineering	2.5	2.1
Elementary education	5.1	4.9
English language and literature	1.0	1.4
General biology	1.8	2.9
Law enforcement	1.7	2.6
Management	4.0	2.9
Marketing	2.5	1.7
Mechanical engineering	2.0	1.9
Nursing	4.2	5.5
Political science	3.0	2.4
Premedical, predental, preveterinary	3.2	4.0
Psychology	4.2	4.5
Secondary education	1.9	1.9
Therapy	2.3	4.9

Stat Fact

It has been predicted that about one-third of all college graduates between 1990 and 2005 will take jobs that do not require their degrees.

a. Which majors rank somewhere in the top three for both years? Which rank somewhere in the bottom three for both years?

b. If "significantly different" is defined as having a data rank difference of at least 3, then which majors, if any, rank significantly different over these two years? If "significantly different" is defined as having a data rank difference of at least 6, then which majors rank significantly different?

c. Identify the major(s) that have increased the most and those that have decreased the most in popularity from 1990 to 1993.

GROUP

9. Some sociologists feel that there is a positive association between the property crime of burglary and the violent crime of murder.[10]

[10] As the rate of one crime increases, so does that of the other crime.

The following table lists the occurrences of these two crimes, per 100,000 of the population, for each of the 50 states and the District of Columbia.

Occurrences of Crimes per 100,000 of the Population

	Murder	Burglary		Murder	Burglary		Murder	Burglary
AL	11.6	1103	AK	7.5	894	AZ	7.7	1670
AR	10.3	1211	CA	11.9	1345	CO	4.2	1250
CT	5.1	1228	DE	5.0	971	DC	77.8	1983
FL	10.7	2171	GA	11.8	1619	HI	4.0	1228
ID	2.7	813	IL	10.3	1063	IN	6.2	943
IA	1.9	808	KS	4.0	1167	KY	7.2	767
LA	17.2	1438	ME	2.4	823	MD	11.5	1120
MA	4.0	1113	MI	10.4	1143	MN	2.7	907
MS	12.2	1251	MO	8.8	1066	MT	4.9	709
NE	2.7	724	NV	9.7	1367	NH	1.9	736
NJ	5.6	1017	NM	9.2	1739	NY	14.5	1161
NC	10.7	1530	ND	0.8	427	OH	6.1	983
OK	8.0	1448	OR	3.8	1135	PA	6.7	729
RI	4.8	1271	SC	11.2	1380	SD	2.0	527
TN	10.5	1264	TX	14.1	1852	UT	3.0	881
VT	2.3	1087	VA	8.8	731	WA	4.9	1263
WV	5.7	657	WI	4.6	751	WY	4.9	631

Source: *The Universal Almanac, 1993*, p. 234.

a. What percentage of states (and D.C.) rank similarly for both crimes? A state "ranks similarly" if the difference of its data ranks for the two crimes is less than 10. (If there are ties in data ranks, use the smallest possible difference.)

b. Does this data seem to support the claim that the two crimes are positively associated? Why or why not?

GROUP   10. The following chart shows the percentages of the popular vote, by state, for the Republican candidates for president in the years 1980, 1984, and 1988.

State	1980	'84	'88	State	1980	'84	'88	State	1980	'84	'88
AL	48.8	60.5	59.2	AR	48.1	60.5	56.4	AK	54.3	66.7	59.6
AZ	60.6	66.4	60.0	CA	52.7	57.5	51.1	CT	48.2	60.7	52.0
CO	55.1	63.4	53.1	DE	47.2	59.8	55.9	FL	55.5	65.3	60.9
GA	41.0	60.2	59.8	HI	42.9	55.1	44.8	ID	66.5	72.4	62.1
IL	49.6	56.2	50.7	IN	56.0	61.7	59.8	IA	51.3	53.3	44.5
KY	49.1	60.0	55.5	KS	57.9	66.3	55.8	LA	51.2	60.8	54.3
MA	41.9	51.2	45.4	ME	45.6	60.8	55.3	MS	49.4	61.9	59.9
MT	56.8	60.5	52.1	MI	49.0	59.2	53.6	MN	42.6	49.5	45.9

State	1980	'84	'88	State	1980	'84	'88	State	1980	'84	'88
MO	51.2	60.0	51.8	MD	44.2	52.5	51.1	NH	57.7	68.6	62.4
NC	49.3	61.9	58.0	NY	46.7	53.8	47.5	NJ	52.0	60.1	56.2
NM	54.9	59.7	51.9	ND	64.2	64.8	56.0	NV	62.5	65.8	58.9
NE	65.9	70.6	60.2	OR	48.3	55.9	46.6	OH	51.5	58.9	55.0
OK	60.5	68.6	57.9	PA	49.6	53.3	50.7	RI	37.2	51.7	43.9
SD	60.5	63.0	52.8	SC	49.4	63.6	61.5	TN	48.7	57.8	57.9
TX	55.3	63.6	56.0	UT	72.8	74.5	66.2	VT	44.4	57.9	51.1
VA	53.0	62.3	59.7	WI	47.9	54.2	47.8	WY	62.6	70.5	60.5
WA	49.7	55.8	48.5	WV	45.3	55.1	47.5				

Source: *Presidential Elections Since 1789*, 1991, 5th ed., pp. 139–141.

Identify the states that have been in the group made up of the 10 highest popular voting percentages in all three elections. Also identify the states that have been in the 10 lowest in all three elections.

Case Study   *School Report Cards*

Can the gap between low- and high-achieving public schools be traced to factors such as teacher and administrator salaries and spending per pupil? *(Jeff Dunn/Picture Perfect ™)*

In 1985 the General Assembly of the State of Illinois adopted a statewide testing program as part of its Education Reform Act. The law required each public elementary school and high school to make an annual report card available to the public. This report card contains standardized test scores as well as demographic and financial information about the school.

On November 30, 1992, the *Chicago Tribune* published some findings based on the most recent set of school report cards. Part of those findings follow.

ELEMENTARY SCHOOL FINDINGS
Per Pupil Spending

Top 10	Annual Average	Bottom 10	Annual Average
Avoca, Wilmette,	$8589	Homer, Lockport	$2945
Glencoe	8536	Kirby, Tinley Park	3213
Golf	8749	McHenry	3206
Kenilworth	8418	Mokena	3133
Niles Dist. 71	8706	Richmond	3214
Northbrook Dist. 28	9553	Sandridge, Lynwood	3037
Rhodes, River Grove	8204	Taft, Lockport	2849
Rosemont	9287	Union, Joliet	2916
Roundout, Lake Forest	9994	Will County, Lockport	3160
W. Northfield, Northbrook	8397	Winthrop Harbor	2959

Teacher Salaries

Top 10	Annual Average	Bottom 10	Annual Average
Butler, Oak Brook	$46,352	Alden-Hebron	$24,165
Deerfield	46,096	Elwood	24,882
Elmhurst	46,857	Fox River Grove	24,981
Elmwood Park	48,172	Hazel Crest	26,646
Glencoe	48,213	Millburn, Wadsworth	26,951
Hinsdale	48,149	Richmond	24,386
Kenilworth	52,228	Sandridge, Lynwood	21,462
Skokie Dist. 68	49,528	Spring Grove	23,909
Wilmette	46,688	Willow Springs	22,799
Winnetka	49,528	Union, Joliet	24,079

I. Find the median per-pupil spending for the elementary schools whose spendings rank in the top 10 and the median per-pupil spending for the elementary schools whose spendings rank in the bottom 10. Then find the median teacher salary for schools with salaries in the top 10 and the median teacher salary for schools with salaries in the bottom 10.

II. Using the medians determined in Exercise I above, approximately how many times greater is the median per-pupil spending for the schools in the top 10 than for the schools in the bottom 10? Does this ratio match that of median teacher *salaries* in the top 10 and bottom 10 schools? If not, which of the two categories (per-pupil spending or teacher salaries) shows a greater difference?

III. The above mentioned findings appeared in Section 2 of the November 30, 1992, issue of the *Chicago Tribune,* on pp. 5–8. Read the article at your library and find the median administrator's salary for the elementary schools that have the top 10 administrator salaries. Also find the median administrator's salary for the elementary schools that have the bottom 10 administrator salaries. Approximately how many times greater is the median administrator salary for schools in the top 10 than that for schools in the bottom 10?

IV. The *Chicago Tribune* article also contains information on the top 15 and bottom 15 Chicago-area public high schools in terms of average scores on the ACT test. Find the median average ACT score for the top 15 schools and the median average ACT score for the bottom 15 schools. Approximately how many times greater is the median for the top 15 schools than the median for the bottom 15 schools?

V. Do you think that the gap between the top and bottom schools, as measured by per-pupil spending, teacher and administrator salaries, and ACT test results, has narrowed or widened since 1992? Research more up-to-date data and write a

CHAPTER SUMMARY

The field of statistics deals with the analysis of data. "Data" is a plural noun, and individual items of data in this text are often referred to as "scores." In this chapter, the data are ungrouped and are recorded in arbitrary forms, or arrays. Such data must be transformed into more orderly structures and are usually sorted for further examination.

A stem-and-leaf display is an efficient method of sorting data by hand and can provide a good picture of the distribution of scores. Also, graphics calculators and many computer software programs, such as spreadsheets and databases, can be used to sort data. In addition, most statistical software programs and some word processing programs can sort data.

Data rank is the position of any given score from the least score, or "bottom." The median of a set of data is the score that is ranked in the middle; there are as many scores ranked below it as above it. The median is found by examining the stem-and-leaf display or sorted list and counting off scores.

The kth percentile of a set of scores, denoted by P_k, is a number that, ideally, divides the data into two parts in such a way that k% of the data are less than P_k and $(100 - k)$% of the data are greater than P_k. As the number of scores in the data set increases, percentiles can usually be determined with more accuracy.

The percentile to which a given score, S, corresponds can be found by the formula

$$\frac{\text{Number of scores less than } S}{\text{Total number of scores in data}}.$$

This fraction is then written as a percentage and rounded accordingly.

Finding the score that corresponds to a given percentile is more difficult. Given P_k, the first thing that must be calculated is its data rank, R, by the formula

$$R = \frac{k}{100} \cdot n,$$

where n is the total number of scores in the data. If R is not a whole number, then R is always rounded up, and P_k is equal to the score with data rank R. If R is a whole number, then P_k is the mean of the scores with data ranks R and (R + 1).

Deciles, quintiles, and quartiles are percentiles that are multiples of 10, 20, and 25, respectively. Stanines divide the data into nine equal parts. The range is the difference of the highest score and the lowest score; the interquartile range (IQR) is $P_{75} - P_{25}$.

Key Concepts and Terms

Videotape Suggestions

Program 1 from the series *Against All Odds: Inside Statistics* is an overview of some of the applications of statistics. *Program 2* from the same series provides more information on stem-and-leaf displays and could be a preview of future work in picturing distributions.

Information about acquiring these programs can be obtained by calling 1-800-LEARNER (1-800-532-7637).

(Courtesy of NASA)

Graphic methods
facilitate the visual
interpretation of multiple
data sets, as
represented here by the
rings of Saturn.

C h a p t e r

2 Comparing Several Sets of Data

Often it is necessary to compare two or more sets of data side by side in order to study differences and trends. Achievement test scores of men and women, reaction times of different drugs, profit margins of several companies, crime rates in different communities, and automobile driving records for various age groups are just a few examples of comparisons that are made every day.

This chapter explains several graphic methods that make the visual interpretation of data easier and more efficient. Box plots, confidence intervals for the median, and other devices are introduced, and the **case study** at the end of the chapter uses these tools to investigate the ages and salaries of the nation's top executives.

2.1 BIVARIATE AND MULTIVARIATE DATA

A set of data may contain observations about one, two, or any number of different variables. Graphic methods of comparing data are especially helpful when two or more variables are involved.

Definition ▶

A set of data is said to be **bivariate** if it contains information on two variables or **multivariate** if it contains information on more than two variables.

For example, Exercise 10 (in Exercises 1.6) on page 34 contains a data set that may be considered multivariate since it contains information on three variables: the Republican voting percentages for 1980, 1984, and 1988. This chapter concentrates on the comparison of quantitative variables. Chapter 11 will describe methods of comparing qualitative variables.

Consider the following situation. A medical researcher believes that she has developed a screening device that can detect teenagers who have higher-than-normal cholesterol counts. Forty teenagers were identified by this device and had their cholesterol counts taken. Then 40 teenagers were chosen at random and had their cholesterol counts taken. The following table shows the results. Group H is the group of teenagers who were identified as possibly having higher-than-normal cholesterol counts, and Group R is the group chosen at random.

Cholesterol Counts

Group H				Group R			
139	157	160	201	160	198	140	130
122	188	194	240	138	85	98	108
200	180	230	179	126	208	178	170
194	190	180	188	175	208	161	195
179	194	140	136	130	120	110	128
180	193	159	152	223	98	163	165
176	157	178	206	150	186	248	83
160	120	176	192	186	161	165	188
203	210	176	188	163	145	126	195
194	189	163	196	210	123	116	164

The next section explores several different graphic approaches to comparing these two sets of data. Although graphics calculators and software were used to construct some of the following displays, all of the work in this chapter (and later ones) can be done by hand.

2.2 GRAPHIC REPRESENTATIONS OF DATA

There are several ways to illustrate the distribution of scores in a set of data. This chapter will study dot plots, five-number summaries, and box plots. Box plots will be used in a variety of ways, including an early introduction to confidence intervals. Additional graphic representations, histograms and frequency curves, will be studied in Chapter 3 and applied through the remainder of the text.

Dot Plots

A simple way to graphically compare two sets of data is through the construction of separate dot plots.

Definition ▶   A **dot plot** is a display in which each score is shown as a dot along a horizontal axis.

These dot plots were drawn with the statistical software program Minitab, which will be featured throughout this text. Minitab was originally developed at The Pennsylvania State University in 1972 to teach college-level basic statistics courses. It is now used for teaching and research at more than 2000 institutions around the world.

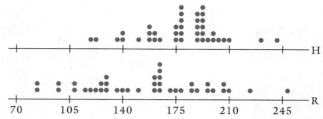

Figure 2-1 Dot plots of cholesterol data.

S t a t   F a c t

A recent research project aimed at identifying causes of high cholesterol found the typical cholesterol level of Japanese living in Japan to be 180. However, Japanese living in Hawaii had levels near 220, and Japanese living in San Francisco had levels close to 230.

Because they show a great amount of detail, dot plots are most useful when the data set is not very large. Histograms, to be studied in the next chapter, are better for treating larger amounts of data.

On a dot plot, each score is represented by a dot. When several sets of data are being compared, their dot plots should all be drawn to the same scale. Figure 2-1 shows two dot plots for the cholesterol data. An examination of the dot plot for the random group (R) shows a slight clustering of scores around 165, with the rest fairly evenly spread between about 100 and 200. There are a few extremes on each side. The dot plot for the group of teenagers identified as possibly having high cholesterol counts (H) is different. This group's scores cluster about two points, each of which is greater than the 165 for the random group. Also, the rest of the scores seem to be distributed less evenly than those of the random group, and there are no low extremes.

Before drawing any conclusions from these dot plots, a statistician might want to examine some key statistics[1] provided by the data on which the plots were based.

The Five-Number Summary

When comparing sets of data, it is often useful to quickly pick out some key scores. The minimum and maximum scores, the median, and the lower and upper quartiles help to define the distribution of scores in a set of data. These key scores are written in a five-number summary and graphically illustrated in a box plot.

Definition ▶   A **five-number summary** is a display that quickly shows the minimum and maximum scores, the median (P_{50}), the lower quartile (P_{25}), and the upper quartile (P_{75}).

These data are usually presented in rectangular form, sometimes with a box drawn around them as in Figure 2-2.

---

[1] For now, the word "statistic" will refer to any measure that can be used to describe the data set. Thus, range, IQR, median, and so on, will be called statistics. Chapter 7, however, will refine this definition.

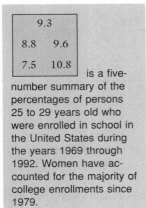 is a five-number summary of the percentages of persons 25 to 29 years old who were enrolled in school in the United States during the years 1969 through 1992. Women have accounted for the majority of college enrollments since 1979.

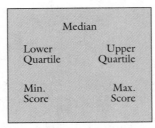

Figure 2-2 The five-number summary format.

Figure 2-3 Five-number summaries of cholesterol counts.

Five-number summaries are useful for comparing the distributions of scores from several sets of data. Notice that the range (the difference of the maximum and the minimum scores) and the IQR (the difference of the third and first quartiles) are easily computed from a five-number summary. Figure 2-3 shows five-number summaries for the two cholesterol groups.

The information presented in five-number summaries can be obtained in several different ways:

- A stem-and-leaf display can be constructed by hand and then used to determine the percentiles, as explained in Chapter 1.
- A spreadsheet program can be used to sort the data, and then the percentiles can be determined by hand.
- A graphics calculator can be used.
- A statistical software program can be used.

2.3 *Technology* TOOLS

Using a Graphics Calculator to Determine a Five-Number Summary

As an example, consider using a TI-82 graphics calculator to find all the information needed for the five-number summary of Group H.

- Clear a list—say, L1.

This is accomplished with the keystroke sequence

$\boxed{\text{STAT}}$, $\boxed{4}$, $\boxed{\text{2nd}}$, $\boxed{1}$, $\boxed{\text{ENTER}}$.

- Enter all of the data from Group H in list L1. (Press $\boxed{\text{STAT}}$, $\boxed{1}$ to begin entering scores.)

Next, go back to the menu by pressing $\boxed{\text{STAT}}$:

```
EDIT CALC
1: Edit...
2: SortA(
3: SortD(
4: ClrList
```

Move the active prompt to the calculation menu (CALC).

- Press the right arrow key, $\boxed{\blacktriangleright}$.
- Type $\boxed{1}$ to choose 1-variable statistics.

This screen then appears:

```
1-Var Stats
```

The user is now being asked to name the list for which 1-variable statistics are to be calculated. The data in list L1 are specified by typing $\boxed{\text{2nd}}$ followed by $\boxed{1}$. After the $\boxed{\text{ENTER}}$ key is pressed, the down-arrow key, $\boxed{\blacktriangledown}$, can be used to scroll down the screen and show these calculations:

```
1-Var Stats
↑ n=40
  minX=120
  Q1=160
  Med=180
  Q3=194
  maxX=240
```

The information shown is exactly what is needed to fill out a five-number summary.

Using Statistical Software to Determine a Five-Number Summary

The statistical software program Minitab is very versatile, powerful, and easy to use.[2] It can perform all of the calculations necessary for a five-number summary. First enter all of the scores from Group H in column C1, and all of the scores from Group R in column C2, of a Minitab worksheet. Figure 2-4 shows how to begin.

Next, pulling down the **Stat** menu and choosing **Basic Statistics** with the **Descriptive Statistics** option results in the following calculations:

Minitab: Descriptive Statistics Calculations

	N	MEAN	MEDIAN	TRMEAN	STDEV	SEMEAN
H	40	178.98	180.00	179.08	25.99	4.11
R	40	155.60	161.00	155.14	39.39	6.23

	MIN	MAX	Q1	Q3
H	120.00	240.00	160.00	194.00
R	83.00	248.00	126.00	186.00

The DOS version of Minitab opens onto a Session window. The data screen is then entered through the Edit menu or by pressing ALT + D. Minitab's worksheets, or data screens, resemble spreadsheets.

	C1 Group H	C2 Group R	C3
1	139	160	
2	122	138	
3	200	126	
4	194	175	
5	179	130	
6	180	223	
7	176	150	
8	160	186	

Figure 2-4 A Minitab worksheet.

[2] Minitab Release 8, the Macintosh version, has been used for illustrations in this text, but for these applications other releases are extremely similar. Minitab is also available in DOS and Windows versions for PCs.

Every version of Minitab has an interactive feature with pull-down menus and can be used with macro commands (which are nearly the same in all versions) to automate repetitive chores or extend Minitab's capabilities. The output of the Macintosh version and the outputs of the PC versions are practically identical.

Unless otherwise specified, all the features illustrated with the Macintosh version in this text can be accomplished with a minimal number of keystrokes in the DOS version of Minitab.

The rows H and R indicate the two groups, and N is the number of scores in each group. The statistics MEAN and TRMEAN (trimmed mean) will be explained in the next chapter, STDEV in Chapter 7, and SEMEAN (standard error of the mean) in Chapter 8. The remaining statistics, MEDIAN, MIN, MAX, Q1, and Q3 are exactly the ones needed for a five-number summary.[3]

2.4   BOX PLOTS

Although five-number summaries give a sense of the overall distribution of scores in a set of data, they are not visually effective.

Definition ▶   A **box plot** is a display that presents the information from a five-number summary in graphic form.

Box plots are often preferred over five-number summaries as a first step in data analysis because of the ease with which they can be read. Figure 2-5 illustrates a box plot.

Box plots in this text use the quartiles as dividing points. Unfortunately, agreement on this is not universal, and some statisticians use, instead of Q_1 and Q_3, two other points called the upper and lower **hinges,** which represent the medians of the upper half and lower half of the scores, respectively. In nearly all practical situations, however, the hinge values are so close to Q_1 and Q_3 that there is little or no difference. Therefore, in the interest of simplicity, this text uses Q_1 and Q_3 to construct box plots.

Box plots are sometimes called box-and-whisker displays.

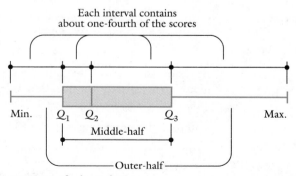

Figure 2-5 The structure of a box plot.

[3]Percentiles computed by hand and those obtained with a computer do not always agree exactly, as these do. Minitab uses a rule that involves interpolation to find percentiles, whereas the manual method shown in Chapter 1 does not. In general, Minitab provides more refined statistics than it would be practical to try to obtain by hand.

Box plots may also be drawn vertically, with one placed next to the other.

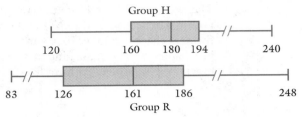

Figure 2-6 Box plots of cholesterol counts.

The visual form of a box plot makes the information on a five-number summary easier to visualize. It is important to remember that the four intervals defined on a box plot have about the same numbers of scores. This implies that Figure 2-5 on page 47 shows data that has a clustering of scores somewhere between Q_1 and Q_2, with the higher scores (those above Q_3) more spread out than the lower scores.

When using box plots to compare several sets of data, it is important to draw both to the same scale and to align them properly relative to each other. Hand-drawn box plots for the information in the five-number summaries of the cholesterol groups would resemble Figure 2-6. These box plots supply some of the same information contained in the dot plots. The smaller box for Group H signifies more clustering in the middle half. The extreme low scores for Group R are readily seen. Quartiles can easily be compared.

These two box plots could also be drawn with a TI-82 graphics calculator, using the following steps.

- Clear lists L1 and L2.

This is accomplished with the keystroke sequence

[STAT], [4], [2nd], [1], [ENTER],
[STAT], [4], [2nd], [2], [ENTER].

- Enter all of the data from Group H in list L1 and all of the data from Group R in list L2. (Press [STAT], [1] to begin entering scores.)
- Press [2nd], [Y=] to gain access to the statistical plotting capabilities. Then press [1].

By using the arrow ([▶]) and [ENTER] keys, turn on plot 1, choose the box plot graph icon, specify the location of the data ("Xlist") as list L1, and specify an individual score frequency of 1. The final screen appears at left in Figure 2-7.

- Repeat the preceding step for the data in list L2.

That is, press [2nd], [Y=], [2]. Use the [▶] and [ENTER] keys as before, but in this case specify Xlist as L2.

- Return to the statistical plot menu by [2nd], [Y=] to double-check for accuracy.

The screen should appear as in Figure 2-8.

Notice that two plots are turned on (since two box plots are desired) and the rest are turned off. Also verify that box plots are chosen, with the correct data location specified.

More than one list of data may also be cleared by simply using commas to separate list names. In this case, [STAT], [4], [2nd], [1], [,], [2nd], [2], [ENTER] will clear data lists L1 and L2.

Figure 2-7

Figure 2-8

The specified values for Xscl, Ymin, Ymax, and Yscl are graphing parameters and are not statistical.

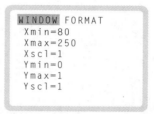

Figure 2-9 Limits of the viewing window for box plots.

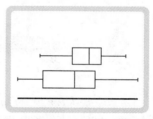

Figure 2-10 Box plots of cholesterol data produced by a TI-82 graphics calculator.

- Press [WINDOW] and set reasonable limits for the display (Figure 2-9).
- Press the [GRAPH] key to obtain the screen in Figure 2-10.

Pressing the [TRACE] key will cause a cursor to appear on the box plot and a statistic in the lower left corner of the screen. The arrow keys may then be used to move the cursor to different locations on the box plot.

Although the values of the minimum, maximum, and quartiles do not appear on the box plot screen, they can easily be shown on the screen by using the [TRACE] and arrow keys.

The median cholesterol count of the group identified by the screening device is definitely greater than that of the group chosen at random. Does this imply that the device actually does what it claims to do? The results may be due merely to luck—who was picked and who wasn't. If a different group were chosen at random, would the results, with reasonable certainty, be the same? Section 2.4 will examine more closely the influence of chance on the analysis of data.

Exercises 2.1–2.4

1. Name some qualitative variables that would yield bivariate or multivariate data.

2. In what ways is a dot plot similar to a stem-and-leaf display?

3. Each of the four intervals of a box plot contains about one-fourth of the scores. Under what conditions would each interval contain *exactly* one-fourth of the scores?

4. Which statistic gives the length of the box part of a box plot? Which one gives the length of the entire box plot?

GROUP   5. Sketch a box plot to correspond to each of the following dot plots.

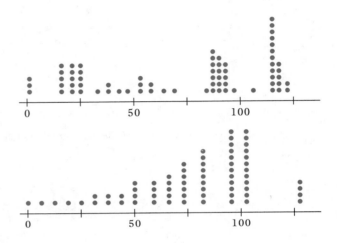

The five-number summaries required for Exercises 6–10 can easily be obtained with a graphics calculator, as shown at the start of Section 2.3.

GROUP   6. In 1991 a survey was conducted to compare the hourly wages of newly licensed registered nurses (RNs) with those of experienced RNs. The following table summarizes some of the results.

Stat Fact

A July 1994 survey of salaries by the College Placement Council found that pay for new nursing graduates declined by more than 4% last year. However, the same survey found that starting salaries in nursing are still higher than those in most other fields.

City	Newly Licensed RNs	Experienced RNs
Atlanta	$10.34 to $12.00	$15.00 to $18.50
Boston	$11.05 to $15.93	$15.15 to $26.93
Chicago	$10.66 to $13.56	$14.50 to $20.48
Denver	$11.00 to $12.61	$13.24 to $19.95
Detroit	$10.34 to $13.82	$11.37 to $18.38
Houston	$9.05 to $15.02	$12.05 to $19.90
Los Angeles	$11.25 to $16.60	$16.12 to $24.60
Miami	$9.16 to $13.52	$13.74 to $20.30
Milwaukee	$9.67 to $11.54	$12.47 to $18.13
Minneapolis–St. Paul	$12.09 to $12.48	$16.84 to $17.89
New York City (Manhattan)	$13.70 to $15.87	$16.10 to $21.39
Philadelphia	$10.00 to $10.52	$13.26 to $15.34
Washington, D.C.	$13.96 to $18.85	$16.76 to $22.06

Source: *Nurse's FactFinder,* Springhouse Corporation, 1991, p. 7.

a. Use the mean of the two pay extremes indicated in each category to construct two five-number summaries that compare the wages of new and experienced RNs.

b. Construct box plots to illustrate graphically the five-number summaries from Part a.

c. State some reasons why the pay for experienced RNs seems to be more variable.

GROUP

7. Exercise 13 (in Exercises 1.1–1.3) on page 13 gave the ages of the last 45 actresses to receive the Oscar for best actress. Following are the ages of the last 45 actors to receive the Oscar for best actor:

Stat Fact

The Academy Awards draws one of the largest audiences for television advertising, particularly female-directed advertising. The 1994 Academy Awards viewers were estimated to be 65% female, with over half of them individually earning more than $40,000 per year. 70 million people watched the Academy Awards and each minute of advertising cost an average of $1.37 million.

Actor	Age	Actor	Age
Broderick Crawford	38	Marlon Brando	48
Jose Ferrer	38	Jack Lemmon	48
Humphrey Bogart	52	Art Carney	56
Gary Cooper	51	Jack Nicholson	38
William Holden	35	Peter Finch	60
Marlon Brando	30	Richard Dreyfuss	32
Ernest Borgnine	38	Jon Voight	40
Yul Brynner	41	Dustin Hoffman	42
Alec Guinness	43	Robert De Niro	37
David Niven	49	Henry Fonda	76
Charlton Heston	35	Ben Kingsley	39
Burt Lancaster	47	Robert Duvall	55
Maximilian Schell	31	F. Murray Abraham	45
Gregory Peck	46	William Hurt	35
Sidney Poitier	39	Paul Newman	61
Rex Harrison	56	Michael Douglas	33
Lee Marvin	41	Dustin Hoffman	51
Paul Scofield	44	Daniel Day-Lewis	32
Rod Steiger	42	Jeremy Irons	42
Cliff Robertson	43	Anthony Hopkins	54
John Wayne	62	Al Pacino	52
George C. Scott	43	Tom Hanks	37
Gene Hackman	40		

Source: *The World Almanac and Book of Facts, 1995*, pp. 327–329, 357–370.

Draw separate dot plots, five-number summaries, and box plots for the data in this exercise and for the data in Exercise 13 (in Exercises 1.1–1.3) on page 13. Use them to answer the following questions.

a. What is the difference in median ages?

b. What percentage of the actors were younger than the median age of the actresses when they received their Oscars?

c. What percentage of the actresses were younger than Q_3 for the actors?

d. Identify the actresses whose ages were closest to the median age of the actors.

e. What ages are in the middle half (second and third quartiles) of both groups, and to whom do they belong?

8. In 1993 the 15 states with the largest populations responded to a survey that asked for the percentages of their populations completing different levels of schooling. Some of the data in this survey were broken down as follows:

State	Less than High School	At Least a Bachelor's Degree
CA	20.3	25.0
FL	20.4	19.8
GA	25.3	21.1
IL	20.8	22.1
IN	20.8	14.1
MD	17.4	26.1
MA	15.4	30.0
MI	18.5	19.1
NJ	17.9	27.9
NY	19.3	24.2
NC	25.2	18.5
OH	17.2	19.5
PA	20.2	18.7
TX	22.7	22.1
VA	19.3	25.8

Source: *Digest of Education Statistics, 1994*, p. 23.

a. Construct two five-number summaries for these data, one in terms of less than a high school education and the other in terms of at least a bachelor's degree.

b. On the basis of these summaries, might any general conclusions be drawn that could be valid for the entire country? If so, what are they?

c. Four states rank the same from the top of one category as they rank from the bottom of the other. Name those states and explain why this "inverse ranking" occurs.

GROUP   9. Following are the 15 top-rated TV shows for the decades of the 1950s, 1960s, 1970s, and 1980s, along with their average ratings in each decade.

	1950s		1960s	
Program		*Average Rating*	**Program**	*Average Rating*
1. "Arthur Godfrey Talent Scouts"		32.9	1. "Bonanza"	29.6
2. "I Love Lucy"		31.6	2. "The Red Skelton Show"	26.4
3. "You Bet Your Life"		30.1	3. "The Andy Griffith Show"	22.4

Stat Fact

Males		Females	
5.4		5.2	
4.7	6.5	4.6	5.9
3.8	7.5	3.8	6.7

These two five-number summaries represent annual American high-school drop-out rates for those in grades 10–12 during the years 1967 through 1992.

1950s		1960s	
Program	*Average Rating*	*Program*	*Average Rating*
4. "Dragnet"	24.6	4. "Beverly Hillbillies"	21.9
5. "The Jack Benny Show"	22.3	5. "The Ed Sullivan Show"	21.7
6. "Arthur Godfrey and Friends"	19.5	6. "The Lucy Show"	21.3
7. "Gunsmoke"	15.6	7. "The Jackie Gleason Show"	16.5
8. "The Red Skelton Show"	15.2	8. "Bewitched"	14.8
9. "December Bride"	13.8	9. "Gomer Pyle"	13.4
10. "I've Got a Secret"	12.9	10. "Candid Camera"	11.2
11. "$64,000 Question"	11.2	11. "The Dick Van Dyke Show"	11.1
12. "Disneyland"	10.8	12. "The Danny Thomas Show"	10.7
13. "The Ed Sullivan Show"	10.6	13. "Family Affair"	9.8
14. "Have Gun—Will Travel"	10.3	14. "Laugh-In"	7.9
15. "The Danny Thomas Show"	9.9	15. "Rawhide"	7.5

1970s		1980s	
Program	*Average Rating*	*Program*	*Average Rating*
1. "All in the Family"	23.1	1. "60 Minutes"	23.5
2. "M*A*S*H"	17.6	2. "Dallas"	21.0
3. "Hawaii Five-O"	16.5	3. "The Cosby Show"	16.9
4. "Happy Days"	15.9	4. "Dynasty"	14.5
5. "The Waltons"	14.0	5. "Knots Landing"	14.2
6. "The Mary Tyler Moore Show"	13.7	6. "Cheers"	14.0
7. "Sanford & Son"	13.4	7. "Magnum, P.I."	13.7
8. "One Day at a Time"	11.4	8. "Murder, She Wrote"	12.9
9. "Three's Company"	10.8	9. "Who's the Boss?"	12.2
10. "60 Minutes"	10.0	10. "Family Ties"	11.8
11. "Maude"	9.8	11. "Falcon Crest"	11.6
12. "Gunsmoke"	9.7	12. "The Golden Girls"	11.0
13. "Charlie's Angels"	9.6	13. "Kate & Allie"	10.6
14. "The Jeffersons"	9.4	14. "Night Court"	9.8
15. "Laverne & Shirley"	9.3	15. "Newhart"	8.7

a. Compare the data for these four decades by constructing a box plot for each decade.

b. Are the medians, ranges, and IQRs remaining fairly constant? If not, how are they changing and what conclusions, if any, can be drawn?

GROUP   10. The following array shows net energy consumption in the United States (in quadrillions of BTUs), broken down in terms of residential and commercial, industrial, and transportation.

Year	Residential and Commercial	Industrial	Transportation
1973	15.766	25.917	18.584
1974	15.246	24.994	18.095
1975	15.200	22.737	18.219
1976	15.997	24.038	19.076
1977	15.828	24.593	19.794
1978	16.023	24.637	20.589
1979	15.709	25.679	20.447
1980	15.075	23.854	19.669
1981	14.541	22.533	19.480
1982	14.629	20.020	19.043
1983	14.395	19.401	19.109
1984	14.964	21.183	19.773
1985	14.839	20.520	20.036
1986	14.791	20.102	20.781
1987	15.152	21.113	21.415
1988	16.012	22.082	22.269
1989	16.270	22.292	22.524
1990	15.636	22.813	22.497
1991	16.089	22.674	22.254

a. Construct three box plots and three five-number summaries that illustrate these data.

b. Which of these three users has the greatest median consumption?

c. Rank these users from least to greatest in terms of their IQRs. What would a relatively small or a relatively large IQR imply about possibilities for future energy conservation by that user?

Excursion 2.4A • USING BOX PLOTS TO COMPARE DIFFERENT AMOUNTS OF DATA

The material and exercises that use box plots to compare different amounts of data are optional. They may be covered at any time or omitted.

In actual practice, different amounts of data are often available for two variables that are to be compared. Fortunately, the numbers of scores represented by the box plots do not necessarily have to be the same. Different amounts of data for two variables can be indicated by adjusting the heights of the boxes in proportion to the relative numbers of scores in the corresponding sets of data, as Figure 2-11 shows.

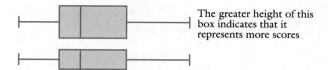

The greater height of this box indicates that it represents more scores

Figure 2-11

E x a m p l e
1

The following arrays show the grade-point averages (GPAs) of 25 students enrolled in an elective freshman humanities course and the GPAs of 40 students enrolled in a required freshman mathematics course, chosen at random.

GPAs for Humanities and Mathematics Students

Humanities				Mathematics				
3.32	3.19	2.97	3.11	3.05	2.11	2.35	3.11	3.17
3.30	3.45	2.75	3.05	3.02	3.45	2.50	2.43	2.05
2.75	2.99	3.72	3.66	3.37	3.15	2.75	2.11	2.98
3.21	3.33	3.71	3.11	3.15	3.17	3.88	3.05	3.52
2.88	2.77	3.51	2.53	2.87	2.91	3.12	3.13	3.56
3.47	1.94	2.77	3.55	2.15	3.90	3.47	2.16	2.72
3.77				2.35	2.41	2.94	3.81	2.75
				2.44	2.65	3.50	2.15	2.33

a. Construct two box plots to compare these data.
b. State three observations that seem to follow from the box plots in Part a.

Solution a. The first step in constructing box plots is to determine the information in a five-number summary. This can be done in several different ways, such as by making separate stem-and-leaf displays, by using a graphics calculator, or by using Minitab.

This example uses the TI-82 graphics calculator as shown on page 44–45 to obtain the five-number summaries.

Humanities

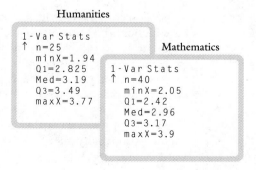

```
1-Var Stats
↑ n=25
  minX=1.94
  Q1=2.825
  Med=3.19
  Q3=3.49
  maxX=3.77
```

Mathematics

```
1-Var Stats
↑ n=40
  minX=2.05
  Q1=2.42
  Med=2.96
  Q3=3.17
  maxX=3.9
```

These screens show exactly the information needed to complete the following five-number summaries.

GPAs for Humanities
and Mathematics Students

Humanities		Mathematics	
3.19		2.96	
2.83	3.49	2.42	3.17
1.94	3.77	2.05	3.90

Because there are almost twice as many mathematics scores as humanities scores, the height of the mathematics box plot should be drawn about twice that of the humanities box plot.

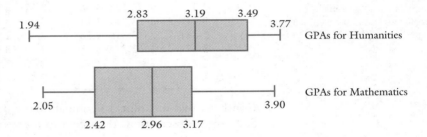

b. One could make three observations by examining these two box plots:

• About twice as many mathematics scores are being examined as humanities scores.
• The upper half of the humanities scores are about as high as the upper fourth of the mathematics scores.
• The scores in the lower fourth of humanities and the upper fourth of mathematics are more spread out than the rest of the scores, suggesting the possibility of a few extreme scores in those areas.

These observations may lead one to suspect that, in general, humanities students have higher GPAs than mathematics students. However, such a conclusion, which would be based on relatively small samples that were not even the same size, could easily be incorrect. It is possible that most of the weaker humanities students were, for some reason, not included. Also, such a conclusion would not take into account certain realities, such as the fact that one course is an elective and the other is required.

Drawing general conclusions from sample observations requires some knowledge of probability theory, which will be explored thoughtfully in Chapter 5A and beyond. •

Exercises 2.4A

1. There seems to be a difference between the lowest recorded temperatures in the western United States and those in the eastern United States. As of 1993, the lowest recorded (Fahrenheit) temperatures for states in these two areas were as follows:

Stat Fact

The lowest recorded temperature at a permanently inhabited location on Earth was $-92°F$, in Siberia.

Western		Eastern			
WA	-48	ME	-48	NC	-34
OR	-54	NH	-46	SC	-19
CA	-45	MA	-35	GA	-17
NV	-50	RI	-23	FL	-2
AZ	-40	CT	-32	AL	-27
UT	-69	NY	-52	MS	-19
ID	-60	PA	-42	LA	-16
MT	-70	MD	-40	TX	-23
		WV	-37		

   a. Draw box plots to compare the temperatures of the two areas.

   b. What western state has a lowest recorded temperature closest to one of the lowest recorded quartile temperatures of the eastern states?

2. In general, do chief justices of the U.S. Supreme Court serve longer than associate justices? To help answer this question, 10 chief justices and 22 associate justices were chosen at random. Their names and the lengths of their terms follow.

Stat Fact

John Rutledge served as an associate justice from 1790 to 1791. President George Washington nominated him for chief justice in 1795. In its only rejection of a chief justice nominee up to that time or since, the Senate refused to confirm him. (Some later chief justice nominees withdrew before the Senate could take action.)

Chief Justice	Term (years)	Associate Justice	Term (years)	Associate Justice	Term (years)
John Marshall	34	David Davis	14	Edward White	16
Earl Warren	16	Potter Stewart	23	John Campbell	8
Melville Fuller	21	Abe Fortas	4	Horace Gray	20
Roger Taney	28	John Blair	6	Gabriel Duval	22
Morrison Waite	14	Levi Woodbury	5	Ward Hunt	9
William Taft	8	Hugo Black	34	William Day	19
Harlan Stone	5	William Brennan	33	William Moody	3
Warren Burger	17	John Rutledge	1	Charles Hughes	5
Fred Vinson	7	Henry Brown	15	Wiley Rutledge	6
John Jay	5	Joseph Lamar	5	Tom Clark	18
Louis Brandeis	22	Bushrod Washington	31		

   a. Draw box plots to compare the length of terms of these chief justices and associate justices.

   b. What conclusions, if any, can be drawn from the box plots in Part a?

   c. What would need to be done in order to answer the question of relative term length with absolute certainty?

GROUP   3. The following list shows average life expectancies for females born in 1994 in selected countries of the world, roughly grouped into four geographic regions.

	Life Expectancy	Population (millions)
North, Central, and South America		
United States	78.60	258.3
Canada	79.79	28.1
Mexico	66.00	90.0
Brazil	67.50	152.0
Chile	75.05	13.5
Costa Rica	77.04	3.0
Ecuador	67.59	10.3
Guatemala	59.43	10.0
Panama	74.10	2.4
Peru	66.58	22.9
Uruguay	74.88	3.0
Venezuela	72.80	20.7
Europe		
Austria	79.02	7.6
Belgium	76.79	10.1
Cyprus	78.13	0.7
Czechoslovakia	75.81	10.3
Denmark	77.70	5.1
Finland	78.90	5.0
France	80.94	57.7
Greece	76.35	10.5
Hungary	73.71	10.3
Italy	79.70	57.8
Netherlands	79.88	15.2
Norway	79.81	4.2
Poland	75.49	38.5
Portugal	77.17	10.2
Spain	79.69	39.1
Sweden	80.41	8.4
Switzerland	80.00	6.5
United Kingdom	78.03	58.0
Yugoslavia	74.48	23.8
Asia		
Bangladesh	55.97	113.9
India	55.67	897.4
Iran	55.04	62.8
Israel	78.09	4.5
Japan	81.81	124.8
Pakistan	59.20	122.4
South Korea	74.96	44.6
Sri Lanka	71.66	17.8
Syria	68.05	13.5

	Life Expectancy	Population (millions)
Africa and Oceania		
Egypt	60.30	58.3
Kenya	60.50	27.7
South Africa	63.50	39.0
Australia	80.01	17.8
New Zealand	77.96	3.4

Compare life expectancies by drawing box plots of different heights to reflect the different numbers of countries in the four groups.

GROUP

4. Adjust the heights of the box plots in Exercise 3 so that they reflect the total populations of the groups of countries.

5. The following list contains a sample of passenger-car production by the three major U.S. auto manufacturers in 1991.

General Motors		Ford Motor Company	
Sunbird	118,615	Cougar	52,157
Grand Am	147,467	Topaz	49,493
Camaro	52,177	Continental	52,406
Corvette	21,082	Sable	110,619
Grand Prix	114,718	Town Car	126,354
Toronado	5,934	Taurus	317,810
Skylark	61,537	Mustang	81,594
Riviera	10,314	Escort	188,450
Century	98,670	Thunderbird	71,395
Cavalier	252,857		
Eldorado	16,276	**Chrysler Corporation**	
Calais	34,616	Sundance	61,605
Ciera	111,628	Daytona	11,492
Delta 88	105,151	Dynasty	114,439
Geo Prizm	94,927	LeBaron	29,388
Reatta	1,210	Viper	3
Bonneville	86,355	Spirit	80,216

Construct three box plots that illustrate these data. Adjust the heights of the boxes to reflect the relative number of cars manufactured by each corporation.

2.5   NOTCHED BOX PLOTS AND CONFIDENCE INTERVALS

Consider again the question of whether one can be reasonably certain that the cholesterol screening device mentioned at the start of this chapter does indeed identify those with high cholesterol counts. Although the median cholesterol counts

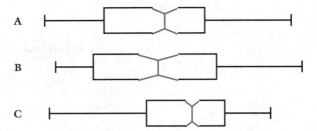

Figure 2-12

of the two samples differ, this difference might be due solely to the chance selection of certain individuals in each group. Are the medians far enough apart to discount the influence of chance? For now, this question will be answered by the use of notched box plots.[4]

The "notching" of box plots is a graphic method of widening the location at which the median is graphed. One reason for notching is to compare several box plots in order to determine whether a difference in medians is due only to the chance selection of data that made up the samples. For many different reasons, unusual or nonrepresentative scores may have been included in the sample data,[5] and the factor of chance selection of scores needs to be considered. It is very unlikely that any two data sets taken from two relatively large groups would have exactly the same median, but the question that must be answered is: Are the medians of the samples different enough to provide strong evidence that the larger groups from which they were taken must have different medians as well?

If it is determined that the difference between the sample medians is great enough to not be due to chance, then that difference is said to be **statistically significant** and implies that the medians of the larger groups from which the samples were taken are also different. The criterion for statistical significance is a very important concept, and it will be developed more throughout the course of this text. For now, however, a difference between two medians is statistically significant if the notched part of one box plot is not vertically in line with any part of the notch of the other box plot. In Figure 2-12, the placement of the medians in box plots A and B could be attributed to chance, since the notches around the medians overlap. However, the difference between the median of the data represented in box plot C and the median shown in either box plot A or B is statistically significant and is most likely not due to chance.

The width of the notch in a given box plot depends upon the IQR, the number of scores in the data set, and the degree of confidence in, or the certainty of, the final conclusion.

[4]Chapter 10 will re-examine this same question by a different method.

[5]Chapter 7 will examine this difficulty more closely.

Formula for Notch Width

$$\text{Notch width} = \frac{1.57 \cdot \text{IQR}}{\sqrt{n}} \qquad\qquad (2\text{-}1)$$

In Formula 2-1, IQR stands for the interquartile range, and $\sqrt{n}$ is the square root of the number of scores in the set of data. The constant 1.57 is a factor that results in a notch centered about the median such that 95% of the time medians whose differences are due to chance result in overlapping notches.[6] The notch forms a confidence interval for the median.

Definitions ▶   A **confidence interval** for a median is an interval constructed in a specific manner, in which the true median can be expected to lie a certain percentage of the time.

The **true median** is the median of the larger group, or population, from which a sample was taken.

Formula 2-1 for determining notch width is adequate for reaching some preliminary conclusions. Many statistical software programs use more sophisticated formulas that provide more refined versions and allow for different percentages to determine the width of the confidence interval.

E x a m p l e 2   Refer to the information in this chapter on the comparison of the cholesterol counts of a group chosen at random and a group identified by a screening device. Determine whether the difference in the medians is statistically significant or due to chance.

---

[6] Exactly how one obtains 1.57 is beyond the scope of this text and is a topic of order statistics.

Solution For convenience, the five-number summaries computed earlier are repeated:

Cholesterol Data

Group H		Group R	
180		161	
160	194	126	186
120	240	83	248

Using Formula 2-1 for notch width,

$$\text{Notch width for Group H} = \frac{1.57 \cdot 34}{\sqrt{40}} \approx 8.44;$$

$$\text{Notch width for Group R} = \frac{1.57 \cdot 60}{\sqrt{40}} \approx 14.89.$$

Next, center the appropriate notch width about each median by marking half of each notch width on each side of the median, obtaining the following diagram:

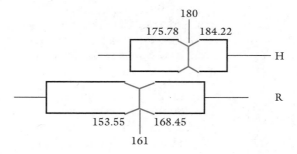

Since the notches do not overlap, then there is a 95% likelihood that the difference in the medians of the two groups is not due to chance and is statistically significant. ●

This result implies that, most likely, the two large groups from which these samples were taken (people who would be identified by this device and people chosen at random) have different median cholesterol counts, with those identified by the device having the higher counts.

**E x a m p l e
3**

A high-school counselor randomly chose 40 scores from the verbal and mathematics sections of the SAT test taken by 80 students of the class of 1993.

SAT Scores

Verbal					Mathematics				
540	580	480	370	460	760	620	590	620	750
660	510	510	530	730	660	430	600	800	620
510	510	390	700	550	780	590	440	380	540
520	510	530	590	410	670	750	600	390	600
500	520	510	460	430	710	680	600	600	420
690	530	580	500	650	740	600	660	530	500
500	480	520	640	480	730	670	500	500	430
630	580	520	540	560	670	500	480	590	680

a. Construct two box plots and compare the relative distributions of these scores.
b. Use the method of notched box plots to see whether the difference in median values is statistically significant or due to chance.

Solution   Minitab will be used for this example, although all of the work could be done by hand. Also, the use of Minitab's macro command BOXPLOT, with "By" and "Notches" subcommands will be illustrated here. Macro commands are typed directly onto the Session window after the **MTB** > prompt. A semicolon (;) must follow every macro command that precedes a subcommand. A period (.) must follow the last subcommand.

As can be seen from the following diagrams, box plots obtained by a macro command are represented by dotted lines, with the location of the median at the plus sign. Quartiles are indicated by "I" symbols.[7] (The asterisks indicate scores that are unusually extreme in comparison with the rest of the scores in the data set, which are discussed in the next section, on outliers.)

a. All of the data are entered in column C1 of a Minitab worksheet. Next to each score, in column C2, a 1 is typed if the score is a verbal score, or a 2 is typed if it is a mathematics score. The "By C2" subcommand separates the scores.

The box plots for the verbal scores are shown on line 1, and those for the mathematics scores, on line 2.

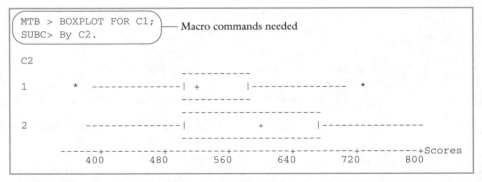

---

[7]Technically, the "I" represents the lower and upper hinges, which were defined earlier. Recall that their values are very nearly the same as the first and third quartiles and will be treated as the same in this text.

Several comparisons can be made at a glance. The range of the math scores is greater than that of the verbal scores; the IQR of the math scores is approximately twice that of the verbal scores. The unusual placement of the median of the verbal scores suggests that a real clustering of those scores occurs just below their median.

b. The "Notches 95%" subcommand is used to provide the following graphs. The parentheses indicate the ends of the notches. Notice that, unlike the notches obtained with the manual method, these notches are not centered about the median. That is because Minitab uses a more sophisticated method for locating the confidence interval about the median.

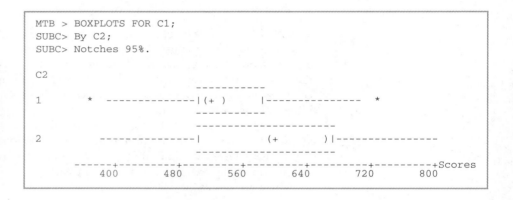

```
MTB > BOXPLOTS FOR C1;
SUBC> By C2;
SUBC> Notches 95%.

C2
                                          -----------
1          *    --------------|(+  )      |--------------    *
                                          -----------
                                      ----------------------
2          ---------------|              (+       ) |----------------
                                      ----------------------
          ------+---------+---------+---------+---------+---------+Scores
              400       480       560       640       720       800
```

Since these notches do not overlap, the difference in the median scores for the verbal and mathematics sections is statistically significant and not due to chance. ●

**E x a m p l e
4**

Refer to the cholesterol data of Example 2. Use Minitab (or other statistical software) to find the greatest degree of certainty that the difference in medians of these samples is statistically significant.

Solution   The difference in medians is statistically significant if the confidence intervals (notches) fail to overlap.

By hand, notches were constructed at the 95% confidence level and they did not overlap. Thus, the degree of certainty that the difference in medians is statistically significant is at least 95%. Minitab's "Notch" subcommand used at the 96%, 97%, and 98% confidence levels produces the following graphs:

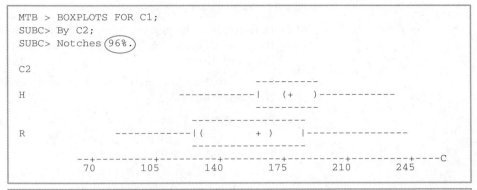

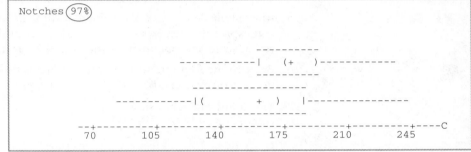

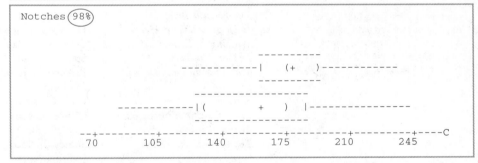

On the basis of these graphs, one may conclude with 97% certainty that the difference in medians of the samples is statistically significant and implies a difference in the medians of the larger groups, or populations, from which these samples came. A 98% certainty, however, is not justified since the notches overlap at that confidence level.

Notice that, as the confidence level increases, the notches widen.[8] This should seem reasonable since the true median is more likely to lie in a wider interval. ●

---

[8]Although the notch width for Group H is also widening, it is hard to notice. That is because of its smaller IQR and the limitations of the printing process.

Exercises 2.5

1. In Example 2, the phrases "most likely" and "95% likely" were used to describe the confidence in the conclusion. Why can't one be absolutely certain of the conclusion?

2. Under what conditions would it be unnecessary to notch box plots when comparing medians?

3. As the number of scores in a set of data increases, what happens to the width of the confidence interval for the median?

4. If one could construct notched box plots in which medians whose differences were due to chance overlapped 90%, 80%, and 70% of the time, how would the widths of their confidence intervals compare?

5. In Example 3 on page 62, 80 different students' scores are represented. Would you expect different results if 40 students contributed two scores each—one verbal score and one mathematics score? Explain.

6. Refer to the footnote to Example 3. What is the maximum number of scores the first quartile can be away from the lower hinge?

7. The following table shows the ages of some of the best-selling American authors of 1992.

Title	Author	Age in 1992
The Pelican Brief	John Grisham	37
It Doesn't Take a Hero	H. Norman Schwarzkopf	58
Waiting to Exhale	Terry McMillan	41
Dolores Claiborne	Stephen King	45
The Road to Omaha	Robert Ludlum	65
Sex	Madonna	34
Sahara	Clive Cussler	61
More Wealth Without Risk	Charles Givens	50
Mixed Blessings	Danielle Steel	45
The Stars Shine Down	Sidney Sheldon	75
Scruples Two	Judith Krantz	65
Tale of the Body Thief	Anne Rice	51
Creating Love	John Bradshaw	59
Mexico	James Michener	85
Every Living Thing	James Herriot	76

Source: *The World Almanac and Book of Facts 1994*, page 289; *Contemporary Authors Cumulative Index;* and *Who's Who in America 1994.*

a. Construct a box plot that illustrates the ages of these authors.

b. Notch the box plot from Part a to determine a 95% confidence interval for the median age of best-selling American authors of 1992.

8. The following table shows accidental deaths per 100,000 of the population in selected nations.

Nation	Deaths	Nation	Deaths
Australia	34.8	Austria	49.7
Bulgaria	43.0	Canada	35.3
Costa Rica	31.5	Cuba	81.5
Czechoslovakia	56.1	Denmark	50.7
Ecuador	48.7	England, Wales	22.8
France	59.2	Greece	42.4
Hungary	76.9	Iceland	33.6
Japan	25.4	Netherlands	23.5
Norway	46.4	Poland	54.1
Portugal	45.0	Switzerland	49.7
United States	39.5	West Germany	32.3
Yugoslavia	39.6		

Source: *The Universal Almanac 1993*, p. 354.

a. Construct a notched box plot for this data, and state the 95% confidence interval for the median accidental death rate worldwide.

b. What countries from this sample actually have accidental death rates within the confidence interval found in Part a?

c. Do you think these countries constitute a typical, or valid, sample for use in finding the worldwide confidence interval? Why or why not? If not, do you think the worldwide confidence interval should be higher or lower?

 Exercises 9–12 can be efficiently worked with a graphics calculator by first entering all of the data in separate lists.

9. The following array shows percentages of school enrollment in 30 randomly selected countries of the world for the years 1970 and 1990.

Country	School Enrollment 1970	1990	Country	School Enrollment 1970	1990
Malta	78	93	Argentina	81	96
Denmark	88	100	Iran	52	82
Chad	19	29	Finland	92	100
Japan	92	98	Austria	84	87
United States	100	100	Spain	88	100
Togo	39	64	Sweden	90	95
Hungary	84	88	Egypt	55	81
Thailand	58	58	Guinea	24	20
Bolivia	61	76	Turkey	67	78
Belgium	93	99	Netherlands	89	100
Syria	61	87	Peru	72	98
Israel	83	92	Greece	85	98
Canada	86	100	Nigeria	21	47
Kuwait	75	86	Costa Rica	76	76
Mexico	67	84	Ethiopia	11	27

a. Draw two notched box plots that compare the data from these years, and state a conclusion based on whether or not the notches overlap.

b. What country had both a 1970 school enrollment that fell within the confidence interval for the median that year and a 1990 enrollment that fell within the confidence interval for 1990?

c. If the range and IQR are much smaller for one set than another, what are the implications?

GROUP

10. The following array shows the average cost per day of hospital care in 24 randomly selected states, in the years 1980, 1985, and 1990.

State	Average Cost per Day 1980	1985	1990	State	Average Cost per Day 1980	1985	1990
AL	$209	$389	$588	MT	$160	$312	$405
AZ	290	591	867	NH	203	422	671
CA	362	654	939	NJ	212	400	613
DE	238	474	771	NY	257	419	641
GA	218	386	630	ND	177	322	427
ID	208	373	547	OK	239	455	632
IL	277	498	717	PA	234	468	662
KS	207	401	532	SC	186	358	590
ME	217	394	574	SD	189	282	391
MA	294	500	788	TX	226	461	752
MI	267	507	716	VA	211	399	635
MS	174	319	439	WI	218	392	554

a. Draw three notched box plots that compare the median costs of a hospital stay for these years.

b. What is the greatest value in the confidence interval for the median cost in 1985? What percentage of the costs reported in 1980 are less than this value? What percentage of the costs in 1990 are less than this value?

11. Some argue that the Eastern Division of the American League was stronger than the Western Division. One way to settle this question is to show that usually the Eastern Division champions had a higher winning percentage than the Western Division champions. Fifteen years have been chosen at random. Draw notched box plots to determine whether the difference in the median winning percentages for this sample is statistically significant. What do the results seem to imply?

Year	East Winner	Percentage	West Winner	Percentage
1969	Baltimore Orioles	.673	Minnesota Twins	.599
1971	Baltimore Orioles	.639	Oakland Athletics	.627
1972	Detroit Tigers	.551	Oakland Athletics	.600
1974	Baltimore Orioles	.562	Oakland Athletics	.556
1975	Boston Red Sox	.594	Oakland Athletics	.605

	East			West	
Year	*Winner*	*Percentage*	*Winner*	*Percentage*
1977	New York Yankees	.617	Kansas City Royals	.630
1978	New York Yankees	.613	Kansas City Royals	.568
1979	Baltimore Orioles	.642	California Angels	.543
1980	New York Yankees	.636	Kansas City Royals	.599
1982	Milwaukee Brewers	.586	California Angels	.574
1983	Baltimore Orioles	.605	Chicago White Sox	.611
1984	Detroit Tigers	.642	Kansas City Royals	.519
1985	Toronto Blue Jays	.615	Kansas City Royals	.562
1986	Boston Red Sox	.590	California Angels	.568
1988	Boston Red Sox	.549	Oakland Athletics	.611

12. Refer to Exercise 11. Consult an almanac and find the median winning percentage of each division, using all of the years from 1969 through 1993. Use both half-season statistics for 1981. How do these results compare to the results from Exercise 11?

13. Refer to Exercise 11. If one division has a higher median winning percentage than the other, does it necessarily follow that one is stronger? What are some other possibilities?

GROUP

14. The following table is a list of the birth rates (births per thousand of the population) in 30 selected countries for the years 1980, 1985, and 1990.

Stat Fact

The United States birth rate in 1950 was 24.1, and the projected United States birth rate for the year 2050 is 13.2.

Country	Birth Rate			Country	Birth Rate		
	1980	*1985*	*1990*		*1980*	*1985*	*1990*
Australia	15.3	15.7	15.4	Luxembourg	11.5	11.2	13.3
Austria	12.0	11.4	11.6	Malta	16.0	14.2	15.2
Belgium	12.7	11.5	12.6	Mauritius	27.0	18.8	21.0
Canada	15.4	14.8	5.2	Netherlands	12.8	12.3	13.3
Cuba	14.1	18.0	17.6	Norway	12.5	12.3	14.3
Former				Panama	26.8	26.6	23.9
Czechoslovakia	16.4	14.5	13.4	Poland	19.5	18.2	14.3
Denmark	11.2	10.6	12.4	Portugal	16.4	12.5	11.8
Finland	13.1	12.8	13.2	Singapore	17.3	16.6	17.0
France	14.8	13.9	13.5	Sweden	11.7	11.8	14.5
Greece	15.4	11.7	10.2	Switzerland	11.3	11.6	12.5
Hungary	13.9	12.2	12.1	United Kingdom	13.5	13.3	13.9
Ireland	21.9	17.5	15.1	United States	16.2	15.7	16.7
Israel	24.1	23.5	22.2	West Germany	10.0	9.6	11.4
Italy	11.2	10.1	9.8	Former Yugoslavia	17.0	15.9	14.0
Japan	13.7	11.9	9.9				

Source: *The 1994 Information Please Almanac*, p. 135.

Use box plots to determine whether there seems to be any pattern in the median birth rate. Notch the medians to determine whether a statistically significant change occurred.

GROUP 15. Refer to Exercise 14.

   a. What country had the highest birth rate for two of the three years shown?
   b. What country had the lowest birth rate for two of the three years shown?
   c. Does the United States rank generally high or generally low for the years shown?

2.6 IDENTIFYING OUTLIERS: UNUSUAL AND RARE SCORES

In addition to their use to compare different sets of data, box plots can be used, with a few modifications, to determine whether any scores within a single set of data are unusually large or small.

Definition ▶ | An **outlier** is a score that is either so large or so small in relation to the rest of the scores in the data that it seems unbelievable or questionable.

Outliers can be identified by locating lines, or "fences," at different places on the box plot. Consider Figure 2-13. Inner fences are located at 1.5·(IQR) from the first and third quartiles and outer fences are 1.5·(IQR) beyond the inner fences. **Adjacent scores,** one on each side of the median, come closest to but do not reach the two inner fences. Scores between the adjacent scores are **ordinary scores.**

Definitions ▶ | A **possible outlier,** or **unusual score,** is a score between the inner and outer fences.
A **probable outlier,** or **rare score,** is a score beyond the outer fence.

Stat Fact

In a nonmathematical sense, an outlier is an outlying portion or member of anything, detached from the main system to which it belongs. It may also be a person who lodges away from the place with which he or she is connected by business or otherwise.

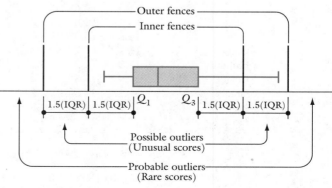

Figure 2-13 Locations of fences and outliers.

E x a m p l e 5

The infant mortality rate is the number of deaths of infants under 1 year old per 1000 live births in a given year. The following array shows 1993 infant mortality rates in selected countries of the world.

S t a t F a c t

In the United States in 1990, the state of Maine had the lowest infant mortality rate, 6.2, and Georgia had the greatest rate, 12.4.

Infant Mortality Rates

Country	Rate	Country	Rate
Afghanistan	162	Japan	5
Australia	7	Jamaica	14
Austria	9	Kuwait	15
Belgium	8	Malaysia	20
Bangladesh	108	Poland	17
Cambodia	116	Singapore	8
Canada	7	Syria	39
Costa Rica	17	Thailand	24
Cuba	13	Uganda	94
Denmark	6	United States	8
Ethiopia	122	Uruguay	20
France	7	Venezuela	33
Hungary	17	Vietnam	54
Ireland	8	Yugoslavia	21
Israel	10	Zimbabwe	55

Find adjacent scores and determine which countries, if any, have infant mortality rates that can be considered outliers; if there are any, classify them as possible outliers (unusual) or probable outliers (rare).

Solution   The first step is to sort the data with a stem-and-leaf display, a graphics calculator, or computer software. For illustration purposes in this example, a stem-and-leaf display has been constructed by hand, as shown:

```
Stem │ Leaf
  16 │ 2
  12 │ 2
  11 │ 6
  10 │ 8
   9 │ 4
   5 │ 4 , 5
   3 │ 3 , 9
   2 │ 0 , 0 , 1 , 4
   1 │ 0 , 3 , 4 , 5 , 7 , 7 , 7
   0 │ 5 , 6 , 7 , 7 , 7 , 8 , 8 , 8 , 8 , 9
```

Next, the quartiles are determined with the help of the stem-and-leaf display. Q_1 is the eighth score, which is 8. Q_2 is the mean of the 15th and 16th scores, which is 17. Q_3 is the 23rd score, which is 39.

The following box plot can now be drawn.

The information in a five-number summary is the starting point for determining outliers.

In order to locate the fences, the IQR must be found.

$$IQR = Q_3 - Q_1$$
$$= 39 - 8$$
$$= 31.$$

Also,

$$1.5(IQR) = 1.5 \cdot 31$$
$$= 46.5.$$

The coordinates of the fences at the right side of the box plot, or upper end of the scores, is now determined. The inner fence is located at $39 + 46.5 = 85.5$, and the outer fence at $85.5 + 46.5 = 132$.

Clearly, there are no negative mortality rates, so fences are left off of the lower, or left, side of the box plot. (If it is obvious that outliers occur on only one side of a box plot, then only one set of fences need be drawn.)

The following diagram shows the completed box plot—with fences, the adjacent score, and possible and probable outliers indicated.

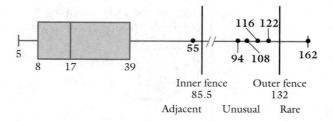

The infant mortality rate of 55 (Zimbabwe) can be termed adjacent. Infant mortality rates of 94 (Uganda), 108 (Bangladesh), 116 (Cambodia), and 122 (Ethiopia) are possible outliers and may be considered unusual. The infant mortality rate of 162 (Afghanistan) is a probable outlier and is considered rare. All of the other mortality rates can be considered ordinary. ●

If this same analysis were conducted using Minitab, **Boxplot** would be chosen from the **Graph** menu. The format of Figure 2-14 differs a little from that drawn by hand. The ends of the lines attached to the box plot are connected to the adjacent values instead of the minimum and maximum scores. Also, note that

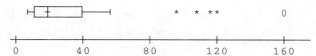

Figure 2-14 Identification of outliers using Minitab.

possible outliers are indicated by an asterisk, and probable outliers are shown with a zero.

E x a m p l e
6

At the beginning of Chapter 1, the following array showed the numbers of calls received by a 911 dispatcher for 36 days chosen at random:

$$\begin{array}{cccccc}
22 & 76 & 6 & 23 & 54 & 31 \\
30 & 27 & 36 & 19 & 71 & 48 \\
17 & 30 & 48 & 28 & 105 & 22 \\
63 & 41 & 26 & 37 & 36 & 44 \\
11 & 41 & 64 & 65 & 52 & 63 \\
8 & 34 & 38 & 32 & 43 & 30
\end{array}$$

For planning purposes, the likelihood that a score of about 105 will occur again must be determined. Future staffing will have to reflect that likelihood. Classify the score of 105 as ordinary, adjacent, a possible outlier, or a probable outlier.

Solution   Use Minitab to easily solve the problem.

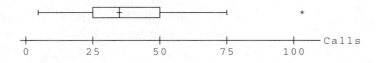

The score of 105 is shown to be a possible outlier but not a probable one—unusual but not unbelievable. The remote possibility of another score as large should be taken into account when planning for future staffing needs. •

E x e r c i s e s **2 . 6**

1. What is the significance of the adjacent score?

2. As the IQR increases, what happens to the fences?

3. Does the location of the median affect the placement of the fences? Explain.

4. Outliers are more likely to occur at the right side, or toward the higher values, of the box plot. Why?

5. In Example 6, the score of 105 is identified as a possible outlier. Name some ways to account for a score that large.

6. If the score of 105 in Example 6 had been identified as a probable outlier instead of a possible one, what difference, if any, would that make in future staffing plans?

7. The occurrence of an outlier can also be a desired result. Name some practical situations in which outliers would be welcome.

In Exercises 8–15, the locations of fences to determine outliers depend upon the IQR. A five-number summary obtained from a graphics calculator is an efficient method of determining the IQR.

8. The following table shows some estimated health care costs for 1994, in billions of dollars.

Type of Health Care	1994 Expenditures
Home health care	22.2
Hospitals	408.8
Dentists' services	47.5
Physicians' services	194.9
Other professional services	54.0
Durable medical equipment	15.5
Nondurable medical equipment	80.2
Nursing home care	85.5
Other personal health care	26.2

Source: *RN*, September 1994, p. 11.

Draw a box plot with fences to help determine what health care costs, if any, are possible or probable outliers.

9. The following table lists cholesterol content, in milligrams, for single servings of different types of foods.

Food	Cholesterol (mg)	Food	Cholesterol (mg)
Lean roast beef	77	Plain yogurt	14
Lean ground beef	80	Creamed cottage cheese	31
Lean roasted pork rib	67	Macaroni and cheese	32
Fried beef liver	372	Vanilla ice cream	30
Chicken without skin	76	Vanilla ice milk	9
Broiled halibut	48	Butter	31
Steamed hard-shell crab	96	Margarine	0
Peanut butter	0	Mayonnaise	8
Large cooked egg	274	Frosted devil's food cake	50
Whole milk	33	Apple pie	2

Which, if any, foods have cholesterol counts that are so high that they would be considered possible or probable outliers? What are the adjacent scores?

10. The following array shows the infant mortality rates for most of the countries in Europe.

Country	Rate	Country	Rate
Albania	32	Italy	9
Austria	9	Netherlands	7
Belgium	8	Norway	6
Bulgaria	14	Poland	17
Czechoslovakia	13	Romania	19
Denmark	6	Spain	9
Finland	5	Sweden	6
France	7	Switzerland	7
Greece	13	Turkey	62
Hungary	17	United Kingdom	8
Ireland	8	Yugoslavia	21

Source: *The Universal Almanac 1993*, pp. 350–351.

With respect to Europe, find the adjacent rate, and determine whether any of the rates are possible or probable outliers.

11. In the preceding exercise, why is the phrase "With respect to Europe" necessary, and what is its significance?

GROUP

12. Exercise 9 (in Exercises 1.6) on page 33 gave the occurrences of murder and burglary per 100,000 of the population for each of the 50 states and the District of Columbia. For convenience, these data are repeated here.

	Murder	Burglary		Murder	Burglary		Murder	Burglary
AL	11.6	1103	**AK**	7.5	894	**AZ**	7.7	1670
AR	10.3	1211	**CA**	11.9	1345	**CO**	4.2	1250
CT	5.1	1228	**DE**	5.0	971	**DC**	77.8	1983
FL	10.7	2171	**GA**	11.8	1619	**HI**	4.0	1228
ID	2.7	813	**IL**	10.3	1063	**IN**	6.2	943
IA	1.9	808	**KS**	4.0	1167	**KY**	7.2	767
LA	17.2	1438	**ME**	2.4	823	**MD**	11.5	1120
MA	4.0	1113	**MI**	10.4	1143	**MN**	2.7	907
MS	12.2	1251	**MO**	8.8	1066	**MT**	4.9	709
NE	2.7	724	**NV**	9.7	1367	**NH**	1.9	736
NJ	5.6	1017	**NM**	9.2	1739	**NY**	14.5	1161
NC	10.7	1530	**ND**	0.8	427	**OH**	6.1	983
OK	8.0	1448	**OR**	3.8	1135	**PA**	6.7	729
RI	4.8	1271	**SC**	11.2	1380	**SD**	2.0	527
TN	10.5	1264	**TX**	14.1	1852	**UT**	3.0	881
VT	2.3	1087	**VA**	8.8	731	**WA**	4.9	1263
WV	5.7	657	**WI**	4.6	751	**WY**	4.9	631

Answer the following questions in relation to the 50 states and the District of Columbia.

a. Which murder rates are possible or probable outliers?

b. Which burglary rates are possible or probable outliers?

c. What are the adjacent scores for both categories?

GROUP
13. Exercise 13 (in Exercises 1.1–1.3) on page 13 and Exercise 7 (in Exercises 2.1–2.4) on page 51 gave the ages of those who have received the Oscar for best actress and the Oscar for best actor, respectively. Use this information to answer the following questions.

  a. Which actresses, if any, won the Oscar at an age that was unusual or rare?

  b. Which actors, if any, won the Oscar at an age that was unusual or rare?

  c. In the combined group of actors and actresses, which individuals won Oscars at ages that were unusual or rare for the combined group? Were these people also cited in Parts a and b?

14. The following table shows the average annual percent increases in hospital care expenditures from 1988 through 1991, in 20 states selected at random.

State	Percent Increase	State	Percent Increase
DE	11.6	MA	9.0
CA	9.7	MT	11.1
MD	10.4	OR	10.2
ND	8.7	IL	9.6
NM	9.5	MN	11.0
SC	15.9	MS	10.8
NY	10.3	UT	11.1
IA	10.0	TX	12.7
VA	11.7	MI	8.3
TN	12.4	OH	9.2

Source: *Health United States 1993,* U.S. Department of Health and Human Services, p. 251.

  a. Find a 95% confidence interval for the median percent increase in hospital care expenditures for all 50 states.

  b. Is South Carolina's increase of 15.9% a possible outlier, a probable outlier, or neither?

GROUP
15. A typing class was taught using three different types of experimental typewriters: X, Y, and Z. At the end of the semester, 36 students were chosen at random from each group and given a speed test. The following arrays show the numbers of words per minute that were typed by the members of each group:

Group X			Group Y			Group Z		
30	21	19	41	26	28	58	26	26
52	24	43	52	60	16	15	36	38
71	18	32	30	26	44	59	77	17
42	18	25	21	34	34	18	59	26
71	21	8	52	31	33	26	43	47
52	31	39	60	40	31	43	27	29
42	9	46	63	16	40	43	15	27
46	21	19	58	40	17	43	39	19

Group X			Group Y			Group Z		
19	31	48	41	18	32	17	44	44
22	26	38	43	21	39	27	44	49
28	47	26	34	45	48	28	28	39
23	26	28	31	38	38	45	45	36

a. Construct a five-number summary for each group.

b. Construct a box plot to the same scale for each group.

c. Notch the box plots in Part b and decide whether the difference in medians is due to chance or is statistically significant. If any median differences are due to chance, clearly state which ones.

d. Add fences to the box plots from Part b and identify the scores, if any, that are possible or probable outliers.

Case Study   *The Corporate Elite*

Background information on top corporate executives may provide a clue to their success. *(Pictured: Robert J. Eaton, Chairman and CEO, Chrysler Corporation)*

What is the typical CEO like? How well educated are most CEOs? Which university did most of the corporate elite attend? What is their mean, or median, income? In what field did most of them start out? On average, how many years did it take them to reach the top? What percentage are foreign-born?

The chief executives of the Business Week 1000, the most valuable publicly held U.S. companies, are profiled on pp. 60–106 of the October 11, 1993, issue of *Business Week*. The article provides vital statistics on these leaders along with some background information on their educations and career paths. A random sample of $n = 30$ of these chief executives reveals the following information.

Company	Chief Executive Officer	Birth Date	Annual Pay
AFLAC	Daniel P. Amos	8-13-51	$1,332,000
AMOCO	H. Laurence Fuller	11-8-38	1,223,000
Ashland Oil	John Richard Hall	11-30-32	728,000
Bank of Boston	Ira Stepanian	11-14-36	1,297,000
Borden	Anthony S. D'Amato	6-22-30	735,000
CBS	Laurence Alan Fisch	3-15-23	1,573,000
Citicorp	John Shepard Reed	2-7-39	2,185,000
Conrail	James A. Hagen	3-27-32	607,000
Dean Foods	Howard M. Dean	6-27-37	748,000
Ecolab	Pierson M. Grieve	12-5-27	992,000
First Security	Spencer Fox Eccles	8-24-34	750,000
Ford Motor	Harold A. Poling	10-14-25	1,183,000
Genentech	G. Kirk Raab	9-27-35	1,056,000
GTE	Charles R. Lee	2-15-40	1,851,000

Company	Chief Executive Officer	Birth Date	Annual Pay
Home Depot	Bernard Marcus	5-12-29	2,667,000
Intel	Andrew S. Grove	9-2-36	1,351,000
Kerr-McGee	Frank A. McPherson	4-29-33	525,000
MAPCO	James E. Barnes	3-2-34	671,000
McDonald's	Michael R. Quinlan	12-9-44	1,580,000
Nevada Power	Charles Albert Lenzie	9-5-37	389,000
Northern Trust	David W. Fox	8-29-31	971,000
Paychex	B. Thomas Golisano	11-14-41	375,000
Raytheon	Dennis J. Picard	8-25-32	1,280,000
Scholastic	Richard Robinson	5-15-37	553,000
St. Paul	Douglas W. Leatherdale	12-6-36	628,000
Trinova	Darryl F. Allen	9-7-43	573,000
Unisys	James A. Unruh	3-22-41	1,801,000
Walgreen	Charles R. Walgreen, III	11-11-35	1,194,000
Whirlpool	David R. Whitwam	1-30-42	1,985,000
Xerox	Paul Arthur Allaire	7-21-38	1,525,000

I. Construct a five-number summary for the ages of these 30 chief executives. Compute their ages as of January 1, 1994, and round all ages to the nearest half-year.

II. Use the five-number summary from Part I to construct a 95% confidence interval for the median age of all 1000 CEOs in the *Business Week* article.

III. Construct a 95% confidence interval for the median annual pay of this sample of 30 CEOs.

IV. How reliable is the confidence interval found in Part III? Obtain a copy of the October 11, 1993, *Business Week* at your library and form a sample of 60 CEOs in the article who are listed directly before and after those shown here. Construct a 95% confidence interval for the median annual pay based on this new sample of 60. Write a paragraph comparing the characteristics of the two confidence intervals for the median annual pay. Do the confidence intervals overlap?

V. Use statistical software to find the actual median pay for all 1000 CEOs in this article. (This will take about 45 minutes.) Is the median within either of the confidence intervals you found earlier?

CHAPTER SUMMARY

This chapter concentrated on ways to compare several sets of data. Beyond using simple dot plots for comparison, data sets can be compared via their five-number summaries. A five-number summary for a set of data lists its low score, high score, and first, second, and third quartiles. A box plot is a graphic representation of a five-number summary that can be adapted for several special purposes, as a first step in data analysis.

Two or more sets of data can be compared by drawing their box plots to the same scale and placing one beneath (or next to) the other. The heights of the box plots can be varied to reflect two (or more) data sets that have different numbers of scores.

A box plot is notched about the median to fix the median's location within an interval. Such notching can be carried out on box plots that represent samples taken from larger data groups in order to determine whether any difference of medians in the samples is great enough to not be due merely to the chance selection of the data. If the difference is not due to chance, then it is said to be statistically significant, implying a difference in the medians of the larger groups from which the samples came. The width of the notch of a box plot is a function of the IQR, the number of scores, and the degree of confidence that is desired in the resulting conclusion.

Box plots may also be used to identify scores that are exceptionally high or low, called outliers. The identification of outliers is a function of the IQR.

Key Concepts, Terms, and Formulas

Videotape Suggestion

Program 3 from the series *Against All Odds: Inside Statistics* provides additional information on box plots and five-number summaries and would be a preview of future work with the mean and the standard deviation.

The shape of the Sears
Tower in Chicago
resembles a histogram.

3 Measures of Central Location and Dispersion: Histograms

In Chapters 1 and 2, percentiles were used to identify important data scores. In some cases the goal was to find a number that best reflected the data as a whole: an "average" score. However, the word "average" can be used loosely and have many meanings. It is often preferable to use some specific measure of central location.

Until now, the range and the IQR have been used as the measures of dispersion, or spread of scores, in a set of data. Other measures of dispersion often prove more useful, and this chapter describes three of them, then introduces alternative types of graphic displays for data. The **case study** examines the relationship between crime rates and imprisonment rates.

3.1 MEASURES OF CENTRAL LOCATION

Definition ▶ A **measure of central location** (or **central tendency**) is a single number that represents the typical score in the data.

There are several measures of central location: the median, the mean, the trimmed mean, the midrange, and the mode. The nature of the data usually determines which of these is used, but there can be advantages and disadvantages to each. For small data sets, it is possible that none of these measures of central location will render a truly typical score.

The median, mean, and mode are the three principal measures of central location. The first two chapters dealt with the first of these, the **median**—the score that has equal numbers of scores above it and below it. The median is not always the best measure of central location. Consider the following data.

Seven Salaries

$76,800

76,000

75,800

74,600

12,800

10,200

9,500

The median salary is $74,600, but that value may not best represent the typical salary, since nearly half the salaries are substantially less than it and the others are about equal to it. Therefore, $74,600 may not be the best measure of central location for this small data set.

A second measure of central location, the mean, would probably be more representative of the typical salary in this case. The **mean** of a set of scores, symbolized by μ, is found by adding the scores and then dividing that sum by the number of scores, N:

$$\mu = \frac{x_1 + x_2 + x_3 + \cdots + x_N}{N}.$$

In this formula, x_1 represents the first score, x_2 the second score, and so on until the last score, x_N, is added to the sum.

Because this formula is cumbersome to write, many statisticians use a shorthand notation, called **sigma notation,** for the numerator, so that

$$x_1 + x_2 + x_3 + \cdots + x_N = \sum_{i=1}^{N} x_i.$$

The formula for the mean may thus be written more compactly as

$$\mu = \frac{\sum_{i=1}^{N} x_i}{N}. \qquad (3\text{-}1)$$

Note that the use of a Greek letter, μ (instead of $\overline{X}$), for the mean and an uppercase N (instead of a lowercase n) is significant; it implies that the data set is the whole population and that all of its scores (not just a sample of them) are used in the computation.[1]

---

[1] Most calculator displays, however, do not make this distinction and always label the mean $\overline{X}$. The discussion of sampling in Chapter 7 will explain this topic fully.

Applying the formula for the mean to the salary illustration,

$$\mu = \frac{(76{,}800 + 76{,}000 + 75{,}800 + 74{,}600 + 12{,}800 + 10{,}200 + 9500)}{7}.$$

Rounded to the nearest penny, μ = \$47,957.14. In some ways, this is not the typical salary either; after all, no one has that salary or one even close to it. However, if only one salary is to be named as typical, this one is probably less misleading than the median since it takes into account the relative size of each salary in the data set.

A refined version of the mean is the trimmed mean. The **5% trimmed mean** of a set of data is the mean of all the scores except the smallest 5% (rounded to the nearest integer) and largest 5% of them. (If there are duplicate scores at the ends, then 5% of the repetitions are removed.) This measure of central location is often used when the data contain an extreme score or two at one end and the inclusion of those scores in the computation of the mean would produce misleading results. However, the trimmed mean may not be much more effective than the mean at providing a good measure of central location if extreme scores occur at both ends or if the number of scores is very small. In the preceding salary illustration, the trimmed mean would be the same as the mean since 5% of 7 = .35, which, rounded to the nearest integer, is 0. Thus, no scores would be "trimmed" from the ends.

The **midrange** is the mean of the smallest and largest scores in the data set. The midrange is usually used to provide a quick estimate of the mean and can be fairly accurate in cases of relatively small data sets or data sets with scores that are spread, or distributed, fairly evenly throughout their ranges. In the salary illustration, the midrange would be $\frac{76{,}800 + 9500}{2}$ = \$43,150, which is within approximately 10% of the actual mean.

The last measure of central location to discuss is the **mode** of a set of data, which is the most-often-repeated score. If two (or more) scores are repeated the most often, then the data have a **bimodal** (or **multimodal**) distribution. If all of the scores repeat the same number of times, then the data are **nonmodal.** Classifications of five sets follow.

> It is also common practice to compute 10% and 20% trimmed means.

Stat Fact

In another, related usage, the word "mode" denotes a temporarily prevailing fashion or convention, especially one characteristic of a particular place or period. For example: "Long, flowing hair was the mode for men and women in the hippie era of the 1960s."

Possible Mode Classifications

A	B	C	D	E
45, 40, 38, 38, 35, 35, 35, 30, 28, 26, 26, 20	78, 50, 41, 40, 39, 37, 36, 36, 33, 30, 20, 20	80, 79, 68, 66, 59, 58, 49, 30, 29, 28, 25, 20	59, 58, 40 40, 37, 37, 36, 32, 32, 30, 22, 20	41, 41, 37, 37, 24, 24, 22, 22, 18, 18, 10, 10
Mode = 35.	Bimodal; 20 and 36 are the modes.	Nonmodal.	Multimodal; 32, 37, and 40 are the modes.	Nonmodal.

In summary,

Definitions ►

> The **median** is the score that has equal numbers of scores above it and below it.
> The **mean** is the arithmetic average of the scores.
> The **mode** is the most-often-repeated score.
> The **5% trimmed mean** is the mean of the middle 90% of the scores.
> The **midrange** is the mean of the largest and smallest scores.

The mode is the only measure of central location that may be used with qualitative data.

Example 1

The combined SAT scores of 40 college freshmen enrolled in an introductory statistics class are as follows:

720	610	700	520	760	440	380	600	720	600
610	800	320	460	580	610	780	770	760	570
640	520	600	480	500	520	680	780	780	680
660	580	490	620	520	780	700	680	760	520

Stat Fact

6.2 million students attended community colleges in 1992—an increase of about 2 million since 1985.

a. For these data, determine the five measures of central location presented thus far.
b. Briefly discuss the advantages and disadvantages of each measure of central location for these data, and then decide which measure or measures best represent the typical score.

Solution a. The *mean* is the sum of all 40 scores divided by 40, or $\frac{24{,}800}{40}$, which is 620.

A stem-and-leaf display is helpful for finding the median, trimmed mean, midrange, and mode. From Figure 3-1, the *mode* is easily identified as 520.

Stem	Leaf
80	0
78	0,0,0,0
77	0
76	0,0,0
72	0,0
70	0,0
68	0,0,0
66	0
64	0
62	0
61	0,0,0
60	0,0,0
58	0,0
57	0
52	0.0,0,0,0
50	0
49	0
48	0
46	0
44	0
38	0
32	0

Mean = 620
Mode = 520
Median = 610
Trimmed mean ≈ 626
Midrange = 520

Figure 3-1

The *median* is the mean of the scores with data ranks 20 and 21, which is score 610.

To find the *trimmed mean,* first find 5% of 40, which is 2. Next, the two highest scores (800 and one of the 780s) and the two lowest scores (320 and 380) are excluded from the data set. The mean of the remaining scores, $\frac{22,520}{36} \approx 626$, is the trimmed mean.

$$\text{The } \textit{midrange} \text{ is } \frac{800 + 320}{2} = 560.$$

b. *Mode.* Unlike the other measures of central location, the mode is guaranteed to be an actual score and thus is potentially more typical than the other measures. In this example, however, the mode does not seem very typical. It is greater than only seven scores and is very much less than either the mean or the median. Indeed, there is no safeguard against extreme values when dealing with the mode. Although there may be situations in which a measure of central location that is an actual score is of primary importance, the mode is not commonly used as a measure of central location.

Median. The median is a "democratic" measure of central location in the sense that it gives each score, regardless of its size, one "vote" in determining the typical score. The median thus moderates the presence of a few extreme (and possibly misleading) scores.

Mean. The mean is a usually good measure of central location that, unlike the median or mode, takes the relative value of every single score into account. Because of this, however, an outlier can greatly affect the mean. In this example, the stem-and-leaf display does not indicate the presence of an outlier.

Midrange. The midrange is simply a quick estimate of the mean. It is used most effectively in large data sets with scores that are fairly evenly spaced out. Since the scores in this example are not evenly spaced and tend to stretch toward the lower values, the midrange underestimates the mean.

Trimmed mean. In the data of Figure 3-1, the two lowest scores are more extreme in relation to the rest of the data than are the two highest scores. As a result, the elimination of all four of these scores raises the mean of the remaining scores. This trimmed mean may indeed be most typical of the bulk of the scores.

Considering all the advantages and disadvantages of each measure of central location for this example, most statisticians would probably use either the mean, the trimmed mean, or the median as the typical score. •

The mean will be of fundamental importance throughout the rest of this text. Although the mean can take a long time to compute, commercial software and hand-held calculators make the work much easier, as the next examples show.

Minitab can be used to find the mean, median, and 5% trimmed mean. It can also be used to produce a stem-and-leaf display from which the mode can easily be identified.

S t a t   F a c t

A 1994 study found that 34% of Americans are at least 20% heavier than their ideal weights, compared with 26% in the late 1970s.

Compare the median, mode, mean, trimmed mean, and midrange for each group of 40 teenagers who had their cholesterol counts examined in the last chapter. Use the results to decide whether there is further evidence that the screening device used to identify people with high cholesterol counts is effective. The data for the two groups are repeated here:

Cholesterol Counts

Group H (high cholesterol)				Group R (randomly selected)			
139	157	160	201	160	198	140	130
122	188	194	240	138	85	98	108
200	180	230	179	126	208	178	170
194	190	180	188	175	208	161	195
179	194	140	136	130	120	110	128
180	193	159	152	223	98	163	165
176	157	178	206	150	186	248	83
160	120	176	192	186	161	165	188
203	210	176	188	163	145	126	195
194	189	163	196	210	123	116	164

Solution   As shown in Section 2.3 of the last chapter, Minitab may be used to obtain the following descriptive statistics:

Minitab: Descriptive Statistics Calculations

```
MTB > Describe C1 C2
                N       MEAN    MEDIAN    TRMEAN     STDEV    SEMEAN
   H           40      178.98   180.00    179.08     25.99     4.11
   R           40      155.60   161.00    155.14     39.39     6.23

               MIN       MAX        Q1        Q3
   H         120.00    240.00    160.00    194.00
   R          83.00    248.00    126.00    186.00
```

From this Minitab display, the mean, median, and 5% trimmed mean (TRMEAN) can be read as follows:

Group H	Group R
Mean = 178.98	Mean = 155.60
Median = 180.00	Median = 161.00
Trimmed mean = 179.08	Trimmed mean = 155.14

The midrange for each group is easily found by determining the mean of the minimum and maximum scores shown in the Minitab summary.

$$\text{Midrange for Group H} = \frac{120 + 240}{2}$$
$$= 180;$$
$$\text{Midrange for Group R} = \frac{83 + 248}{2}$$
$$= 165.5.$$

Finally, the mode can be determined from a stem-and-leaf display, either by hand or by the choice of **Stem-and-Leaf** from Minitab's **Graph** menu:

Minitab: Stem-and-Leaf Displays

```
                                 Stem-and-leaf of R
                                 Leaf Unit = 1.0
                                    2      8 35
                                    4      9 88
                                    5     10 8
                                    7     11 06
       Stem-and-leaf of H         12     12 03668
       Leaf Unit = 1.0           15     13 008
          2     12 02            17     14 05
          4     13 69            18     15 0
          5     14 0             (8)    16 01133455
          9     15 2779          14     17 058
         12     16 003           11     18 668
         18     17 666899         8     19 558
         (7)    18 0008889        5     20 88
         15     19 02344446       3     21 0
          7     20 0136           2     22 3
          3     21 0              1     23
          2     22                1     24 8
          2     23 0
          1     24 0
```

Notice that Minitab prints the stem-and-leaf displays in ascending rather than descending order and does not show commas between scores.

Also, there is an extra column at the left side of each of these stem-and-leaf displays. It indicates how many scores are on a given line and beyond, in the direction of the closest "end" of the list. The line with parentheses contains the median, and the number in the parentheses is the number of scores on that line.

Upon close examination of these stem-and-leaf displays, it can be seen that the mode for Group H is 194. Group R is multimodal, with modes of 98, 126, 130, 161, 163, 165, 186, 195, and 208—too many to be of much help in determining a typical score.

With the exception of some of the modes of Group R, every measure of central location was higher for Group H than for Group R. This is further evidence (along with Example 2 of Chapter 2) that the screening device identifies people with high cholesterol counts. Chapter 10 will state this claim more precisely and examine it more closely. ●

3.3 MEASURES OF DISPERSION: THE STANDARD DEVIATION

The **dispersion** of a set of data is a measure of how spread out, or scattered, the scores are. There are several measures of dispersion (or variability, or spread), and two of them have already been covered: the range and the IQR. Both of these measures, however, are rough and are sometimes of limited value. Consider two sets of data:[2]

A	B
1, 2, 3, 40, 40, 40, 40, 40, 53, 54, 101	1, 2, 3, 10, 20, 30, 40, 50, 53, 90, 101
($\mu_A = 37.64$)	($\mu_B = 36.36$)

For each of these data sets, the range is 100 and the IQR is 50. It is clear, however, that an alternative measure of dispersion is needed to express the different patterns, or distributions, of scores in the two sets. This measure is called the standard deviation.

Definition ▶ The **standard deviation** is a measure of dispersion that indicates how closely the scores are clustered about the mean.

Unlike the range and the IQR, the standard deviation takes every score into account. It is a function of the mean and the square of the difference, or deviation, between each score and the mean. The standard deviation has the same units as the mean.

The symbol that will be used for standard deviation is the Greek letter sigma (σ) (pronounced sig′-ma). The standard deviation of a set of data with N scores is as follows.

The use of the Greek letter σ indicates that the formula is used with all the data from an entire population, and not just a sample of the population.

Definition Formula for Standard Deviation

$$\sigma = \sqrt{\frac{\sum\limits_{i=1}^{N} (x_i - \mu)^2}{N}}$$

(3-2)

In this formula, $(x_i - \mu)$ is the difference between an individual score, x_i, and the mean, μ, and is called the score's **deviation** from the mean. The numerator inside the radical is thus the sum of the squared differences. This sum is divided by N, and then the square root of the quotient is found. That number is the standard deviation of the data.

---

[2]These data sets are small so that all the scores can be taken in at a glance. In reality, there is often no need to examine dispersion or compute other statistics for sets this small.

A shortcut formula for the standard deviation can be derived algebraically from the definition formula.

Shortcut Formula for Standard Deviation

$$\sigma = \sqrt{\dfrac{\sum\limits_{i=1}^{N} x_i^2 - \dfrac{\left(\sum\limits_{i=1}^{N} x_i\right)^2}{N}}{N}}$$

(3-3)

Even though this formula may not look any easier to use, it is, because it does not require that the mean be computed first. Also, the shortcut formula does not depend on any possibly rounded value of the mean, which makes it potentially more accurate than the definition formula.

The standard deviation is very useful in comparing the relative amounts of variability for several sets of data. In many practical situations, however, the standard deviation is computed from a slightly different formula than either of those just shown. After an introduction to sampling and estimation in Chapter 7, another method for finding the standard deviation will be presented along with guidelines about when to use it.

E x a m p l e 3

Apply Formulas 3-2 and 3-3 to find the standard deviation for each of the two sets of data already illustrated, which are repeated below. Use the results to decide which set has scores that are more clustered about the mean.

A	B
1, 2, 3, 40, 40, 40, 40, 40, 53, 54, 101	1, 2, 3, 10, 20, 30, 40, 50, 53, 90, 101
(μ_A = 37.64)	(μ_B = 36.36)

Solution Upon careful examination of these two groups of data, the scores appear to be more clustered about the mean in Group A than in Group B. The standard deviation for Group A, σ_A, should therefore be less than the standard deviation for Group B, σ_B. The following calculations should confirm this.

For the sake of illustration, the definition formula (3-2) will be used for Group A, and the shortcut formula (3-3), for Group B.[3]

Group A: Definition Formula (3-2)

Step 1 Find the mean.
For these data, μ = 37.64.

---

[3] In general, all nonexact answers in this text are rounded to the nearest hundredth, and all intermediate calculations are rounded to the nearest thousandth.

Step 2 *Find the deviation of each score from the mean.*

The deviation of score x from the mean, μ, is denoted by d and is found by $d = (x - \mu)$. A score's deviation may be a positive number, a negative number, or zero.

Step 3 *Square each deviation found in Step 2.*

Step 4 *Add the squares of the deviations found in Step 3.*

For efficiency, the following table is used to complete Steps 2 through 4.

x	$d = x - \mu$	$d^2 = (x - \mu)^2$
1	$1 - 37.64 = -36.64$	1342.490
2	$2 - 37.64 = -35.64$	1270.210
3	$3 - 37.64 = -34.64$	1199.930
40	$40 - 37.64 = 2.36$	5.570
40	$40 - 37.64 = 2.36$	5.570
40	$40 - 37.64 = 2.36$	5.570
40	$40 - 37.64 = 2.36$	5.570
40	$40 - 37.64 = 2.36$	5.570
53	$53 - 37.64 = 15.36$	235.930
54	$54 - 37.64 = 16.36$	267.650
101	$101 - 37.64 = 63.36$	4014.490
		Sum $= 8358.550$

Step 5 *Divide the sum of the squared deviations by the number of scores.*

$$\frac{8358.55}{11} = 759.868.$$

Step 6 *Find the standard deviation.*

The standard deviation for Group A, σ_A, is the square root of the number obtained in Step 5:

$$\sigma_A = \sqrt{759.868}$$
$$\approx 27.57.$$

Group B: Shortcut Formula (3-3)

Step 1 *Find the sum of the squares of each score.*

$$1^2 + 2^2 + 3^2 + 10^2 + 20^2 + 30^2 + 40^2 + 50^2 + 53^2 + 90^2 + 101^2 = 26{,}624.$$

Step 2 *Find the sum of the scores, square that sum, and then divide by the number of scores.*

$$\frac{(1 + 2 + 3 + 10 + 20 + 30 + 40 + 50 + 53 + 90 + 101)^2}{11} = 14{,}545.455.$$

Step 3 *Subtract the computation obtained in Step 2 from the computation in Step 1.*

$$26{,}624 - 14{,}545.455 = 12{,}078.545.$$

Step 4   Divide the difference obtained in Step 3 by the number of scores.

$$\frac{12{,}078.545}{11} = 1098.050.$$

Step 5   Find the standard deviation.
The standard deviation of Group B, σ_B, is equal to the square root of the number obtained in Step 4:

$$\sigma_B = \sqrt{1098.050}$$
$$\approx 33.14. \quad \bullet$$

The fact that the standard deviation of Group A is less than that of Group B ($27.57 < 33.14$) implies that the scores in Group A are clustered more about their mean than the scores in Group B are clustered about their mean. The dot plots for these two data sets shown in Figure 3-2 reinforces this fact.

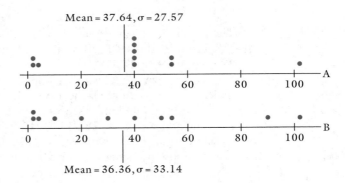

Figure 3-2

The formulas used for Example 3 are appropriate for work done essentially by hand. If a calculator with a statistics mode (a scientific calculator or a graphics calculator) is available, its use will save time because it has these and other frequently used formulas built into it.

E x a m p l e
4

Using a graphics calculator, compute the standard deviation for the data in Group A of Example 3, on page 89.

Solution   The steps are identical to those for using a graphics calculator to construct a five-number summary, illustrated in Chapter 2.

• Clear list L1.

This is accomplished with the keystroke sequence

(STAT), (4), (2nd), (1), (ENTER).

• Enter all of the data from Group A in list L1. (Press (STAT), (1) to begin entering scores.)

Next, go back to the menu by pressing (STAT) (Figure 3-3).

```
EDIT CALC
1: Edit...
2: SortA(
3: SortD(
4: ClrList
```

Figure 3-3

```
1-Var Stats
```

Figure 3-4

```
1-Var Stats
 x̄=37.63636364
 Σx=414
 Σx²=23940
 Sx=28.91114915
 σx=27.56569913
↓ n=11
```

Figure 3-5

Move the active prompt to the calculation menu (CALC).

- Press the right arrow key, ▶.
- Type 1 to choose 1-variable statistics as in Figure 3-4.
- Type 2nd, 1 to specify the data in list L1.
- Press ENTER.

The calculations will appear as in Figure 3-5.

From Figure 3-5, the standard deviation (labeled σx) can be seen as $27.56569913 \approx 27.57$.[4] •

This screen contains other important information as well, such as the mean (labeled $\bar{x}$), the sum of all the scores (Σx), the sum of the squares of all the scores (Σx^2), and the number of scores ($n = 11$).

The statistic $Sx = 28.91114915$ is found by a formula identical to that for σx except that the divisor is $(n - 1)$ instead of n. As already mentioned, Chapter 7 will discuss the situations in which it is correct to use this latter formula. For now, the standard deviation will be obtained by use of a divisor of n, and care must be taken to read the correct statistic from the calculator display.[5]

Exercises 3.1–3.3

If a graphics calculator is used for the following exercises, it will be helpful to remember that "1-Var Stats" may be scrolled down to reveal a five-number summary.

1. Why would it not be appropriate to use the midrange as a measure of central location for a data set with an outlier?

2. The **midquartile** is also a measure of central location and is the mean of the first and third quartiles—that is, $\dfrac{Q_1 + Q_3}{2}$. Under what conditions would the midquartile be a good measure of central location?

---

[4]As a rule of thumb, answers will contain only one or two more decimal places than were present in the original data. All of the decimal places in this display are misleading and imply an accuracy that is not present.

[5]It will be important to remember that the statistic provided in Minitab's "Describe" command, labeled STDEV, is Sx and not σx. However, it will always be the case that $\sigma x = Sx \cdot \sqrt{\dfrac{n-1}{n}}$.

3. Consider the following three sets of data, in the form of stem-and-leaf displays:

A			B			C	
Stem	Leaf		Stem	Leaf		Stem	Leaf
14	0		47	0		69	5 , 5
13	8 , 8		13	5 , 9		13	0
12	5 , 6		12	6		12	8
11	0 , 0 , 0 , 5		11	0		11	0 , 0 , 0 , 8
10	0 , 0 , 7 , 8		10	0 , 0 , 5		10	0 , 0 , 8
9	0 , 3 , 7 , 8		9	0 , 6 , 7 , 8 , 9		9	0 , 3 , 5 , 5
8	3 , 8		8	2 , 5 , 8 , 8		8	0 , 5
7	5		7	6 , 8		7	3 , 6 , 8
			0	2			

Using estimation only, which set of data would probably have the largest midrange? The smallest midrange? Which set would have the largest trimmed mean? The smallest trimmed mean? Which set would have the largest mean? The smallest mean?

GROUP

4. Calculate the actual midranges, trimmed means, and means for the three sets of data in Exercise 3, and check the estimates.

5. In a comparison of two sets of data, which of the three principal measures of central tendency (mean, median, mode) seems least likely to provide a true picture of the difference?

6. Following is a list of the 20 longest-running daytime network TV game shows ever, as of January 1993, and their lengths of runs.

Stat Fact

A prospective contestant must do 24 chin-ups in 30 seconds, run a 40-yard dash in 5 seconds, climb a 20-foot rope in 10 seconds, and run 15 feet back and forth—sideways—eight times in 12 seconds to qualify as a contestant on "American Gladiators."

Show	Length of Run
"The Price Is Right" (CBS version)	15 years, 6 months
"Concentration"	14 years, 7 months
"Hollywood Squares"	13 years, 8 months
"Wheel of Fortune"	12 years, 9 months
"Let's Make a Deal"	12 years, 6 months
"Jeopardy!"	10 years, 9 months
"Family Feud"	8 years, 11 months
"The Price Is Right" (NBC/ABC version)	8 years, 9 months
"Queen for a Day"	8 years, 9 months
"The Newlywed Game"	8 years, 5 months
"The Big Payoff"	7 years, 10 months
"The Dating Game"	7 years, 6 months
"The Match Game"	6 years, 9 months
"Strike It Rich"	6 years, 8 months
"You Don't Say"	6 years, 6 months
"To Tell the Truth"	6 years, 3 months
"Who Do You Trust?"	6 years, 3 months
"The $10,000/$20,000 Pyramid"	6 years, 1 month
"Password"	6 years, 0 months
"Truth or Consequences"	5 years, 11 months

Source: David Schwartz, Steve Ryan, and Fred Wostbrock, *The Encyclopedia of TV Game Shows*, 1987, p. 585.

Counting the two versions of "The Price Is Right" as two separate shows, determine the mean, median, and mode. Also indicate the names of the shows that correspond most closely to these measures of central location.

Hint: *First convert years and months to years; for instance, 15 years, 6 months = 15.50 years. Use two decimal places.*

GROUP

7. Make up a data set with seven scores (positive or negative) such that:

   a. Mean < median < mode.   b. Median < mean < mode.

   c. Mode < mean < median.   d. Mode < median < mean.

   e. Mean < mode < median.   f. Median < mode < mean.

8. Which is least likely to be an actual score in the data set—mean, median, or mode?

GROUP

9. The overall lengths, wheelbases, and overall widths (in inches) of 24 small cars follow. (Each dimension is rounded to the nearest whole number.)

Stat Fact

1993 model cars emit just 1% of the pollutants generated by cars sold in 1973.

Car	Overall Length	Wheelbase	Overall Width
Acura Integra 4-door	178	103	67
Dodge Colt 4-door	174	98	66
Dodge Shadow 4-door	172	97	67
Eagle Summit sedan	174	98	66
Ford Escort 4-door	171	98	67
Ford Aspire 5-door	156	94	65
Geo Metro sedan	151	93	63
Geo Prizm sedan	173	97	66
Honda Civic 4-door	173	103	67
Hyundai Elantra 4-door	173	98	66
Hyundai Excel	161	94	63
Infiniti G20 sedan	175	100	67
Mazda Protege	172	98	66
Mitsubishi Eclipse 2-door	173	97	67
Mitsubishi Mirage 4-door	172	98	67
Nissan Sentra 4-door	170	96	66
Saturn sedan	176	102	67
Subaru Justy 3-door	146	90	60
Subaru Impreza 4-door	172	99	67
Toyota Celica LB	174	100	69
Toyota Corolla 4-door	172	97	66
Toyota Tercel 4-door	162	94	65
Volkswagen Golf 4-door	161	97	67
Volkswagen Jetta 4-door	173	97	66

Sources: *Edmund's 1994 New Car Prices,* November 1993, pp. 230–241, and *Edmund's 1994 Import Car Prices,* April 1994, pp. 243–256.

Find the mean overall length, mean wheelbase, and mean overall width. Determine which single car comes closest to the means of all three categories.

* 10. For data set A of Exercise 3, find the mean of the deviations of all the scores. Then repeat the calculation for data sets B and C. State a rule for the sum of the deviations of any set of data.

GROUP 11. Find the standard deviation (in terms of months) for the data in Exercise 6 in three different ways:

    a. Use the definition formula (3-2).

    b. Use the shortcut formula (3-3).

    c. Use a hand-held calculator with a statistics mode, or use computer software.

* 12. Use the definition formula (3-2) for σ to derive its shortcut formula (3-3).

13. Specify whether each of the following is true or false. If a statement is false, explain why. The standard deviation:

    a. indicates how closely the scores are clustered about the mean.

    b. increases as the IQR increases.

    c. is the only useful measure of dispersion.

    d. is a precise measure of dispersion.

    e. and the range take every single score into account.

GROUP 14. Consider the following six sets of data, presented in stem-and-leaf displays.

A	
Stem	Leaf
12	0
10	0
9	5,5,6
8	2,5,7,8,9
6	0,2
3	3,2
1	1

B	
Stem	Leaf
5	0
4	5,5
3	3,3,5,5,6,8
2	2,2,4
1	2,4
0	5

C	
Stem	Leaf
5	0,5,5,6,9
4	2,3,5,7
3	1,2,7
2	5
1	7
0	8

D	
Stem	Leaf
5	5
4	2,3
3	6
2	2,4,6,8
1	3,3,3,4,5,6
0	7

E	
Stem	Leaf
16	7
7	5,7
6	2,3,7
5	0,0,5,8
3	3,4,7
2	4,8

F	
Stem	Leaf
50	0
45	0,0
33	0,0,0
31	0,0,0,0,0
28	0,0,0
10	0

    a. Compute the range and standard deviation for each set of data.

    b. For each of the six sets of data, divide the range by the standard deviation. Use the six results to state a "rule of thumb" for the approximate number of standard deviations that are equal to the range in a distribution without obvious outliers.

15. The following array shows the ages of National Football League head coaches when they were hired, as of January 31, 1993.

Team	Coach	Age
Atlanta	Jerry Glanville	38
Buffalo	Marv Levy	58
Chicago	Dave Wannstedt	40
Cincinnati	Dave Shula	32
Cleveland	Bill Belichick	38
Dallas	Jimmy Johnson	45
Denver	Dan Reeves	37
Detroit	Wayne Fontes	49
Green Bay	Mike Holmgren	43
Houston	Jack Pardee	53
Indianapolis	Ted Marchibroda	60
Kansas City	Marty Schottenheimer	45
L.A. Raiders	Art Shell	42
L.A. Rams	Chuck Knox	59
Miami	Don Shula	40
Minnesota	Dennis Green	42
New England	Dick McPherson	60
New Orleans	Jim Mora	50
N.Y. Giants	Ray Handley	46
N.Y. Jets	Bruce Coslet	43
Philadelphia	Rich Kotite	38
Phoenix	Joe Bugel	49
Pittsburgh	Bill Cowher	34
San Francisco	George Seifert	49
San Diego	Bobby Ross	55
Seattle	Tom Flores	54
Tampa Bay	Sam Wyche	47
Washington	Joe Gibbs	40

Source: *USA Today,* January 20, 1993, p. 4C.

Find the mean, standard deviation, and range of these ages. Do the standard deviation and range follow the rule of thumb from Part b of Exercise 14?

3.4   OTHER MEASURES OF DISPERSION

The standard deviation is a common and useful measure of dispersion, but there are situations when it is not the best measure of spread. Two of these situations will now be mentioned, and a measure of dispersion will be introduced which is more appropriate in those instances than the standard deviation.

The Coefficient of Variation

The coefficient of variation, denoted by v, is unlike the mean and standard deviation in that it lacks units, allowing comparison of the variability of scores in two data sets that do not necessarily measure the same thing. The coefficient of variation is also useful for comparing the variability of data sets that do measure the same thing but have scores of very different magnitudes.

Definition ▶ The **coefficient of variation,** v, is defined as

$$v = \frac{\sigma}{\mu}. \tag{3-4}$$

E x a m p l e
5

Following are the numbers of hospitals classified as adult or pediatric in nine geographic divisions of the United States.

Geographic Division	Adult	Pediatric
New England	216	2
Middle Atlantic	513	3
South Atlantic	749	6
East North Central	771	8
East South Central	419	3
West North Central	717	6
West South Central	691	7
Mountain	355	2
Pacific	533	6

Source: *American Hospital Association Hospital Statistics,*
1993–94, Table 13B.

Use the coefficient of variation to determine whether the number of adult hospitals or the number of pediatric hospitals is more variable.

Solution   The coefficient of variation, v, is determined by the formula $v = \frac{\sigma}{\mu}$. Therefore, the standard deviation and mean of each of the two sets of data must first be computed. The following table shows the results of these calculations.

	Adult Hospitals	Pediatric Hospitals
σ	183.97	2.15
μ	551.56	4.78

The coefficient of variation for adult hospitals can then be calculated as $v_A = \frac{183.97}{551.56} \approx .33$, and the coefficient of variation for pediatric hospitals as $v_P = \frac{2.15}{4.78} \approx .45$. This implies that the number of pediatric hospitals is more variable than the number of adult hospitals.   ●

The Variance

Another measure of dispersion, to be used more often, later in the text, is the variance.

Definition ▶ The **variance** of a set of data is the square of its standard deviation. That is, variance $= \sigma^2$.

Like the standard deviation, the variance is used to compare the variability of scores from two or more sets of like data. Chapter 13, on ANOVA testing, will examine the variance further.

Exercises 3.4

If a graphics calculator is used for the following exercises, "1-Var Stats" will show the mean and standard deviation. These values can then be used to determine the variance, or coefficient of variation.

1. A small computer firm has six employees. Classified by years of experience and annual salary, they are as follows:

Number of Years Experience	Salary
2	$43,000
2	44,500
3	52,000
5	50,000
7	61,700
15	92,600

   Use the coefficient of variation to determine whether the years of experience or the salaries are more variable.

2. Following are the prices of 17 models of men's running shoes along with their weights, in ounces.

Brand	Price	Weight
Adidas Tube 2	$130	10.8
Asics Gel-Kayano	130	13.0
Asics Gel-122 Cross	75	11.5
Asics Gel-Saga	60	10.0
Asics GT-2000	85	12.8
Avia Lantra	80	12.0
Brooks Allegro	65	12.6
Brooks Solaris	55	11.2
Etonic StableAir Strata	75	11.5
L.A. Gear Lightspeed Tech	80	14.5
New Balance 485	60	12.5
New Balance 865	80	10.5
New Balance 555	75	12.5
Nike Air Hurricane Plus	115	11.0
Nike Air Pegasus RD	75	12.0
Nike Air Structure II	110	14.5
Saucony Grid Sensation	85	11.0

Source: *Runner's World,* October 1993, pp. 54–72.

   Use the coefficient of variation to determine whether shoe price or shoe weight is more variable among these models.

3. Does there seem to be any advantage to using the variance instead of the standard deviation to compare variability of scores? For the same two sets of data, could a comparison of the variances ever yield a conclusion different from the one yielded by a comparison of the standard deviations?

4. Following are the prices of 16 models of women's running shoes along with their weights, in ounces.

Brand	Price	Weight
Adidas Equipment Running	$100	9.9
Adidas Torsion Revenge	90	14.2
Asics Gel-121	70	9.5
Asics Gel-Eagle	65	10.5
Avia 2075	90	13.5
Avia 3020	60	9.0
Avia Meridian	80	8.9
Brooks Addition	70	11.6
Brooks Eldorado	70	11.3
Brooks Prodigy	45	8.6
Converse Ellipse	70	11.7
New Balance 530	70	9.0
Nike Air/Atlantis	80	11.5
Puma Terrain	120	13.0
Saucony Grid Courageous	95	10.7
Saucony Jazz 3000	68	9.3

Using the above data and the data in Exercise 2, and using the variance as the measure of variability, determine whether shoe prices for men or for women are more variable.

5. The following table shows the campaign spending (in millions of dollars) of winning congressional candidates.

Year	Senate	House of Representatives
1977–78	42.3	55.6
1979–80	40.0	78.0
1981–82	68.2	114.7
1983–84	97.5	127.0
1985–86	104.3	154.9
1987–88	123.6	171.0
1989–90	115.4	179.1
1991–92	124.3	239.8

Source: *Congressional Digest,* April 1994, p. 103.

Use the variance to determine whether Senate or House of Representatives campaign spending was more variable in these years.

6. The following array shows the median sale prices (in thousands) for existing

one-family houses in selected metropolitan areas in the year 1988 and in the first quarter of 1992.

City	1988	1992	City	1988	1992
Baltimore, MD	$ 88.7	$111.5	Milwaukee, WI	$ 74.5	$ 96.1
Boston, MA	181.2	168.2	Minneapolis–St. Paul, MN	85.2	94.8
Chicago, IL	98.9	132.9	Philadelphia, PA	102.4	120.7
Cincinnati, OH	69.7	87.5	Phoenix, AZ	80.0	84.7
Cleveland, OH	69.2	88.1	Pittsburgh, PA	63.2	74.8
Dallas, TX	90.8	90.8	San Diego, CA	147.8	182.7
Denver, CO	81.8	91.5	San Francisco Bay area, CA	212.6	243.9
Detroit, MI	73.1	77.5	Seattle–Tacoma, WA	88.7	141.3
Houston, TX	61.8	78.2	Washington, DC	132.5	152.5
Los Angeles area, CA	179.4	218.0			

a. Use both the coefficient of variation and the variance to determine in which year these median prices were more variable.

b. Do the two results from Part a agree? If not, which measure of dispersion would seem more valid to use in order to draw a conclusion? Why?

7. Following are the winning times (in seconds) for men's and women's 100-meter dashes in previous Olympic games.

Men

Year	Winner	Country	Time
1896	Thomas E. Burke	United States	12.00
1900	Francis W. Jarvis	United States	10.80
1904	Archie Hahn	United States	11.00
1908	Reginald E. Walker	South Africa	10.80
1912	Ralph C. Craig	United States	10.80
1920	Charles W. Paddock	United States	10.80
1924	Harold M. Abrahams	Great Britain	10.60
1928	Percy Williams	Canada	10.80
1932	Eddie Tolan	United States	10.30
1936	Jessie Owens	United States	10.30
1948	Harrison Dillard	United States	10.30
1952	Lindy J. Remigino	United States	10.40
1956	Bobby J. Morrow	United States	10.50
1960	Armin Hary	Germany	10.20
1964	Robert L. Hayes	United States	10.00
1968	James Hines	United States	9.90
1972	Valery Borzov	USSR	10.14
1976	Hasely Crawford	Trinidad and Tobago	10.06
1980	Allan Wells	Great Britain	10.25
1984	Carl Lewis	United States	9.99
1988	Carl Lewis	United States	9.92
1992	Liford Christie	Great Britain	9.96

Women

Year	Winner	Country	Time
1928	Elizabeth Robinson	United States	12.20
1932	Stanislawa Walasiewicz	Poland	11.90
1936	Helen Stephens	United States	11.50
1948	Francina Blankers-Koen	Netherlands	11.90
1952	Marjorie Jackson	Australia	11.50
1956	Betty Cuthbert	Australia	11.50
1960	Wilma Rudolph	United States	11.00
1964	Wyomia Tyus	United States	11.40
1968	Wyomia Tyus	United States	11.00
1972	Renate Stecher	East Germany	11.07
1976	Annegret Richter	West Germany	11.01
1980	Lyudmila Kondratyeva	USSR	11.06
1984	Evelyn Ashford	United States	10.97
1988	Florence Griffith-Joyner	United States	10.54
1992	Gail Devers	United States	10.82

Source: *The Universal Almanac 1994*, pp. 692–693.

Are the winning times for men or the winning times for women more variable? Choose the measure of variability, and state the reasons for your choice.

8. Carefully examine the data of Exercise 7.

   a. Does it appear that the winning times of one gender are decreasing more consistently than the winning times of the other? Why or why not?

   b. Examine just the years in which winning times were recorded for both men and women. By how much has the difference between men's and women's winning times decreased? What does this decrease imply?

   c. Do you think a woman's winning time for this event will ever be less than a man's in the same year? Why or why not?

GROUP 9. The following array shows crude oil production (in millions of barrels per day) by OPEC and non-OPEC countries in selected years.

OPEC Countries

Country	1960	1970	1980	1990
Indonesia	.41	.85	1.58	1.40
Iran	1.07	3.83	1.66	3.09
Iraq	.97	1.55	2.51	2.01
Nigeria	.02	1.08	2.06	1.83
Saudi Arabia	1.31	3.80	9.90	6.48
Venezuela	2.85	3.71	2.17	2.14
Other OPEC	2.07	8.59	7.11	6.76

Non-OPEC Countries

Country	1960	1970	1980	1990
Canada	.52	1.26	1.44	1.53
China	.10	.60	2.11	2.77
Mexico	.27	.49	1.94	2.55
United States	7.04	9.64	8.60	7.30
USSR	2.91	6.97	11.46	10.68
Other non-OPEC	1.42	3.50	5.20	9.72

Use the coefficient of variation to accomplish the following.

a. Rank the years from most variable to least variable for OPEC production.

b. Rank the years from most variable to least variable for non-OPEC production.

c. Does there seem to be any relation between variability of production in OPEC countries and in non-OPEC countries? Explain.

* 10. The following array shows three-year averages for rates of first marriages, divorces, and remarriages. (First marriages are per 1000 single women, divorces are per 1000 married women 15 to 44 years old, and remarriages are per 1000 widowed and divorced women 15 to 54 years old.)

Period	First Marriage	Divorce	Remarriage
1921–23	99	10	98
1924–26	95	11	99
1927–29	94	12	84
1930–32	81	10	61
1933–35	92	11	69
1936–38	98	13	83
1939–41	106	14	103
1942–44	108	17	139
1945–47	143	24	163
1948–50	134	17	135
1951–53	122	16	136
1954–56	120	15	129
1957–59	112	15	129
1960–62	112	16	119
1963–65	109	17	143
1966–68	107	20	166
1969–71	109	26	152
1972–74	103	32	151
1975–77	85	37	134
1978–80	83	40	134
1981–83	84	39	125
1984–86	80	38	115
1987–89	76	37	109

Source: *U.S. Bureau of the Census, Current Population Reports: P23-180, Marriage, Divorce, and Remarriage in the 1990's,* U.S. Government Printing Office, Washington, D.C., 1992.

a. A **three-year average** is a mean computed over three years and is often used to "smooth" data. Explain what you think "smooth" means in this application.

b. Which of the three measures of dispersion studied in this chapter (standard deviation, coefficient of variation, and variance) do you believe would be the best measure for comparing the variability of marriage, divorce, and remarriage? Why?

c. Based on the measure of dispersion chosen in Part b, is the marriage rate, divorce rate, or remarriage rate most variable? Which is least variable?

3.5   HISTOGRAMS AND FREQUENCY CURVES

Histograms

Box plots have been shown to be fairly effective in describing the distribution of scores in a data set by showing the relative locations of the quartiles and the extreme scores. However, it is often difficult to see the pattern made by all the scores. Dot plots show this pattern very well. With increasingly great numbers of scores, however, dot plots would require larger and larger pieces of paper, and thus their use is usually restricted to relatively small data sets.

Definition ▶   A **histogram** is a bar chart in which there is no space between successive bars, or intervals. It graphically displays the data's distribution in such a way that the frequency of each score contributes to its shape.

A histogram can show the distribution of a great number of scores.

E x a m p l e 6   Draw a histogram illustrating the distribution of SAT scores for the 40 students listed in Example 1, repeated here:

720	610	700	520	760	440	380	600	720	600
610	800	320	460	580	610	780	770	760	570
640	520	600	480	500	520	680	780	780	680
660	580	490	620	520	780	700	680	760	520

Solution

Step 1   Decide on the number of intervals, or classes, into which to divide the data.

A histogram divides the data into any chosen number of intervals, determined by the amount of detail that is desired. More intervals show more detail, but after a certain point they add no more information than a stem-and-leaf display would provide. The goal is to attain a good compromise.

In the absence of any other directions, use the following rule of thumb. **Sturgess' Rule** recommends that the number of intervals be approximately $(1 + 3.3 \cdot \log(N))$, where $\log(N)$ is the logarithm base 10 of the number of scores in the data. In this case, $N = 40$, so

$$1 + 3.3 \cdot \log(40) \approx 1 + 3.3 \cdot (1.602)$$
$$\approx 6.28.$$

Therefore, according to Sturgess' Rule, the histogram will be drawn with about six intervals. Sturgess' Rule is only a starting point; the actual number of intervals in the final histogram could be a few more or less, depending upon the results of later steps.

Step 2   Determine the interval, or class width.

First find the range—in this case, $800 - 320$, or 480. Next, divide the range by the desired number of intervals as calculated in Step 1, and round the answer to the nearest whole number. In this case, no rounding is necessary, and the interval width is $\frac{480}{6}$, or 80.[6]

Step 3   Set up the intervals.

From Step 2 the determination was made to sort the data into intervals of width 80. The greatest score is 800, so, working backwards, the last interval includes scores greater than $(800 - 80)$ and less than or equal to 800. Continuing to work backwards in steps of 80, a total of seven intervals is obtained as follows:

The interval notation $(720, 800]$ indicates all numbers greater than 720 but less than or equal to 800; $[720, 800)$ would indicate all numbers greater than or equal to 720 but less than 800.

Interval 1: Greater than 720 and less than or equal to 800 $(720, 800]$

Interval 2: Greater than 640 and less than or equal to 720 $(640, 720]$

Interval 3: Greater than 560 and less than or equal to 640 $(560, 640]$

Interval 4: Greater than 480 and less than or equal to 560 $(480, 560]$

Interval 5: Greater than 400 and less than or equal to 480 $(400, 480]$

Interval 6: Greater than 320 and less than or equal to 400 $(320, 400]$

Interval 7: Greater than 240 and less than or equal to 320 $(240, 320]$

The end points of the intervals are treated very carefully. It is important that end points be included on only one end of each interval. The end on which they are included is completely arbitrary. In this case they are included on the right end, and hence this is called a **right-end histogram.**[7] (If end points had been included on the left end, this would have been called a **left-end histogram.**)

---

[6] If the quotient in this step is a whole number or if it is rounded down to arrive at a whole number, then there will be one more interval than planned. If the quotient is rounded up, there will be exactly as many intervals as planned.

[7] The fact that the score 320 can not be included in the interval of scores greater than 320 and less than or equal to 400 forces the "extra" interval from $(240, 320]$.

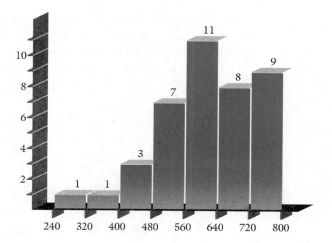

Figure 3-6

Step 4   *Mark the intervals along a line and tally the number of scores in each interval.*
The number of tally marks is the frequency of the interval.

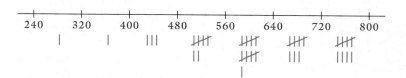

Step 5   *Add a vertical axis to the line from Step 4, and construct bars with heights*
equal to corresponding frequencies.

The interval with the most scores in it is called the **modal interval.** In Figure 3-6,
the modal interval is (560, 640]. Changing a histogram from right-end to left-end
can sometimes change the modal interval and result in a slightly different appear-
ance, although the left-end version of the histogram in this example would look
the same as Figure 3-6.   ●

E x a m p l e
7

Use a graphics calculator to draw a histogram for the data of Example 6. Use an
interval width of 40.

Solution   Enter the data in the same manner as shown earlier.

• Clear list L1.

This is accomplished with the keystroke sequence

STAT, 4, 2nd, 1, ENTER.

• Enter all of the SAT scores in list L1. (Press STAT, 1 to begin entering scores.)
• Press 2nd, Y= to access the statistical plotting capabilities. Then press 1.

By using the arrow (▶) and ENTER keys, turn on plot 1; choose the histogram icon; specify the location of the data (Xlist) as list L1; and specify an individual score frequency of 1. The final screen should appear as follows:

- Press the WINDOW key to set reasonable limits for the histogram.[8]

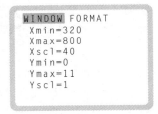

- Press the GRAPH key to obtain the histogram shown in Figure 3-7.

It is interesting to compare Figure 3-7 with histograms that contain twice as many intervals (*Xscl* = 20, Figure 3-8) and half as many intervals (*Xscl* = 80, Figure 3-9).

These figures show a histogram has an optimum number of intervals; too many intervals may be confusing, and too few intervals may fail to show important details.

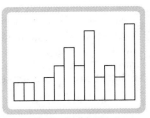

Figure 3-7 A histogram produced by a TI-82 graphics calculator.

[8]At some point it should be verified that the other two statistical plots, Plot2 and Plot3, are turned off. The keystrokes 2nd, Y= will produce a screen indicating whether these plots are on or off. If they are on, turn them off by typing each one's number and then using the arrow keys and ENTER.

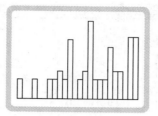

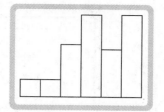

Figure 3-8 *Xscl* = 20.   **Figure 3-9** *Xscl* = 80.   ●

E x a m p l e
8

Use Minitab to produce histograms that compare the cholesterol counts studied in the last chapter and in Example 2 of this chapter. The data are repeated here:

Cholesterol Counts

	Group H				Group R		
139	157	160	201	160	198	140	130
122	188	194	240	138	85	98	108
200	180	230	179	126	208	178	170
194	190	180	188	175	208	161	195
179	194	140	136	130	120	110	128
180	193	159	152	223	98	163	165
176	157	178	206	150	186	248	83
160	120	176	192	186	161	165	188
203	210	176	188	163	145	126	195
194	189	163	196	210	123	116	164

Solution   As in any application of Minitab, the first step is to enter the data into a worksheet shown in Figure 3-10. Next, the **Histogram** command from the **Graph** menu is chosen, and information about the histogram—basically the midpoints of the first and last intervals along with the desired interval width—is supplied in the

	C1	C2	C3
↓			
1	139	160	
2	122	138	
3	200	126	
4	194	175	
5	179	130	
6	180	223	

Figure 3-10

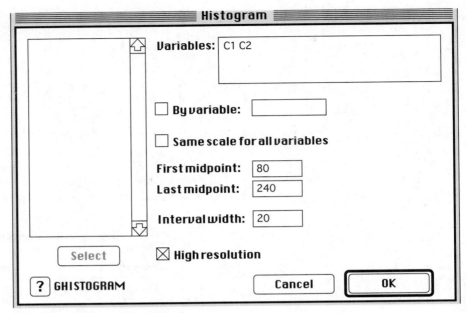

Figure 3-11 A histogram dialog box.

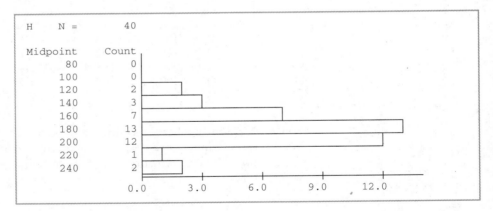

Figure 3-12

resulting dialog box. In this example, an interval width of 20 is specified. Figure 3-11 shows the completed dialog box. The dialog box for the DOS version of Minitab is similar.

Minitab histograms show several slight differences from histograms produced with calculators. First, the midpoints (rather than the end points) of each interval are identified. Second, the histograms are produced "sideways," with the high scores at the bottom instead of the top.[9] Also, all Minitab histograms are right-ended.

Figure 3-12 shows a histogram for Group H (high cholesterol group). A

---

[9] Not all statistical software programs produce histograms with this orientation. Other programs produce histograms that look exactly like the ones shown in Example 5.

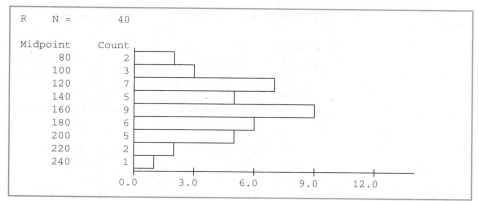

Figure 3-13

histogram for Group R (randomly selected) would look like Figure 3-13. From a comparison of these two histograms the trend toward higher cholesterol counts in Group H is easily seen.   ●

The word "histogram" refers to a display similar to those in the preceding examples. Histograms are sometimes called **absolute frequency histograms** (or **frequency histograms**) because the height (or length) of each bar corresponds to the number of scores in the interval. A **relative frequency histogram** is a histogram whose vertical scale is graduated in percentages of the total number of scores in the data. Example 6 completed as a relative frequency histogram would look like Figure 3-14. This histogram and the one from Example 6 on page 105 look exactly the same. The only difference is that percentages are indicated instead of frequencies. The first class, for example—(240, 320]—has a frequency of 1; 1 out of 40 scores is $\frac{1}{40}$, or 2.5%.

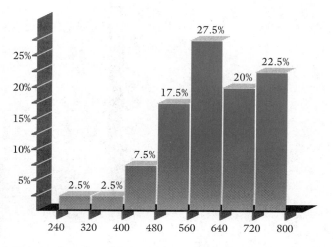

Figure 3-14 A Relative frequency histogram.

Cumulative Frequency Histograms

Definition ▶ | A **less-than cumulative frequency histogram** is a histogram in which the frequency of each interval is equal to the frequency of scores in that interval plus all the frequencies to the left of it.

Figure 3-15 shows a less-than cumulative frequency histogram for the SAT scores of Example 6 on page 103. The vertical scale had to be adjusted to fit a cumulative frequency of 40 onto the page. The frequency of each interval tells how many scores are equal to or less than the upper value in that interval. For example, the frequency of 5 in the interval (400, 480] indicates that five scores are 480 or less. For the construction of a less-than cumulative frequency histogram, these cumulative counts must be obtained from the data, either by examining the frequency histogram, if one is constructed, or by using a stem-and-leaf display.

Definition ▶ | A **greater-than cumulative frequency histogram** is a histogram in which the frequency of each interval is equal to the frequency of scores in that interval plus all the frequencies to the right of it.

If a greater-than cumulative frequency histogram were constructed for the SAT scores, the frequency of (400, 480] would be 38, indicating that 38 scores were greater than 400.

Both types of cumulative frequency histograms can be converted to relative-frequency cumulative histograms by replacing interval frequencies with percentages.

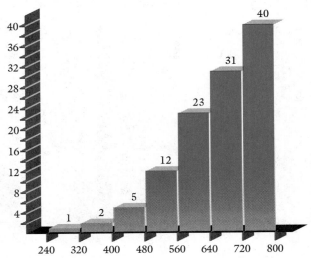

Figure 3-15 A less-than cumulative frequency histogram.

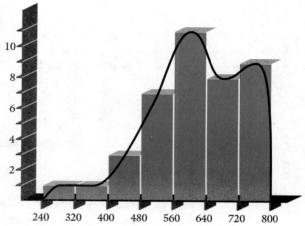

Figure 3-16 A frequency curve constructed from a histogram.

Frequency Curves

Definitions ▶ A **frequency polygon** is a straight-line graph that is formed by connecting the midpoints of the tops of the bars of a histogram.

A **frequency curve** is a "smoothed-out" frequency polygon.

Frequency curves will be very important and will be used often throughout the rest of this text. Figure 3-16 is an example of a frequency curve for the SAT scores. In the drawing of frequency curves, the bars are not usually drawn; often the frequencies of interval midpoints are indicated instead, as shown in Figure 3-17.

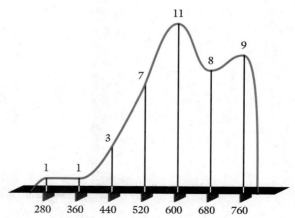

Figure 3-17 A frequency curve with interval midpoints shown.

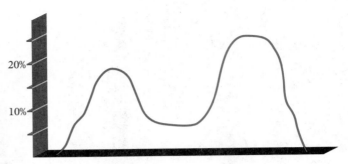

Figure 3-18

The frequency curve is a very informative display from which the distribution of data can be described and inferences drawn. For example, consider the **relative frequency curve** in Figure 3-18. This figure shows a nearly bimodal distribution of scores that could represent many different situations. One might be the shoe sizes of a random selection of people, with the scores clustering around two sizes. Sizes would be expected to tend toward one of two values: the modal size of men and the modal size of women. If more men than women were selected, then a relative frequency curve similar to this one could result.

Ogives

An **ogive** is a frequency curve formed from a cumulative frequency histogram by connecting the corners of the bars with a smooth curve. Using the SAT scores shown in the less-than cumulative frequency histogram in Figure 3-15, a less-than ogive would appear as shown in Figure 3-19. The bars have been included but

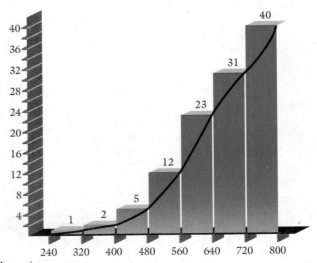

Figure 3-19 An ogive.

can be deleted. By examining the shape and steepness of an ogive, intervals where the frequency of scores increase significantly can be identified. Also, comparison of the heights of the curve at specific scores with the vertical scale can allow one to determine how much of the data are less than or equal to that score. Greater-than ogives can also be constructed and used.

Relative frequency polygons, relative frequency curves, and relative frequency ogives use percentages in the frequency scale.

Exercises 3.5

1. Name some situations that would generate data with a bimodal distribution of scores.

2. Explain an approximate method to obtain a percentile for a given score from a less-than cumulative frequency histogram. Why is the method only approximate?

3. Very often, what effect does increasing the number of intervals have on a data set's histogram?

Definition ▶   A **left-end/minimum-score histogram** is a left-end histogram in which the first interval begins with the smallest score in the data.

4. In May 1994 a check of supermarket shelves revealed the information in the table at the top of page 114. Use Sturgess' Rule to construct a left-end/minimum-score histogram for these data. Round the interval width to the nearest whole number.

Stat **F**act

Since 1987 Americans have eaten more than $3.5 billion worth of cold cereal each year.

(*Marc Sherman*)

Amount of Sodium (mg) per Serving in Selected Low-Fiber Cereals

Puffed Rice	0	Puffed Wheat	0
Kix	260	Honey-Comb	190
Corn Flakes	330	Kellogg's 19	330
Rice Chex	230	Apple Jacks	135
Cocoa Puffs	190	Cookie Crisp Chocolate Chip	110
Corn Chex	270	Crispix	230
Corn Pops	95	Lucky Charms	200
Froot Loops	150	Rice Krispies	360
Golden Crisp	40	Special K	250
Alpha-Bits	210	Cheerios	290
Total Corn Flakes	200	Wheaties Honey Gold	200
Fruity Pebbles	150	Cocoa Pebbles	160
Cocoa Krispies	190	Golden Grahams	280
Frosted Flakes	200	Cap'n Crunch	220
Triples	190	Cinnamon Toast Crunch	210
Cap'n Crunch's Crunch Berries	210	Smacks	40

Definition ▶ A **right-end/maximum-score histogram** is a right-end histogram in which the last interval ends with the largest score in the data.

5. In May 1994 a check of supermarket shelves revealed the following information. Use Sturgess' Rule to construct a right-end/maximum-score histogram for these data.

Calories per Serving in High-Fiber Ready-to-Eat Cereals

100% Bran	70
All-Bran Original	80
All-Bran with Extra Fiber	50
Basic 4	210
Bran Buds	70
Natural Bran Flakes	90
Cracklin' Oat Bran	230
Crunchy Corn Bran	140
Fiber One	60
Frosted Mini Wheats	190
Fruit & Fibre Peaches, Raisins, Almonds & Oat Clusters	120
Grape-Nuts	100
Multi Bran Chex	220
Oatmeal Crisp with Raisins	210
Post Banana Nut Crunch	120
Post Great Grains, Double Pecan	120
Raisin Bran	170
Shredded Wheat 'N Bran	90
Shredded Wheat	80
Team Flakes	110

6. A survey was conducted to compare food prices at grocery stores. A market basket of basic items was purchased at 40 supermarkets, and these are the prices paid:

$ 88.13	$ 95.99	$84.58	$ 85.85	$99.28
90.19	92.18	86.08	105.72	98.77
97.63	99.58	98.13	94.30	89.86
91.43	92.68	88.74	99.83	95.11
100.69	96.32	96.48	94.92	85.07
106.20	99.99	95.82	86.48	88.51
85.12	93.85	96.47	88.87	96.18
90.07	100.01	89.17	93.05	94.32

   a. Construct a left-end/minimum-score histogram to illustrate these data. Use an interval width of $3.61.

   b. What is the modal interval? That is, which interval contains the most scores?

7. Construct a less-than cumulative frequency histogram for the data of Exercise 6.

8. Answer the following questions by consulting the cumulative frequency histogram constructed in Exercise 7.

   a. What percentage of stores charged less than $91.80 for the basket?

   b. What percentage of stores charged at least $88.19 but less than $99.02?

   c. What percentage of stores charged at least $95.41?

   d. What percentage of stores charged either less than $88.19 or at least $102.63?

9. Consider the following data on the median ages for people in different occupations:

Occupation	Age	Occupation	Age
Millwright	43.1	Elementary school teacher	41.3
Dentist	40.5	Civil engineer	42.0
Tool and die maker	39.5	Locomotive operator	46.5
Barber	51.4	Geologist	40.8
Stationary engineer	42.2	Pharmacist	39.1
Electrician	38.3	Secondary school teacher	42.0
Composer	46.5	Veterinarian	39.5
Airplane pilot	36.2	Telephone installer	40.3
Farmer	48.0	Brickmason	38.2
Railroad conductor	43.1	Structural metal worker	36.5
Crane operator	41.9	Police supervisor	41.5
Machinist	39.3	Plumber	38.4
Dental technician	39.9	Operating engineer	37.8

Source: *Statistical Abstract of the United States: 1994*, p. 415.

Draw a left-end/minimum-score frequency histogram with eight intervals for these data. Use the histogram to answer the following questions.

a. Which occupations are in the youngest interval? Which are in the oldest interval?

b. Identify the modal intervals (the intervals with the most scores recorded).

c. Use the histogram to make an educated guess about the values of the median and the mean.

GROUP   10. Use the frequency histogram of Exercise 9 to construct a frequency curve, a less-than ogive, and a greater-than ogive.

Although the following exercises are specific to the TI-82, other graphics calculators have some of the same features.

11. Describe an easy method of using the graphics calculator to determine $Xmin$ and $Xmax$ for setting WINDOW parameters.

12. What is displayed when the TRACE key is used with a histogram?

13. Refer to the histogram in Figure 3-7 on page 106. For the same data, find a value of $Xscl$ that produces a histogram containing two intervals in which there are no scores.

14. Clear list L1 and enter the data from Exercise 5. By trial and error, determine the value of $Xscl$ that will produce a histogram with six intervals. Let $Xmin$ and $Xmax$ be the minimum and maximum scores, respectively.

15. Clear list L1 and enter the data from Exercise 9. The goal is to draw both a box plot and a histogram for these data on the same screen. What should be the value of the WINDOW parameters so that the two displays do not overlap?

3.6 MISLEADING HISTOGRAMS AND OTHER TYPES OF GRAPHS

A graph can unintentionally misrepresent data in many ways. For instance, in the drawing of histograms, ogives, and other graphs, it is often tempting to show more detail where changes occur than where there is little change.

As an example, consider the (left-ended) histogram in Figure 3-20, which shows the average length of a hospital stay in selected community hospitals across the country in 1993. This histogram clearly summarizes the data. For all hospitals in the selection, the average stay was at least 5 days; most average stays were between 6 and 7 days, but some hospitals reported average stays of more than 10 days.

It would be misleading to present the following histogram, which shows only a portion of the data, in an effort to focus on the majority of the cases. Figure 3-21 is very misleading in several ways:

• The absence of the longer stays gives the false impression that 8 days is the longest average stay reported. Even if the accompanying text reports the occurrence of longer stays, that fact may easily be overlooked or forgotten during the study of this graph.

- This histogram in Figure 3-21 is actually the upper part of the histogram presented in Figure 3-20 and is an example of a **truncated** graph. Truncated graphs are always misleading because they are out of proportion. For instance, the scale of Figure 3-21 suggests that there are about twice as many hospital stays between 6 and 7 days as there are between 5 and 6 days, even though the actual frequencies are noted at the tops of the bars.

Although truncated graphs should be avoided, they are sometimes necessary due to space limitations. The use of slash marks (//) in a truncated graph helps the reader remember that the graph is out of proportion, as demonstrated by Figure 3-22. The use of slash marks is a compromise between showing every detail

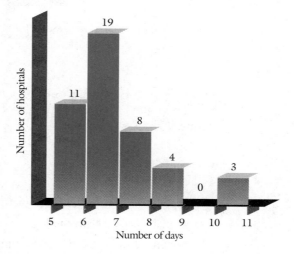

Figure 3-20

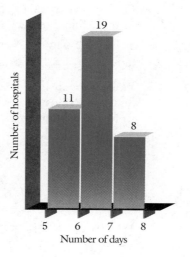

Figure 3-21

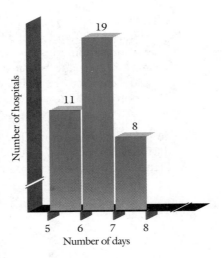

Figure 3-22

in a graph and focusing only on a specific part of it. The nature of the data being presented and the experience of the intended readers determine whether the compromise is satisfactory.

E x e r c i s e s 3 . 6

1. State two general rules to keep in mind when drawing histograms and other graphs, so that the finished picture does not mislead the reader.

2. Explain what a truncated graph is and why such a graph may be misleading.

3. Sometimes truncated graphs cannot be avoided. What are some valid reasons for using truncated graphs?

4. Find several examples of misleading or inaccurate graphs in a recent newspaper or magazine. For each example:

   a. Explain why you believe it is misleading or inaccurate.

   b. Suggest a better way of presenting the data.

5. When icons are used to form graphs, the resulting display is called a **pictograph.** Suppose an illustrator wished to show that the value of a certain stock doubled from the previous year, and she produced the following pictograph:

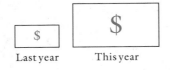

What is misleading or inaccurate about this graph? How should the problem be corrected?

Case Study

Analyzing crime rates and imprisonment rates helps us to understand patterns of crime over time. (*Alcatraz, National Park Service*)

The data presented in this case study are examples of **time series data.** Time series data record the behavior of a variable, such as the national crime rate, over time. When plotting time series data it is standard practice to place values of the variable along the vertical axis and time along the horizontal axis.

Crime and Punishment

Is crime in America increasing or decreasing? Is the prison population increasing or decreasing? Is the crime rate or the imprisonment rate more variable? One way to answer such questions is to examine some statistics. Page 33 of the November 13, 1993, issue of *The Economist* (London) contains information from which the following data were estimated.

Year	National Crime Rate (per 100,000 of the population)	Imprisonment (rate per 100,000 of the population)
1980	11,600	140
1981	12,000	150
1982	11,600	165
1983	10,700	182
1984	10,300	190
1985	10,000	200
1986	9,500	215
1987	9,700	225
1988	10,000	250
1989	9,800	275
1990	9,300	290
1991	9,200	310

I. Find the standard deviation of the crime rates and the standard deviation of the imprisonment rates. Then find the coefficient of variation for each.

II. Is the crime rate or the imprisonment rate more variable? Write a paragraph to support your choice, using one of the measures of variability from Part I to support your argument.

III. A **line graph** for this data would place years along the horizontal axis and rates along the vertical axis. Each year and its corresponding rate could then be plotted and the points connected. Draw separate line graphs for the crime rates and the imprisonment rates. What seems to be the relationship between the two graphs?

IV. The national crime rates in this case study are based on a national survey of 100,000 people. Another index, the **Uniform Crime Reports** (UCR), charts only crimes for which official police reports have been made. Do you think the UCR index for the same years would show higher or lower rates than those above? Explain your answer.

V. Obtain a copy of the *Economist* article from your library, and give some reasons why the imprisonment rate could be unrelated to the crime rate.

CHAPTER SUMMARY

The mean, median, and mode are three principal ways of arriving at a typical score for a set of data. The trimmed mean and midrange are also used under special circumstances. These statistics are referred to as measures of central location or central tendency. The preferred measure in a given situation depends on the nature and intended use of the data.

Dispersion, or variability, is a measure of the scattering of the scores in a data set. Earlier chapters presented the range and the IQR as two measures of dispersion. However, a better measure of dispersion is often the standard deviation. Unlike the range and the IQR, the standard deviation indicates scattering about the mean and takes every score into account. A relatively small standard deviation indicates more clustering of scores about the mean. The standard deviation, coefficient of variation, and variance can be used to compare the dispersion of different sets of data.

A histogram is a graphic method of showing the pattern, or distribution of scores, in a set of data. As in a dot plot, each score contributes to the overall shape, but the size of a histogram is more easily managed. Since every score contributes to the shape of a histogram, it is usually more informative than a box plot.

There are many special types of histograms; relative frequency histograms, less-than cumulative frequency curves, and frequency polygons are just a few. The word "relative" implies the placement of a percentage scale on the histogram.

Key Concepts, Terms, and Formulas

Mean: $\mu = \dfrac{\sum\limits_{i=1}^{N} x_i}{N}$ (3-1)   *Page 82*

5% trimmed mean, midrange, mode   *Page 83*

Standard deviation: $\sigma = \sqrt{\dfrac{\sum\limits_{i=1}^{N}(x_i - \mu)^2}{N}}$ (3-2)   *Page 88*

Definition formula

$$\sigma = \sqrt{\dfrac{\sum\limits_{i=1}^{N} x_i^2 - \dfrac{\left(\sum\limits_{i=1}^{N} x_i\right)^2}{N}}{N}}$$ (3-3)   *Page 89*

Shortcut formula

Coefficient of variation: $v = \dfrac{\sigma}{\mu}$ (3-4)   *Page 97*

Variance   *Page 99*

Histogram   *Page 103*

Cumulative frequency histogram   *Page 110*

Frequency curve   *Page 111*

Misleading graphs   *Page 116*

Videotape Suggestions

Programs 2 and 3 from the series *Against All Odds: Inside Statistics,* recommended in earlier chapters, would provide a good perspective on some of the material in this chapter, as well. *Program 4* in the same series gives additional information on frequency (density) curves as smoothed histograms.

The St. Louis skyline
resembles a distribution
of data.

C h a p t e r

4 Interpreting Histograms and Frequency Curves

One often needs to interpret data that are summarized in a table, histogram, or other display. Newspaper and magazine writers frequently try to make their points with convincing graphs that show the overall distribution of scores but not the individual score themselves. As an example, the histogram from this chapter's **case study,** shown in Figure 4-1, gives the percentages of people at least 25 years old who hold bachelor's degrees or higher degrees in each of California's 52 congressional districts.

To further examine the situation in California, it might be helpful to compute an estimated mean percentage for the entire state. This chapter explains how to estimate and examine measures of central location and variability for data presented in this way, without individual scores.

4.1 ESTIMATING THE MEAN AND STANDARD DEVIATION FROM GROUPED DATA

When applying the following techniques for estimating the mean, median, mode, and standard deviation, it is important to remember that the measures can only be estimates when actual scores are not given.

123

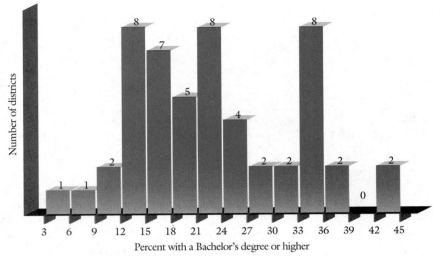

Figure 4-1

Definitions ▶ | **Grouped data** are data presented in the form of a histogram, a frequency curve, an interval tally, or a similar display.
An **interval tally** is a list of intervals and their frequencies of scores.

Chapter 7 will describe how
to estimate the mean and
standard deviation of a population by using a random
sample.

**E x a m p l e
1**

Following is an interval tally of last year's orders to a dealer in rare coins. Estimate the mean price paid for a rare coin from this dealer last year.

Price Range	Interval Tally (frequency)
$0 to under $25	7
$25 to under $50	24
$50 to under $75	20
$75 to under $100	9
$100 to under $125	5
$125 to under $150	10
$150 to under $200	15
$200 to under $300	30
$300 to under $500	6
$500 to under $750	4

Solution   Notice that the interval widths are *not* all the same. They range from $25 to $250. Assume that the prices are scattered throughout each interval and that the mean interval price of interval number i, denoted by m_i, is located at the midpoint of the interval. Without information to the contrary, these assumptions should seem reasonable.

For example, in the first interval, $[0, 25)$, the mean interval price is $\dfrac{0 + \$24.99}{2} \approx \$12.50$. An easier way to find m_i is to simply take half the sum of two consecutive end points:[1] $\dfrac{0 + \$25.00}{2} = \$12.50$. The estimated price of each coin in that interval is $12.50, and the total sales of the seven coins in that interval are estimated at $7 \cdot \$12.50$, or $87.50. This may be more or less than the actual value of all prices in the interval, but it is the best estimate that can be made without the actual prices paid.

More symbolically, if m_i stands for the mean interval price and f_i for the frequency of scores in interval number i, then the amount of money represented in this particular interval is $m_i \cdot f_i$.

If there are k such intervals, then the total sales for all k intervals is found as follows:

$$\text{Estimated total sales (score total)} = \sum_{i=1}^{k} m_i \cdot f_i$$
$$= m_1 \cdot f_1 + m_2 \cdot f_2 + m_3 \cdot f_3 + \cdots + m_k \cdot f_k.$$

In this case, $k = 10$, and total sales are

$$12.50 \cdot 7 + 37.50 \cdot 24 + 62.50 \cdot 20 + \cdots + 625 \cdot 4 = \$19{,}987.50.$$

This is the best possible estimate of the total sales.

To find the estimated mean sale price, divide the total sales by the total number of sales. The total number of sales, or scores, is denoted by N and is found by the following formula.

$$N = \sum_{i=1}^{k} f_i.$$

In this case,

$$N = \sum_{i=1}^{10} f_i$$
$$= 7 + 24 + 20 + \cdots + 4$$
$$= 130.$$

The estimated mean sale price is $\frac{19{,}987.50}{130} = \$153.75.$ ●

[1] Taking half the sum of the least and greatest *possible* scores in an interval to determine the value of m_i is an unnecessary complication.

The preceding example suggests the following formula.

$$\text{Estimated value of } \mu = \frac{\text{Score total}}{\text{Total number of scores}}$$

$$= \frac{\sum_{i=1}^{k} m_i \cdot f_i}{N} \tag{4-1}$$

Once again, notice that this is an *estimated* value of μ. Unless the actual scores are given, there is no way of knowing the true value of μ.

Analogously, the standard deviation, σ, can be estimated from grouped data by the following formula.

$$\text{Estimated } \sigma = \sqrt{\frac{\sum_{i=1}^{k} f_i \cdot (m_i - \mu)^2}{N}} \tag{4-2}$$

As in the estimation of μ, k = the number of intervals, f_i is the absolute frequency of scores in each interval, m_i represents each interval's midpoint, μ is the estimated mean, and N is the total number of scores. This formula is very similar to the definition formula from Chapter 3, except for two features. First, interval midpoints replace individual scores. This has to be so, since individual scores are not given and the midpoints are used as the best estimates of them. Second, the presence of the frequency f_i means that each interval midpoint is counted the same number of times the actual interval scores would be counted if they were known.

A shortcut formula to estimate the standard deviation can be derived algebraically from the preceding formula. It follows.

$$\text{Estimated } \sigma = \sqrt{\frac{\sum_{i=1}^{k} f_i \cdot m_i^2 - N \cdot \mu^2}{N}} \tag{4-3}$$

Shortcut formula

A school profile may comprise a list of grade-point averages, a summary of SAT or ACT scores, a list of typical college acceptances, the number of National Merit finalists, the number of honors and AP courses offered, and other data.

As part of a school profile that a local high school sends to colleges, the grade point averages (GPAs) of all 180 graduates from last year are grouped and presented as follows:

GPA Interval	Number of Students in Interval
[3.20, 4.00)	32
[2.50, 3.20)	58
[1.20, 2.50)	70
[0.50, 1.20)	12
[0.00, 0.50)	8

The college counselor thinks that it would be useful to also indicate the estimated mean GPA and standard deviation. Estimate those measures from the grouped data shown.

Solution When working with grouped data, it is often helpful to organize them in a table.

Interval	Midpoint, m_i	Frequency, f_i	$m_i \cdot f_i$
[3.20, 4.00)	3.60	32	115.2
[2.50, 3.20)	2.85	58	165.3
[1.20, 2.50)	1.85	70	129.5
[0.50, 1.20)	0.85	12	10.2
[0.00, 0.50)	0.25	8	2.0
		$\sum f_i = 180$	$\sum m_i f_i = 422.2$

By the formula given earlier,

$$\text{Estimated value of } \mu = \frac{\sum\limits_{i=1}^{k} m_i \cdot f_i}{N} \tag{4-1}$$

$$= \frac{422.2}{180}$$

$$\approx 2.35.$$

A few more columns in the table would be helpful to estimate the standard deviation using the shortcut formula.

Interval	Midpoint, m_i	Frequency, f_i	$m_i \cdot f_i$	$(m_i)^2$	$f_i \cdot (m_i)^2$
[3.20, 4.00)	3.60	32	115.2	12.9600	414.720
[2.50, 3.20)	2.85	58	165.3	8.1225	471.105
[1.20, 2.50)	1.85	70	129.5	3.4225	239.575
[0.50, 1.20)	0.85	12	10.2	.7225	8.670
[0.00, 0.50)	0.25	8	2.0	.0625	.500
					$\sum f_i m_i^2 = 1134.57$

Using the shortcut formula,

$$\text{Estimated } \sigma = \sqrt{\frac{\sum\limits_{i=1}^{k} f_i \cdot m_i^2 - N \cdot \mu^2}{N}},$$

the following estimated standard deviation is obtained.

$$\text{Estimated } \sigma = \sqrt{\frac{1134.57 - 180 \cdot (2.35)^2}{180}}$$

$$\approx 0.88. \quad \bullet$$

**E x a m p l e
3**

To convert a relative frequency histogram to an absolute frequency histogram, the total number of scores represented must be known.

Figure 4-2 is a relative frequency histogram. *Percentages* are indicated at the tops of the bars. Assume that this histogram was formed from 80 scores, and calculate an estimated mean and an estimated standard deviation of those scores, to the nearest hundredth.

Solution   It is possible to slightly modify the formulas for estimated mean and estimated standard deviation so that they use relative frequencies instead of absolute frequencies. (See Exercise 3 [in Exercises 4.1–4.3] on page 135.)

But for illustration, this example will be completed using absolute frequencies, and so the first task is to convert the relative frequency histogram in Figure 4-2 to an absolute frequency histogram. If the total number of scores, N, is 80, then the absolute frequency of each interval is found by multiplying N by the relative frequency of that interval. The first interval, for instance, has a frequency of $80 \cdot (.25) = 20$. All of the remaining interval frequencies are found the same way. Figure 4-3 shows the resulting histogram.

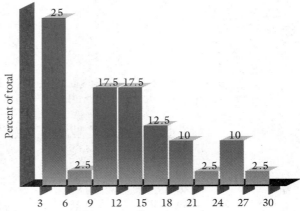

Figure 4-2 A relative frequency histogram.

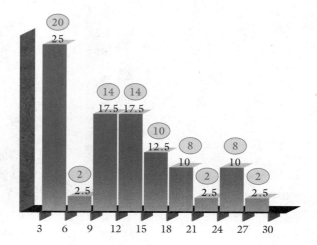

Figure 4-3

For a determination of the estimated mean, all data are arranged in tabular form to prevent confusion. It does not matter whether the histogram is left-ended or right-ended when the mean and standard deviation are estimated. However, for correct definition of the intervals, either left- or right-ended intervals must be assumed. A right-ended interval is used arbitrarily here. The completed table follows.

Interval	Midpoint, m_i	Frequency, f_i	Midpoint · Frequency, $m_i \cdot f_i$
(3, 6]	4.5	20	$4.5 \cdot 20 =$ 90
(6, 9]	7.5	2	$7.5 \cdot 2 =$ 15
(9, 12]	10.5	14	$10.5 \cdot 14 =$ 147
(12, 15]	13.5	14	$13.5 \cdot 14 =$ 189
(15, 18]	16.5	10	$16.5 \cdot 10 =$ 165
(18, 21]	19.5	8	$19.5 \cdot 8 =$ 156
(21, 24]	22.5	2	$22.5 \cdot 2 =$ 45
(24, 27]	25.5	8	$25.5 \cdot 8 =$ 204
(27, 30]	28.5	2	$28.5 \cdot 2 =$ 57
		$N = \sum_{i=1}^{k} f_i = 80$	$\sum_{i=1}^{k} m_i \cdot f_i = 1068$

$$
\text{Estimated mean} = \frac{\sum_{i=1}^{k} m_i \cdot f_i}{N}
$$

$$
= \frac{1068}{80}
$$

$$
= 13.35.
$$

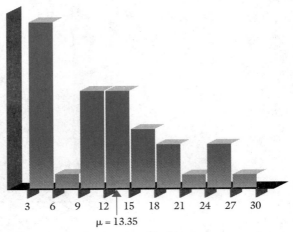

3   6   9   12 | 15   18   21   24   27   30
$\mu = 13.35$

Figure 4-4 The mean as the balance point of a histogram.

Figure 4-4 shows how the estimated mean can also be interpreted as the balance point of a histogram.

Next, using the shortcut formula (4-3) to determine the standard deviation, tabulate the intermediate calculations.

m_i	m_i^2	f_i	$f_i \cdot m_i^2$
4.5	20.25	20	$20 \cdot 20.25 =$ 405.0
7.5	56.25	2	$2 \cdot 56.25 =$ 112.5
10.5	110.25	14	$14 \cdot 110.25 =$ 1,543.5
13.5	182.25	14	$14 \cdot 182.25 =$ 2,551.5
16.5	272.25	10	$10 \cdot 272.25 =$ 2,722.5
19.5	380.25	8	$8 \cdot 380.25 =$ 3,042.0
22.5	506.25	2	$2 \cdot 506.25 =$ 1,012.5
25.5	650.25	8	$8 \cdot 650.25 =$ 5,202.0
28.5	812.25	2	$2 \cdot 812.25 =$ 1,624.5

$$\sum_{i=1}^{k} f_i \cdot m_i^2 = 18,216$$

Finally, substitute into the shortcut formula for the estimated standard deviation.

$$\text{Estimated } \sigma = \sqrt{\frac{\sum_{i=1}^{k} f_i \cdot m_i^2 - N \cdot \mu^2}{N}} \qquad (4\text{-}3)$$

$$= \sqrt{\frac{18,216 - 80 \cdot (13.35)^2}{80}}$$

$$\approx 7.03. \quad \bullet$$

Estimating the mean and standard deviation of grouped data can be tedious, as Example 3 illustrated. Appropriate statistical software or a calculator into which interval frequencies can be keyed can make the task much easier. Graphics calculators, illustrated in the next example, have that capacity.

E x a m p l e
4

Using a TI-82 graphics calculator, estimate the mean and standard deviation of the data presented in the histogram of Example 3. Figure 4-5 repeats the histogram, with the necessary frequencies and interval midpoints indicated. (Recall that the circled numbers are absolute frequencies that were computed from the uncircled relative frequencies.)

Solution   Use two lists of data: L1 for interval midpoints and L2 for the corresponding frequencies.

• Clear lists L1 and L2.

Accomplish this with the keystroke sequence

STAT, 4, 2nd, 1, ENTER,
STAT, 4, 2nd, 2, ENTER.

• Enter all the midpoints in list L1 and all the frequencies in list L2. (Press STAT, 1 to begin entering scores.)

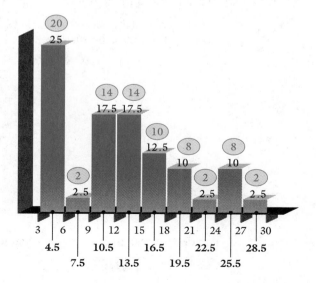

Figure 4-5

When both lists of data are complete, the screen should appear as follows:

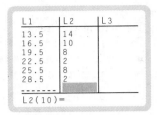

- Next, press STAT, ▶ to obtain the following menu.

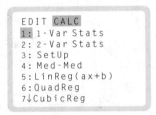

- Type ③ to set up the data calculations in such a way that scores are taken from L1 and score frequencies from L2.

Use the arrow (▶) and ENTER keys to position the cursor over L1 and L2 as shown in the following diagram.

The desired results can now be obtained with the keystrokes

- STAT, ▶, ①, ENTER.

Estimated mean of histogram

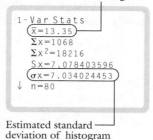

Estimated standard ───
deviation of histogram

4.3 THE WEIGHTED MEAN

Estimating the value of μ from grouped data involves the same procedure followed to compute a weighted mean.

Definition ▶ A **weighted mean** of a group of scores is a mean computed in such a way that the frequency, or relative importance, of each score is taken into consideration.

In Example 4 on page 131, the (nonweighted) mean of the midpoints is

$$\frac{4.5 + 7.5 + 10.5 + 13.5 + 16.5 + 19.5 + 22.5 + 25.5 + 28.5}{9} = 16.5.$$

This mean, however, does not take the frequency, or relative importance, of each midpoint into consideration. The weighted mean 13.35, which is the estimated mean of the data in the histogram, is a more accurate measure of central location.

Most people have experience with weighted means, even though the terminology may be unfamiliar. For instance, suppose a small business pays its president $200,000 per year and pays each of ten salesmen $50,000 per year. The mean of these two salaries is

$$\frac{\$200,000 + \$50,000}{2} = \$125,000.$$

No one's salary comes close to this. On the other hand, the weighted mean of these salaries is

$$\frac{\$200,000 + 10 \cdot \$50,000}{11} \approx \$63,636.36.$$

Although still higher than nearly all the salaries, this value is clearly more typical than the unweighted mean and is therefore a better measure of central location.

E x a m p l e
5

The following array shows eight major retailers with their 1992 estimated sales and percents of growth in sales.

S t a t   F a c t

General Motors has the largest earnings of any American corporation, with revenues of $138 billion. By comparison, the combined revenue of all the teams in major league baseball was $1.80 billion.

Retailer	1992 Estimated Sales	Percent of Growth in Sales
Circuit City	$ 3.2 billion	22.5%
Costco	6.6 billion	47.4%
Dillard's	4.8 billion	16.9%
Home Depot	7.1 billion	38.4%
K-Mart	39.0 billion	7.7%
Target	10.0 billion	15.7%
Toys 'R' Us	7.2 billion	20.2%
Wal-Mart	55.0 billion	29.8%
	$132.9 billion	

Source: *Business Week*, December 21, 1992, pp. 68–69.

Use a weighted mean to determine the mean percent of growth of these eight giant retailers.

Solution   As indicated, the total sales of these eight retailers is $132.9 billion. The relative contribution of each retailer to this total is

$$\frac{\text{Retailer's sales (in billions of dollars)}}{132.9}.$$

This relative contribution can be used to weight each retailer's percent of growth by the formula

Weighted % of growth = (% of growth) · (relative contribution).

Following this formula for Wal-Mart's weighted percent of growth, for instance, one would obtain

$$(29.8\%) \cdot \left(\frac{55.0}{132.9}\right) \approx 12.33\%.$$

The weighted mean percent of growth is the sum of each retailer's weighted percent of growth:

In each case, the percent growth in sales is multiplied by a weighting factor.

$$(22.5\%) \cdot \frac{3.2}{132.9} + (47.4\%) \cdot \frac{6.6}{132.9} + (16.9\%) \cdot \frac{4.8}{132.9} + (38.4\%) \cdot \frac{7.1}{132.9} +$$

$$(7.7\%) \cdot \frac{39.0}{132.9} + (15.7\%) \cdot \frac{10.0}{132.9} + (20.2\%) \cdot \frac{7.2}{132.9} + (29.8\%) \cdot \frac{55.0}{132.9} \approx 22.43\%. \quad \bullet$$

In the last example above, the nonweighted mean would be approximately 24.83%, implying nearly 2.5% more mean growth than would be justified.

Exercises 4.1–4.3

1. In an estimation of the mean or standard deviation of an interval tally, do all intervals need to be the same width? Why or why not?

* 2. Algebraically derive the shortcut formula (4-3) for the estimated standard deviation.

* 3. Explain how Example 3 on page 128 can be completed using only relative frequencies, and then work it that way.

4. A 1993 survey of the annual incomes of nurses on salary is the basis for the following histogram.

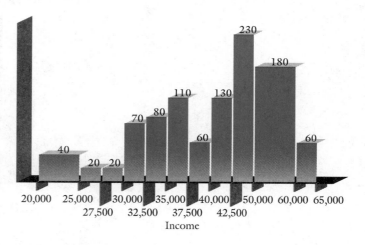

Source: *RN*, October 1993, p. 55.

Estimate the mean annual income of nurses on salary.

GROUP 5. A 1993 survey of the annual incomes of nurses on hourly pay is the basis for the following histogram.

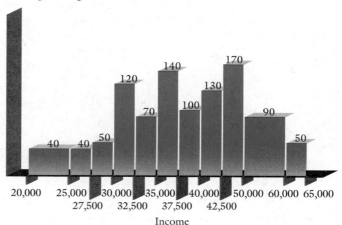

Stat Fact

For full-time nurses, the number of years of nursing experience affects pay more than the service performed or the shift worked. However, for part-time nurses who are hired to fill gaps, the service performed is more likely to affect pay.

a. Estimate the mean annual income of nurses on hourly pay.

b. By having two bars in each interval, a **double-bar histogram** allows for the easy comparison of two separate histograms with identical intervals. Construct a double-bar histogram for the data presented in this exercise and Exercise 4. Shade the bars for the Exercise 4 data, and leave the bars representing the data from this exercise unshaded.

GROUP

6. The histogram completed as Example 6 on page 103 in Chapter 3 is repeated here.

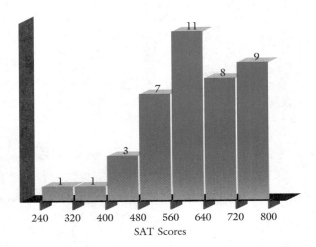

SAT Scores

Estimate the mean and standard deviation on the basis of this histogram. Then see how close those estimates are to the actual values, which can be computed from the data on page 103.

7. All 337 of the people in a small town responded to a questionnaire about the amounts of their telephone bills for the month of December. The following interval tally summarizes the results.

Telephone Bills

Interval	Tally
$80 to under $110	66
$50 to under $80	75
$35 to under $50	80
$20 to under $35	65
$12 to under $20	35
$8 to under $12	7
$0 to under $8	9

Find the estimated mean and estimated standard deviation of all telephone bills in that town for the month of December.

* 8. Refer to Exercise 7. Find the largest and smallest possible values of the mean

of all telephone bills if interval midpoints are not used to estimate scores within an interval.

Hint: *Use the left end point of each interval as the typical score for that interval to obtain the smallest possible mean.*

9. The following absolute frequency histogram shows the average annual pay, in thousands of dollars, received by workers in the states and the District of Columbia in 1988.

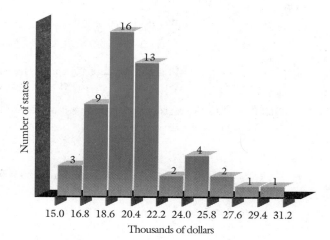

a. Estimate the mean state pay and the standard deviation of state pays from this histogram.

b. The mean state pay from Part a is not the same as the mean pay for the nation. What do you think the difference is, and what additional information would be needed to find the mean pay for the nation?

10. The following absolute frequency histogram shows the distribution of per-capita income, in thousands of dollars, among the 50 largest U.S. cities in 1987.

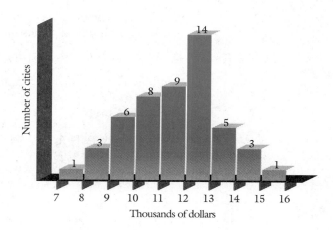

    a. Estimate the mean per-capita income and the standard deviation for these 50 cities.

    b. Approximately what percentage of the area under the histogram appears to lie within one estimated standard deviation of the estimated mean? That is, what percentage of the histogram seems to be in an interval that has left end point

$$L = \text{mean} - \text{standard deviation}$$

and right end point

$$R = \text{mean} + \text{standard deviation?}$$

What percentage appears to lie within two standard deviations of the mean? Within three standard deviations of the mean?

GROUP   11. A car rental company supplied the following mileage information on its fleet of cars.

Mileage of Rental Cars

Accumulated Mileage (thousands)	Tally
50 to under 60	8
40 to under 50	16
20 to under 40	79
15 to under 20	85
10 to under 15	75
5 to under 10	30
1 to under 5	30
0.5 to under 1	22
0.25 to under 0.5	8
Under 0.25	10

    a. Find the estimated mean mileage and estimated standard deviation for this fleet of autos.

    b. What percentage of the fleet appears to have mileages within two estimated standard deviations of the estimated mean?

    Hint: See Part b of Exercise 10.

    c. Combine pairs of consecutive intervals to form an array with only half as many intervals. Do you think the estimated mean increased or decreased? Check your answer.

* 12. The following histogram is a less-than cumulative frequency histogram. It shows the numbers of families (in millions) living below the poverty level during 23 selected years between 1959 and 1989. (For example, this display indicates that for 21 of those years, less than 8.1 million families lived below the poverty level.)

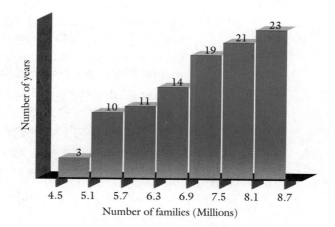

Estimate the mean number of families living below the poverty level in those 23 years.

Hint: *First form an interval tally as follows.*

Interval	Tally
[4.5, 5.1)	③
[5.1, 5.7)	10 − 3 = ⑦
[5.7, 6.3)	11 − 10 = ①
•	•
•	•
•	•

For Exercises 13 through 15, the large screen of a graphics calculator will allow you to "proofread" your weighted mean calculation before entering it.

GROUP

13. The following list shows the numbers of job openings in certain fields in 1990 and their gains or declines from 1980.

Job Openings from 1980 to 1990

More Openings	Fewer Openings
Computer programmers: 663,000 jobs, up 162%	Elevator operators: 11,000 jobs, down 48%
Insurance adjusters: 345,000 jobs, up 106%	Chief executives: 19,000 jobs, down 47%
Economists: 152,000 jobs, up 59%	Telephone operators: 233,000 jobs, down 24%
Financial managers: 636,000 jobs, up 55%	Garbage collectors: 60,000 jobs, down 22%
Lawyers: 747,000 jobs, up 49%	Barbers: 85,000 jobs, down 20%
Police and detectives: 822,000 jobs, up 45%	
Airline pilots: 110,000 jobs, up 44%	
Social workers: 659,000 jobs, up 43%	

Source: *Chicago Sun-Times,* January 30, 1993, p. 6.

Using the number of jobs cited as the base for the relative importance of each percentage, accomplish the following.

a. Compute the weighted mean percentage of the jobs that declined and the weighted mean percentage of the jobs that increased.

b. Putting all the jobs into one group, find the weighted mean. Does this indicate a net increase or net decrease of jobs?

c. Describe a shortcut to calculate Part b.

14. The following array shows the job cuts at ten large U.S. defense contractors.

Job Cuts at U.S. Defense Contractors

Company	Initial Number Employed	Percent Change
General Dynamics	102,220	−40%
GE Aerospace	41,000	−28%
McDonnell Douglas	128,000	−26%
Grumman	28,900	−25%
Northrop	41,000	−15%
Raytheon	77,600	−14%
Lockheed	82,500	−14%
Martin Marietta	65,000	−11%
United Technologies	201,400	−11%
GM Hughes	66,600	−0.6%

Source: *Business Week*, October 19, 1992, p. 44.

Use a weighted mean to determine the percentage change for these ten contractors.

15. The following graph shows the changes in the unemployment rates in five areas of the country.

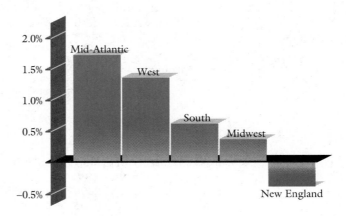

Sources: *Business Week*, December 7, 1992, p. 93, and *Statistical Abstract of the United States*, 1993, p. 26.

Use your best estimates of the percentages indicated on the graph to answer the following questions.

a. If all five of these areas employ about the same number of people, what is the estimated mean change in the unemployment rate?

b. Use a weighted mean to determine the change in the unemployment rate if the 1992 total resident populations of these five areas, in thousands, were as follows:

New England	13,200
Middle Atlantic	37,900
South	88,100
Midwest	60,700
West	55,100

4.4   THE EMPIRICAL RULE AND CHEBYSHEV'S THEOREM

The Empirical Rule

The Empirical Rule is also known as the 68%–95%–99% rule. Chapter 6 will show in much greater detail how to determine areas under mound-shaped frequency curves.

The **Empirical Rule** states a special relationship between the mean and the standard deviation of data whose histograms or frequency curves are roughly bell-shaped, or in other words, mound-shaped.

a. About 68% of the scores fall within one standard deviation of the mean. For example, if the mean of some mound-shaped distribution (such as the results of several types of IQ tests) is 100 and the standard deviation is 10, then about 68% of the scores will fall in the interval from $[100 - 10]$ to $[100 + 10]$— that is, in the interval $[90, 110]$.

b. About 95% of the scores fall within two standard deviations of the mean in the interval $[80, 120]$.

c. Practically all of the scores fall within three standard deviations of the mean in the interval $[70, 130]$.

Obviously, the more exact and balanced the bell shape of the distribution, the closer these three rules will come to reflecting the actual percentages.

The Empirical Rule (and Chebyshev's Theorem, to be stated shortly) can be used to construct an interval that contains a specified percentage of the scores. The procedure for finding this interval is similar to that which will be shown in Chapter 8 for constructing a confidence interval for the mean. (Loosely speaking, a **confidence interval for the mean** is an interval in which the actual mean is likely to lie a certain percentage of the time.)

Refer to the school profile of Example 2 on page 127. Assuming that the distribution of the GPAs is considered bell-shaped, determine the following.

a. An interval, centered about the estimated mean, in which approximately 68% of the GPAs should lie.

b. An interval, centered about the estimated mean, in which approximately 95% of the GPAs should lie.

Solution   a. Using the Empirical Rule for bell-shaped distributions, approximately 68% of the GPAs should be in an interval that extends one standard deviation on each side of the mean. Using the estimated mean and estimated standard deviation from Example 2 on page 127, the required interval is

$$[2.35 - .88, 2.35 + .88] = [1.47, 3.23].$$

That is, approximately 68% of the GPAs represented in the school profile should be at least 1.47 but no more than 3.23.

b. Furthermore, the Empirical Rule states that 95% of the GPAs should be in an interval within two standard deviations of the mean. This is the interval

$$[2.35 - 2 \cdot .88, 2.35 + 2 \cdot .88] \approx [0.59, 4.11].$$

Since no one can have a GPA of 4.11, the right end of this interval may be adjusted to 4.00 and written as

$$[0.59, 4.00]. \quad \bullet$$

An administrator at a small bottling company constructed the frequency curve shown in Figure 4-6, which shows the number of bottles broken in shipping during

Very few frequency curves based on actual data are perfectly mound-, or bell-shaped. This one is reasonably close.

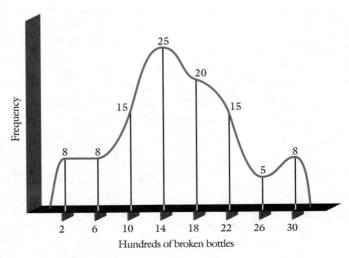

Figure 4-6

each of the last 104 weeks (2 years). Assume that this frequency curve is roughly bell-shaped.

a. Find the estimated mean and standard deviation for the last 2 years.
b. Use the answers to Part a to construct an interval, centered about the estimated mean, in which the weekly number of broken bottles occurred about 68% of the time. Repeat the construction for 95% of the time. Round your final answers in terms of the nearest bottle.
c. If each broken bottle costs 17 cents, determine the minimum amount of money lost during one of the days in which so many bottles broke that such a day occurred only about $2\frac{1}{2}$% of the time.

Solution   a. Frequency curves and histograms are closely related. The histogram that is related to the given frequency curve follows.

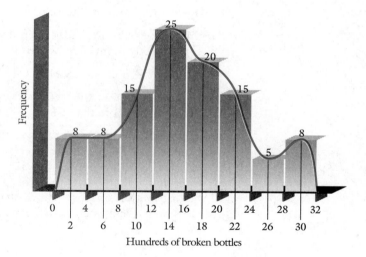

Use the method for finding the estimated mean and standard deviation of any histogram. A graphics calculator or other calculator with the option of entering frequencies is the easiest to use for this type of problem. For the sake of illustration, however, this example is solved with a calculator which does not have that capacity.

$$\text{Estimated value of } \mu = \frac{\sum\limits_{i=1}^{k} m_i \cdot f_i}{N} \qquad (4\text{-}1)$$

$$= \frac{2 \cdot 8 + 6 \cdot 8 + 10 \cdot 15 + \cdots + 30 \cdot 8}{8 + 8 + 15 + \cdots + 8}$$

$$= \frac{1624}{104}$$

$$\approx 15.62.$$

Using the shortcut formula,

$$\text{Estimated } \sigma = \sqrt{\frac{\sum\limits_{i=1}^{k} f_i \cdot m_i^2 - N \cdot \mu^2}{N}} \tag{4-3}$$

$$= \sqrt{\frac{(2\cdot 8^2 + 6\cdot 8^2 + 10\cdot 15^2 + \cdots + 30\cdot 8^2) - 104\cdot(15.615)^2}{104}}$$

$$= \sqrt{\frac{31040 - 25358.135}{104}}$$

$$\approx 7.39.$$

Since both the estimated mean and standard deviation are in terms of hundreds of bottles, the estimated mean number of bottles broken each week for the last 2 years is $15.615 \cdot (100)$, or about 1562 bottles per week, with a standard deviation of $7.391 \cdot (100)$, or about 739 bottles.

b. Assuming a bell-, or mound-shaped distribution of broken bottles, an interval in which the weekly number of broken bottles would lie about 68% of the time is

$$[1562 - 739, 1562 + 739] = [823, 2301].$$

That is, for about 68% of the weeks surveyed, the number of broken bottles was at least 823 but no more than 2301.

c. The Empirical Rule states that 95% of the scores are within two standard deviations of the mean—that is, in the interval

$$[1562 - 2 \cdot 739, 1562 + 2 \cdot 739] = [84, 3040].$$

Since the distribution of scores is bell-shaped, the remaining 5% of the scores is regarded as split evenly between the two "tails" indicated in Figure 4-7. The

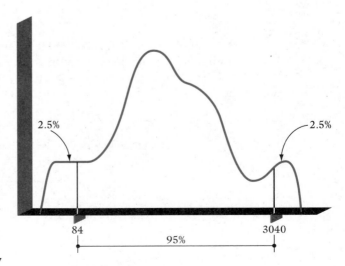

Figure 4-7

minimum number of broken bottles included in the upper tail is 3040. At 17 cents each, this amounts to $516.80 for the cheapest week in the upper tail.   •

Chebyshev's Theorem

Chebyshev's Theorem is used with frequency curves that are not reasonably close to mound-shaped.

Chebyshev's Theorem states the relationship between the mean and standard deviation of any set of data, regardless of how irregular its frequency curve or histogram might be.

Theorem ▶

> **Chebyshev's Theorem**
>
> At least $\left(1 - \dfrac{1}{k^2}\right)$ of the scores will be within k standard deviations of the mean.

Chebyshev's Theorem implies the following statements for k equal to 1, 2, 3, and 4, respectively.

$k = 1$ It is possible that none of the scores will fall within one standard deviation of the mean.

$k = 2$ At least $1 - \dfrac{1}{2^2} = \dfrac{3}{4}$ of the scores will fall within two standard deviations of the mean.

$k = 3$ At least $1 - \dfrac{1}{3^2} = \dfrac{8}{9}$ of the scores will fall within three standard deviations of the mean.

$k = 4$ At least $1 - \dfrac{1}{4^2} = \dfrac{15}{16}$ of the scores will fall within four standard deviations of the mean.

**E x a m p l e
8**

Refer to Example 3 on page 128. Use Chebyshev's Theorem to determine an interval that contains at least $\frac{3}{4}$ of the scores represented. Figure 4-8 shows the related absolute frequency histogram, estimated mean, and standard deviation.

S t a t   F a c t

Pafnuty Lvovich Chebyshev (1821–1894) was one of the most distinguished mathematicians Russia has ever produced. His interests and writings ranged over a wide variety of subjects, including probability theory, calculus, gearings, and the construction of geographic maps. He also spent part of his time on theoretical mechanics and worked on the problem of obtaining rectilinear motion from rotary motion.

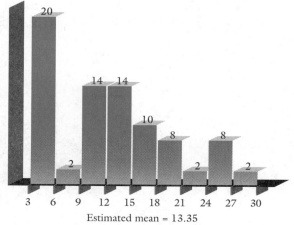

Estimated mean = 13.35
Estimated standard deviation = 7.03

Figure 4-8

Solution   The distribution of these scores does not appear to be bell-shaped. Chebyshev's Theorem, however, applies to all distributions, not just bell-shaped ones. Therefore, by Chebyshev's Theorem, at least $\frac{3}{4}$ of the scores represented in this histogram fall within two standard deviations of the mean, implying that the required interval can be found as follows:

$$[\mu - 2 \cdot \sigma, \mu + 2 \cdot \sigma = [13.35 - 2 \cdot 7.03, 13.35 + 2 \cdot 7.03]$$
$$= [-.71, 27.41].$$

Since there are no scores less than 3, this interval may be stated as $[3, 27.41]$.

In the following diagram, the interval $[3, 27.41]$ is shaded. It appears that at least $\frac{3}{4}$ of the scores are certainly in that interval.

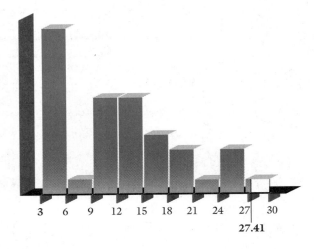

E x a m p l e
9

Use the histogram in Figure 4-9, Chebyshev's Theorem, and a graphics calculator to find the specified intervals (round all answers to the nearest hundredth).

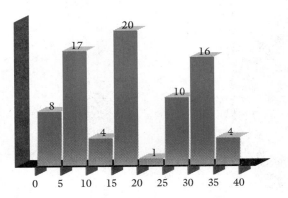

Figure 4-9

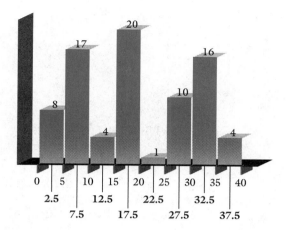

Figure 4-10

a. An interval, centered about the estimated mean, in which at least $\frac{3}{4}$ of the scores must lie

b. An interval, centered about the estimated mean, in which at least $\frac{1}{2}$ of the scores must lie

Solution To use Chebyshev's Theorem in this example, first determine the estimated values of σ and μ. Then it will be necessary to find a value of k such that the interval $[\mu - k \cdot \sigma, \mu + k \cdot \sigma]$ contains the desired number of scores.

The estimated values of μ and σ will be found with a graphics calculator. Figure 4-10 repeats the histogram of Figure 4-9 with interval midpoints indicated. Enter the midpoints and their frequencies in the graphics calculator in exactly the same manner as in Example 4 on page 131. When the data have been entered and the appropriate functions selected, the final screen should appear as follows:

```
1-Var Stats
 x̄=18.9375
 Σx=1515
 Σx²=38350
 Sx=11.05778121
 σx=10.98845275
↓ n=80
```

Thus, the estimated mean and standard deviation are approximately 18.938 and 10.988, respectively.

Now determine the value of k from Chebyshev's Theorem.

Although k does not need to be an integer, the expression $1 - \frac{1}{k^2}$ is meaningless in this context if $k < 1$.

a. Solving the equation

$$1 - \frac{1}{k^2} = \frac{3}{4}$$

for k results in

$$1 - \frac{3}{4} = \frac{1}{k^2},$$

$$\frac{1}{4} = \frac{1}{k^2},$$

$$2 = k.$$

The required interval is

$$[\mu - k \cdot \sigma, \mu + k \cdot \sigma] = [18.938 - 2 \cdot 10.988, 18.938 + 2 \cdot 10.988]$$
$$= [-3.04, 40.91].$$

The negative sign for the left end of this interval has no practical application in this particular histogram and can be ignored, with the required interval being rewritten as $[0, 40.91]$. Chebyshev's Theorem claims that at least $\frac{3}{4}$ of the scores will be in this interval. Examination of the histogram verifies that this condition is met: all of the scores are in this interval.

b. In a similar manner, solving the equation $1 - \frac{1}{k^2} = \frac{1}{2}$ for k results in

$$1 - \frac{1}{2} = \frac{1}{k^2},$$

$$\frac{1}{2} = \frac{1}{k^2},$$

$$k = \sqrt{2},$$

$$k \approx 1.414.$$

$$[\mu - k \cdot \sigma, \mu + k \cdot \sigma] = [18.938 - 1.414 \cdot 10.988, 18.938 + 1.414 \cdot 10.988]$$
$$= [3.40, 34.48].$$

A visual check of the original histogram indicates that at least $\frac{1}{2}$ of the scores lie in the shaded region determined by this interval.

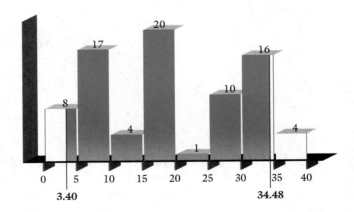

Exercises 4.4

1. Refer to Example 6, Part a, on page 142. Do you think it would be permissible to replace the word "should" with "must"? Justify your answer.

2. Common sense seems to imply what restrictions on *k* in Chebyshev's Theorem?

3. Will the use of the Empirical Rule or of Chebyshev's Theorem result in a wider interval for a given percentage of scores? Explain why.

* 4. In Chebyshev's Theorem, how is *k* related to the fraction of scores, *f*, that must fall within *k* standard deviations of the mean?

If using a graphics calculator for Exercises 5–10, remember to enter interval midpoints in L1 and corresponding frequencies in L2.

5. The following frequency curve shows the percentages of the voting-age populations that voted in the 1992 Democratic presidential primary in the 40 states that held primaries. Consider it bell-shaped.

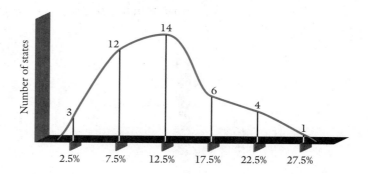

a. Find the estimated mean percentage of voter turnout.

b. Determine an interval which contains approximately 68% of the percentages of voter turnout.

GROUP

6. The following histogram shows the winning times, in minutes and seconds, for Kentucky Derby winners from 1900 through 1992.

Stat Fact

The Santa Anita Derby, held about a month before the Kentucky Derby, has proved to be a good indicator of the Kentucky Derby winner. One-third of the winners at Santa Anita have gone on to finish in one of the first three positions at Churchill Downs.

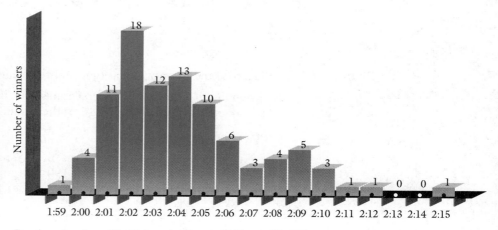

Based on data from *The Universal Almanac 1993*, pp. 674–675.

a. Estimate the mean winning time and the standard deviation.

b. Assuming that this histogram is bell-shaped, determine a time interval that, based on previous results, would be 68% likely to contain the winning time of next year's race.

c. Assuming that this histogram is not considered bell-shaped, determine an interval that, based on previous results, would be at least 75% likely to contain the winning time of next year's race.

Hint: Convert all times to seconds or convert all times to seconds beyond 2 minutes, in which case 1:59 = −1.

GROUP 7. The following frequency polygon shows the winning times, in minutes and seconds, of the Preakness Stakes from 1900 through 1992. (In 1918 there were two winning times, so the sum of the frequencies is 94 instead of 93.)

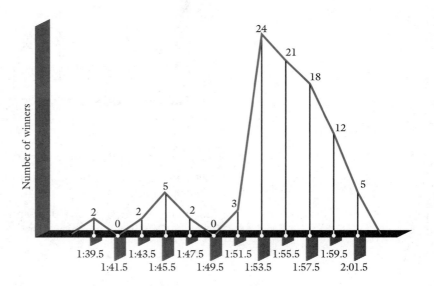

a. Estimate the mean winning time and the standard deviation.

b. Assuming that this frequency polygon is bell-shaped, determine an interval that contains approximately 68% of the winning times.

Hint: See the hint for Exercise 6.

GROUP 8. The following frequency curve shows the winning times, in minutes and seconds, of the Belmont Stakes from 1900 through 1992. (For several years no winning times were recorded, so the sum of the frequencies is less than 93.)

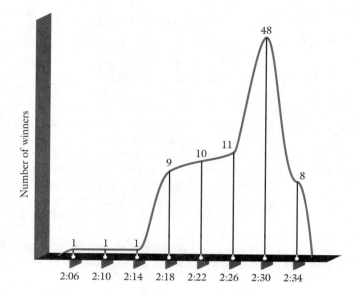

a. Estimate the mean and the standard deviation of the data represented by this frequency curve.

b. Assuming that these data are not bell-shaped, use Chebyshev's Theorem to find an interval that contains at least 50% of these times.

Hint: See the hint for Exercise 6.

9. Refer to Exercise 4 (in Exercises 4.1–4.3) on page 135. Assuming that the histogram is bell-shaped, determine an interval that contains approximately 68% of the annual incomes of nurses on salary.

GROUP   10. Refer to Exercise 5 (in Exercises 4.1–4.3) on page 135, and assume the histogram to be bell-shaped.

a. Determine an interval that contains approximately 68% of the annual incomes of nurses on hourly pay.

b. Using the information in Exercise 9 and Part a of this exercise, what is the difference of the two estimated means?

4.5 ESTIMATING THE MODE AND MEDIAN

It is possible to estimate from grouped data other measures of central location besides the mean. To simplify the following illustrations and exercises, assume that all histograms are left-ended.

Estimating the Mode

Figure 4-11 geometrically illustrates one way of estimating the location of the mode from a given histogram. The interval with the most scores is called the **modal interval.** The following method assumes that there is a true mode and that it lies in the modal interval, even though neither has to be the case.

In Figure 4-11, the estimated mode is slightly to the right of the midpoint of the modal interval (rather than exactly at the midpoint). This location takes into account the fact that the interval to the right of the modal interval has a higher frequency of scores than the interval to the left. If the frequencies of scores in the intervals on both sides of the modal interval were the same, then the midpoint of the modal interval would be used as the estimated mode.

If two or more intervals have the highest frequency, then the distribution is estimated as multimodal, and the following procedure is performed the required number of times.

The following formula is used to find the actual location of the estimated mode.

$$\text{Estimated mode} = L + \left(\frac{a}{a + b}\right) \cdot w \qquad \textbf{(4-4)}$$

Figure 4-11 has been redrawn (Figure 4-12) with indications of the important points and measures that are part of Formula 4-4. In this formula, a and b are the differences in frequencies, L is the lower limit of the modal interval, and w is the

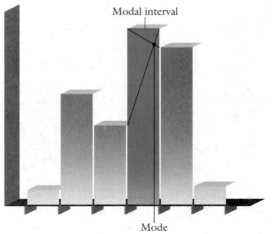

Figure 4-11 Location of the estimated mode.

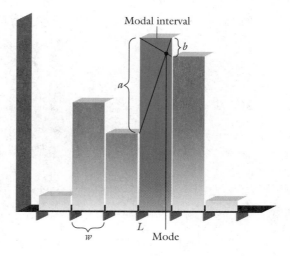

Figure 4-12

modal interval width. All interval widths are assumed to be the same. If they were not, a refinement of the formula would be necessary.[2]

E x a m p l e
10

The histogram in Figure 4-13 shows, as a percentage of income, the estimated state and local taxes paid in 1992 by families of four with $50,000 incomes, in selected large cities. Based on this histogram, estimate the modal percentage of taxes paid.

S t a t **F** a c t

Of all cities included in this histogram, Memphis, Tennessee, had the lowest estimated state and local taxes, 5.8%, and Newark, New Jersey, had the greatest, 22.9%.

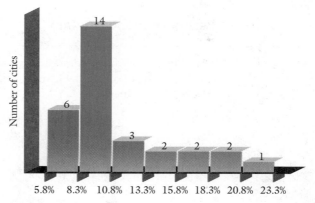

Figure 4-13

Based on data from *Statistical Abstract of the United States 1994*, p. 310.

---

[2]If the interval widths are not all the same, then the **interval density** is the ratio of the interval frequency to the interval width. The modal interval is then the interval with the greatest interval density. L is the same as before, w is the width of the modal interval, and a and b are the differences in interval densities.

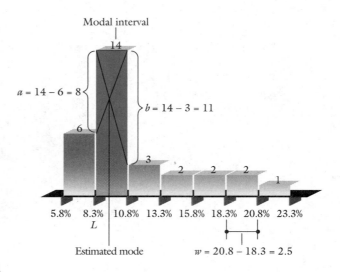

Figure 4-14

Solution   In Figure 4-14, the histogram is redrawn with indications of important features. In this case,

$$L = 8.3,$$
$$a = 14 - 6 = 8,$$
$$b = 14 - 3 = 11,$$
$$w = 2.5.$$

Therefore, the estimated modal tax is $8.3 + \left(\dfrac{8}{8 + 11}\right) \cdot 2.5 \approx 9.35\%$.   •

Estimating the Median

E x a m p l e
11

The following histogram shows, as a percentage of income, the estimated state and local taxes paid in 1992 by families of four with $100,000 incomes, in 30 selected large cities.

S t a t   F a c t

Of all the cities included in this histogram, Memphis Tennessee, had the lowest estimated state and local taxes, 5.7%, and Newark, New Jersey, had the greatest, 23.4%.

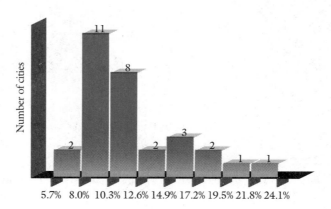

Based on this histogram, estimate the median percentage of taxes paid.

Solution

Step 1 Identify the interval in which the median must lie.

Since $N = 30$, the median is the mean of the 15th and 16th scores from the bottom. Both of these scores are in the interval from 10.3% to 12.6%. The median interval may or may not be the same as the modal interval.

Step 2 Apply the formula for the estimated median.

$$\text{Estimated median} = L + \left(\frac{\frac{N}{2} - F}{f}\right) \cdot w \qquad (4\text{-}5)$$

In this formula, L is the lower limit of the median interval, N is the total number of scores in the data, F is the sum of the frequencies up to but not including the median interval, f is the frequency of the median interval, and w is the width of the median interval.[3]

In this example, $L = 10.3$, $N = 30$, $F = 13$, $f = 8$, and $w = 2.3$. Substituting these values in the expression for the estimated median,

$$\text{Estimated median tax} = 10.3 + \left(\frac{\frac{30}{2} - 13}{8}\right) \cdot 2.3 \approx 10.88\%. \quad \bullet$$

Both the formula for the estimated mode and the formula for the estimated median may be used with frequency curves or interval tallies instead of histograms.

E x a m p l e
12

The frequency curve in Figure 4-15 shows the ages at which a certain disease affected a given population. Estimate the mean, median, and mode. Round all calculations to the nearest hundredth.

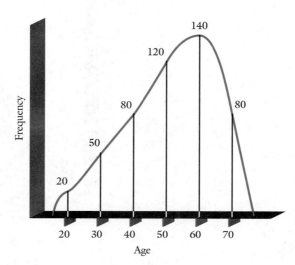

Figure 4-15

[3]In an estimation of the median, it is not necessary for all interval widths to be the same.

Solution   *Mean*. For an estimation of the mean, midpoints of intervals and interval frequencies are needed and are exactly the information given by the frequency curve. Apply the following formula.

$$\text{Estimated value of } \mu = \frac{\sum_{i=1}^{k} m_i \cdot f_i}{N} \tag{4-1}$$

$$= \frac{20 \cdot 20 + 30 \cdot 50 + \cdots + 70 \cdot 80}{20 + 50 + \cdots + 140}$$

$$= \frac{25,100}{490}$$

$$\approx 51.22.$$

Median. To estimate the median (and mode), it is necessary to construct the histogram from which this frequency curve was derived. By centering the intervals about the midpoints indicated in Figure 4-15, the histogram in Figure 4-16 can be sketched.

Now apply the estimated median formula,

$$\text{Estimated median} = L + \left(\frac{\frac{N}{2} - F}{f} \right) \cdot w. \tag{4-5}$$

Since there is a total of 490 scores, the median is the mean of the scores with data ranks 245 and 246. By adding interval frequencies from the left, it can be seen that these two scores occur in the interval from 45 to 55. Thus, $L = 45$.

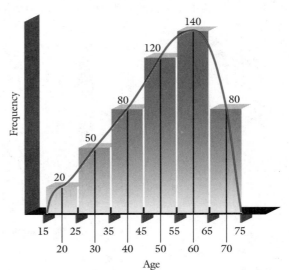

Figure 4-16 A histogram constructed from the frequency curve in Figure 4-15.

$$\text{Estimated median} = 45 + \left(\frac{\frac{490}{2} - 150}{120}\right) \cdot 10$$

$$\approx 52.92.$$

Mode. Since all interval widths are the same, the modal interval is the tallest interval, $[55, 65)$. Applying the formula for estimated mode, with $L = 55$,

$$\text{Estimated mode} = L + \left(\frac{a}{a + b}\right) \cdot w \qquad \textbf{(4-4)}$$

$$= 55 + \left(\frac{20}{20 + 60}\right) \cdot 10$$

$$= 57\frac{1}{2}. \quad \bullet$$

In the last example, mean < median, so the mean is on the left side of the median. It is always true that scores are more widely scattered, or skewed, on the side of the median where the mean occurs. Accordingly, these scores (as evidenced by the frequency curve) are skewed to the left.

Exercises 4.5

In the following exercises, round all calculations to the nearest hundredth.

1. Give an example of ten scores from which a histogram can be constructed in which the true mode is not found within the modal interval.

2. If you were estimating the median from a histogram, under what conditions would it be impossible to identify the interval in which the median must lie? Under these circumstances, what would seem a reasonable estimate for the median?

3. Refer to Example 10 on page 153, and estimate the median and the mean.

4. Refer to Example 11 on page 154, and estimate the mode and the mean.

GROUP 5. Refer to Exercises 3 and 4. Are all three estimates of central location for taxes on \$50,000 incomes less than the corresponding estimates of central location for taxes on \$100,000 incomes? Which measure of central location is lowest for both?

GROUP 6. The following frequency curve shows the U.S. defense budget, in billions of dollars, for 28 selected years between 1945 and 1988.

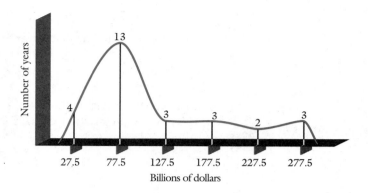

Estimate the mean, median, and mode.

Hint: Draw a histogram from these data in order to estimate the median and mode.

GROUP   7. The following frequency curve shows the U.S. human resources budget, in billions of dollars, for 28 selected years between 1945 and 1988.

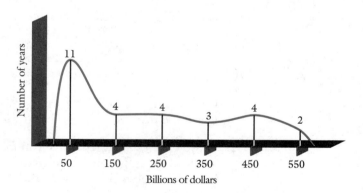

Estimate the mean, median, and mode.

Hint: Draw a histogram from these data in order to estimate the median and mode.

8. Refer to Exercises 6 and 7. What factors could explain why all measures of central location for the human resources budget except one are greater than those for the defense budget?

GROUP * 9. The following histogram shows the estimated numbers of deaths from major earthquakes, worldwide, in the years 1977 through 1991.

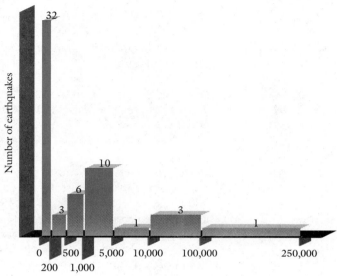

Estimate the mean, median, and modal numbers of deaths that occurred during those years.

10. For each of the following six frequency curves, determine which of these three conditions is most likely true:

$$\text{Mean} < \text{median},$$
$$\text{Mean} > \text{median},$$
$$\text{Mean} \approx \text{median}.$$

Make your determination by sketching some "test" histograms, with arbitrary values, that roughly correspond to the following graphs.

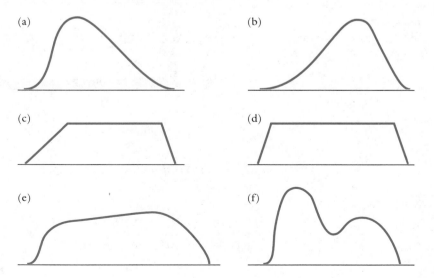

4.6 CONTROL CHARTS

Sometimes data can be graphed over a period of time, and a pattern, or cycle, is discovered.

Definition ▶ A **control chart** is a graph that can be used to indicate how a series of new scores compares with a historically based mean and standard deviation.

Control charts are used most often with sequential data. Like the case study of Chapter 3, page 119, control charts are an application of time series data.

For example, with 1970, 1975, 1980, 1985, and 1990 used as reference years, the mean conventional mortgage rate on new homes was 10.79%, with a standard deviation of 2.04%. A control chart can be constructed to analyze how the interest rate for each month of 1992 compares with this historical base. Some arbitrary amount of deviation from the historical mean is expected—say, one standard deviation. This locates the **upper control limit** and the **lower control limit,** the boundaries for expected scores.[4]

Figure 4-17 is a completed control chart for the historical interest rate, with the interest rate for each month of 1992. This control chart shows that every month, with the exception of March, had interest rates below what would be expected from the historical data.

3σ control charts, which place the control limits at $\mu \pm 3\sigma$, are very common in industry.

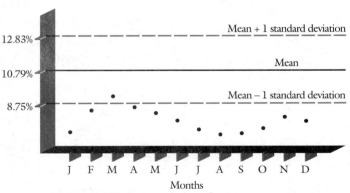

Figure 4-17 A control chart for interest rates.

Source: *Chicago Sun-Times,* January 28, 1993, p. 1, and *The Universal Almanac 1993,* p. 251.

---

[4]In other applications of this procedure, such as quality control of items produced on an assembly line, location of the upper and lower control limits follows a different, less arbitrary procedure. Later courses in statistics will examine control charts more closely.

Exercises 4.6

1. If the historical mean is an accurate and valid predictor of future events, then the locations of the upper and lower control limits determine whether or not the sequential data is "out of control." If the distribution of historical data are roughly mound-shaped, what percentage of scores would not be within control limits placed one standard deviation from the mean? What percentage would not be within two standard deviations of the mean?

2. Why do the data in the interest-rate situation described on page 160 seem not to follow the guidelines of Exercise 1? Name several situations in which the historical mean would possibly be a more valid predictor of future scores than it was with the interest rates.

3. A certain machine produces ball bearings with diameters of 1.25 cm, and from long experience it is known that the standard deviation of ball bearings produced by this machine is 0.08 cm. Ball bearings that are more than three standard deviations from the desired mean cannot be used and must be discarded. The diameters of newly manufactured ball bearings are measured at regular time intervals, with the following results: 1.19   1.22   1.15   1.28   1.28   1.30   1.27   1.35   1.30   1.29   1.41   1.39   1.44   1.44   1.39   1.50   1.44   1.56   1.59   1.55   1.61   1.60   1.66.

   a. Draw a control chart for these sequential data. Indicate the desired mean, μ, and $\mu \pm 3\sigma$ on the control chart. Plot each of these 23 measurements on the control chart.

   b. What does the control chart indicate about the product that is being manufactured?

GROUP

4. The following table lists prices, in dollars per ounce, of gold, silver, and platinum in the years 1981 through 1988.

	1981	'82	'83	'84	'85	'86	'87	'88
Gold	460.00	376.00	424.00	361.00	318.00	368.00	448.00	438.00
Silver	10.50	8.00	11.40	8.10	6.10	5.50	7.00	6.50
Platinum	475.00	475.00	475.00	475.00	475.00	519.00	600.00	600.00

Construct three separate control charts, one for each of these three precious metals. Locate the upper and lower control limits at 1.5 standard deviations from the mean of these prices. Use the control charts to determine, within the following 8 months of 1992, the months during which these approximate prices of precious metals were "out of control."

	Jan.	Feb.	Mar.	May	June	July	Sept.	Nov.
Gold	360.00	350.00	345.00	340.00	345.00	362.00	340.00	338.00
Silver	4.18	4.10	4.12	4.00	4.00	3.80	3.70	3.70
Platinum	360.00	360.00	355.00	373.00	378.00	385.00	365.00	365.00

Based on data from *Statistical Abstract of the United States, 1991*, p. 488, and *Business Week*, December 28, 1992, p. 110.

5. A control chart is one of many types of devices called **statistical process controls (SPC).** Many SPCs use data gathered over a certain time period to check the reliability of a manufacturing process. Besides control charts, another example of an SPC is a **p-chart,** which plots proportions of defective items over time. Describe some situations in which you feel a p-chart would be more appropriate than the control charts described in this section.

Case Study   *Higher Education*

Histograms help illustrate the number of people who hold a bachelor's degree. *(Lisa Martino)*

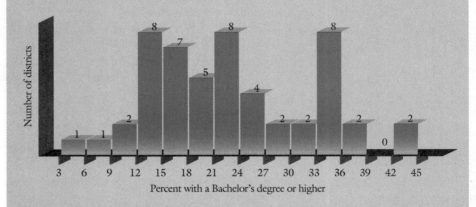

The following left-ended histogram was based on data compiled from Appendix VII of *Statistical Abstract of the United States, 1993.* It represents the percentages of people at least 25 years old who hold bachelor's degrees or higher degrees in each of California's 52 congressional districts.

For example, in eight California congressional districts, at least 33% but fewer than 36% of those who are at least 25 years hold bachelor's degrees or higher degrees.

I. Determine the estimated mean percentage of all 52 districts.

II. Do you predict that the estimated median is greater than or less than the estimated mean? Compute the estimated median, and check your answer.

III. Go to your library and use a copy of *Statistical Abstract of the United States, 1993,* to compute the actual mean and median for these data. How close are the estimated values of the mean and median to the actual values?

IV. Do you think this histogram for California is typical of other states in the country? If not, how do you think it differs?

V. New York and Pennsylvania combined have about the same number of congressional districts as California. Use a copy of *Statistical Abstract of the United States, 1993,* to construct a histogram similar to the one in this case study for the combined districts of New York and Pennsylvania. Does it confirm your answer to Part IV?

CHAPTER SUMMARY

This chapter was concerned with estimating the measures of central location and dispersion from grouped data illustrated by histograms, frequency curves, or similar displays. These graphs give the frequency of scores within a certain interval but not the scores themselves.

In estimations of the mean and standard deviation, all scores that fall in a particular interval are treated as if they fell at the midpoint of the interval. The problem thus becomes one of dealing with estimated scores, and is treated accordingly. Formulas have been developed to find the estimated mean and standard deviation. Calculators into which score frequencies can be entered make such calculations much easier.

The mode can be estimated from grouped data by first identifying the modal interval, which is the interval with the greatest frequency. Rather than the midpoint of that interval simply serving as the estimated mode, the midpoint is adjusted left or right, according to formula, to reflect the relative frequencies of the two intervals on both sides of it.

The median is estimated by first determining the median interval—the interval in which the median, or the two scores that will determine the median, must lie. An appropriate formula is then used to arrive at an estimated value of the median.

A control chart plots sequential data to determine whether new data are following an established or acceptable trend. It is one of many types of devices called statistical process controls.

Key Concepts, Terms, and Formulas

$$\text{Estimated median} = L + \left(\dfrac{\dfrac{N}{2} - F}{f}\right) \cdot w \quad \textbf{(4-5)}$$

Control chart

Statistical process control (SPC)

Videotape Suggestions

Program 4 from the series *Against All Odds: Inside Statistics* provides more information about working with frequency curves. *Program 6* from the same series contains additional information about control charts and patterns over time. *Program 13* has further illustrations of control charts and could serve as a preview of the work in Chapter 7.

| CUMULATIVE REVIEW CHAPTERS 1–4 | ▶ | O v e r v i e w |

The material in this first unit of the text was divided into two parts: ungrouped data (Chapters 1 through 3) and grouped data (Chapter 4).

In Chapter 1 the primary goal was to explore ways of ranking data and using data ranks to compare scores and determine percentiles. Chapter 2 concentrated on graphic ways to illustrate the distribution of scores in a single set of data, as well as comparing one set of data with another. Box plots were one of the most important graphic devices studied, and they were also used to illustrate the idea of a confidence interval and statistical significance. In Chapter 3 measures of central location and dispersion were examined, along with methods of constructing histograms, frequency curves, and other displays.

Chapter 4 concentrated on estimating measures of central location and dispersion from grouped data. The skills learned in Chapter 3, especially working with histograms and frequency curves, were very important for the work in this chapter because scores are often found graphically, in the form of a histogram or other display.

Throughout this unit the importance of technology should have been apparent. The graphics calculator, for instance, can easily sort scores, determine percentiles and five-number summaries, compute the mean and standard deviation, and draw box plots and histograms.

Procedure Index

EXERCISES   ▶

1. Must the median always lie somewhere in the longest row of a stem-and-leaf display? Explain.

2. Name two separate conditions under which deciles would not divide a set of data into exactly ten equal parts.

3. Following are the earned run average (ERA) leaders in the National League from 1970 through 1992.

1970: 2.81 (Seaver)	1982: 2.40 (Rogers)
1971: 1.76 (Seaver)	1983: 2.25 (Hammaker)
1972: 1.98 (Carlton)	1984: 2.48 (Pena)
1973: 2.07 (Seaver)	1985: 1.53 (Gooden)
1974: 2.28 (Capra)	1986: 2.22 (Scott)
1975: 2.24 (Jones)	1987: 2.76 (Ryan)
1976: 2.52 (Denny)	1988: 2.18 (Magrane)
1977: 2.34 (Candelaria)	1989: 2.28 (Garelts)
1978: 2.43 (Swan)	1990: 2.21 (Darwin)
1979: 2.71 (Richard)	1991: 2.39 (Martinez)
1980: 2.21 (Sutton)	1992: 2.08 (Swift)
1981: 1.69 (Ryan)	

Use a stem-and-leaf display to identify the median ERA. Put the unit and tenths places in the stem.

4. The following table shows average weekly amounts of TV viewing in hours (h) and minutes (m), by age.

	November 1992	November 1991
Children 2–5	25 h, 32 m	26 h, 23 m
Children 6–11	21 h, 20 m	21 h, 10 m
Female teens	21 h, 0 m	22 h, 11 m
Male teens	22 h, 39 m	22 h, 41 m
Women 18–24	28 h, 54 m	28 h, 54 m
Women 25–54	31 h, 5 m	32 h, 5 m
Women 55 and older	44 h, 11 m	43 h, 31 m
Men 18–24	23 h, 31 m	23 h, 1 m
Men 25–54	28 h, 44 m	28 h, 4 m
Men 55 and older	38 h, 28 m	39 h, 49 m

Source: *The 1994 Information Please Almanac*, p. 748.

a. Find the median weekly average viewing times for November 1992 and November 1991.

b. The mean of what two classifications of viewers is used to determine both medians from Part a?

Hint: Convert hours to minutes or minutes to decimal parts of an hour.

5. The verbal SAT scores of 65 members of a recent high school class are as follows:

650	480	800	770	650	420	340	660	700	550
520	500	660	680	720	710	350	380	760	800
400	790	680	660	660	740	730	380	460	640
680	580	490	600	770	790	350	560	790	800
600	560	500	680	540	450	480	600	680	780
710	670	730	800	630	570	700	600	570	450
530		580		780		390		470	

Determine the quintiles. That is, find the SAT scores that mark P_{20}, P_{40}, P_{60}, and P_{80}. Also find the range and the IQR.

6. The following table lists the percentages of males and females who have taken the SAT test in 1993, classified by course work taken in high school.

Course Work	% Male	% Female
Music appreciation	43	57
British literature	46	54
Geometry	48	52
Physics	53	47
Ancient history	52	48
German	54	46
Modern Hebrew	47	53
Calculus	53	47
American literature	46	54
Dance	13	87
Composition	45	55
Biology	46	54
Economics	45	55
Sociology	38	62
Data processing	47	53

Source: *The College Board: 1993 Profile of SAT and Achievement Test Takers*, pp. 2–5.

a. Construct two five-number summaries for these data.

b. Use the five-number summaries from Part a to construct two notched box plots that compare these data.

c. Determine whether or not the difference in medians is statistically significant. What are the implications of your answer?

7. A set of data is composed of three scores that repeat over and over. What is the effect of this composition on the box plot for the data? What kind of display would best show the distribution of these scores?

8. Sketch a box plot that might result from the following stem-and-leaf display. Assume that this skeleton contains all different scores.

Stem	Leaf
X	X , X , X , X , X , X
X	X , X
X	X , X , X
X	X
X	X , X , X , X , X , X , X , X , X , X
X	X , X

9. Explain the difference between a hinge and a quartile.

10. In 1991 the monthly average passenger-car production, in thousands, for 20 of the top auto-producing countries was as follows:

Austria	1.2	Australia	25.9
Canada	74.2	Brazil	24.4
France	206.1	Czechoslovakia	14.7
Italy	135.6	India	16.1
Netherlands	7.1	Japan	813.0
Romania	6.2	Mexico	61.1
South Korea	94.3	Poland	13.9
United Kingdom	106.5	Spain	152.5
West Germany	355.8	USSR	98.7
Yugoslavia	18.2	United States	450.6

Find the adjacent scores and determine which countries, if any, have production figures that can be considered outliers; if so, classify them as unusual or rare.

11. Which two measures of central location are easiest to identify from a stem-and-leaf display?

12. Which measures of central location are completely unaffected by the presence of an outlier?

13. The following table shows annual vehicle-miles (in millions) and numbers of fatal-injury accidents in the western states in 1992.

	Annual Vehicle-Miles	Fatal-Injury Accidents
AK	820	8
AZ	12,504	179
CA	147,374	1445
CO	11,426	137
HI	3318	32
ID	1781	19
MT	893	12
NM	4042	72
NV	3218	39
OR	7157	61
UT	5430	39
WA	18,812	127
WY	862	7

Source: *Highway Statistics 1992,* U.S. Department of Transportation, Federal Highway Administration, p. 220.

Use the coefficient of variation to determine whether annual vehicle-miles or the number of fatal-injury accidents is more variable.

14. Identify each of the following statements as true or false. If you answer false, give an example that shows why. Remember, unless a statement is always true, it is false.

a. If the mean of a data set is k and every score is multiplied by 2.7, then the new mean obtained is $2.7 \cdot k$.

b. If the mean of a data set is k and every score is increased by 2.7, then the new mean obtained is $2.7 + k$.

c. Two data sets, A and B, each with the same numbers of scores, are given. If every member of A is less than the smallest member of B, then it must follow that $\sigma_A < \sigma_B$.

d. The standard deviation of a set of data is q. Every score is decreased by 3. The standard deviation of the resulting data is $(q - 3)$.

e. If the variance of the data in one set is greater than the variance of data in another set, then the coefficients of variation of the two data sets must be related in the same way.

15. Sturgess' Rule is only a rough guide for the number of desired intervals in a histogram. The following table lists battle deaths in World War II by country.

Australia	26,976	Austria	280,000
Belgium	8460	Brazil	943
Bulgaria	6671	Canada	42,042
China	1,324,516	Czechoslovakia	6683
Denmark	4339	Finland	79,047
France	201,568	Germany	3,250,000
Greece	17,024	Hungary	147,435
India	32,121	Italy	149,496
Japan	1,270,000	Netherlands	6500
New Zealand	11,625	Norway	2000
Poland	664,000	Romania	350,000
South Africa	2473	U.S.S.R.	6,115,000
United Kingdom	357,116	United States	291,557
Yugoslavia	305,000		

a. Use Sturgess' Rule to construct a left-end/minimum-score histogram, rounding the interval width to the nearest whole number.

b. Redraw the histogram from Part a with twice as many intervals and with interval widths of your own choosing that are not necessarily all the same, so that the display is more informative.

16. The number of freighters in the merchant fleets of some major countries of the world are as follows:

Interval for Number of Vessels	Tally
500 to under 800	3
200 to under 500	13
100 to under 200	11
75 to under 100	3
50 to under 75	3
25 to under 50	18
0 to under 25	5

Based on data from *Statistical Abstract of the United States 1992*, p. 637.

Draw a histogram for these data, and use it to estimate the mean, median, and mode.

Note: *Be careful when estimating the mode, because not all interval widths are the same. See footnote 2, Chapter 4, page 153.*

17. Panama and the former Soviet Union have merchant fleets of 1770 and 1759 freighters, respectively. If the largest interval in Exercise 16 is extended to include fleets of 500 to under 2000, which of the estimated values of mean, median, and mode will change the most? Recompute the estimates to verify your guess.

18. How are the Empirical Rule and Chebyshev's Theorem similar? How are they different? The use of which one would result in a more narrow confidence interval?

19. The following frequency curve could show the numbers of hours the students enrolled in a particular school spent watching TV during the last week of December. Consider it mound-shaped.

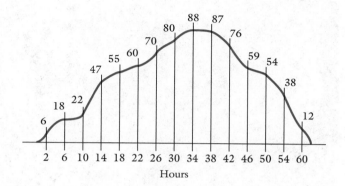

a. Find an interval for the number of hours spent watching TV that week that would apply to 68% of the students.

b. What is the maximum number of hours a student who is ranked in the lowest $2\frac{1}{2}$% of all students could have watched TV that week?

20. A **slugging percentage** in baseball is determined by dividing the total number of bases earned by the total number of at-bats. For example, if a player hit a home run, hit a double, and struck out in his first three at-bats of the season, his slugging percentage would be $\dfrac{4 + 2}{3} = 2.000$, and his batting average would be $\frac{2}{3} = .667$. Explain how a slugging percentage is like the weighted mean of a batting average.

The measure of probability arose largely through games of chance.
(*Courtesy of TropWorld Casino and Entertainment Resort*)

Unit 2
Foundations of Inference

Chapter 5A
Fundamentals of Probability Theory
Key Objectives: Understand that probability theory is used to make decisions in the face of uncertainty; learn how to apply the Fundamental Counting Principle.

Chapter 5B (Optional)
Further Topics in Probability
Key Objectives: Thoroughly understand special counting situations such as permutations, combinations, and permutations of nondistinct objects; determine conditional probabilities using a probability tree; and apply Bayes's Theorem.

Chapter 6
Random Variables and Probability Distributions
Key Objectives: Understand the difference between discrete random variables and continuous random variables; learn when to apply the Binomial Formula and how to use standard normal curve areas.

Chapter 7
Sampling and Estimating Population Parameters
Key Objectives: Understand the difference between working with a sample and working with a population; learn how to use a random number table.

2 Foundations of Inference

Although descriptive statistics may be very informative and may effectively illustrate the distribution of a set of data, the formation of general conclusions and of decisions based on a single set of data often requires more advanced techniques. Such techniques are the basis of inferential statistics, and their mathematical foundation is the theory of probability. Decision-making in the face of uncertainty, when a limited amount of data is available, is the heart of probability.

Chapter 5A approaches the theory of probability as it was originally developed—through games of chance. Definitions and rules are introduced

gradually, and a framework is created to support the study of random variables in Chapter 6.

Chapter 6 describes many current applications of probability and introduces the normal probability distribution, which is of fundamental importance to inferential statistics. That chapter will extend your probability skills.

Finally, Chapter 7 investigates random sampling. It is the last step in preparation for the study of inferential statistics. The case study at the end of the chapter illustrates the fact that even the U.S. government, in several tries, was unable to put the theory of random sampling into practice.

(Garden State Park, New Jersey)

Horseracing, the nation's largest spectator sport, tests people's abilities to predict a specific outcome.

5A Fundamentals of Probability Theory

Blaise Pascal (1623–1662) helped found the theory of probability. (*North Wind Picture Archives*)

Before you advance further in the study of statistics, it is necessary to understand the principal concepts of probability theory. Most of the statistics presented so far can be termed descriptive statistics or data analysis; that is, they are descriptions of given data and the patterns of scores within those data. Some elementary analysis, such as finding rare or unusual scores, has also been performed. These are the most elementary forms of statistical study.

Inferential statistics, on the other hand, involves judgments—conclusions and predictions based on data obtained from relatively small samples of larger groups. This chapter and the next two provide the probability theory on which the concepts of inferential statistics are based. **Probability** is a branch of mathematics that attempts to measure the likelihoods of random events of any kind, ranging from those connected with games of chance to questions of social science. Probability plays a very important role in any study of inferential statistics. The **case study** at the end of this chapter examines the historical foundations of probability.

5A.1 AN ILLUSTRATIVE EXAMPLE

As a hypothetical example of the use of probability in the gathering of statistics, consider a survey that a local medical society plans to send to 1000 doctors. The society wants a sense of the percentage of its doctors who have ever performed an abortion. However, because of the social stigma that some attach to abortion, the society cannot be sure that all the doctors will respond truthfully, even though the survey is to be confidential.

Doctors can be asked to use the following procedure when answering the question.

S t a t   F a c t

In 1990 there were
6,154,000 registered phy-
sicians in the United
States—approximately 40
people per doctor.

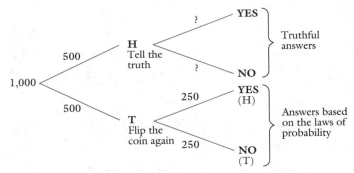

Figure 5A-1

Flip a coin. If it lands heads, then tell the truth; that is, answer either yes or no. However, if the coin lands tails, then flip the coin a second time. If the second flip is heads, then, regardless of the actual truth, answer yes; if it is tails, then answer no.

Figure 5A-1 illustrates this procedure.

If 1000 doctors are surveyed, then about one-half of them, or 500, will flip tails and will have to flip again. On the second flip, about half of these, or 250, will flip heads and about half will flip tails again. The following fraction estimates the percentage of doctors who have performed abortions.

$$\frac{\text{Total number of "Yes" answers} - 250}{\text{Total number of responses} - 500}.$$

The actual performance of an abortion will no longer be the sole criterion for a "Yes"; the doctor's response will also depend on the laws of probability. With this procedure, the survey subjects should be more comfortable answering "Yes" on the questionnaire, and more reliable results should be obtained.

In this situation, there is no guarantee that the coin flips will be split approximately evenly between heads and tails. However, a mathematical principle called the **Law of Large Numbers** assures that the percentages of heads and tails actually flipped will likely be increasingly close to the theoretical percentages as the number of flips increases.

S t a t   F a c t

Simeon Denis Poisson
(1781–1840), a French
mathematician, did im-
portant work in probability
and introduced the term
"Law of Large Numbers."

5A.2 SOME FUNDAMENTAL DEFINITIONS AND PRINCIPLES

Definitions ▶   An **event** is an outcome of an experiment.

A **simple event** is an event that cannot be broken down into smaller parts.

A **compound event** is an event made up of two or more other events.

Events may be as simple as obtaining a number from the throw of a die or as complicated as the election of the president of the United States. For example, the event of throwing the number 4 in a single die toss is a simple event, whereas being dealt a full house (three cards of one value and two cards of another value) in a poker hand depends on the occurrence of many simple events and thus is a compound event.

If E stands for a particular event, then $P(E)$ stands for the probability that event E will occur. The following is always true.

Rule ▶

> For any event, E,
>
> $$0 \le P(E) \le 1.$$

$P(E) = 0$ implies that E cannot happen. $P(E) = 1$ implies that E is certain to happen. $P(E)$ is usually expressed as a fraction, which is determined from the following definition.

Definition ▶

> The **Classical Definition** of the Probability of an Event
>
> $$P(E) = \frac{\text{Number of successful ways for } E \text{ to occur}}{\text{Total number of possible outcomes}}.$$

Some additional assumptions are made in determining probabilities.

- The denominator in the classical definition formula just shown is a count of every possible outcome, with all outcomes being equally likely. The event E is one of the possible outcomes. The denominator is often referred to as the **sample space** of the experiment.
- If $E_1, E_2, E_3, \ldots, E_n, \ldots$ are events that have nothing in common but together make up the set of all possible outcomes, then

$$P(E_1) + P(E_2) + P(E_3) + \cdots + P(E_n) + \cdots = 1.$$

E x a m p l e
1

S t a t   F a c t

Many European roulette wheels have only a single 0.

Roulette, one of the most popular games at gambling casinos, is played with a wheel and a small ball. The wheel has a total of 38 pockets. Thirty-six are alternately black and red and are numbered from 1 through 36, respectively. There are also two white pockets opposite each other on the wheel, numbered 0 and 00. When the wheel is spun, every pocket is equally likely to have the ball stop in it.

a. A **straight bet** is a bet on a single number. What is the probability that a straight bet on the number 7 will win?

b. A **color bet** (red or black) is a bet that the winning number will be a particular color. What is the probability that a red color bet will win?

c. A **square bet** is a bet that one of four numbers that form a square on the betting table will win. What is the probability that a given square bet will win?

Solution   a. The event of the ball's landing in the number 7 pocket is just one of 38 equally likely outcomes of the experiment of spinning the roulette wheel. By the classic definition of the probability of an event, the probability of obtaining a 7 is denoted by $P(7)$ and is

$$P(7) = \frac{1}{38}.$$

b. The event of the winning number's being red can happen in 18 ways. Therefore,

$$P(\text{Red}) = \frac{18}{38}$$

$$= \frac{9}{19}.$$

c.
$$P(\text{Square bet}) = \frac{4}{38}$$

$$= \frac{2}{19}. \quad \bullet$$

Sometimes the probability of an event is more easily computed if the event represents the complement of another event.

Definition ▶

If E is an event, the **complement** of E, denoted by E^c, is the event that
a. has nothing in common with E and
b. must occur if E does not occur.
It must always follow that $P(E) + P(E^c) = 1$.

As an illustration of working with complementary events, consider the roulette wheel again. A **small line bet** is a bet that one of the numbers 0, 00, 1, 2, or 3 will win. If the event E is a small line bet, then the probability of E's occurring is $P(E) = \frac{5}{38}$. The probability that the complement of E will occur (the probability of *not* winning a small line bet) is found by solving the equation $P(E) + P(E^c) = 1$:

$$P(E^c) = 1 - P(E)$$

$$P(E^c) = 1 - \frac{5}{38}$$

$$= \frac{33}{38}.$$

Part b of the next example will illustrate how the probability of an event is easily determined from the probability of its complement.

Example 2 Two dice are tossed. Find the following probabilities.

a. Their sum is 7.
b. Their sum is less than 11.

Solution The ordered pairs shown in Figure 5A-2 indicate the ways in which the indicated sums can occur. In order not to confuse the two dice, think of one of them as red and the other as green. For example, the ordered pair 5, 3 indicates that the sum 8 can be attained with a 5 on the red die and a 3 on the green die. The sum 8 can also be attained with a 3 on the red die and a 5 on the green die, designated 3, 5.

Notice that, since each die has six faces, there are $6 \cdot 6 = 36$ possible outcomes.

a. From the display in Figure 5A-2, it can be seen that, out of 36 equally likely outcomes, there are six ways in which the sum 7 can occur. Thus, the probability of throwing the sum 7, denoted by $P(7)$, is

$$P(7) = \frac{6}{36} = \frac{1}{6}.$$

b. Instead of counting all the ordered pairs that have sums less than 11, notice that only three of the 36 outcomes are not less than 11. Throwing a sum that is not less than 11 is the complement of throwing a sum that is less than 11. Thus,

$$\begin{aligned} P(\text{Sum is less than } 11) &= 1 - P(\text{Sum is not less than } 11) \\ &= 1 - P((\text{Sum is less than } 11)^c) \\ &= 1 - \frac{3}{36} \\ &= \frac{33}{36} \\ &= \frac{11}{12}. \end{aligned}$$

Since there are more ways for two dice to total 7 than any other sum, the sum 7 is the most likely result if two dice are tossed.

					4,3					
				3,3	3,4	4,4				
			3,2	4,2	5,2	5,3	5,4			
		2,2	2,3	2,4	2,5	3,5	4,5	5,5		
	2,1	3,1	4,1	5,1	6,1	6,2	6,3	6,4	6,5	
1,1	1,2	1,3	1,4	1,5	1,6	2,6	3,6	4,6	5,6	6,6
2	3	4	5	6	7	8	9	10	11	12

Figure 5A-2 Possible outcomes when two dice are tossed. ●

Exercises 5A.1–5A.2

1. Refer to the discussion of the abortion survey on pp. 175–76. Describe how to obtain the percentage of doctors who have performed abortions if only $P\%$ of the 1000 doctors returned the survey.

2. Do you think there is a difference between an impossible event and an event with zero probability? If so, try to explain the difference.

3. If the two white pockets on a roulette wheel were next to each other instead of across from each other, would that affect the probability of the ball's landing in one of them?

GROUP

Stat Fact

In 1992 the total amount wagered in American casinos was $329 billion. This is more than the GNPs of Australia and Argentina combined.

4. Determine the probabilities of the following bets in a game of roulette.

   a. An **even bet**—a bet that an even number (not 0 or 00) will win.

   b. A **street bet**—a bet that one of three specified numbers will win.

   c. A **high bet**—a bet that one of the "high" numbers, 19 through 36, will win.

5. Many European roulette wheels contain only a single-zero pocket and no double-zero pocket. If the rest of a European wheel is identical to an American wheel, would the probabilities of Exercise 4 be higher or lower for the European wheel? Why?

6. Two dice are rolled. Determine the following by referring to the solution to Example 2 on page 179.

   a. What is the probability of rolling a sum that is divisible by 5?

   b. What is the probability of rolling a sum less than or equal to 10? Less than 15?

   c. What is the probability of rolling a sum that is either less than 8 or divisible by 5?

7. Two special dice are made. One has eight sides, numbered 1 through 8, and the other has ten sides, numbered 1 through 10. These two dice are tossed.

   a. How many outcomes are possible?

   b. Which two sums have the least probability, and what is that probability?

   * c. What is the probability of rolling a sum of 6?

   * d. What is the probability of rolling a sum of 9?

A standard deck has 52 cards distributed in four suits, with 13 values per suit. The suits are clubs, spades, diamonds, and hearts. Clubs and spades are black; diamonds and hearts are red.

8. A card is chosen at random from a standard deck.

   a. What is the probability that it is black?

   b. What is the probability that it is either a queen or a jack?

   c. What is the probability that it is either an ace or a red numbered card?

9. Answer all parts of Exercise 8 under the condition that a red ace has first been removed from the deck. Which probability decreased?

GROUP   * 10. Which of the following five-card poker hands do you *intuitively* believe is least likely to occur? Rank the rest from least likely to most likely.[1]

    a. A full house (three of one value and two of another value)

    b. A royal flush (10, J, Q, K, and A all of the same suit)

    c. A straight (five cards with consecutive values)

    d. A pair

    e. A flush (five cards of the same suit)

5A.3   THE FUNDAMENTAL COUNTING PRINCIPLE

As Example 2 on page 179 implies, counting the number of ways in which an event can occur is a key part of determining the probability of that event. An important rule to follow when counting is called the Fundamental Counting Principle (FCP).

The Fundamental Counting Principle

Let E be an event. Suppose that E is made up of several distinct steps and that each step has a certain number of distinct choices. The total number of ways in which E can occur is equal to the product of the number of steps and the number of choices.

Tree diagrams are most practical in relatively simple situations.

A **tree diagram** is a simple straight-line sketch that shows all the possible outcomes in a multiple-step experiment. It can logically illustrate the FCP.

As an illustration, consider an event composed of, say, three steps. There are two choices for the first step, three choices for the second step, and two choices for the third step. An identification code composed of the letter X or Y followed by 3, 5, or 7, with either A or B at the end, can represent such an event. The FCP says that there are $2 \cdot 3 \cdot 2 = 12$ possible codes in this case. The tree diagram in Figure 5A-3 verifies this calculation. It has a total of 12 ($2 \cdot 3 \cdot 2$) branches, and the codes at the ends of the branches indicate the sequence of possible choices.

Use of the FCP for the two-dice problem of Example 2 on page 179 implies that there must be $6 \cdot 6$, or 36, possible outcomes for a toss. Note that 36 possible

---

[1]Refer to optional Chapter 5B to see how the actual probabilities of these hands can be determined.

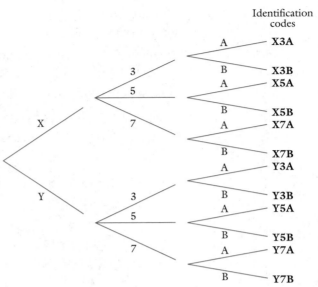

Figure 5A-3 A tree diagram for possible codes.

outcomes do not necessarily mean 36 different sums. Each sum can result from several different outcomes.

The FCP, with or without its accompanying tree diagram, allows very rapid computation of the number of possible outcomes.

E x a m p l e 3

At the racetrack, a **trifecta** is a bet that picks the first three horses in the correct order. Answer the following questions for a six-horse race.

a. How many trifecta bets are possible?
b. How many trifecta bets are possible in which horse number 4 finishes in the second, or "place," position?

Solution   a. Theoretically, any of the six horses could finish in the first, or "win," position. There would then be five horses that could finish in the second, or "place," position and four horses that could finish in the third, or "show," position. By the Fundamental Counting Principle, $6 \cdot 5 \cdot 4 = 120$ trifecta bets would be possible. Obviously, not all of these would be equally likely.

b. This part can be easily visualized with the help of the tree diagram in Figure 5A-4. There are five ways of picking the winning horse, one way of picking the second-place horse, and then four remaining ways of picking the horse that will show. By the Fundamental Counting Principle, $5 \cdot 1 \cdot 4 = 20$ trifecta bets that meet the given conditions are possible.

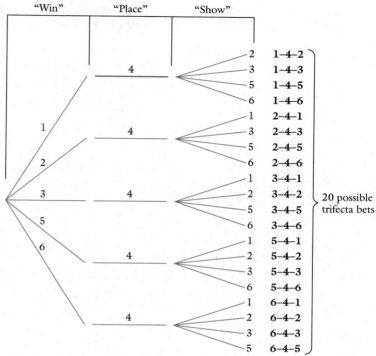

Figure 5A-4   ●

E x a m p l e
4

To win the **daily double,** one must pick the winning horse in the first two races. On a certain day, the first race has six horses in it and the second race has eight horses in it. If all horses are equally likely to win and a daily-double bet is made completely by chance, what is the probability that it will be a winner?

Solution   There are six different horses that could win the first race and eight different horses that could win the second race. By the Fundamental Counting Principle, $6 \cdot 8 = 48$ outcomes are possible for these two races. The probability that a particular daily-double bet is the winning outcome is $\frac{1}{48}$.   ●

E x a m p l e
5

Fifteen children are on a Little League baseball team.

a. How many batting orders of nine players are possible?
b. How many batting orders are possible if the coach's two children are on the team and must bat in the first two spots?
c. What is the probability that a randomly determined batting order would consist of the nine youngest children batting in alphabetical order by last name?

Solution   a. The following diagram shows the number of choices for filling each spot in the lineup. (In this situation the lineup determines who will play.)

$$\underline{15} \cdot \underline{14} \cdot \underline{13} \cdot \underline{12} \cdot \underline{11} \cdot \underline{10} \cdot \underline{9} \cdot \underline{8} \cdot \underline{7}$$

First position   Last position

Example 1 in optional Chapter 5B, on page 211, demonstrates an efficient method of using a graphics calculator to calculate $15 \cdot 14 \cdot 13 \cdot \cdots \cdot 7$. This type of calculation is necessary in the study of permutations and combinations.

There are 15 choices for the first position, 14 choices for the second position, 13 choices for the third position, and so on through 7 choices for the last position. Therefore, by the FCP,

$$15 \cdot 14 \cdot 13 \cdot 12 \cdot 11 \cdot 10 \cdot 9 \cdot 8 \cdot 7 \approx 1{,}816{,}214{,}400$$

batting orders are possible.

b. Study of the following diagram reveals that the number of possible batting orders under these special conditions is

$$2 \cdot 1 \cdot 13 \cdot 12 \cdot 11 \cdot 10 \cdot 9 \cdot 8 \cdot 7 = 17{,}297{,}280.$$

$$\underset{\substack{\text{Coach's}\\\text{children}}}{\underline{2} \cdot \underline{1}} \cdot \underset{\text{Other people's children}}{\underline{13} \cdot \underline{12} \cdot \underline{11} \cdot \underline{10} \cdot \underline{9} \cdot \underline{8} \cdot \underline{7}}$$

c. Only one batting order has the nine youngest players batting in alphabetical order. Using the result of Part a of this example, the required probability is

$$\frac{1}{1{,}816{,}214{,}400} = .0000000005506,$$

which is practically zero, implying that such a coincidence would be a near miracle. As optional Chapter 5B will demonstrate, the chance of winning the Illinois State Lottery is about 70 times greater than the chance of this batting order's occurring at random. ●

E x a m p l e
6

Figure 5A-5 illustrates the face of a three-wheel slot machine. Most slot machines have 20 symbols per wheel, as listed. When the lever is pulled, each wheel spins and stops independently of the others.

a. What is the probability that three bells will appear in the middle, or payoff, row?
b. What is the probability that at least one bar will appear in the payoff row?

Stat   Fact

Charles Fey of San Fran-
cisco, California, built the
first slot machine, the
nickel-operated "Liberty
Bell," in 1895. He rented
out his machine to gam-
bling parlors for 50% of
the profits, and the profits
were large. With three
wheels and 20 symbols
on each wheel, there were
8000 possible results for
each pull of the handle, of
which only 12 paid out.

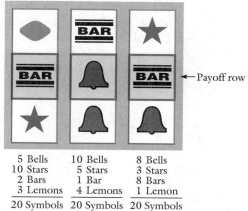

5 Bells	10 Bells	8 Bells
10 Stars	5 Stars	3 Stars
2 Bars	1 Bar	8 Bars
3 Lemons	4 Lemons	1 Lemon
20 Symbols	20 Symbols	20 Symbols

Figure 5A-5 A slot machine.

Solution   a. By the classic definition of the probability of an event,

$$P(3 \text{ bells}) = \frac{\text{No. of successful ways of obtaining 3 bells}}{\text{Total number of possible outcomes}}.$$

First find the denominator of this probability fraction. There are 20 spaces for the first wheel to stop on, 20 spaces for the second wheel, and 20 spaces for the third wheel. Therefore, by the Fundamental Counting Principle, there is a total of $20 \cdot 20 \cdot 20 = 8000$ possible outcomes.

Next determine the numerator. There are five "successful" spaces for the first wheel to stop on, ten "successful" spaces for the second wheel, and eight "successful" spaces for the third wheel. Thus, by the Fundamental Counting Principle, there are $5 \cdot 10 \cdot 8 = 400$ successful combinations of the three wheels. The desired probability is

$$P(3 \text{ bells}) = \frac{400}{8000}$$

$$= \frac{1}{20}.$$

b. Rather than considering all the possible ways in which at least one bar can appear, you can answer this part more easily by computing the probability of the *complement* of at least one bar's appearing, then subtracting that probability from 1.

Define the following events.

$$E = \text{At least one bar appears in the payoff row.}$$
$$E^c = \text{No bars appear in the payoff row.}$$

(These two events are complements because they have nothing in common and one or the other must occur.)

Determine $P(E^c)$ as follows. There are $20 - 2 = 18$ ways for the first wheel to *not* show a bar in the payoff row, 19 ways for the second wheel to *not* show a bar, and 12 ways for the third wheel to *not* show a bar. Thus, by the Fundamental Counting Principle, there are $18 \cdot 19 \cdot 12 = 4104$ ways for the three wheels to stop without showing a bar on any of them. By the classic definition of the probability of an event,

$$P(E^c) = \frac{4104}{8000}$$
$$= .513.$$

Find the desired probability, $P(E)$, by solving the equation

$$P(E) + P(E^c) = 1.$$

Therefore,

$$P(E) = 1 - .513$$
$$= .487. \quad \bullet$$

The Monte Carlo Method

Sometimes it is necessary to compute the likelihood of an event through trial and error, but repeated experiments that would yield the necessary data would be too time-consuming, tedious, or dangerous. In such a case, one might use some procedure that simulates the real-life process of obtaining a historic record of results, called a **Monte Carlo method.** Monte Carlo methods often involve computer simulation, the results of which can be used to indicate probabilities associated with tasks. Monte Carlo methods were first used during World War II for the design of the first atomic reactors.[2]

E x a m p l e
7

Write and run a computer program in the Microsoft Quick BASIC language to simulate the toss of a fair coin.

Solution   The simulation must allow for two equally likely outcomes. It can do this through the random number generating function, RND, which prints random decimals between 0 and 1. There are many ways to proceed.[3] Two possibilities follow.

[2]The term "Monte Carlo method" was selected because such methods use computational statistical games involving sampling and probability theory to help scientists understand chance events. The mathematician S. Ulam assigned the term to the method in the 1940s.

[3]Many versions of Microsoft BASIC do not require line numbers and have some commands that are more sophisticated than the ones shown here.

Version A	Version B
10 Randomize Timer	10 Randomize Timer
20 LET X = RND	20 LET X = RND
30 IF X < .5 THEN PRINT	30 LET Y = 10*X
''Heads''	40 LET Z = INT(Y)
40 IF X > .5 THEN PRINT	50 LET W = Z/2
''Tails''	60 IF W = INT(W) THEN PRINT
50 IF X = .5 THEN PRINT	''Heads''
''Edge''	70 IF W <> INT(W) THEN PRINT
60 END	''Tails''
	80 END

Exercises 5A.3

1. In a seven-horse race, you think that horse number 3 will win and that either horse number 1 or horse number 5 will finish second. To determine the number of possible ways your trifecta bet can be made, complete the tree diagram that has been started below.

2. Complete Exercise 1 with the condition that whichever horse does not place finishes "out of the money"—that is, in fourth place or further.

3. A **perfecta bet** is a bet that picks the winning horse and the second-place horse. How many perfecta bets are possible in a nine-horse race?

4. A **quinella bet** is a bet that picks the first two horses but does not specify which is the winner. How many quinella bets are possible in a nine-horse race?

    *   5. How many times more likely is a daily-double bet made on two five-horse races than a daily-double bet made on two ten-horse races?

Not all tree diagrams have the format and orientation of those in the previous examples and exercises. The following exercise uses a vertical tree with multiple branches to count cumulative exposures.

GROUP *   6. A certain, very contagious virus can be passed from one person to another directly or through a third person. That is, a given person can have the disease, carry it, or both. Assume that John Doe's first possible contact with the virus is at age 16, when he begins to be exposed each year to two different people who either might have it or might be passing it on.

        Complete the following tree diagram to determine the maximum number of exposures John Doe is liable to have by age 19. When completing the tree diagram, assume that each of the people to whom he is exposed has had the same number of exposures as John Doe himself. Boldface letters represent direct exposures, and nonboldface letters signify indirect exposures.

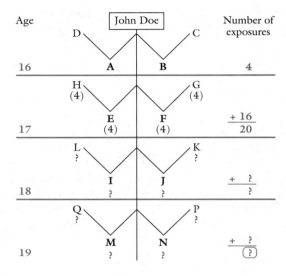

7. Following is a drawing of a dart board. What is the probability that a dart that was thrown and hit the dart board hit in the region labeled 1?

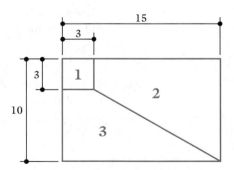

Hint: Obviously, counting the "number of successful ways" in which the dart can hit in region 1 is out of the question. Instead, use a ratio of appropriate areas to find a relative frequency.

8. Refer to the dart board in Exercise 7. If a person threw a dart that hit the dart board but *not* in region 2, what is the probability that the dart hit in region 1?

GROUP

9. In a manner similar to that of Exercise 7, use a ratio of areas to find the following probabilities associated with the following dart board. Assume that

midpoints determine the indicated regions and that a dart is guaranteed to land somewhere on the dart board.

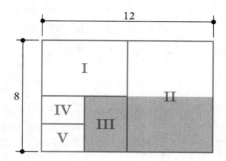

a. The probability that a dart lands in region IV

b. The probability that a dart lands in region I or in the shaded part of the board

c. The probability that the dart is not in region V if it is not in region III

d. The probability that a dart landed in region III if the dart hit the shaded part of the board

e. The probability that a dart landed in the shaded part of the board if it landed in the bottom half of the board

f. The probability that a dart did *not* land in region II if it landed in the shaded part of the board

10. Recall that a slot machine has three wheels with 20 symbols per wheel. Consider a slot machine with the following symbols.

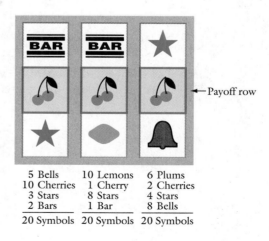

5 Bells	10 Lemons	6 Plums
10 Cherries	1 Cherry	2 Cherries
3 Stars	8 Stars	4 Stars
2 Bars	1 Bar	8 Bells
20 Symbols	20 Symbols	20 Symbols

a. What is the probability of each wheel's stopping on a star?

b. What is the probability of all the wheels' stopping on the same symbol?

c. What is the probability of obtaining at least one star?

d. What is the probability of exactly two wheels' stopping on cherries?

e. What is the probability of at least two wheels' stopping on cherries?

GROUP   11. A certain casino has three different types of slot machines. Type A has two wheels with 50 symbols per wheel, including three stars on one wheel and two stars on the other wheel. Type B has three wheels with 20 symbols per wheel, including four stars on the first wheel, three stars on the second wheel, and two stars on the third wheel. Type C has four wheels with ten symbols per wheel, including three stars on the first wheel, one star on the second wheel, three stars on the third wheel, and four stars on the fourth wheel. If payoffs are made for obtaining all stars, which of the three machines is most likely to provide a winning window? Which is least likely?

* 12. His Serene Highness Cosimo II, Fourth Grand Duke of Tuscany, lived from 1590 to 1621. He was an important figure in foreign relations and a patron of the arts as well as a gambler who faced a paradox. When he rolled three dice at a time, he knew that he could get the sum 9 in six different ways and that there were also six different ways of getting the sum 10. He reasoned that the probabilities of getting the two sums should therefore match. Experience, however, showed him that a 10 was more likely than a 9. Cosimo took this paradox to Galileo, whom he had appointed first professor of philosophy and mathematics at Pisa in 1610. Galileo solved and explained the paradox. Can you?

Galileo Galilei (1564–1642) is known mostly for his work in physics and astronomy, but he was also consulted for his knowledge of probability. (*North Wind Picture Archives*)

13. Give some examples of events that would be examined most effectively by Monte Carlo methods.

14. Write a computer program that simulates the toss of a fair coin a given number of times, n. The program should keep track of the percentage of heads obtained.

a. Let $n = 50$ and run the program. Then run the program again with $n = 500$ and with $n = 5000$.

b. How do the percentages of heads obtained for the three different values of n compare? What seems to be happening, and must it happen?

* 15. The following dart board is exactly 1 foot square. The shaded region is one-fourth of a circle with its center at the lower left-hand corner and with a radius of 1 foot.

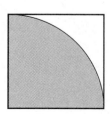

Write and run a computer program in Microsoft BASIC that simulates 5000 throws of a dart. Write the program so that it meets the following two conditions.

• Each dart is guaranteed to hit somewhere on the dart board.
• The program computes the probability of hitting the shaded region.

How do you think the probability obtained should be related to π?

5A.4 VENN DIAGRAMS AND RELATIONSHIPS BETWEEN SETS

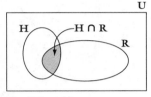

U = {All cards}
H = {All honor cards: jack, queen, king, ace}
R = {All red cards}

Figure 5A-6

A **Venn diagram** is a picture that shows relationships between sets,[4] which is often helpful in work with probabilities.

Recall that a standard deck of playing cards contains four suits, two of which are red (♥ and ♦) and two of which are black (♣ and ♠). Each suit contains 13 values, 2 through ace. The jack, queen, king, and ace are commonly referred to as honor cards.

Consider the Venn diagram in Figure 5A-6, which shows the relationship between two of the subsets in a deck of cards: all red cards (♥ and ♦) and the honor cards. $H \cap R$ (read "H and R" or "H intersection R") is the set consisting of all cards each of which is both a red card and an honor card. This is the **intersection** of the two sets H and R.

On the other hand, the shaded region in Figure 5A-7 shows cards that are either honor cards or red cards (or both). This is the **union** of the two sets H and R and is denoted by $H \cup R$ (read "H or R" or "H union R").

Probabilities can be associated with the set operations of intersection and union. The next sections discuss in detail the formulas that relate such probabilities.

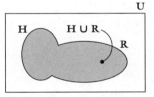

Figure 5A-7

5A.5 ADDITION AND SUBTRACTION OF PROBABILITIES

If A and B are two events, then a mathematical rule called the **addition rule** for the probability of two events can be used to calculate the probability that at least one of them occurs.

Rule ▶

If A and B are two events, then the probability that at least one of them occurs, denoted by $P(A \cup B)$, is found by the formula

$$P(A \cup B) = P(A) + P(B) - P(A \cap B). \qquad \textbf{(5A-1)}$$

[4]See Appendix A-1 for a review of set theory.

The cards represented in Figure 5A-6 and 5A-7 can illustrate the addition rule. Set H contains 16 cards (four honor cards in four suits) and set R (all the red cards) contains 26 cards. There are eight cards that are both honor cards and red cards; thus, the set $H \cap R$ contains eight cards. These numbers imply the following probabilities.

$$P(H) = \frac{16}{52}, \qquad P(R) = \frac{26}{52}, \qquad P(H \cap R) = \frac{8}{52}.$$

The probability of drawing an honor card or a red card, denoted by $P(H \cup R)$, can be found by applying the addition rule:

$$P(H \cup R) = P(H) + P(R) - P(H \cap R)$$
$$= \frac{16}{52} + \frac{26}{52} - \frac{8}{52}$$
$$= \frac{17}{26}$$
$$\approx .65.$$

Notice the importance of subtracting $P(H \cap R)$. Otherwise, it would have been added twice in the determination of $P(H \cup R)$.

E x a m p l e
8

Two dice are tossed. What is the probability of getting a sum of 7 or 11?

Solution   Let S denote the event of throwing a 7 and E the event of throwing an 11. Figure 5A-8 is a Venn diagram for the toss. The two ovals in the figure represent the ways of combining the numbers on the two dice to total either 7 or 11. Remember that there is a total of 36 possible outcomes, which can make 11 different sums.

The fact that $(S \cap E) = \emptyset$ implies that there is no way to make a 7 and an 11 at the same time. Therefore, $P(S \cap E) = 0$. (This is different from the card example, in which it was possible to pick a single card that was both an honor card and a red card.)

Applying the addition rule,

$$P(S \cup E) = P(S) + P(E) - P(S \cap E)$$
$$= \frac{6}{36} + \frac{2}{36} - 0$$
$$= \frac{8}{36}$$
$$= \frac{2}{9}. \quad \bullet$$

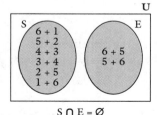

$S \cap E = \emptyset$

Figure 5A-8

A new type of price scanner was tested for use in grocery stores. Test baskets of typical items were assembled and their prices scanned. In 8% of the baskets the scanner overcharged for some items, and in 5% of the baskets the scanner under-charged for some items. Also, in 4% of the baskets both overcharges and under-charges were made. What is the probability that this price scanner will make some type of mistake in a basket of groceries when reading prices?

Solution   Let O be the event of an overcharge and U the event of an undercharge. Interpret the percentages obtained from the test data as the probabilities

$$P(O) = .08, \qquad P(U) = .05, \qquad P(O \cap U) = .04.$$

"Some type of mistake" would consist of an overcharge, an undercharge, or both, in the same basket. The probability of such an event is denoted by $P(O \cup U)$ and can be found by the addition rule:

$$P(O \cup U) = P(O) + P(U) - P(O \cap U)$$
$$= .08 + .05 - .04$$
$$= .09. \quad \bullet$$

E x e r c i s e s 5 A . 4 – 5 A . 5

1. If $P(X \cap Y)$ is not subtracted when the addition rule is applied to find $P(X \cup Y)$, what could be the effect on the calculated value of $P(X \cup Y)$?

2. In the determination of probabilities, are the set operations of intersection and union commutative? That is, for any two sets A and B, will it always be true that

$$P(A \cap B) = P(B \cap A) \qquad \text{and} \qquad P(A \cup B) = P(B \cup A)?$$

3. Find the missing probabilities, if possible.

	$P(X)$	$P(Y)$	$P(X \cap Y)$	$P(X \cup Y)$
a.	$\frac{1}{4}$	$\frac{1}{10}$	?	$\frac{1}{100}$
b.	$\frac{2}{3}$	$\frac{1}{6}$	$\frac{1}{18}$	?
c.	?	.30	.15	.43
d.	.48	?	.10	.33

* 4. Referring to Exercise 3, find the interval of possible values for $P(X \cap Y)$ that would make one of the parts of that exercise possible.

5. A student is taking a statistics examination and a literature examination on the same day. The probability of passing both is .65, and the probability of passing at least one is .90. If the probability of passing the statistics exam is .80, what is the probability of passing the literature exam?

6. A student is taking both a calculus examination and a physics examination on the same day. The probabilities of passing them singly are the same. However,

the probability of passing both is only $\frac{2}{3}$, whereas the probability of passing at least one of them is $\frac{3}{4}$. What is the probability of passing the calculus exam?

7. Insurance companies know that, historically, the probability that a member of a given high-risk group will both receive a speeding citation and be involved in an accident (sometime in his or her driving career) is .65. The probability that the person will experience one or the other of these two events is .95. If the two events are equally likely, find the probability that a member of this group will be involved in an accident.

8. Solve Exercise 7 if $P(\text{Speeding} \cup \text{Accident})$ remains the same but $P(\text{Speeding} \cap \text{Accident}) = .25$, and the probability of receiving a speeding ticket is three times the probability of being involved in an accident.

Use the following information to answer Exercises 9–13.

Leaving genetic factors aside, and if fertility drugs are not used, twins are born about once in every 89 births, triplets about once in every 7700 births, quadruplets about once in every 680,000 births, and quintuplets about once in every 85,000,000 births.

Source: *Collier's Encyclopedia*, 1995, vol. 16, p. 697.

9. Approximately how many times less likely is each successive multiple birth?

10. Find the probability of twins or triplets' resulting from a given pregnancy: $P(\text{Twins} \cup \text{Triplets})$.

11. What is the probability that any one of these four multiple births will occur?

12. In 250,000,000 births, about how many sets of triplets would be expected? How many sets of quintuplets?

13. What are some unmentioned factors that would need to be taken into account to answer Exercises 9–12 more accurately?

* 14. Research multiple births at your library, and find the name of the sextuplets, born in 1974, who are thought to be the first recorded set of sextuplets to survive early infancy. In what town were they born?

* 15. The general formula for the probability of the union of three events is

$$P(A \cup B \cup C) = P(A) + P(B) + P(C) - P(A \cap B)$$
$$- P(A \cap C) - P(B \cap C) + P(A \cap B \cap C).$$

Use the following diagram to illustrate this rule.

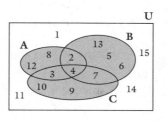

5A.6 MULTIPLICATION OF PROBABILITIES

Besides an addition rule, there is also a multiplication rule for probabilities, and it can be expressed in either of two ways, depending upon whether or not the events involved are independent of each other.

Definitions ▶

Two events, A and B, are **independent events** if the occurrence (or nonoccurrence) of one of them does not in any way affect the probability of the other's occurring.

If two events are not independent, then they are called **conditional events**.

Optional Chapter 5B examines conditional events more closely through an application of tree diagrams.

For example, if a coin is tossed twice, the probability of obtaining heads on the second toss is not in any way affected by whether or not heads was obtained on the first toss. The two events (obtaining heads on the first toss and obtaining heads on the second toss) are therefore independent. The probability of both events' occurring is expressed as $P(A \cap B)$ and can be calculated by the **multiplication rule for independent events**:

Rule ▶

If events A and B are independent, then the probability of both events' occurring, denoted by $P(A \cap B)$, is

$$P(A \cap B) = P(A) \cdot P(B). \qquad (5A\text{-}2)$$

This rule implies that the probability of obtaining two heads on two flips of a fair coin is

$$P(H \cap H) = P(H) \cdot P(H)$$
$$= \frac{1}{2} \cdot \frac{1}{2}$$
$$= \frac{1}{4}.$$

Now consider two events that are *not* independent, called conditional events. The conditional probability of event B, given that event A has occurred, is denoted by $P(B \mid A)$, read "the probability of B given A."

For example, suppose a drawer has two blue socks and four green socks in it. Two socks are to be drawn at random, one after the other, without replacing the first sock before the second sock is drawn. If a blue sock is drawn first, then the probability of drawing a blue sock on the second draw is $\frac{1}{5}$. However, if a green sock is drawn first, then the probability of drawing a blue sock on the second draw is $\frac{2}{5}$. These facts are denoted by $P(\text{blue} \mid \text{blue}) = \frac{1}{5}$ and $P(\text{blue} \mid \text{green}) = \frac{2}{5}$, respectively. The outcome of the first event (the color of the first sock drawn) clearly affects the probability of the second event (the color of the second sock drawn), and the two events are therefore conditional.

The **general multiplication rule** for any two events (independent or conditional) can be stated in terms of conditional probability:

Rule ▶

> If A and B are any two events, independent or conditional, then the probability of both events' occurring, $P(A \cap B)$, is
>
> $$P(A \cap B) = P(A) \cdot P(B \mid A). \tag{5A-3}$$
>
> [This may also be expressed as $P(A \cap B) = P(B) \cdot P(A \mid B)$.]

Notice that if two events, A and B, are independent, then $P(B \mid A) = P(B)$, and the general multiplication rule is identical to the multiplication rule for independent events.

The following examples illustrate applications of the multiplication rule for both independent events and conditional events.

E x a m p l e 10

Two dice are tossed. What is the probability that both dice land showing even numbers?

Solution   Let A be the event of obtaining an even number on one die, and let B be the event of obtaining an even number on the other die. Since there is no way one die can "know" and be influenced by what the other die does, intuitively we know that these two events are independent. Therefore,

This result can also be verified by examining the sample space.

$$P(A \cap B) = P(A) \cdot P(B)$$

$$= \frac{3}{6} \cdot \frac{3}{6}$$

$$= \frac{1}{4}. \quad \bullet$$

E x a m p l e 11

A card was drawn from a standard deck of cards and its color was noted. The card was *not* replaced in the deck, and a second card was drawn. What is the probability that both cards were red?

Solution   Let R_1 denote the event of drawing a red card on the first draw and R_2 the event of drawing a red card on the second draw. Similarly, B_1 is the event of drawing a black card on the first draw and B_2 is the event of drawing a black card on the second draw.

Consider the tree diagram shown in Figure 5A-9. The top branch is the only branch that corresponds to the compound event of drawing two red cards.

Since there are 26 red cards out of a total of 52 cards, the probability of picking a red card on the first draw is $\frac{1}{2}$. If a red card is chosen on the first draw but not replaced in the deck before the second draw, then 25 red cards remain in

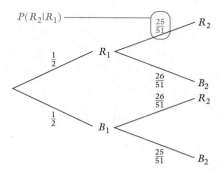

Figure 5A-9

the deck, and the probability of picking a second red card under these conditions is $\frac{25}{52-1} = \frac{25}{51}$, which is slightly less than $\frac{1}{2}$. By the general multiplication rule,

$$P(R_1 \cap R_2) = P(R_1) \cdot P(R_2 \mid R_1)$$

$$= \frac{1}{2} \cdot \frac{25}{51}$$

$$= \frac{25}{102} \approx .245. \quad \bullet$$

**E x a m p l e
12**

What is the probability of a ball's landing in a red pocket of a roulette wheel five times in a row?

Solution   Recall that a roulette wheel has 38 pockets, 18 of which are red. Since the ball is equally likely to land in any of the pockets on any given spin, the outcomes of the individual spins are independent of one another. By an extension of the multiplication rule for independent events,

$$P(R_1 \cap R_2 \cap R_3 \cap R_4 \cap R_5) = P(R_1) \cdot P(R_2) \cdot P(R_3) \cdot P(R_4) \cdot P(R_5)$$

$$= \left(\frac{18}{38}\right)^5$$

$$\approx .024. \quad \bullet$$

**E x a m p l e
13**

A gumball machine contains five green gumballs, three red gumballs, and six white gumballs.

For each dime, two gumballs come out. What is the probability that both gumballs are green?

Solution   Let G_1 be the event that the first gumball out of the machine is green and G_2 the event that the second gumball is green. By the general multiplication rule,

$$P(G_1 \cap G_2) = P(G_1) \cdot P(G_2 \mid G_1).$$

If a package of M&M's has three brown candies, two yellow, two red, one green, one orange, and one tan, then the probability of shaking out two yellow ones is $\frac{2}{10} \cdot \frac{1}{9} = \frac{1}{45}$.

Since 14 gumballs are in the machine at the start of this experiment, 5 of which are green, the probability that the first gumball out is green, $P(G_1)$, is $\frac{5}{14}$. Then $14 - 1 = 13$ gumballs remain, with $5 - 1 = 4$ of them green. Therefore, the conditional probability, $P(G_2 \mid G_1)$, is $\frac{4}{13}$. Substituting these values,

$$P(G_1 \cap G_2) = P(G_1) \cdot P(G_2 \mid G_1)$$

$$= \frac{5}{14} \cdot \frac{4}{13}$$

$$= \frac{10}{91}. \quad \bullet$$

E x a m p l e
14

Two events, X and Y, are such that $P(X) = \frac{4}{7}$, $P(Y) = \frac{1}{2}$, and $P(X \cap Y) = \frac{3}{14}$. Find the following.

a. $P(X \cup Y)$
b. $P(Y \mid X)$
c. $P(X \mid Y)$

Solution   a. Substituting into the addition rule,

$$P(X \cup Y) = P(X) + P(Y) - P(X \cap Y),$$

$$P(X \cup Y) = \frac{4}{7} + \frac{1}{2} - \frac{3}{14}$$

$$= \frac{12}{14}$$

$$= \frac{6}{7}.$$

b. There is no statement of whether events X and Y are independent. However, since $\frac{3}{14} \neq \frac{4}{7} \cdot \frac{1}{2}$, it follows that $P(X \cap Y) \neq P(X) \cdot P(Y)$, and X and Y must not be independent. $P(Y \mid X)$ can be found by substituting into the general multiplication rule:

$$P(X \cap Y) = P(X) \cdot P(Y \mid X),$$

$$\frac{3}{14} = \frac{4}{7} \cdot P(Y \mid X),$$

$$\frac{\frac{3}{14}}{\frac{4}{7}} = P(Y \mid X),$$

$$\frac{3}{8} = P(Y \mid X).$$

c. Solving for $P(X \mid Y)$ the same way as in Part b,

$$P(X \cap Y) = P(Y \cap X) = P(Y) \cdot P(X \mid Y),$$

$$\frac{3}{14} = \frac{1}{2} \cdot P(X \mid Y)$$

$$\frac{\frac{3}{14}}{\frac{1}{2}} = P(X \mid Y)$$

$$\frac{3}{7} = P(X \mid Y). \quad \bullet$$

The result from Part c implies that the formula for conditional probability, $P(A \mid B)$, may be expressed as follows.

Formula for Conditional Probability

$$P(A \mid B) = \frac{P(A \cap B)}{P(B)}. \tag{5A-4}$$

By studying Formula 5A-4 it may be seen that $P(A \mid B) = \dfrac{P(A \cap B)}{P(B)}$ and $P(B \mid A) = \dfrac{P(A \cap B)}{P(A)}$. This implies that if $P(A \cap B) \neq 0$, then $P(A \mid B)$ and $P(B \mid A)$ are equal only if $P(A) = P(B)$.

Exercises 5A.6

1. Which of the following are always true?
   a. $P(A \cap B) = P(B \cap A)$
   b. $P(A \cup B) = P(B \cup A)$
   c. If $P(A) \neq 0$, $P(B) \neq 0$, $P(A \cap B) = 0$, then $P(A \mid B) = P(B \mid A)$.
2. Refer to Example 11 on page 196.
   a. Identify $P(B_2 \mid R_1)$, $P(R_2 \mid B_1)$, and $P(B_2 \mid B_1)$.
   b. What is true about the sum of the probabilities on all three pairs of branches?
   c. If the first card drawn were replaced in the deck before the second card was drawn, would that make $P(R_1 \cap R_2)$ larger or smaller?
3. Refer to Example 12 on page 197. Would a bet on a black pocket be an especially wise bet for the sixth spin of the wheel? Why or why not?

4. A fair coin is flipped three times.

   a. Complete the following tree diagram for this compound event. Indicate probabilities along all branches of the completed tree diagram.

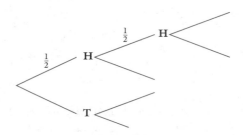

   b. What is the probability of flipping three heads?

      Hint: Multiply the three probabilities along the top branch.

   c. What is the probability of flipping exactly two heads?

      Hint: Multiply the probabilities on each branch with exactly two heads, then add the products.

   d. What is the probability of flipping at least one head?

5. Answer all parts of Exercise 4 under the condition that the coin is not fair and the probabilities of heads and tails are $\frac{3}{5}$ and $\frac{2}{5}$, respectively.

6. A woman has a pocketful of n fair coins. She pulls them all out of her pocket and tosses them on a table. Answer the following questions in terms of n.

   a. What is the probability that all the coins land heads?

   b. What is the probability that at least one of the coins lands heads?

      Hint: At least one coin's landing heads is the complement of all of the coins' landing tails.

GROUP

7. The accuracy of a certain missile is 80%. Four shots are taken at a target, with the result of each attempt being independent of the previous result.

   a. A tree diagram with $2^4 = 16$ branches has been started to help answer Parts b and c. Complete the tree diagram, labeling probabilities along the branches.

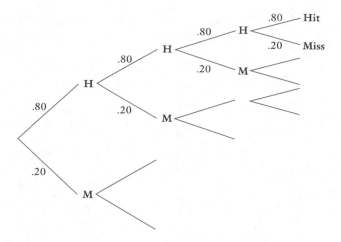

b. What is the probability that the missile hits the target exactly twice?

c. What is the most likely number of hits?

8. A family has three children. Assume that male and female births are equally likely and are independent of each other. Use a tree diagram with $2^3 = 8$ branches to answer the following questions.

What are the chances of having at least one boy among three siblings? (*Joy Small*)

a. What is the probability that the family has all girls?

b. What is the probability that there is at least one boy?

c. What is the probability of more boys than girls?

*

    d. If it is somehow known for certain that the family has at least two boys, then what is the probability that the family has three boys?

    Hint: Study the tree diagram carefully, and start by counting the number of branches that have at least two boys.

*

    e. If it is somehow known for certain that the family has at least one girl, then what is the probability that the family has two boys?

9. Complete the following table.

	$P(A)$	$P(B)$	$P(A \cup B)$	$P(A \cap B)$	$P(A \mid B)$	$P(B \mid A)$
a.	$\frac{3}{4}$	?	$\frac{9}{10}$	$\frac{1}{5}$	?	?
b.	?	?	?	$\frac{2}{3}$	$\frac{8}{9}$	$\frac{4}{5}$
c.	$\frac{1}{5}$	$\frac{1}{2}$	?	?	?	$\frac{1}{20}$
d.	$5k$	$3k$	?	$1k$	?	?

10. In Part d of Exercise 9, what restrictions, if any, should be placed on k?

11. Two people are to be chosen randomly to fill vacancies on a condominium's board of directors. There are four men and six women from which to choose. Draw a tree diagram to answer the following questions.

    a. What is the probability that both people chosen are men?

    b. What is the probability that both people chosen are women?

    c. Must $P(2 \text{ men}) + P(2 \text{ women}) = 1$? Explain why or why not.

    d. Determine the probability that the second person chosen is a woman if a man is chosen first.

    e. Is the answer to Part d the same as the probability of choosing a woman second? Why or why not?

* 12. A gumball machine has five red gumballs, eight green gumballs, three white gumballs, and seven blue gumballs. For each quarter, three gumballs come out of the machine.

    a. What is the probability that all three gumballs are white?

    b. What is the probability that all three gumballs are the same color?

    c. What is the probability that no green gumballs come out?

    d. What is the probability that exactly two red gumballs come out?

5A.7   CROSS-TABS TABLES

A cross-tabs table can also illustrate the idea behind conditional probability.

Definition ▶

A **cross-tabs table** is a table that contains information on two qualitative variables.

A cross-tabs table is also known as a contingency table or a two-way classification. Later chapters of the text will use it more extensively.

In the following cross-tabs table, the two qualitative variables are grades in school and performance on an IQ test.

IQ vs. Grades

Score on IQ Test	Good Grades, G	Poor Grades, P	Total
High, H	240	60	300
Low, L	10	90	100
Total	250	150	400

A cross-tabs table can be used efficiently to determine probabilities. Here, for example, the probability that an individual scored high on the test is $P(H) = \frac{300}{400} = \frac{3}{4}$. The probability that an individual has poor grades, given that he or she scored low on the test, is $P(P \mid L) = \frac{90}{100} = \frac{9}{10}$. Note that the denominator, 100, is the total number of students who scored low, and the numerator, 90, is the number of students who had poor grades and scored low.

Exercises 5A.7

Use the cross-tabs table just presented, which compares school grades to IQs, to determine the probabilities in Exercises 1–3.

1. Find $P(G)$ and $P(G \cap H)$.
2. Find $P(G \mid H)$, $P(H \mid G)$, and $P(P \mid H)$.
3. Find $P(L \cup P)$ by using the addition rule.
4. A certain college has 100 women and 150 men. Twenty women and 25 men are enrolled in a probability course.

   a. Construct a cross-tabs table for the preceding information.

   Use the cross tabs-table from Part a to answer the following two questions.

   b. What is the probability that a student selected at random is a woman?

   c. What is the probability that a student selected at random is a woman, given the fact that the student is enrolled in the probability class?

S t a t   F a c t

McDonald's ranks first in the number of fast food franchises. It has nearly 10,000 outlets in the United States and over another 6000 in 79 countries around the world. Plans are underway to open additional franchises in Bombay and Dehli in early 1996, where beefless burgers will be served.

5. The following cross-tabs table shows the geographic distribution of the locations of a group of fast-food franchises.

| | **Region** | | | |
	SW	NW	SE	NE
Urban	10%	20%	5%	5%
Rural	5%	3%	12%	15%
Suburban	5%	12%	3%	5%

Compute row and column totals for this cross-tabs table, and then answer the following questions.

a. If a franchise is chosen at random, what is the probability that it is in the Northeast?

b. If a franchise is chosen at random, what is the probability that it is in a rural area of the Southeast?

c. What is the probability that a suburban franchise is in the Southwest?

d. What is the probability that a franchise in the Northwest is suburban?

e. What is the probability that a franchise that is not in the Northeast is also not urban?

5A.8 VENN DIAGRAMS AND INDEPENDENT EVENTS

The question of independence is an important one, and it is not always obvious or easy to determine. Recall that if two events are independent, then the occurrence (or nonoccurrence) of one of them does not in any way affect the probability of the other's occurring. For instance, if a king was drawn at random from a deck of cards, that occurrence does not in any way affect the probability that the card drawn was a diamond. Thus, the two events, drawing a king and drawing a diamond, are independent. On the other hand, if a diamond was drawn at random from a deck of cards, that does affect the probability that the card drawn was red, since diamonds are red cards. Thus, the events of drawing a diamond and drawing a red card are not independent.

Erroneously assuming the independence of events and then multiplying their probabilities to arrive at a probability of all the events' occurring is a very common mistake. One of the most famous examples, in which even experts went wrong, involved a criminal trial in California. A woman was robbed by an interracial couple driving a yellow car; the woman criminal was reported as having blonde hair worn in a ponytail. Such a couple was later found and charged with the crime. The prosecution produced an expert witness from the mathematics department of a nearby college to testify that the probability of finding a couple with all of these characteristics was so small as to ensure their guilt beyond a reasonable doubt. The

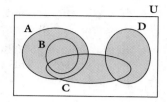

Figure 5A-10

jury agreed and convicted them. Years later, however, the California Supreme Court overturned the conviction. The California Supreme Court heard further (and, this time, correct) testimony that the characteristics were not independent, the prosecution's computation of probability had, therefore been incorrect, and the jury had thus been misled.[5]

Although a Venn diagram cannot identify independent events with certainty, it may be used to identify pairs of events that could *not* possibly be independent. Consider Figure 5A-10. By definition of independent events, the events A and B in this Venn diagram cannot possibly be independent, because if B has occurred, then A must have also, since B is part of A. Thus, the occurrence of event B affected the probability of event A. Events A and B therefore cannot be independent, and it must follow that

$$P(A \mid B) \neq P(A) \qquad \text{and} \qquad P(B \mid A) \neq P(B).$$

Similarly, the events A and D (or B and D) cannot possibly be independent, because if A has occurred, then D definitely has not, since the two sets (events) have nothing in common. In this circumstance, one could say that events A and D (or B and D) are **mutually exclusive.**

It is not possible to draw a conclusion about the independence of events B and C, A and C, or D and C on the basis of Figure 5A-10. More information is needed to determine whether the events in each of these pairs are independent.

In essence, if two sets have nothing in common, or if one lies inside the other, then the events they represent cannot be independent. If the two sets are not related in either of these ways, then there is a possibility (but no guarantee) that the events they represent are independent.

Exercises 5A.8

1. Refer to the opening paragraph of this section on page 204. Rewrite the explanation of independent events in your own words, using the classifications woman, parent, and mother.

2. It is easy to think that if two events are mutually exclusive, they are also independent. However, this could never be the case. Explain why.

---

[5]See "The Laws of Probability," *Time,* January 8, 1965, page 42, and "Trial by Mathematics," *Time,* April 26, 1968, page 41, for more details. The case study in optional Chapter 5B examines this case more closely.

3. If two events are independent, how must their relationship look in a Venn diagram? Are all events pictured in this way independent? Explain.

4. In the following Venn diagram, the labeled sets represent events. Indicate which pairs of events are possibly independent and which pairs are definitely not independent.

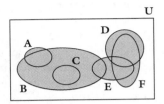

5. A card is drawn from a standard deck of playing cards. Define the following events.

$$D = \text{Draw a diamond},$$
$$C = \text{Draw a club},$$
$$A = \text{Draw an ace},$$
$$K = \text{Draw a king},$$
$$R = \text{Draw a red card}.$$

Draw a Venn diagram that illustrates these events, and use it to determine which pairs of events are possibly independent and which pairs are definitely not independent.

Case Study   *The Historical Foundations of Probability*

Karl Friedrich Gauss (1777–1855), born in Germany, is considered one of the greatest mathematicians of all time. (*Library of Congress*)

It may seem to many students that the contents of mathematics courses have always been as complete and systematic as they appear today. Each course contains a set of rules that, if followed exactly, always lead to the correct answer. This orderly structure was not originally the case. Centuries ago, arguments and even superstitions surrounded many of the most basic mathematical rules and formulas we take for granted today.

Probability is a perfect example. It originated in the society, particularly the gambling parlors, of 17th-century France. Gamblers consulted the renowned mathematicians of that time about how to win the most while risking the least. Luck was often confused with mathematics. Gambling and games gave rise to the theory of probability. **Blaise Pascal** and **Pierre Fermat** quickly moved to the forefront of this new field and corresponded about problems connected with games of chance.

In the early 1700s the theory of probability took some giant leaps forward through the work of **Abraham de Moivre** and **James Bernoulli.** It was de Moivre who first stated the multiplication rule for the theory of probability (which will be a key concept for work with discrete random variables in Chapter 6 and some later applications).

De Moivre was also one of the first mathematicians to explore the normal probability distribution, which is extremely important in statistics as well as in probability. Bernoulli, too, did important work that helped to develop probability as a legitimate branch of mathematics in its own right.

In 1812 the French mathematician **Marque de La Place** wrote a treatise on probability that contained the most systematic and complete treatment of games of chance up to that time. In addition, it described many applications of probability theory to a wide range of scientific and practical problems. As a result, the applications of probability theory expanded rapidly.

Still in the early 19th century, **Karl Friedrich Gauss** and La Place, working independently, applied probability theory to astronomical observations. They laid the foundation for a specialty branch of mathematics called error theory. It, in turn, is related to a mathematical technique called the method of least squares, which is very important in statistics.

Pascal, Fermat, and their contemporaries used the classic definition of the probability of an event employed in this text. This classic definition has liabilities, however. In some situations, not all possibilities are equally likely, or many steps must be taken, each depending upon the outcome of the previous one. These difficulties in combinatorial analysis (counting) and conditional probability make it very hard for even the most logical person to determine the "number of successful ways" and the "total number of possible outcomes" in order to apply the definition.

Another difficulty with the classic definition arises from the inevitable application of probability theory to practical situations such as life expectancy, birth rate, and other social and demographic questions. The probability that a person will live to at least age 95 in a certain community, for example, could be estimated statistically: the number of people who have reached that age could be divided by the total population and the resulting ratio adjusted and refined. But any attempt to apply the classic probability definition literally by counting the number of successful ways in which a person could live to age 95 would clearly fail.

Gradually, it became evident that for more advanced work in probability a revision of the classic definition, or an alternative to it, would be necessary. Since 1850 mathematicians have been seeking such an alternative. You may learn about their efforts in more advanced courses in statistics and probability.

Exercise

The early history of probability is filled with colorful people living in interesting times. Following are the names of a few more of them. Find information in your library about them and their times. Also, if possible, describe their contributions to the theory of probability.

Galileo (1564–1642)

Chevalier de Mere (a French knight and an ardent dice gambler of the 17th century, also known as Antoine Gombauld)

Christiaan Huygens (1629–1695)

Daniel Bernoulli (1700–1782)

Thomas Bayes (1702–1761)

Leonhard Euler (1707–1783)

Thomas Simpson (1710–1761)

Jean le Rond d'Alembert (1717–1783)

Marquis de Condorcet (1743–1794)

Simeon Denis Poisson (1781–1840)

Andree Andreevich Markov (1856–1922)

CHAPTER SUMMARY

The probability that an event, E, will occur is defined as

$$P(E) = \frac{\text{Number of successful ways for } E \text{ to occur}}{\text{Total number of possible outcomes}}.$$

This is the classical definition of the probability of an event. Along with a few logical restrictions, it is the formula used to determine an event's probability.

Use of the classic definition requires certain techniques of counting. Critical to any method of counting is the Fundamental Counting Principle (FCP):

Let E be an event. Suppose that E is made up of several distinct steps and that each step has a certain number of distinct choices. Then the total number of ways in which E can occur is equal to the product of the number of steps and the number of choices.

A tree diagram is a graphic representation of the FCP and is often helpful in problems requiring careful analysis.

The addition rule for the probability of two events, A and B, is

$$P(A \cup B) = P(A) + P(B) - P(A \cap B).$$

The union symbol ($\cup$) is equivalent to the word "or," and the intersection symbol ($\cap$) is equivalent to the word "and."

$P(B \mid A)$ denotes the probability of the occurrence of event B, given that event A has happened (or is guaranteed to happen). $P(B \mid A)$ is referred to as a conditional probability.

Two events, A and B, are independent if the likelihood of one's occurring is not affected by whether or not the other has occurred—that is, if $P(B \mid A) = P(B)$.

The general multiplication rule for probabilities is

$$P(A \cap B) = P(A) \cdot P(B \mid A).$$

If the events A and B are independent, then the general multiplication rule is simply $P(A \cap B) = P(A) \cdot P(B)$.

Key Concepts, Terms, and Formulas

Videotape Suggestions

Program 15 in the series *Against All Odds: Inside Statistics* provides additional information on the theory of probability and could serve as a preview of the topic of randomness in Chapter 7. *Program 16* in the same series includes the topics of independence and the multiplication rule for independent events as well as a preview of some of the work in Chapter 6.

Applying knowledge of
permutations and
combinations helps
ensure the security of
computer passwords,
bank customers' PIN
numbers, and other
ordered items.

5B Further Topics in Probability

(The material in this chapter is optional.)

Chapter 5A showed that counting usually plays an important part in determining the probability of an event. However, counting the number of ways in which an event can occur is not always easy. The **case study** at the end of this chapter examines a famous criminal case in which even the experts counted incorrectly.

5B.1 PERMUTATIONS AND COMBINATIONS

Two sophisticated methods of counting, forming permutations and forming combinations, can shed light on the procedure of counting.

Permutations

Definition ▶

A **permutation** is an arrangement of r objects chosen from n objects, the order of which is important. The possible number of such arrangements is denoted by $_nP_r$.

S t a t   F a c t

Another, less common meaning of "permutation" is a change from one position to another. Charles Darwin referred to the permutation of the continents in his *Life and Letters*.

In a permutation, no object may be used more than once; that is, objects are counted, chosen, or arranged **without replacement.**

The batting order example on page 183 in Chapter 5A involved permutations of players. Recall that 15 children were on a Little League baseball team, and the goal was to find how many batting orders made up of 9 of the 15 children were possible. With the FCP, it was determined that the number of possible batting orders was the product of the numbers of choices possible at all the positions:

1,816,214,400. In mathematical notation, 9 players chosen from 15 players can be arranged in $_{15}P_9$ ways.

$$n = 15 \text{ and } r = 9$$

$$\underbrace{\underset{n}{15} \cdot \underset{(n-1)}{14} \cdot \underset{(n-2)}{13} \cdot \underset{(n-3)}{12} \cdot \underset{(n-4)}{11} \cdot \underset{(n-5)}{10} \cdot \underset{(n-6)}{9} \cdot \underset{(n-7)}{8} \cdot \underset{(n-8)}{7}}_{r \text{ factors}}$$

First position ↓ ... Last position ↓

Many calculators have a key that calculates $_nP_r$ directly from a given n and r. If they do not, they usually have a factorial key,[1] and Formula 5B-1 can be applied to determine $_nP_r$.

$$_nP_r = \frac{n!}{(n-r)!}. \qquad \text{(5B-1)}$$

In other words, $n \cdot (n-1) \cdot (n-2) \cdot \cdots \cdot (n-r+1) = \frac{n!}{(n-r)!}$. Also note that

$$_nP_n = \frac{n!}{(n-n)!}$$

$$= \frac{n!}{0!}$$

$$= \frac{n!}{1}$$

$$= n!.$$

In the batting order example, $n = 15$ and $r = 9$. Therefore,

$$_{15}P_9 = \frac{15!}{(15-9)!}$$

$$= \frac{15!}{6!}$$

$$= \frac{15 \cdot 14 \cdot 13 \cdot 12 \cdot 11 \cdot 10 \cdot 9 \cdot 8 \cdot 7 \cdot 6 \cdot 5 \cdot 4 \cdot 3 \cdot 2 \cdot 1}{6 \cdot 5 \cdot 4 \cdot 3 \cdot 2 \cdot 1}$$

$$= 15 \cdot 14 \cdot 13 \cdot 12 \cdot 11 \cdot 10 \cdot 9 \cdot 8 \cdot 7$$

$$= 1,816,214,400.$$

Most graphics calculators can determine the value of $_nP_r$, as the next example illustrates.

---

[1] Recall from algebra that n factorial, written $n!$, is equal to

$$n \cdot (n-1) \cdot (n-2) \cdot \cdots \cdot 1,$$

and 0! is defined as 1.

E x a m p l e
1

Use a TI-82 graphics calculator to determine the value of $_{15}P_9$.

Solution Make the following keystrokes.

• Enter the value n, which is 15.
• Press the [MATH] button.
• Press the right-arrow key, [▶], three times until the PRB menu is selected.
• Press [2] to activate the permutation function.
• Enter the value of r, which is 9.

The screen should appear as shown in Figure 5B-1. When the [ENTER] key is pressed, the value of $_{15}P_9$ is computed as 1,816,214,400, which matches the result obtained by hand.

```
15 nPr 9
```

Figure 5B-1 •

E x a m p l e
2

Many banks require each customer using an automatic teller machine to enter a personal identification code made up of three digits followed by three letters.

a. How many codes are possible if neither digits nor letters may be used more than once?
b. How many codes are possible if digits and letters may be used more than once?

Solution a. The order in which the digits (or letters) are entered is important. For instance, 573MNY is different from 735MYN. Since order is important, and since no digit or letter may be used more than once, these arrangements are permutations.

$$10 \cdot 9 \cdot 8 \cdot 26 \cdot 25 \cdot 24$$

$$\underbrace{}_{\substack{\text{3 digits} \\ _{10}P_3}} \cdot \underbrace{}_{\substack{\text{3 letters} \\ _{26}P_3}}$$

As the diagram indicates, three digits chosen from ten digits may occur in $_{10}P_3$ ways.

$$_{10}P_3 = \frac{10!}{(10-3)!}$$

$$= 10 \cdot 9 \cdot 8, \quad \text{or} \quad 720 \text{ ways.}$$

Three letters chosen from 26 letters may occur in $_{26}P_3$ ways.

$$_{26}P_3 = \frac{26!}{(26 - 3)!}$$
$$= 26 \cdot 25 \cdot 24, \quad \text{or} \quad 15{,}600 \text{ ways.}$$

By the FCP, $_{10}P_3 \cdot {}_{26}P_3 = 720 \cdot 15{,}600$, or $11{,}232{,}000$, identification codes are possible.

b. Since the digits and letters may be used more than once, the different arrangements are not permutations, but the number of possible arrangements can still be found by application of the FCP.

$$\underbrace{10 \cdot 10 \cdot 10}_{\substack{\text{3 digits} \\ 10^3}} \cdot \underbrace{26 \cdot 26 \cdot 26}_{\substack{\text{3 letters} \\ 26^3}}$$

As the diagram shows, if digits and numbers may be used more than once, there are 10 ways to pick the first digit, 10 ways to pick the second digit, and 10 ways to pick the third digit. Then there are 26 ways to pick the first letter, 26 ways to pick the second letter, and 26 ways to pick the last letter. Thus, $10^3 \cdot 26^3 = 17{,}576{,}000$ identification codes are possible under these conditions. ●

E x a m p l e 3

THE BIRTHDAY PROBLEM

Forty people are chosen at random.

a. What is the probability that none have the same birthday?
b. What is the probability that at least two of the people have the same birthday?

(Include February 29, and use 366 as the number of days in the year.)

Technically, not all 366 birth dates are equally likely. Besides the obvious exception of February 29, fewer babies are born in the United States during the month of April than in any other month.

Solution   a. Let E be the event that no two people have the same birthday. Choose any person to start with, and note his or her birthday. Choose a second person. The probability that the second person does not have the same birthday as the first person is $\frac{365}{366}$, since there are 365 successful ways to not have the same birth date out of 366 possible birth dates.

Choose a third person. The probability that the third person does not have one of the other two people's birthdays is

$$\frac{(366 - 2)}{366} = \frac{364}{366},$$

since there are 364 successful ways not to have the same birth date out of 366 possible birth dates. Continuing in this manner, the probability that the 40th person does not have the same birthday as one of the first 39 people is

$$\frac{(366 - 39)}{366} = \frac{327}{366}.$$

Note that each of these is a conditional probability determined by the preceding ones.

With repeated application of the general multiplication rule, these individual conditional probabilities are multiplied together to find the probability of event *E*.

$$P(E) = \frac{365}{366} \cdot \frac{364}{366} \cdot \frac{363}{366} \cdot \cdots \cdot \frac{327}{366}$$

$$= 1 \cdot \frac{365}{366} \cdot \frac{364}{366} \cdot \frac{363}{366} \cdot \cdots \cdot \frac{327}{366}$$

$$= \frac{366}{366} \cdot \frac{365}{366} \cdot \frac{364}{366} \cdot \frac{363}{366} \cdot \cdots \cdot \frac{327}{366}$$

$$= \frac{_{366}P_{40}}{(366)^{40}}$$

$$\approx .109455.$$

In other words, about 11% of the time, a randomly chosen group of 40 people would be made up of people who do not share a birth date with anyone else in that group.

b. The probability that at least two people have the same birthday is the complement of no two having the same birthday. Thus, the desired probability for this part is

$$1 - \frac{_{366}P_{40}}{(366)^{40}} = 1. - .109455 = .890545. \quad \bullet$$

This result includes the probability that two, three, four, or any number up to and including 40 have the same birthday. It is not the answer one would generally expect.

Combinations

S t a t   F a c t

Hundreds of years ago, the word "combination" was almost always used in a negative sense, implying a conspiracy.

Definition ▶

A **combination** is a set (not an arrangement) of *r* objects chosen from *n* objects. The order in which the objects are chosen, listed, or arranged in the set is not important. The number of possible sets is denoted by $_nC_r$.

The notation $_nC_r$ is read "*n* choose *r*," or "*n* combination *r*." Like permutations, combinations are formed without replacement.

There are just three possible combinations of two of the first three letters of the alphabet: {(*A*, *B*), (*A*, *C*), and (*B*, *C*)}. The order in which the letters of any particular combination are listed is not important. There is no need to count (*A*, *B*) as separate from (*B*, *A*), as one would when counting the number of possible permutations.

To develop a formula for $_nC_r$, first consider the number of possible ways of arranging *r* objects chosen from *n* objects when the order *is* important: $_nP_r$. Each

The procedure for using the TI-82 graphics calculator to determine $_nP_r$, described in Example 1 on page 213, may be modified to determine $_nC_r$ by pressing ③ instead of ② in the fourth step.

arrangement is of length r; therefore, by the FCP, each contains $r!$ different orderings. Thus, if the number of choices in which the order does matter ($_nP_r$) is divided by the number of orderings possible for each choice ($r!$), the result is a formula for $_nC_r$, the number of possible choices in which the order does not matter.

$$_nC_r = \frac{_nP_r}{r!} = \frac{n!}{(n-r)! \cdot r!} \qquad \text{(5B-2)}$$

Many calculators have a key to compute $_nC_r$ directly. If not, $_nC_r$ can be calculated from Formula 5B-2.

E x a m p l e 4

Stat Fact

California's lottery originally had each player choose six numbers from 49, but then 49 was changed to 53 to reduce the expected payoff.

Stat Fact

It is more than 23 million times more likely that two statistics students in a class of 40 will share the same birthday than win the State Lottery.

The Illinois State Lottery sponsors Lotto, a game with enormous payoffs in which a person chooses six numbers from 1 through 54.

a. In how many ways can these six numbers be chosen?
b. What is the probability that someone will pick all six correctly?

Solution   a. There are 54 numbers from which to choose six, and the order in which the six are chosen is not important. The number of ways in which six numbers can be chosen from 54 without replacement is $_{54}C_6$.

$$_{54}C_6 = \frac{54!}{6! \cdot 48!} = 25,827,165 \text{ ways.}$$

b. The probability that someone will pick all six numbers correctly is, by the classic probability definition,

$$\frac{\text{Number of ways to pick all 6 numbers correctly}}{\text{Total number of ways of picking 6 numbers}}.$$

Since the order of the winning numbers is not important, there is only one correct choice of six numbers. The required probability is therefore $\frac{1}{25,827,165}$, or about 0.00000003871.   ●

E x a m p l e 5

A full house is a five-card hand that includes three cards (a triple) of one value and two (a pair) of another. How many different full houses would have the triple in queens?

Solution   Recall that a deck of cards has four suits with 13 values in each suit. The values are {2, 3, 4, 5, 6, 7, 8, 9, 10, J, Q, K, A}. Each value has four suits. For example, the four cards having the value of queen appear as the queen of ♣, the queen of ♦, the queen of ♥, and the queen of ♠.

Three of the four possible queens must be chosen, which can happen in $_4C_3$ ways. To obtain the final two cards, which are a pair, first the value of that pair must be determined from the remaining 12 values (which can be accomplished in

$_{12}C_1$ ways), and then two of the four possible suits of that value must be chosen (in one of $_4C_2$ possible ways). By the FCP, this hand (**QQQ**xx) can occur in the following number of ways:

$$\underbrace{\underset{\left(\substack{\text{choose 3 of 4}\\\text{possible queens}}\right)}{_4C_3} \cdot \underset{\left(\substack{\text{choose 1 of 12}\\\text{remaining values}}\right)}{_{12}C_1} \cdot \underset{\left(\substack{\text{choose 2 of 4}\\\text{possible suits}}\right)}{_4C_2}}_{\text{Ways to get the pair}}$$

Now, substituting in the actual numbers,

$$_4C_3 \cdot {_{12}C_1} \cdot {_4C_2} = 4 \cdot 12 \cdot 6$$
$$= 288. \quad \bullet$$

Example 5 shows that there are 288 five-card hands that are full houses with the triple in queens. On the other hand, the number of full houses with any value, not just queens, as the triple is

$$_{13}C_1 \cdot {_4C_3} \cdot {_{12}C_1} \cdot {_4C_2} = 3744.$$

Example 6

Stat Fact

It can be proved mathematically that the longer a player gambles, the better the house's chances of winning, and if a given casino can keep its gamblers playing for just 5 more minutes a night, it adds millions to its annual gross. Thousands of dollars are spent on interior design elements, ranging from lighting to special scents sprayed in the air, that help to keep players at the tables longer. Also, there is often a long, circuitous route to the payoff windows.

Find the probability of being dealt four-of-a-kind (four cards of the same value in a five-card hand).

Solution Let F be the event of being dealt four-of-a-kind. By the classic definition of the probability of an event,

$$P(F) = \frac{\text{Number of ways of being dealt four-of-a-kind}}{\text{Total number of hands possible}}$$

First determine the numerator. In order for four-of-a-kind to be dealt, 1 of 13 values must be chosen, which can happen in $_{13}C_1 = 13$ ways. Next, four suits with that value must be chosen, which can happen in $_4C_4 = 1$ way. Finally, the last card must be chosen. Forty-eight cards remain, and one of them can be chosen in $_{48}C_1 = 48$ ways. By the FCP, the numerator is

$$_{13}C_1 \cdot {_4C_4} \cdot {_{48}C_1} = 13 \cdot 1 \cdot 48$$
$$= 624 \text{ ways.}$$

The denominator is the number of different five-card hands that can be pulled from 52 cards without replacement, when the order of the cards in the hand is not important. There are $_{52}C_5 = 2{,}598{,}960$ such hands. Therefore,

$$P(F) = \frac{624}{2{,}598{,}960}$$
$$\approx .00024. \quad \bullet$$

Exercises 5B.1

A TI-82 graphics calculator may be used for the following exercises involving combina-tions. Employ the method for permutations shown in Example 1 on page 213, but press ③ *to activate the combination function.*

1. When answering Exercise 1, assume that a telephone number may start with any of the ten digits.

    a. Can telephone numbers be considered permutations?

    b. How many three-digit area codes are possible if no area code may start with 0 or 1 but repetitions of digits are allowed?

    c. How many different seven-digit telephone numbers are possible within a single area code if no telephone number may start with 0 or 1 but repetitions of digits are allowed?

2. Using the information in Exercise 1, determine how many telephone numbers are possible if the three-digit area code is used.

3. Refer to Exercise 2. The three-digit area code 555 is reserved for some special uses. How many area-code-plus-seven-digit telephone numbers are possible that do not use the 555 area code?

* 4. The numerator of Example 6 on page 217 could also be written as

$$_{13}C_1 \cdot {}_4C_4 \cdot {}_{12}C_1 \cdot {}_4C_1.$$

Explain.

5. Identify each of the following as true or false.

    a. Permutations and combinations are methods of counting.

    b. There are fewer permutations of four items than there are combinations of those four items. (All four items must be used.)

    c. A winning Lotto number is a permutation.

    d. Sometimes the FCP requires that combinations be multiplied together.

6. Refer to the Birthday Problem, Example 3, on page 214. By trial and error, determine the minimum number of people who must be chosen at random so that the probability that at least two of them have the same birthday is at least $\frac{1}{2}$.

* 7. Refer to the Birthday Problem, Example 3, on page 214. By trial and error, determine the minimum and maximum numbers of people who must be chosen at random so that the probability that at least two of them have the same birthday is at least $\frac{1}{3}$ but less than $\frac{2}{3}$.

* 8. If there are about 40 common first names for males, what is the probability that at least two of 15 males at a party have the same first name? Assume that all names are equally likely to be represented.

    Hint: This exercise is modeled on the Birthday Problem from page 214.

9. Example 4 on page 216 shows that the probability of winning the Illinois State Lottery is practically zero. Yet people do win it. Explain why.

* 10. How many lottery tickets would one need to buy in order to have an "even chance" (a probability of $\frac{1}{2}$) of winning the Illinois State Lottery? What conditions need to be satisfied?

GROUP

11. Match each five-card poker hand in column A with the correct calculating formula—the formula used to determine the numerator of the probability fraction—in column B.

A	B
Three-of-a-kind (*xxxyz*)	$_{13}C_2 \cdot {}_4C_2 \cdot {}_4C_2 \cdot {}_{11}C_1 \cdot {}_4C_1$
Flush (all the same suit)	$_4C_3 \cdot {}_4C_2$
Two pairs (*xxyyz*)	$_{10}C_1 \cdot ({}_4C_1)^5$
Two aces, two kings, one queen	$_{13}C_1 \cdot {}_4C_3 \cdot {}_{12}C_2 \cdot {}_4C_1 \cdot {}_4C_1$
One pair (*xxwyz*)	$_{26}C_5$
Three aces and two kings	$_4C_1 \cdot {}_{13}C_5$
All red cards	$_{13}C_1 \cdot {}_4C_2 \cdot {}_{12}C_3 \cdot ({}_4C_1)^3$
All cards of one color	$_4C_2 \cdot {}_4C_2 \cdot {}_4C_1$
Straight (consecutive values; ace high or low)	$_2C_1 \cdot {}_{26}C_5$

12. Many automatic teller machines require the user to key in a four-digit identification code after inserting the bank card.

a. How many different identification codes are possible if any digit may be used more than once?

b. How many different identification codes are possible if no digit may be used more than once?

* c. If a person loses his or her bank card, what is the probability that the person who finds the card can access the account in three or fewer tries? Assume that any digit may be used more than once in the identification code.

13. A state has automobile license numbers made up of three digits followed by four letters.

a. How many different license numbers are possible if no digit or letter may be used more than once?

b. How many different license numbers are possible if any digit or letter may be used more than once but the digit portion of the plate may not begin with 0?

c. Answer Part b under the condition that license numbers may contain no vowels other than Y.

*The "odds" of an event are closely related to, but not the same as, the probability of that event. If E is an event and P(E) is the probability of that event, then the **odds** in favor of event E occurring, or the odds for E, denoted by O(E), are defined as*

$$O(E) = \frac{P(E)}{1 - P(E)}.$$

This definition implies that the odds for an event, O(E), are the ratio of the probability of the event's happening to the probability of the event's not happening. Unlike probabilities, odds are often written in mathematical notation as P(E):(1 − P(E)).

14. The probability of rolling two dice and obtaining a sum of 7 is $\frac{1}{6}$. What are the odds for doing this?

15. If the odds for an event are known, then that event's probability can be determined. All that is missing is the denominator of the probability fraction. Explain how to find that denominator from a given odds ratio.

16. Unlike probabilities, odds may be greater than 1. Under what circumstances are the odds for an event greater than 1?

17. What do you think is meant by betting in such a way as to "even the odds" or "beat the odds?"

18. The odds for flipping a head on an unfair coin are 5:3. That is,

$$O(\text{Head}) = \frac{5}{3}.$$

   a. What is the probability of flipping a head?

   b. What are the odds *against* flipping a head?

   Hint: *For any event E in which O(E) ≠ 0, it will always be true that (Odds for E) · (Odds against E) = 1.*

19. A **trifecta box** is a bet that picks the first three finishers in a horse race; they do not have to be in correct order.

   a. How many trifecta boxes are possible in an eight-horse race?

   b. What are the odds of winning a trifecta box in an eight-horse race if the bet is made at random and every horse is equally likely to win?

20. An **across-the-board bet** in a horse race is a bet that a horse finishes in one of the first three positions. What are the odds of winning an across-the-board bet in a six-horse race if the bet is made at random and every horse is equally likely to win?

5B.2   PERMUTATIONS OF NONDISTINCT OBJECTS

The number of ways of arranging all n objects of a set of size n has been shown to be $_nP_n$, or $n!$ For example, the number of ways of permuting all four letters (objects) of the word MATH is 4!, or 24:

MATH	MAHT	MTHA	MTAH	MHAT	MHTA
AHMT	AHTM	ATMH	ATHM	AMHT	AMTH
THMA	THAM	TAMH	TAHM	TMHA	TMAH
HMTA	HMAT	HTMA	HTAM	HATM	HAMT

However, if some of these objects are not distinct, then the total number of distinguishably different permutations (permutations that look different from one another) is less. Consider all permutations of the letters in the word STAT that look different:

STAT	STTA	SATT			
TATS	TAST	TTAS	TTSA	TSAT	TSTA
ASTT	ATST	ATTS			

It is not difficult to develop a formula for the number of distinguishably different permutations. If all n objects did look different, then there would be a total of $n!$ distinguishably different permutations. A subset of size r of these objects can be arranged $r!$ ways and thus contributes a factor of $r!$ to this total. Therefore, if r objects all look the same, then the number of distinguishably different permutations must equal the number that would have been possible, divided by $r!$.

In the example of permutations of letters in the word STAT, $n = 4$ and $r = 2$. Thus, the number of distinguishably different permutations is

$$\frac{n!}{r!} = \frac{4!}{2!}$$

$$= 12.$$

By application of the FCP, the following formula can be stated:

As a general rule, if n objects include r_1 objects of type 1, r_2 of type 2, r_3 of type 3, . . . , r_k of type k, then

Number of distinguishably different permutations

$$= \frac{n!}{r_1! \cdot r_2! \cdot r_3! \cdot \cdots \cdot r_k!}. \qquad \text{(5B-3)}$$

E x a m p l e 7

Point B is 7 blocks north and 4 blocks east of point A. How many different paths are there from point A to point B? Movement must be directly north or east in 1-block steps.

Solution The following map shows one possible path.

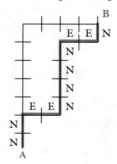

This path, NNEENNNNEEN, can be thought of as one possible permutation of seven N's and four E's. By Formula 5B-3, just developed, there are

$$\frac{11!}{7! \cdot 4!} = 330$$

possible permutations of these letters and therefore 330 possible paths.

Exercises 5B.2

1. Find the number of distinct permutations of the letters in the following words:

   a. MOM

   b. STATS

   c. STATISTICS

2. Using only the digits 1 and 2, how many different four-digit numbers can be formed that have:

   a. Two 1's and two 2's?

   b. At least one 1?

3. Some ships at sea use flag signals—arrangements of flags on a vertical flagpole—to communicate with one another. How many different signals can be displayed using eight flags that are identical except for color, if four are white, two are green, one is red, and one is blue? (Assume that all signals are read from the top down.)

GROUP * 4. Refer to Exercise 3. How many signals can be displayed if either seven or eight flags may be used?

   Hint: Consider several different cases.

GROUP   5. Point B is 5 blocks east and 6 blocks north of point A. Point C is 8 blocks north and 4 blocks east of point B.

   a. How many possible paths are there from point A to point C?

   b. What is the probability of choosing a path from point A to point C that passes through point B?

   c. What is the probability of walking from point A to point C and then back to point A by passing through point B twice? (Assume that all paths in both directions are equally likely.)

5B.3 CONDITIONAL PROBABILITY

Many (though not all) of the probabilities studied thus far have been **noncondi- tional;** that is, they were probabilities of independent events, and the likelihoods of those events were not linked. Many events, however, are not independent. In order to proceed further with the study of probability, the important notion of **conditional probability** demands close examination.

As discussed in Chapter 5A, the mathematical notation $P(A \mid B)$ is read "the probability of A's occurring given that B has occurred or is guaranteed to occur."[2] In some instances, this may or may not be the same as $P(A)$. If $P(A \mid B) = P(A)$, then A and B are said to be independent of each other, and it will always follow that $P(B \mid A) = P(B)$, as well.

Consider the following cross-tabs table of starts and failures for two types of businesses.[3]

Business Starts and Failures

	Business Starts, S	Business Failures, F	Total
Auto repair and services, A	5997	1335	7332
Legal services, L	1296	191	1487
Total	7293	1526	8819

If one of the 8819 businesses is randomly selected, the probability that it is a legal service is $P(L) = \frac{1487}{8819}$, or about 17%. On the other hand, the conditional probability that the selected business is a legal service given that it is a business failure is $P(L \mid F) = \frac{191}{1526}$, or about 13%. Note that in this conditional probability the business is randomly selected not from the sample of 8819 businesses but rather from a part of that sample, the 1526 that are failures. Also, only 191 (instead of 1487) businesses meet the condition of being a legal service *and* a business failure. The Venn diagram in Figure 5B-2 illustrates these ideas.

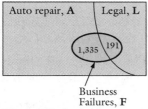

Auto repair, **A**   Legal, **L**

1,335   191

Business Failures, **F**

Figure 5B-2

Notice that the conditional probability, $P(L \mid F)$, was computed by dividing the number of businesses in the intersection of sets L and F by the number of businesses in F. This is different from $P(L \cap F)$, which has the same numerator but a different denominator. The denominator of $P(L \cap F)$ is the total number of businesses in the survey. Thus, $P(L \cap F) = \frac{191}{8819}$, and $P(L \mid F) = \frac{191}{1526}$.

A probability such as $P(L \cap F)$ in this illustration is called a joint probability.

Definitions ▶   A **joint probability** is the probability of two (or more) events' occurring.

A cross-tabs table cell that shows the frequency of the intersection of two events is called a **joint event cell.**

On the other hand, the probability that a randomly selected business is a failure is $\frac{1526}{8819}$ and is referred to as a marginal probability.

Definitions ▶   A **marginal probability** is the probability of an event that is made up of several parts.

A cross-tabs table cell that shows the sum of the frequencies of a row or column is called a **marginal event cell.**

[2]This can be read more simply as "the probability of A given B."

[3]Based on data from *Statistical Abstract of the United States, 1991*, p. 537.

In a cross-tabs table, a marginal probability is the sum of any row or column, divided by the grand total.

As mentioned earlier, the conditional probability that a randomly selected business is a legal service, given that it is a failure, is $\frac{191}{1526}$. Note that

$$\frac{191}{1526} = \frac{\dfrac{191}{8819}}{\dfrac{1526}{8819}}.$$

In other words.

$$\text{Conditional probability} = \frac{\text{Joint probability}}{\text{Marginal probability}}.$$

The formula for computing conditional probabilities from a cross-tabs table is simply

$$\frac{\text{Frequency in joint event cell}}{\text{Frequency in marginal event cell}}.$$

If a cross-tabs table is not used to determine probabilities, then recall from Chapter 5A that a conditional probability may be determined from the formula $P(A \mid B) = \frac{P(A \cap B)}{P(B)}$. (This formula was developed in the last chapter by rearranging the terms in the general multiplication rule.)

5B.4   FURTHER WORK WITH PROBABILITY TREES

A graphic display called a **probability tree diagram** can illustrate marginal and joint probabilities and can make it possible to compute conditional probabilities. Consider the following cross-tabs table, which compares data on the methods of getting to work of the 500 employees of a company.

Transportation Survey

	Driving, *D*	Public Transportation, *PT*	Total
Under 35	100	50	150
35 and older	125	225	350
Total	225	275	500

A probability tree can illustrate these data more graphically. To make the tree, it is first necessary to convert tabled frequencies to probabilities. The frequency in each cell is divided by the grand total, 500.

Transportation Survey

	Driving, D	Public Transportation, PT	Total
Under 35	0.20	0.10	0.30
35 and older	0.25	0.45	0.70
Total	0.45	0.55	1.00

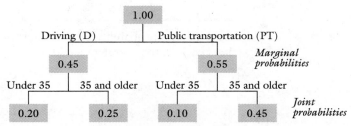

Figure 5B-3 A probability tree for the transportation survey.

The probability tree can now be constructed as shown in Figure 5B-3. Notice that the first row in Figure 5B-3 contains marginal probabilities and the bottom row contains joint probabilities.

The choice of placing the marginal probabilities, P(Drive) and P(Public transportation), in the first row instead of the other marginal probabilities, P(Under 35) and P(35 and older), was arbitrary. Using different marginal probabilities in the first row would have resulted in a slightly different version of this tree diagram. Regardless of how the versions of a probability tree are started, all contain equivalent amounts of information.

Probabilities not explicitly shown, such as the other marginal probabilities and all conditional probabilities, can be determined from this tree diagram. The other marginal probabilities can be found easily, since they are the sums of joint probabilities. For instance, Figure 5B-4 shows the joint probabilities that add to the marginal

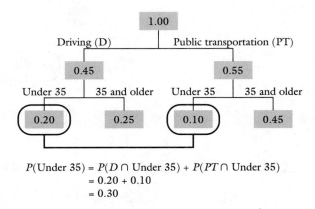

Figure 5B-4

probability $P(\text{Under } 35)$. The conditional probability $P(\text{Under } 35 \mid D)$ can be found by substituting known values into the following relationship:

$$P(\text{Under } 35 \mid D) = \frac{P(D \cap \text{Under } 35)}{P(D)}.$$

$$= \frac{.20}{.45}$$

$$\approx 0.44.$$

E x a m p l e
8

The following probability tree contains data from a survey in which children and adults each stated a preference for one of two brands of cereal.

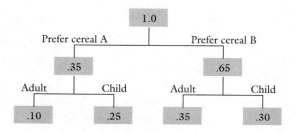

a. Find the percentage of respondents who were children.
b. Find the conditional probability $P(\text{Prefer cereal A} \mid \text{Child})$.
c. Find the conditional probability $P(\text{Child} \mid \text{Prefer cereal A})$.

Solution   a. The percentage of respondents who were children is the same as the probability that a child was surveyed. $P(\text{Child})$ is a marginal probability and can be found by adding the joint probabilities indicated on the following tree diagram.

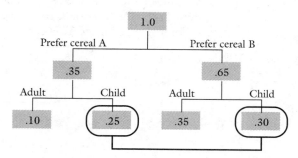

$P(\text{Child}) = P(\text{Prefer cereal A} \cap \text{Child}) + P(\text{Prefer cereal B} \cap \text{Child})$
$= .25 + .30$
$= .55$

Thus, 55% of the respondents were children.

b. $$P(\text{Prefer cereal A} \mid \text{Child}) = \frac{P(\text{Prefer cereal A} \cap \text{Child})}{P(\text{Child})}.$$

Substituting the given joint probability, $P(\text{Child} \cap \text{Prefer cereal A})$, and the marginal probability found from Part a,

$$P(\text{Prefer cereal A} \mid \text{Child}) = \frac{.25}{.55}$$

$$\approx .45.$$

c. $$P(\text{Child} \mid \text{Prefer cereal A}) = \frac{P(\text{Child} \cap \text{Prefer cereal A})}{P(\text{Prefer cereal A})}.$$

Using the known values,

$$P(\text{Child} \mid \text{Prefer cereal A}) = \frac{.25}{.35}$$

$$\approx .71. \quad \bullet$$

Coherency of Estimated Probabilities

A probability tree can be used to conveniently find errors in estimated or intuitive probabilities. Failure of joint probabilities to add to corresponding marginal probabilities indicates subtle mistakes in reasoning. Such probability estimates are said to lack **coherency.**

As an example, suppose that a Chicago White Sox baseball fan believes that Frank Thomas (FT) will stay healthy through the baseball season with probability .90.

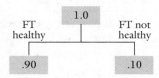

Furthermore, the fan believes that the White Sox have an 85% chance of winning the pennant whether Frank Thomas is healthy or not, but if Frank Thomas is healthy, the probability of the White Sox winning the pennant increases to 95%. These intuitive probabilities are now examined for coherency.

From the marginal probability

$$P(\text{Win pennant}) = .85$$

and the conditional probability

$$P(\text{Win pennant} \mid \text{FT healthy}) = .95,$$

it is possible to compute the joint probability

$$P(\text{FT healthy} \cap \text{Win}) = P(\text{FT healthy}) \cdot P(\text{Win} \mid \text{FT healthy})$$
$$= (.90) \cdot (.95)$$
$$= .855.$$

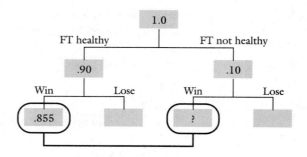

Figure 5B-5

Figure 5B-5 shows how the probability tree can be filled in further. The circled joint probabilities must add to the marginal probability of winning the pennant. That is,

$$P(\text{Win}) = P(\text{FT healthy} \cap \text{Win}) + P(\text{FT not healthy} \cap \text{Win})$$
$$.85 = .855 + P(\text{FT not healthy} \cap \text{Win})$$
$$-.005 = P(\text{FT not healthy} \cap \text{Win})$$

Since probabilities can never be negative, the intuitive probabilities of this fan cannot be correct and are said to lack coherency.

Exercises 5B.3–5B.4

1. When joint and marginal probabilities are determined from a cross-tabs table, what will the denominator always be? What will the denominator always be for conditional probabilities?

2. Refer to the cross-tabs table of the transportation survey on page 224. Find the following probabilities.
    a. $P(D \mid \text{Under 35})$
    b. $P(PT \mid \text{35 and older})$
    c. $P(\text{35 and older})$
    d. $P(D \cap \text{35 and older})$

3. The following cross-tabs table compares the numbers of certain professional degrees conferred in the United States in 1990.

Professional Degrees Conferred

	M.D.	D.D.S.	LL.B. or J.D.	Total
Men	9977	2830	21,059	33,866
Women	5138	1263	15,378	21,779
Total	15,115	4093	36,437	55,645

In relation to the data in this table, find the following probabilities.

a. P(M.D. | Woman)

b. P(Woman | M.D.)

c. P(LL.B. or J.D.)

d. P(Man | M.D. or D.D.S.)

e. P(Man)

f. P(M.D. or D.D.S. | Woman)

g. P(LL.B. or J.D. | Woman)

h. P(Woman | LL.B. or J.D.)

Based on data from *Statistical Abstract of the United States, 1993*, p. 185.

4. Refer again to the cross-tabs table of the transportation survey on page 224. Draw a probability tree diagram for those data in which the indicated marginal probabilities refer to the ages of the respondents instead of the methods of transportation used.

5. Refer to Exercise 3 about numbers of professional degrees conferred. Finish the following probability tree for those data. Express all probabilities as decimals rounded to the nearest thousandth.

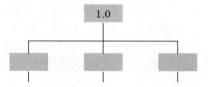

GROUP

6. The following probability tree indicates consumer preference for three types of running shoes—brands A, B, and C—broken down by gender.

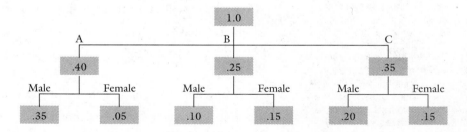

a. Find the marginal probabilities not indicated, P(Male) and P(Female), which are the percentages of males and females in the survey.

b. Find P(Preferred by a male | Brand A).

c. Find P(Brand C | Preferred by a female).

    d. Redraw the probability tree so that the marginal probabilities shown are the percentages of males and females in the survey. Round all decimals to the nearest hundredth.

    e. Using the redrawn probability tree from Part d, determine which conditional probability is equal to $\frac{3}{5}$.

GROUP

7. Refer to Exercise 6. Two tree diagrams are possible for the data: the given one and the redrawn one in Part d. How would one be more useful than the other?

8. It is believed that 80% of the women who use home pregnancy tests are indeed pregnant. A new home pregnancy test has been developed that claims to give correct results 90% of the time. Research showed that this test also indicates a pregnancy 95% of the time, regardless of the woman's true condition. Draw a probability tree to determine whether these probability claims are coherent.

9. It is believed that the probability of inflation next year is 70%. The probability that gasoline will cost more than $1.50 per gallon if inflation does not occur is estimated at 25%. However, some economists are 80% certain that gasoline prices will exceed $1.50 per gallon regardless of inflation. Draw a probability tree to determine whether these estimated probabilities are coherent.

* 10. Refer to Exercise 9. If the estimates are not coherent, then is the stated conditional probability, $P(\text{Gas} > \$1.50 \mid \text{No inflation})$, too high or too low? Explain.

5B.5 BAYES'S THEOREM

Section 5B.4 showed how a marginal probability can be determined by adding two joint probabilities. The relationship between the sums of probabilities determined by intersections is now explored a bit further.

    Consider a small office that has three photocopy machines.

    Machine 1 makes 60% of all copies.
    Machine 2 makes 30% of all copies.
    Machine 3 makes 10% of all copies.

The three machines do not all perform at the same level of quality.

    Machine 1 produces 10% of its copies defectively.
    Machine 2 produces 20% of its copies defectively.
    Machine 3 produces 40% of its copies defectively.

What percentage of all copies are produced defectively?

    In terms of probability, the answer to this question is the same as the probability of using one of the copy machines and producing a defective copy. If M_1, M_2, and

M_3 stand for the copy machines, and D represents the event of producing a defective copy, the given information can be summarized as follows:

$$P(M_1) = .60 \qquad P(M_2) = .30 \qquad P(M_3) = .10,$$
$$P(D \mid M_1) = .10 \qquad P(D \mid M_2) = .20 \qquad P(D \mid M_3) = .40.$$

The Venn diagram in Figure 5B-6 illustrates this information. Each of the events A, B, and C is the intersection of two events. For example, event A is the intersection of the event of machine 1's being used and the event of the production of a defective copy. That is,

$$A = M_1 \cap D, \qquad B = M_2 \cap D, \qquad C = M_3 \cap D.$$

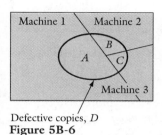

Defective copies, D
Figure 5B-6

The production of a defective copy is analogous to hitting the shaded area in the "dart board" of Figure 5B-6. The probability of hitting the shaded area can be determined by finding the probability of hitting each of the smaller regions—A, B, and C—by the general multiplication rule, then adding those probabilities together. For instance,

$$
\begin{aligned}
P(A) &= P(M_1 \cap D) \\
&= P(M_1) \cdot P(D \mid M_1) \\
&= (.60) \cdot (.10) \\
&= .06.
\end{aligned}
$$

In a similar way, $P(B) = (.30) \cdot (.20)$, or .06, and $P(C) = (.10) \cdot (.40)$, or .04. Thus,

$$
\begin{aligned}
P(D) &= P(A) + P(B) + P(C) \\
&= .06 + .06 + .04 \\
&= .16.
\end{aligned}
$$

That is to say, 16% of all copies made are defective, or the probability of choosing a defective copy from a stack of all the copies made in a day is $P(D) = .16$.

In general, if an event can be broken down into n nonoverlapping parts, its probability can be found as the sum of n joint probabilities. Consider the Venn diagram shown in Figure 5B-7.

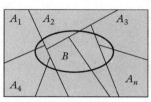

Figure 5B-7

$$
\begin{aligned}
P(B) &= P(A_1 \cap B) + P(A_2 \cap B) + P(A_3 \cap B) + \cdots + P(A_n \cap B) \\
&= P(A_1) \cdot P(B \mid A_1) + P(A_2) \cdot P(B \mid A_2) + P(A_3) \\
&\quad \cdot P(B \mid A_3) + \cdots + P(A_n) \cdot P(B \mid A_n).
\end{aligned}
$$

In sigma notation, this becomes

$$
P(B) = \sum_{i=1}^{n} P(A_i \cap B)
$$

$$
= \sum_{i=1}^{n} P(A_i) \cdot P(B \mid A_i).
$$

These ideas can now be formalized as a theorem.

Theorem ▶ | **Bayes's Theorem**

$$P(A \mid B) = \frac{P(A \cap B)}{\displaystyle\sum_{i=1}^{n} P(A_i) \cdot P(B \mid A_i)}. \qquad (5B\text{-}4)$$

Notice that the denominator of this fraction is $P(B)$, and the equation itself comes from the formula for conditional probability.

E x a m p l e
9

Suppose that a blood test for a certain disease is 95% accurate and that 3% of the population at large has this disease. A person is chosen *at random* from the population, and a blood test indicates that he or she has the disease. What is the probability that the person does, in fact, have the disease?

Solution   The symbol " $+$ " will denote a positive test result (indicating the presence of the disease), and " $-$ " will denote a negative test result. The fact that the test is 95% **accurate** means that $P(+ \mid \text{Disease}) = .95$ and $P(- \mid \text{Not have the disease}) = .95$. The conditional probability, $P(\text{Disease} \mid +)$, is to be determined.

By the formula for conditional probability,

$$P(\text{Disease} \mid +) = \frac{P(\text{Disease} \cap +)}{P(+)}.$$

Using the general multiplication rule, the numerator can be replaced with $P(\text{Disease}) \cdot P(+ \mid \text{Disease})$ to become

$$P(\text{Disease} \mid +) = \frac{P(\text{Disease}) \cdot P(+ \mid \text{Disease})}{P(+)}.$$

The numerator can be computed as

$$P(\text{Disease}) \cdot P(+ \mid \text{Disease}) = (.03) \cdot (.95)$$
$$= .0285.$$

Now all that is missing is the denominator, $P(+)$. To determine it, consider the following probability tree.

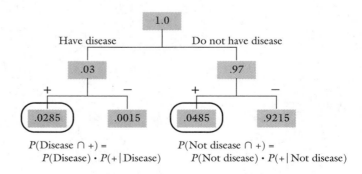

$P(\text{Disease} \cap +) =$ $P(\text{Not disease} \cap +) =$
   $P(\text{Disease}) \cdot P(+ \mid \text{Disease})$ $P(\text{Not disease}) \cdot P(+ \mid \text{Not disease})$

As this tree indicates, there are two ways of getting a positive (+) result on the test—one that correctly identifies the disease, with probability $(.03) \cdot (.95) = .0285$, and the other that incorrectly identifies the presence of the disease, with probability $(.97) \cdot (.05) = .0485$.

All that remains is to substitute into the formula

$$P(\text{Disease} \mid +) = \frac{P(\text{Disease}) \cdot P(+ \mid \text{Disease})}{P(+)}$$

$$= \frac{.0285}{.0285 + .0485}$$

$$= \frac{.0285}{.077}$$

$$\approx .370. \quad \bullet$$

In other words, there is only a 37% chance that a randomly selected person who tests positive actually has the disease. This surprising result dramatizes the dilemma of a **false positive,** which is one of the reasons tests are often repeated many times and additional information gathered before the formation of any definitive conclusions. Figure 5B-8 illustrates why the dilemma occurs in tests of a population of 200 million people.

As can be seen from Figure 5B-8, if people are chosen at random to take the test, many more people test positive who do not have the disease than test positive who do have the disease. Note that the conditional probability found, .370, is identical to the fraction $\dfrac{5700000}{5700000 + 9700000}.$

The difficulty of false positives is one of the arguments against testing for AIDS on a more random basis.

The number of false positives can be very significant if people are selected at random for the testing of a characteristic that affects only a very small part of the population.

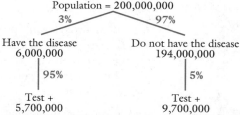

Figure 5B-8 The distribution of positive test results in a population.

Exercises 5B.5

1. Refer to the example of the three photocopy machines on page 230. Explain how the probability of .16 may be interpreted as a weighted mean probability.

2. Refer again to the case of the three photocopy machines on page 230. Assume that a total of 5000 copies are made. Fill in the following table to verify that $P(D) = .16$.

Good and Defective Photocopies by Machine

	Copies per Machine	Good	Defective, D
Machine 1	?	?	?
Machine 2	?	?	?
Machine 3	?	?	?
Total	5000	?	?

3. Refer to Example 9 on page 232. Explain why randomness in the selection of a patient is such an important factor to consider when determining the probability of a false positive.

4. Refer again to Example 9 on page 232. Would the probability of a false positive be lessened more by a test that is 98% (instead of 95%) accurate or by the knowledge that, instead of 3%, the disease affects 10% of the population? Rework the example with the new percentages to test your belief.

*   5. A chest of drawers contains three identical drawers. One drawer contains three gold coins, another contains two gold coins and two silver coins, and another contains one gold and two silver coins. A drawer is chosen at random, and a coin from the drawer is chosen at random. What is the probability that the coin chosen is gold?

   Hint: $P(G) = P(D_1) \cdot P(G \mid D_1) + P(D_2) \cdot P(G \mid D_2) + P(D_3) \cdot P(G \mid D_3)$.

*   6. A chest of drawers contains three identical drawers. One drawer contains two gold coins, and each of the other two drawers contains one gold coin and one silver coin. A drawer is chosen at random, and a coin from that drawer is chosen at random. If a gold coin is chosen, what is the probability that the other coin in the drawer is also gold?

   Hint: *Find* $P(G \mid G) = \dfrac{P(G \cap G)}{P(G)}$, *and the answer may surprise you.*

7. A student took a multiple-choice test with five choices per answer. Assume that the student knew the correct answer to 70% of the questions and guessed on the rest of them. If an answer is returned marked correct, what is the probability that the student really knew the answer, given each of the following conditions?

   a. On the questions that the student did not know, he or she guessed completely at random.

   b. On the questions the student did not know, he or she was able to narrow each answer down to three choices.

   c. Explain how the results of Parts a and b are related.

GROUP   8. Seven people, identified as A through G, read and correct a standardized test. One and only one of these people corrects each test that arrives. The percentage of the tests corrected by each one is proportional to his or her share of the following rectangle. Assume that points V and Z are midpoints and that points W, X, and Y form equal divisions.

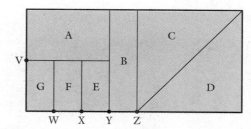

Experience has shown that the probability of reader A's making a mistake is $\frac{1}{10}$. All other readers have a probability of $\frac{1}{20}$ of making a mistake. Determine the following.

a. The percentage of tests corrected by each of the seven people; that is, $P(A), P(B), \ldots, P(G)$

b. The probability that a randomly selected test was corrected by reader B and graded correctly; that is, $P(B \cap \text{Correct})$

c. The probability that any test, chosen at random, is graded correctly; that is, $P(\text{Correct})$

d. $P(D \mid \text{Correct})$

e. $P(\text{Not graded by } B \mid \text{Correct})$

f. $P(D \mid \text{Not graded correctly})$

9. If a person is vaccinated properly, the probability of his or her getting a certain disease is .05. Without a vaccination, the probability of getting the disease is .35. Assume that $\frac{1}{3}$ of the population is properly vaccinated.

a. If a person is randomly selected from the population, what is the probability of that person's getting the disease?

b. If a person gets the disease, what is the probability that he or she was vaccinated?

10. A certain home pregnancy test claims to be 95% accurate. It is estimated that one-third of all women who take this test at home are indeed pregnant. Find the probability that a woman is pregnant if the test indicates that she is.

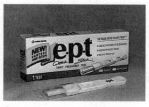

What is the probability that a woman is indeed pregnant if a home pregnancy test indicates a positive result? (*George Semple*)

11. The accuracy of some lie detector tests ranges from 92% to 99%. Furthermore, it is "believed" to be a fact that 15% of the employees of a large automobile manufacturer would lie if asked the brand of car they prefer. An employee is chosen at random and asked for his or her car preference while the lie detector test is administered. If the lie detector indicates that the employee is lying, determine the probability that he is indeed lying, if:

a. The test is 92% accurate.

b. The test is 99% accurate.

12. Refer to Exercise 11. Briefly explain why you think some courts of law do not admit the results of lie detector tests as evidence.

Case Study   *Mistaken Identity?*

Trials are sometimes conducted without clear evidence to acquit an innocent party. (*Courtesy of David Sherman*)

Chapter 5A briefly described a famous case of mistaken identity in which an interracial couple was charged with robbery. The couple was convicted chiefly by the prosecutor's convincing argument, which involved the laws of probability. However, the conviction was later overturned when the prosecutor's mathematics was demonstrated to be incorrect. Information about this case and its appeal can be found in "The Laws of Probability," *Time*, January 8, 1965, page 42, and "Trial by Mathematics," *Time*, April 26, 1968, page 41. Obtain copies of those articles from your library, and use them to answer the following questions.

I. Both articles mention "statistical probability." Do you think that is the same as probability? Explain your answer.
II. The 1965 article discusses the odds against certain events' occurring. In this context the word "odds" is not used correctly. Explain why.
III. How did the public defender argue against the prosecutor's claim that the odds were one to 12 million that another couple fitting the defendants' descriptions could have been at the scene of the crime.
IV. Refer to the 1968 article, and explain the role of conditional probability in the overturning of the verdict.
V. What were the appeal judge's reasons for reversing the conviction?

CHAPTER SUMMARY

Permutations and combinations are the bases of sophisticated methods of counting that arise from applications of the Fundamental Counting Principle, first covered in Chapter 5A. A permutation is an arrangement of r objects chosen from n objects in which order is important. The number of permutations possible is denoted by $_nP_r$. A combination is a collection of r objects chosen from n objects in which the order of choice does not matter. The number of possible combinations is denoted by $_nC_r$. Permutations and combinations are important in many special counting situations and are always formed without replacement. The relationship between permutations and combinations is $_nP_r = (r!) \cdot {}_nC_r$.

Conditional, joint, and marginal probabilities can be found from a cross-tabs table. A probability tree diagram shows marginal and conditional probabilities. It may be drawn in different ways, depending upon the probabilities one wishes to emphasize.

A probability may be determined as the sum of two or more joint probabilities by repeated use of the general rule for the multiplication of probabilities. In sigma notation this becomes

$$P(B) = \sum_{i=1}^{n} P(A_i \cap B) = \sum_{i=1}^{n} P(A_i) \cdot P(B \mid A_i).$$

Bayes's Theorem, which has many practical applications, states that

$$P(A \mid B) = \frac{P(A \cap B)}{\sum_{i=1}^{n} P(A_i) \cdot P(B \mid A_i)}.$$

One example of a false positive is a test that incorrectly indicates the presence of a disease in an individual. Probabilities associated with Bayes's Theorem can determine a false positive.

Key Concepts, Terms, and Formulas

Permutation: $_nP_r = \dfrac{n!}{(n-r)!}$ (5B-1)   *Page 212*

Combination: $_nC_r = \dfrac{n!}{(n-r)! \cdot r!}$ (5B-2)   *Pages 215–216*

Odds: $O(E) = \dfrac{P(E)}{1 - P(E)}$ *Page 219*

Number of permutations of nondistinct objects

$$= \frac{n!}{r_1! \cdot r_2! \cdot r_3! \cdot \cdots \cdot r_k!} \quad \text{(5B-3)}$$

Page 221

Joint, marginal, and conditional probabilities   *Pages 222–223*
Probability trees   *Page 224*
Coherency of estimated probabilities   *Page 227*
Bayes's Theorem:

$$P(A \mid B) = \frac{P(A \cap B)}{\sum_{i=1}^{n} P(A_i) \cdot P(B \mid A_i)} \quad \text{(5B-4)}$$

Page 230

Videotape Suggestions

Program 16 in the series *Against All Odds: Inside Statistics,* suggested in Chapter 5A, may be viewed again for a fuller understanding of the theory of independence and the multiplication rule for independent events. *Program 18* in the same series presents stories about gambling casinos. Both programs are previews of future work with random variables and the Central Limit Theorem.

(Picture Perfect ™)

Trading success on the
American Stock
Exchange hinges on the
ability to predict the
behavior of rising and
falling stocks.

Chapter

6 Random Variables and Probability Distributions

Probability often serves as a tool for the statistical examination of everyday events. Statisticians often determine theoretical probabilities from the statistics of past events and then use those probabilities to construct graphs called probability distributions. The graphs help statisticians examine and predict a wide variety of random occurrences. The case study at the end of this chapter concerns a special application of this technique with some of its uses, ranging from predicting stock market behavior to forecasting the weather.

The following discussion of statistics about traffic citations will introduce many of the important concepts of this chapter.

6.1 DISCRETE RANDOM VARIABLES

A group for the betterment of government wishes to determine the cost of a typical traffic citation in a certain community. It will then use that information for comparison with other communities and to project future revenues. After choosing a random sample of $n = 137$ citations from the past several years, the group compiles the information in the following table.

Revenue from Traffic Citations

Nature of Citation	Number Issued	Monetary Fine, F (dollars)
Parking	38	15
Speeding	22	125
Reckless driving	15	80
Drunken driving	8	200
Driving with expired plates	54	25
	$n = 137$	

239

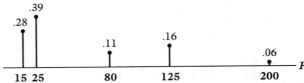

Figure 6-1 The probability density function for the discrete random variable *F.*

The category "Monetary Fine," or *F*, is a variable that can take five different values, depending upon the nature of the citation. For example, *F* = $15 if the citation is for parking, and the probability that *F* = $15 depends upon the probability of choosing a parking citation from the sample. Accordingly, *F* is considered to be a discrete random variable.

Definitions ▶

> A **random variable** is a variable whose values are determined by some type of probability function.
>
> A **discrete random variable** is a random variable that has a countable number of different possible values.
>
> A **statistical probability** is a probability determined by the relative frequency of past events.

S t a t   F a c t

For a week in April 1995, *The Rainmaker* by John Grisham sold 10 copies for every 1 copy sold of *The Hot Zone* by Richard Preston. A mathematician might say the statistical probability that someone bought a copy of *The Rainmaker* that week was 10 times that of *The Hot Zone.*

In the traffic citation study, 38 of the 137 citations sampled were for parking. Thus, the statistical probability[1] of choosing a parking citation from the sample is $P(\text{Parking}) = \frac{38}{137} \approx .28$. Notice that this means $P(F = \$15) = .28$.

A graph showing each possible value of the variable and the associated likelihood of its occurrence can effectively illustrate the values of a discrete random variable. Such a graph is called a **probability density function** or **probability distribution** of the discrete random variable. Figure 6-1 shows the probability density function for the monetary fine variable *F.* Notice that the boldface numbers along the horizontal axis in Figure 6-1 are the possible values of *F.* The height of each line represents the probability (based on sample frequencies) that *F* could be that particular value. As in the determination of *P*(Parking) earlier, each probability is determined by the fraction

$$\frac{\text{Number of citations in the category}}{\text{Total number of citations sampled}}.$$

The Mean and Standard Deviation of a Discrete Random Variable

Earlier chapters showed that the mean and standard deviation of a set of data are useful in describing the data and can be employed to compare one set of data with another. It is also possible to compute the mean and standard deviation of a discrete

[1]The terms "statistical probability" and "probability" will be used interchangeably.

random variable. The **mean value** of a discrete random variable, also called its **average value** or its **expected value,** is the weighted mean of all the possible values of the variable. The probabilities of the values of the random variable are used to weight the mean.

Continue to consider the discrete random variable F in relation to traffic citations. As its probability density function indicates, this variable can take any of five different values with different probabilities. For instance,

$$P(F = 80) = P(\text{Reckless driving})$$
$$= \frac{15}{137}$$
$$\approx .11.$$

For this traffic citation study, the average, mean, or expected value of F, denoted by μ_F, is found by the formula

$$\mu_F = 15 \cdot (.28) + 25 \cdot (.39) + 80 \cdot (.11) + 125 \cdot (.16) + 200 \cdot (.06)$$
$$= \$54.75.$$

Notice that this is the mean of all the values that are possible for F. It is not necessarily an obtainable value for F.

The mean value can also be interpreted as the balance point of the probability density function of F as shown in Figure 6-2.

The general formulas for the mean value and standard deviation of a discrete random variable, X, which could have the values $x_1, x_2, x_3, \ldots, x_n$, are as follows.

The **Mean** and **Standard Deviation** of a Discrete Random Variable, X

If X is a discrete random variable, then

$$\mu_X = \sum_{i=1}^{n} x_i \cdot P(X = x_i). \qquad (6\text{-}1)$$

$$\sigma_X = \sqrt{\sum_{i=1}^{n} (x_i - \mu_X)^2 \cdot P(X = x_i)}. \qquad (6\text{-}2)$$

The probability of the occurrence of any value other than 15, 25, 80, 125, and 200 is zero, as demonstrated by the fact that the only line segments drawn on the probability density function are at these five values.

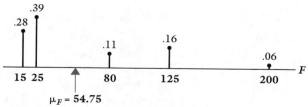

Figure 6-2 The balance point of a probability density function.

On a yearly average, an insurance company for a certain type of medical malpractice knows that a jury could choose three different monetary awards: $100,000 with probability .02, $250,000 with probability .015, and $1,000,000 with probability .005. The only other possibility is that the jury might fail to convict the defendant and thus make no award. What is the minimum annual premium that the insurance company should charge just to break even on all policies of this type?

Solution   The probability that no money will be awarded is

$$1 - [.02 + .015 + .005] = .96.$$

If the discrete random variable A represents the monetary award, then the expected value of A for a given year is its mean value, which can be found as follows:

$$\mu_A = \$100,000 \cdot (.02) + \$250,000 \cdot (.015) + \$1,000,000 \cdot (.005) + \$0 \cdot (.96)$$
$$= \$10,750.$$

Thus, the insurance company would need to charge at least $10,750 each year for each policy of this type just to break even.   •

E x e r c i s e s 6 . 1

1. Refer to the discussion of the discrete random variable F on pages 239–40. Do the data indicate that 28% of the drivers in that community receive parking citations? Explain.

2. Refer again to the random variable F. Make the following two calculations.

   a. (Mean value of F) · (Total number of observations)

   b. Σ ((Individual values of F) · (Number of individual observations))

   Why don't these two calculations have exactly the same value?

3. Refer to Example 1 above. If the insurance company needs to make a profit of 15% on each policy in order to cover its expenses, what should be the company's minimum annual premium?

4. For a certain neighborhood in the city, the probability that a home will be totally destroyed by fire in a given year is .001, and the probability that a 50% fire loss will be suffered is .01.

   a. Ignoring all other possible partial losses, what annual mean payout should an insurance company expect on a $150,000 fire insurance policy?

   b. It is determined that the insurance premium should cover the expected losses due to total damages and 50% damages, plus an additional 10% of those expected losses. What should be the minimum premium on a fire insurance policy of $275,000?

5. A large insurance company has kept yearly records and sampled 750 policies at random for a given year. Ninety-two holders of policies in the sample filed claims that required payouts of $10,000. Furthermore, 68 policies required payouts of $25,000, 51 required payouts of $60,000, and 17 required payouts of $90,000. No other policies required payouts.

   a. Draw the probability density function for the discrete random variable M, which represents the amount of money paid out.

   b. What is the mean value, μ_M, of the yearly payout?

6.2 THE BINOMIAL RANDOM VARIABLE

The binomial random variable is a special discrete random variable with many important applications. However, to understand exactly what a binomial random variable is, one must first understand the definition of a Bernoulli process.

Definition ▶

An experiment is termed a **Bernoulli process** if it defines a random variable in such a way that

   a. only one of two outcomes is possible and
   b. the outcome at each trial is independent of the outcome at any previous trial.

Stat Fact

Jacob Bernoulli (1654–1705) was one of eight world-class mathematicians produced by the Bernoulli family in just three generations. His book on the theory of probability, written in Latin and published posthumously in 1713, was the first such book by a mathematician.

This definition implies that the experiment of rolling a die, for example, is not a Bernoulli process because more than two outcomes are possible. However, the sex of a newborn child is considered a Bernoulli process because only one of two outcomes (boy or girl) is possible, and the outcome of a previous birth does not affect future outcomes.

Definition ▶

A **binomial random variable** is a discrete random variable that keeps score of the number of successes in a Bernoulli process.

Which outcome is successful in a Bernoulli process is arbitrarily decided; then a binomial random variable, X, must meet the following three conditions, which follow from the fact that X represents the number of successes in a Bernoulli process.

a. X is defined in terms of an experiment that is repeated a certain number of times, n. Each repetition, or trial, of the experiment has only two possible outcomes: success and failure.

b. The probability of success is the same for all trials in the experiment. That is, the result of each trial is independent of previous trial results.
c. The value of the random variable is defined as the total number of successes obtained from all the trials. Therefore, a binomial random variable always takes the values 0, 1, 2, 3, . . . , n.

Binomial random variables have many important practical applications because the conditions necessary to define them exist naturally in many situations, such as sampling for defective parts in quality control, ascertaining candidate preferences in pre-election polls, and consumer choices.

Some examples of a binomial random variable are:

- The number of heads obtained in five flips of a coin. (Success is heads; failure is tails.)
- The number of times a sum of 7 is rolled with two dice. (The sum of 7 is a success, and any other sum is a failure.)
- The number of defective parts a machine produces in a batch of 500 parts. (A defective part is a success; a good part is a failure.)

Notice that in each of these examples it is assumed that the probability of success does not change from one trial to the next, and the value of the binomial random variable is the number of successes obtained.

Examples of discrete random variables that are not binomial random variables are:

- The sum of the numbers thrown on two dice
- The final grade-point average in a class, rounded to the nearest hundredth
- The number of red cards drawn from a deck of cards in five tries without replacement

The first two examples are not binomial random variables, because each has more than two possible outcomes. Although in the third example each draw has only two possible outcomes, red and not red, and the value of the variable is the number of successes, the probability of success (drawing a red card) changes from draw to draw if the cards are not replaced in the deck. This means that the result of each trial is not independent of previous results, and therefore the conditions of a binomial random variable are not satisfied.

The following example of a binomial random variable will be developed to suggest a general formula for finding probabilities associated with binomial random variables.

E x a m p l e 2

A coin that is not necessarily fair is tossed four times. The probability of heads is p, and the probability of tails is q, where $q = 1 - p$. The event of obtaining heads is considered a success, and the binomial random variable B represents the number of heads obtained in the four tosses.

a. Draw a tree diagram that shows branch probabilities and the possible values of B.
b. Find the probability of obtaining exactly three heads in these four flips. That is, find $P(B = 3)$.

Solution   a. A tree diagram would appear as follows:

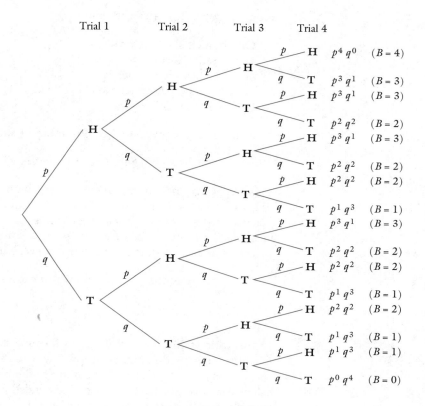

Since the results of the trials (tosses) are clearly independent, the probabilities shown for each trial are multiplied (by the multiplication rule for independent events) to find the probability of obtaining the value of B at the end of each branch. For example, as shown at the top branch of the tree diagram, the probability of tossing four heads, or $P(B = 4)$, is $p \cdot p \cdot p \cdot p = p^4 q^0$ or simply p^4.

b. To find $P(B = 3)$, look for all the branches where $B = 3$ and then add their branch probabilities together. Notice that each branch probability is the product of the probability of success in a single trial raised to a power equal to the number of successes (p^3) and the probability of failure in a single trial raised to the number of failures (q^1), to obtain a branch probability of $p^3 q^1$. Thus,

$$P(B = 3) = \text{Sum of branch probabilities on the second, third, fifth,}$$

and ninth branches

$$= p^3 q^1 + p^3 q^1 + p^3 q^1 + p^3 q^1$$
$$= 4p^3 q^1. \quad \bullet$$

The Binomial Formula

From Example 2, notice the following.

> One branch shows 4 heads.
> Four branches show 3 heads.
> Six branches show 2 heads.
> Four branches show 1 head.
> One branch shows 0 heads.

As Example 2 suggests, if r is the number of successes desired,

$$P(B = r) = \text{(Number of branches with } r \text{ successes)} \cdot p^r q^{n-r}.$$

This means that the number of "successful" branches is important to the determination of probabilities related to binomial random variables. Since it is not always practical to draw tree diagrams to determine the number of branches with r successes in n trials, a more general formula is needed for the number of branches with r successes in an experiment repeated n times. That formula is

$$\text{Number of branches with } r \text{ successes} = {}_nC_r,$$

$$\text{where} \quad {}_nC_r = \frac{n!}{(n-r)! \cdot r!}.$$

In this expression, ${}_nC_r$ is called the **binomial coefficient;** it will be seen again in Chapter 7 from a different point of view. Graphics calculators and many scientific calculators have an ${}_nC_r$ key that computes the binomial coefficient quickly. If your calculator does not have this function, find the value of ${}_nC_r$ from the preceding formula, remembering that $n! = n \cdot (n-1) \cdot (n-2) \cdot \cdots \cdot 1$ and that 0! is equal to 1.

Using the formula for ${}_nC_r$ to obtain the number of branches in Part b of Example 2 (on page 245),

$$\text{Number of branches with three successes in four trials} = {}_4C_3$$

$$= \frac{4!}{(4-3)! \cdot 3!}$$

$$= \frac{4!}{1! \cdot 3!}$$

$$= \frac{4 \cdot 3 \cdot 2 \cdot 1}{1 \cdot 3 \cdot 2 \cdot 1}$$

$$= 4,$$

which is the same result that was obtained by counting the number of "successful" branches in the tree diagram.

For convenience, Table 6-1 lists commonly used values of ${}_nC_r$.

It is now possible to state the binomial formula, which calculates the probability of r successes of a binomial random variable in n trials.

Table 6-1 Table of Binomial Coefficients $_nC_r$

n \ r	0	1	2	3	4	5	6	7	8	9	10
1	1	1									
2	1	2	1								
3	1	3	3	1							
4	1	4	6	4	1						
5	1	5	10	10	5	1					
6	1	6	15	20	15	6	1				
7	1	7	21	35	35	21	7	1			
8	1	8	28	56	70	56	28	8	1		
9	1	9	36	84	126	126	84	36	9	1	
10	1	10	45	120	210	252	210	120	45	10	1

The **Binomial Formula**

Suppose X is a binomial random variable with a probability of success at each trial equal to p and with $q = 1 - p$. If there are n trials, then the probability that there will be exactly r successes is

$$P(X = r) = {}_nC_r \cdot p^r \cdot q^{n-r}.$$ (6-3)

E x a m p l e 3

Construct a probability density function for the following values of p and q in the experiment of flipping a coin four times (Example 2, page 244).

a. $p = .5,$ $q = .5$
b. $p = .7,$ $q = .3$
c. $p = .4,$ $q = .6$

Solution

If $p > .5$, then the probabilities are concentrated to the right of the density function. If $p < .5$, then the probabilities are concentrated to the left.

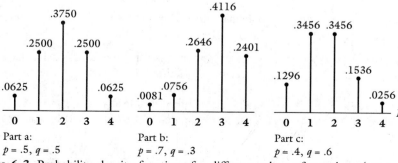

Part a:
$p = .5, q = .5$

Part b:
$p = .7, q = .3$

Part c:
$p = .4, q = .6$

Figure 6-3 Probability density functions for different values of p and q when a coin is flipped four times.

Explanation   The boldface numbers along the horizontal axes in Figure 6-3 are the possible values for B, the number of heads, in four flips. The height of each line is the value of $_nC_r \cdot p^r \cdot q^{n-r}$ for the corresponding values of n (equal to 4), r, p, and q. For instance, the probability that one head will be obtained in four flips of a coin, when the probability of heads is .7, is

$$P(B = 1) = {}_4C_1 \cdot (.7)^1 \cdot (.3)^{4-1}$$
$$= .0756.$$

This appears on the second line of the middle probability density function in Figure 6-3. Notice that, in each of the three probability density functions, the sum of the five probabilities is 1. ●

E x a m p l e 4

A new type of blood pressure medication claims to be effective 75% of the time. The drug is administered to ten patients. Assume that the conditions of a binomial random variable exist, and determine the probability of each of the following events.

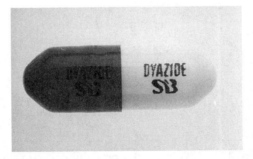

Courtesy SmithKline Beecham

a. Exactly seven patients are helped by the drug.
b. No patients are helped by the drug.
c. At least seven patients are helped by the drug.

Solution   An assumption that the conditions of a binomial random variable exist implies the following.

- An experiment is repeated, with each trial having only two possible outcomes: success and failure. In this case, either the blood pressure medication works or it does not work.
- The probability of success is the same for every trial; that is, every person has the same chance of being helped by the drug.[2]

---

[2]Although this may not actually be true, assume for illustrative purposes that all patients were screened carefully.

- The value of the random variable is the number of successes obtained in all the trials—that is, the number of people helped by the drug.

Let X represent the binomial random variable, with value equal to the number of people who are helped by the new drug.

a. Using the binomial formula with $p = .75$, $q = .25$, $n = 10$, and $r = 7$,

$$P(X = r) = {}_nC_r \cdot p^r \cdot q^{n-r},$$
$$P(X = 7) = {}_{10}C_7 \cdot (.75)^7 \cdot (.25)^{10-7}.$$

According to Table 6-1 on page 247, ${}_{10}C_7 = 120$, so

$$P(X = 7) = 120 \cdot (.75)^7 \cdot (.25)^3$$
$$\approx .2503.$$

Thus, there is about a 25% chance that exactly seven patients are helped by the drug.

b. If no patients are helped by the drug, then $r = 0$ in the binomial formula used for Part a.

$$P(X = 0) = {}_{10}C_0 \cdot (.75)^0 \cdot (.25)^{10}$$
$$\approx 1 \cdot 1 \cdot 0$$
$$\approx 0.$$

This means that the chances that no one will benefit from the drug are virtually nil.

c. The probability that at least seven patients are helped by the drug can be found as follows:

$$P(X \geq 7) = P(X = 7) + P(X = 8) + P(X = 9) + P(X = 10)$$
$$= {}_{10}C_7 \cdot (.75)^7 \cdot (.25)^3 + {}_{10}C_8 \cdot (.75)^8 \cdot (.25)^2$$
$$+ {}_{10}C_9 \cdot (.75)^9 \cdot (.25)^1 + {}_{10}C_{10} \cdot (.75)^{10} \cdot (.25)^0$$
$$\approx .2503 + .2816 + .1877 + .0563$$
$$\approx .7759. \quad \bullet$$

6.3 Technology Tools

It is possible to use statistical software to determine probabilities connected with binomial random variables as well as many other kinds of random variables.

Alternative Solution to Example 4, Using Minitab

Minitab will compute probabilities for binomial random variables if the macro command PDF (for probability density function) is typed, followed by the value of r in an active Session window. The appropriate subcommand is "Binomial" followed by the values of n and p. Figure 6-4 illustrates this process for the first two parts of Example 4 (page 248). Notice that the semicolon (;) typed after the value of r in Figure 6-4 activates the subcommand prompt, **SUBC>**.

```
┌─────────────────────────────────────────────────┐
│ ☐ ══════════════  Session  ══════════════       │
├─────────────────────────────────────────────────┤
│  Worksheet size: 38000 cells                     │
│  ⎛ MTB > PDF 7;                    ⎞              │
│  ⎝ SUBC> Binomial  10    0.75.     ⎠   Macro commands to be typed.
│        ⎛   K          P( X = K)  ⎞               │
│        ⎝  7.00          0.2503   ⎠     Minitab output for Part a.
│  MTB >                                           │
│  MTB >                                           │
│  MTB >                                           │
│  MTB >                                           │
│  ⎛ MTB > PDF 0;                    ⎞             │
│  ⎝ SUBC> Binomial  10    0.75.     ⎠   Macro commands.
│        ⎛   K          P( X = K)  ⎞               │
│        ⎝  0.00          0.0000   ⎠     Minitab output for Part b.
│  MTB >                                           │
└─────────────────────────────────────────────────┘
```

Figure 6-4

Minitab will also compute binomial probabilities with the CDF (for cumulative density function) command. For a given value of r, Minitab will compute the probability that the binomial random variable is less than or equal to r. Part c of Example 4, however, is concerned with the probability of the random variable's being equal to or greater than r. This probability can be determined by using the fact that, for a given value of r,

$$P(X \geq r) = 1 - P(X \leq (r - 1)).$$

Thus, the complement of the desired probability can be found, and that probability subtracted from 1, to obtain the desired results.[3] Figure 6-5 shows the necessary commands.

```
┌─────────────────────────────────────────────────┐
│ ☐ ══════════════  Session  ══════════════       │
├─────────────────────────────────────────────────┤
│  Worksheet size: 38000 cells                     │
│  MTB > CDF 6;                                     │
│  SUBC> Binomial  10    0.75.                      │
│        K   P( X LESS OR = K)                      │
│       6.00          0.2241  ◄───── Complement of Part c
│  MTB >                              1 − .2241 = .7759
└─────────────────────────────────────────────────┘
```

Figure 6-5

[3]There are other ways of using Minitab to carry out calculations like these. Consult the Minitab User's Guide (Appendix A-8 or A-9) or a Minitab manual.

E x a m p l e
5

In a family of eight children, what is the probability that there are at least two girls?

Solution   Let the random variable G represent the number of girls in the family. Assuming that girl and boy births are equally likely at any time, and counting the birth of a girl as a "success," G is a binomial random variable with $n = 8$ and $p = \frac{1}{2}$.

Rather than compute

$$P(G \geq 2) = P(G = 2) + P(G = 3) + P(G = 4) + P(G = 5)$$
$$+ P(G = 6) + P(G = 7) + P(G = 7) + P(G = 8),$$

it is much easier to compute the probability of the event that is complementary to having at least two girls, which is having one or fewer girls. That is,

$$P(G \geq r) = 1 - P(G \leq (r - 1)).$$
$$P(G \geq 2) = 1 - P(G \leq 1)$$
$$= 1 - [P(G = 0) + P(G = 1)]$$
$$= 1 - \left[{}_8C_0 \cdot \left(\frac{1}{2}\right)^0 \left(\frac{1}{2}\right)^8 + {}_8C_1 \cdot \left(\frac{1}{2}\right)^1 \left(\frac{1}{2}\right)^7 \right]$$
$$\approx 1 - [.0039 + .0313]$$
$$\approx 1 - .0352$$
$$\approx .9648. \quad \bullet$$

Alternative Solution to Example 5, Using Minitab

Minitab may also be used to solve the preceding example.

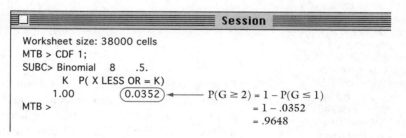

Exercises 6.2 – 6.3

1. In your own words, explain how a Bernoulli process is related to a binomial random variable.

GROUP

2. Give three examples of experiments that may be classified as Bernoulli processes and three that are not Bernoulli processes.

GROUP

3. Identify which of the following discrete random variables are, and which are not, binomial random variables. For those that are not, explain why.

   a. The sum of the numbers that are shown in ten tosses of a single die

b. The number of 6's that come up in ten tosses of a single die

c. The number of traffic tickets a person receives for every 500 trips in his or her car

d. The number of times the ball on a roulette wheel lands in a red pocket in 50 spins

e. The mean of the numbers of the pockets in which the ball of a roulette wheel stops in 50 spins

f. The number of defective parts out of each thousand made by a certain machine

4. A coin is flipped n times. If a tree diagram (similar to that shown in Example 2 on page 245) were drawn to keep track of the possible outcomes, how many branches, in terms of n, would it have?

5. Refer to Table 6-1 on page 247. Explain why not all the boxes contain numbers. What is the interpretation of $_5C_8$?

* 6. Refer to Table 6-1 on page 247. Notice that $_8C_2 = {_8}C_6$, $_5C_3 = {_5}C_2$, and $_{10}C_7 = {_{10}}C_3$. Are these coincidences, or does a general rule apply? If it does, state the general rule and try to prove it.

7. The entries in Table 6-1 on page 247 can also be obtained with a device called **Pascal's triangle.**

$$
\begin{array}{ccccccccc}
 & & & & 1 & & & & \\
 & & & 1 & & 1 & & & \leftarrow n = 1 \\
 & & 1 & & 2 & & 1 & & \leftarrow n = 2 \\
 & 1 & & 3 & & 3 & & 1 & \leftarrow n = 3 \\
1 & & 4 & & 6 & & 4 & & 1 \leftarrow n = 4
\end{array}
$$

$$_4C_0 \quad {_4}C_1 \quad {_4}C_2 \quad {_4}C_3 \quad {_4}C_4$$

Every number in the triangle is the sum of the two numbers above it to the left and right; this pattern continues indefinitely. The numbers across any row of the triangle are the binomial coefficients $_nC_r$ for values of $r = 0$ to $r = n$. Use Pascal's triangle to determine the number of branches in a tree diagram that has eight successes in ten trials.

8. Refer to Example 3 on page 247. Let X be a binomial random variable. For given values of n and r, a whole **family** of probability distributions is possible, depending upon the value of p.

a. What conditions on p would result in a symmetrically shaped distribution?

b. What conditions on p would result in a distribution skewed (more spread out) to the left?

9. Refer to Example 4 on page 248. What conditions might exist that would not justify the use of a binomial random variable as a model for this example?

Use of a TI-82 graphics calculator for the following exercises will reduce many calculations to a single step. For example, Part a of Exercise 10 can be solved by keying in the following, then pressing ENTER.

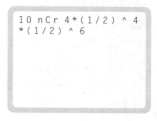

```
10 nCr 4*(1/2)^4
*(1/2)^6
```

10. A fair coin is flipped ten times.

   a. What is the probability of obtaining exactly four heads?

   b. What is the probability of obtaining at most four heads?

11. A fair die is rolled eight times.

   a. What is the probability of rolling a 6 exactly twice?

   b. What is the probability of rolling a 6 at least three times?

   c. What is the probability of rolling a 6 fewer than seven times?

* 12. A coin is bent as in the following sketch, so that it is more likely to land heads than tails:

   The coin is to be tossed eight times. If the binomial random variable H represents the number of heads obtained in these eight flips, then what is the most likely value for H if $p = \frac{3}{4}$?

   Hint: *The most likely value has the longest line on the probability density function.*

* 13. Refer to Exercise 12. Intuitively determine a value for p if a different, unfair coin is flipped 36 times and the most likely value for H is 20.

14. The probability that the engine of a new battle tank will survive a desert sandstorm is .80. Six such tanks are sent on a mission. For the mission to be successful, at least four of the tanks must survive a sandstorm. What is the probability of success of the mission if a sandstorm takes place?

GROUP * 15. In the situation described in Exercise 14, how many tanks should be sent if a sandstorm is certain and at least a 99% chance of completing the mission successfully is desired?

   Hint: *Make a table and use trial and error.*

6.4 THE MEAN AND STANDARD DEVIATION OF A BINOMIAL RANDOM VARIABLE

If a discrete random variable is a binomial random variable, then its mean value and standard deviation can be found quickly with the following special formulas.

The **Mean** and **Standard Deviation** of a Binomial Random Variable

If B is a binomial random variable based on n repetitions of an experiment in which the probability of success in a single trial is p, with $q = 1 - p$, then

$$\mu_B = n \cdot p \qquad\qquad\text{(6-4)}$$

$$\sigma_B = \sqrt{n \cdot p \cdot q}. \qquad\qquad\text{(6-5)}$$

Depending upon the values of n and p, the calculated mean (the expected number of successes) may or may not be a whole number. If a whole number, the mean will be the most likely value of the random variable.

E x a m p l e 6

An experimental coating for children's teeth is claimed to be 70% effective in preventing cavities for 2 years after its application. Let the random variable C represent the number of children who, after receiving this coating, remain cavity-free for 2 years.

a. If a dentist applies the coating to 20 of her young patients, and if the claimed 70% effectiveness is assumed to be true, then what is the mean, or the expected number of children who will be cavity-free for 2 years?

b. What is the probability that the number of children who actually are cavity-free is the expected number from Part a?

Solution   a. C is a binomial random variable, since it keeps track of the number of successes (children without cavities); there are only two possible outcomes for each child (the child has a cavity or doesn't); and the probability of success (not having a cavity) is the same for all children.[4]

Using $p = .70$ as the probability of success and substituting into the formula $\mu = n \cdot p$, the expected number of children who will be cavity-free, denoted by μ_C, is

$$\mu_C = 20 \cdot (.70)$$
$$= 14.$$

---

[4]In actuality, this may or may not be the case. As in Example 4, assume that all patients are screened carefully.

That is, although the random variable C can take any one of the values $\{0, 1, 2, 3, 4, \ldots, 20\}$, its expected, or mean, value for all groups of this size should be 14 if p is indeed equal to .70 for each child.

b. The probability that C actually equals μ_C is

$$P(C = 14) = {}_{20}C_{14} \cdot (.70)^{14} \cdot (.30)^{6}$$
$$= .1916.$$

Although this value may not seem very great, it is the greatest probability of all the 21 possible values for C. If it is possible for a random variable, X, to be equal to its mean, μ_X, then $P(X = \mu_X)$ will always be the most likely of all the obtainable values of X.   ●

Exercises 6.4

1. Refer to Example 3 on page 247. Compute the mean value for each part in two different ways. Then locate the mean value on each probability density function.

2. Refer to Example 6 on page 254. If the dental covering were applied to 21 patients instead of 20, the expected value of C would be 14.7. Since $P(C = 14.7) = 0$, what does intuition seem to imply about the most likely obtainable value of C?

In Exercises 3–8, treat each random variable as a binomial random variable even though, technically, the probabilities of success for the trials are not all the same. The pool is so large that this technicality will be ignored.

3. A mail-order company claims that 90% of its orders are mailed on time. The records of ten customers are chosen at random and examined. When answering the following questions, assume that the company's claim is true.

   a. What is the probability that all ten orders were sent out on time?

   b. What is the probability that at most four orders were sent out on time?

   c. What is the probability that at least five orders were sent out on time?

   d. What is the expected number of orders sent out on time?

4. A new flu virus is said to affect 20% of the population of the United States. Fifty people are chosen at random from the population. Let the binomial random variable F be the number of people in the sample who actually have the flu.

   a. How many would be expected to have the flu? That is, what is the mean number of people who have the flu, taken in groups of 50?

   b. Determine the standard deviation of F.

   c. Round the standard deviation from Part (b) to the nearest whole number to determine an interval with end points $\mu_F - \sigma_F$ and $\mu_F + \sigma_F$. Use the Empirical Rule to determine the probability that an actual value of F is in this interval.

5. A company that produces exploding bolts for spacecraft claims that its bolts function properly (that is, explode when triggered) 95% of the time. A batch of 15 is used on a new shuttle. Assume that the company's claim is true.

   a. What is the expected number of bolts that will function properly during the flight of the shuttle?

   b. What is the probability that at least half (eight or more) of them will explode when triggered?

   c. What is the probability that at most two will not explode?

GROUP

6. Refer to Exercise 5. Suppose one of ten bolts on an escape hatch is known to be defective. In an emergency, at least eight of the ten bolts must explode in order for the hatch to open. What is the probability that the hatch will open in an emergency?

7. Two hundred people are selected at random, and 150 of them own VCRs. Based on this sample, what is the probability that eight out of ten people in a different group own VCRs?

8. Suppose long experience has shown that 23% of all persons who suffer from a particular disease and take no medication recover from the disease.

   a. If 15 patients have the disease, what is the probability that at least 13 of them recover if they take no medication?

   b. A drug company administered an experimental drug to another group of 15 patients and 13 of them recovered. What do these results imply about the effectiveness of the new drug? Explain.

* 9. Refer to Example 2 on page 244. Show that $\sum_{s=0}^{4} P(B = s) = 1$, using only p and q.

   Hint: *Use the fact that the* **binomial theorem** *states that* $(a + b)^n = \sum_{i=0}^{n} {}_nC_i \cdot a^i \cdot b^{n-i}$.

* 10. As you know, the expected value of a binomial random variable is equal to $n \cdot p$. Verify this formula by using the definition of expected value for any (not necessarily binomial) discrete random variable, which is

$$\mu_X = \sum_{i=0}^{n} x_i \cdot P(X = x_i).$$

   Hint: $\sum_{i=0}^{n} x_i \cdot P(X = x_i) = \sum_{i=0}^{n} i \cdot {}_nC_i \cdot p^i \cdot q^{n-i}.$

Excursion 6.4A • THE POISSON RANDOM VARIABLE

*The material and exercises on the Poisson random variable are optional.
They may be covered at any time or omitted.*

Although binomial random variables have wide application, many situations arise in which they cannot serve as models. Conditions in real life are not often structured as experiments in which the same conditions recur over and over again.

The number of people waiting in line at a bank, the number of dents your car is likely to receive in the next year, the number of false alarms phoned in to the fire station next week, and the number of accidents each year on a particular stretch of highway are all examples of discrete random variables for which the conditions of a binomial random variable are not met. In each of these examples, the results of the trials may be independent. However, the probabilities of "successes," such as dents and false alarms, can change from day to day. Much work has been done to adapt the theory of probability to these types of situations, in which success may occur at random in a defined time or space.

A **Poisson process** is the random occurrence of a series of events over time (or space) in such a way that:

- The number of successes can be any whole number.[5]
- The probability of a given number of successes in a specific time or space matches the probability in any equal time or equivalent space. In other words, the probability of success is equally spread out.

When the conditions for a Poisson process are satisfied, then a Poisson random variable can be defined, and the following formula applies.

The **Poisson Formula**

Suppose X is a Poisson random variable. If λ is the mean number of successes in a stated time, t (or space), and r is the desired number of successes, then

$$P(X = r) = \frac{e^{-\lambda} \cdot \lambda^r}{r!}. \tag{6-6}$$

In this formula, e represents an irrational number, which is approximately 2.71828. Your calculator should have an e^X button to help you with these calculations. The solution to the next example will show how the graphics calculator and Minitab can be used to find probabilities for a Poisson random variable.

[5]From a practical point of view, there is usually some limit.

E x a m p l e
7

On an average weekday, three Canadian quarters are deposited at a certain pay telephone. Answer the following questions in regard to the next 5 weekdays.

a. What is the probability that 12 Canadian quarters will be deposited?
b. What is the probability that four or fewer will be deposited?
c. What is the probability that at least ten but no more than 20 will be deposited?

S t a t   F a c t

The letter *e* is used in honor of the Swiss mathematician Leonhard Euler (1707–1783). The value of *e* is exactly equal to the limit of the sequence of numbers generated by $(1 + \frac{1}{N})^N$ as the integer *N* increases. In 1735 the Academy of Sciences in St. Petersburg held a contest that involved solving a difficult mathematical problem, which the eminent mathematicians of the day said would take at least several months. Euler solved it in 3 days, but the effort made him ill, and he subsequently lost the use of one eye. He later lost the use of the other eye as well and learned to perform extensive mental calculations. Euler did some of his finest work while almost totally blind. He is also credited with the introduction of the symbol Σ to represent a sum.

Solution   This solution will illustrate both the graphics calculator and Minitab.

Let the random variable Q be the number of Canadian quarters that will be deposited in the pay telephone in the next 5 weekdays. The variable Q can take on any whole-number value, and in the absence of information to the contrary, it will be assumed that coin deposits are spread out evenly over the next 5 weekdays.

The required value of λ is the mean number of coins deposited over the next 5 days, because that is the time interval stated in the question. Thus, $\lambda = t \cdot 5 = 3 \cdot 5 = 15$.

a. Substituting the variable name Q for X in the Poisson formula,

$$P(Q = r) = \frac{e^{-\lambda} \cdot \lambda^r}{r!}.$$

Also, $r = 12$ and $\lambda = 15$. Therefore,

$$P(Q = 12) = \frac{e^{-15} \cdot 15^{12}}{12!}.$$

All of the quantities in this expression are either very large or very small and are difficult to work with. One way to proceed is to first compute $12! = 479001600$, then let the calculator complete the work.[6] These calculations can be keyed into a TI-82 graphics calculator, and after the (ENTER) key is pressed, the screen appears as follows:

```
e^-15*15^12/4790
01600
           .0828592344
```

Thus, $P(Q = 12) \approx .0829$.

b. The probability that four or fewer quarters will be deposited can be expressed as

$$P(Q \le 4) = P(Q = 0) + P(Q = 1) + P(Q = 2) + P(Q = 3) + P(Q = 4).$$

These probabilities can be worked out by hand as in Part a and then added together, or some software may be used. Using Minitab as in the alternative solution to

---

[6]It is also possible to key in the expression $e^{\wedge}15*15^{\wedge}12/12!$.

Example 4 on page 249 (except applying the "Poisson" instead of the "Binomial" subcommand) yields $P(Q \leq 4) = .0009$ for this part. Note that only the mean value λ (which is 15) needs to be specified in the subcommand.

```
                                            Session
Worksheet size: 38000 cells
MTB > CDF 4;
SUBC> Poisson 15.
       K   P( X LESS OR = K)
      4.00        0.0009
MTB > I
```

c. The probability that at least ten but no more than 20 quarters will be deposited can be written as follows:

$$P(10 \leq Q \leq 20) = P(Q \leq 20) - P(Q \leq 9).$$

Minitab will be used to obtain these values.

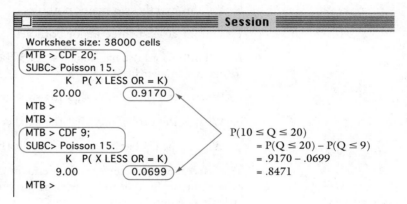

```
                                            Session
Worksheet size: 38000 cells
MTB > CDF 20;
SUBC> Poisson 15.
       K   P( X LESS OR = K)
      20.00        0.9170
MTB >
MTB >
MTB > CDF 9;
SUBC> Poisson 15.
       K   P( X LESS OR = K)
      9.00        0.0699
MTB >
```

$P(10 \leq Q \leq 20)$
$= P(Q \leq 20) - P(Q \leq 9)$
$= .9170 - .0699$
$= .8471$

Exercises 6.4A

1. The mean value of a binomial random variable is found by the formula $\mu = n \cdot p$. The mean value of a Poisson random variable, however, is simply λ. Explain why.

2. Name two significant differences between binomial random variables and Poisson random variables.

For Exercises 3–8, assume that the conditions of a Poisson process apply.

3. Each year in routine maintenance, repair crews on a rail commuter line replace, on the average, about four ties every half mile. Answer the following for a 6-mile stretch of this track.

   a. What is the average number of ties that would need replacement?

   b. What is the probability of replacing 40 ties?

c. If ties cost $37 each, what is the probability that replacement costs will exceed $1850 over a 6-mile stretch of track?

4. The number of customers per day who come to a certain service station for oil changes averages four. The following probabilities pertain to the next 2 days.

a. What is the probability that the service station will perform exactly five oil changes?

b. What is the probability that the service station will perform at least five oil changes?

c. What is the probability that the service station will perform no more than nine oil changes?

5. "Major" potholes are hit on "main" streets about every $2\frac{1}{4}$ miles. If you travel 7 miles down a main street, find the following probabilities.

a. You hit at least one major pothole.

b. You hit exactly two major potholes.

c. You hit at least three major potholes.

6. An old 500-foot hose has 40 leaks. A 25-foot section of it is chosen at random to make a smaller hose.

a. What is the probability that the smaller hose will have no leaks?

b. What is the probability that the smaller hose will have exactly two leaks?

c. What is the probability that the smaller hose will have more than two leaks?

7. The "average" car receives three dents in every 20,000 miles of travel. If you drive your car 50,000 miles, what is the probability that your car will acquire at least one dent?

8. A major carpet manufacturer produces carpets that, according to experience, contain three defects per 90 square yards. What is the probability that a carpet order of 120 square yards will have at least two defects?

GROUP   9. Assume that the average person receives one wrong-number telephone call per month. Make a probability table for the number of wrong-number calls a person is likely to receive in a year. List only probabilities of at least .0650.

   Hint: The highest probabilities for the values of a random variable occur near the average value.

GROUP   10. Refer to Example 3 on page 247. Draw a set of probability density functions for a Poisson random variable in which $\lambda = 0.1, 1.0,$ and 10.0. Use values of r from 0 through 5.

6.5 CONTINUOUS RANDOM VARIABLES

As an introduction to another type of random variable called a continuous random
variable, consider the following traffic survey. The times spent waiting for a particular traffic signal to turn from red to green are measured for a random sample of
$n = 32$ automobiles. The following table contains the results, in seconds rounded
to the nearest tenth.

4.7	17.3	37.2	14.3	24.1	14.0	37.6	15.5
39.8	8.1	21.9	33.2	25.3	38.1	5.3	15.9
26.7	29.0	1.7	31.9	38.6	4.2	7.3	1.8
11.6	18.7	21.3	32.8	12.7	18.0	28.1	32.1

Figure 6-6 shows a (left-ended) histogram constructed for these waiting times,
with interval widths of 5 seconds. From Figure 6-6 it appears that the waiting
times are almost evenly, or uniformly, distributed over the 40-second time limit.
If the random variable W is defined as the waiting time at the traffic signal, this
histogram may be used to determine the probability that W is within a given
time interval.

For example, the probability of waiting at least 10 but no more than 25
seconds, denoted by $P(10 \le W \le 25)$, can be found through a ratio of areas.
That is,

$$P(10 \le W \le 25) = \frac{\text{Area of shaded region}}{\text{Area of entire histogram}}.$$

Each of the areas in this probability fraction can easily be found as the sum of the
areas of rectangles in Figure 6-7.

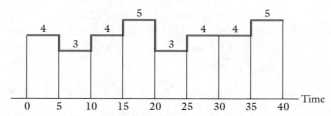

Figure 6-6 A histogram for waiting times.

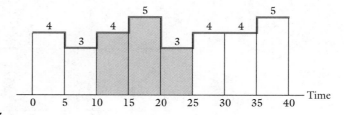

Figure 6-7

$$\text{Area of shaded region} = 5 \cdot 4 + 5 \cdot 5 + 5 \cdot 3$$
$$= 60.$$
$$\text{Area of entire histogram} = 5 \cdot 4 + 5 \cdot 3 + 5 \cdot 4 + 5 \cdot 5 + 5 \cdot 3 + 5 \cdot 4$$
$$+ 5 \cdot 4 + 5 \cdot 3$$
$$= 160.$$

This means that

$$P(10 \le W \le 25) = \frac{\text{Area of shaded region}}{\text{Area of entire histogram}}$$
$$= \frac{60}{160}$$
$$= .375.$$

From this sample, an ideal model may be pictured for a random sample of much larger size, with the characteristics of the smaller sample histogram but with more refined features. Assuming that waiting times, W, are evenly distributed over the 40-second time limit, Figure 6-8 illustrates the large-sample histogram. The shape of this histogram, or its distribution, is a rectangle, and the ratio

$$\frac{\text{Area of shaded region}}{\text{Area of entire histogram}}$$

is the same for any height, h. Furthermore, if the area of the entire histogram is set equal to 1, then the probability that the random variable W is within a given interval is simply the area of the shaded region.

The type of even distribution in Figure 6-8 is called a **uniform probability distribution.** All that is needed to determine the probability that the random variable W is between 10 and 25 seconds is the height, h, of the shaded rectangle. The area of the large rectangle (1) and its base (40) can be used to determine h. That is,

$$A = b \cdot h,$$
$$1 = 40 \cdot h,$$
$$\frac{1}{40} = h.$$

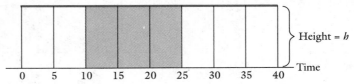

Figure 6-8

This means that

$$P(10 \le W \le 25) = (25 - 10) \cdot \frac{1}{40}$$
$$= \frac{15}{40}$$
$$= .375,$$

which agrees with the earlier answer.

The following and subsequent definitions summarize the ideas behind this traffic survey.

Definition ▶ A **continuous random variable** is a random variable with an uncountable number of different possible values.

It is important to remember that a continuous random variable can take any value (not just whole-number or fractional values) in an interval of values. The number of possible values it can assume cannot be counted. For instance, assuming that the waiting times at the traffic signal can be measured with any degree of accuracy, W can be any one of an uncountable number of different times. Rather than the probabilities of the random variable being concentrated at a certain number of points, as was the case for a discrete random variable, the probabilities associated with a continuous random variable are spread out in some pattern over an interval.

Definitions ▶ The **probability density function,** also known as the **probability distribution,** of a continuous random variable is the frequency curve obtained from the histogram of possible values of the variable.

The area under the probability density function of a continuous random variable is always equal to 1, regardless of its shape.

In the traffic survey, the probability density function of the continuous random variable of waiting times, W, was a horizontal line, as shown in Figure 6-9. The equation $y = \frac{1}{40}$ shown in the figure is the equation of the probability density

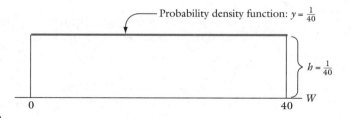

Figure 6-9

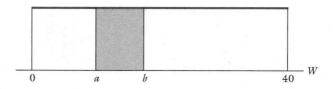

Figure 6-10

function for this random variable, W. (Recall that it was determined by the condition that the rectangle with a base of 40 must have an area of 1.)

In general, the probability that the random variable W is between two times, a and b, can be determined by finding the shaded area. In this case, the shaded area is equal to $(b - a) \cdot \frac{1}{40}$, which means that $P(a \le W \le b) = (b - a) \cdot \frac{1}{40}$.

Notice from Figure 6-10 that, as the end points a and b come closer together, the base of the shaded area decreases and $P(a \le W \le b)$ approaches zero. In fact, the probability that any continuous random variable is equal to one specific value must be zero, because there can be no area above a single point.

E x a m p l e 8

Following is the probability density function of a continuous random variable, T.

a. Sketch a histogram that could serve as a model for this probability density function.
b. Determine the height of this density function at $T = 0$, which is also called the **density of the probability distribution** at $T = 0$.
c. Determine $P(5 \le T \le 10)$.
d. Determine $P(T = 6)$.

Solution a. A histogram that reflects numerous scores near $T = 0$ with steadily fewer until $T = 10$ could serve as a model for the probability distribution. Such a histogram would appear as follows:

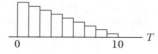

b. The area under the probability distribution of any continuous random variable must be equal to 1. Using the formula for the area of a triangle,

$$A = \frac{1}{2} \cdot b \cdot h,$$

$$1 = \frac{1}{2} \cdot (10) \cdot h,$$

$$\frac{1}{5} = h.$$

This means that the density of the probability distribution at $T = 0$ is equal to $\frac{1}{5}$.

c. $P(5 \leq T \leq 10)$ is equal to the area of this shaded region under the probability distribution.

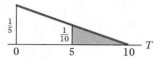

This shaded area is a triangle[7] with base 5 and height $\frac{1}{10}$.

$$P(5 \leq T \leq 10) = \text{Area of shaded triangle}$$

$$= \frac{1}{2} \cdot b \cdot h$$

$$= \frac{1}{2} \cdot 5 \cdot \frac{1}{10}$$

$$= \frac{1}{4}.$$

d. $P(T = 6) = 0$. The probability that any continuous random variable assumes one specific value is always 0 because there is no area under the probability distribution at a single point.   ●

Exercises 6.5

1. Refer to the traffic-signal data on page 261. Using the 32 actual waiting times, find $P(10 \leq W \leq 25)$. How does this compare with the result obtained earlier?

GROUP

2. Classify the following random variables as discrete or continuous.

   a. I = A person's IQ score

   b. R = A real number chosen at random between 0 and 1

   c. T = The low temperature in Chicago on February 2, 1996, if temperatures may be measured only to the nearest hundredth of a degree

   d. T = The low temperature in Chicago on February 2, 1996, if temperatures may be measured to any accuracy desired

   e. The number of heads obtained in six flips of a fair coin.

3. A continuous random variable, T, has a uniform (rectangular) probability distribution over the interval $[2, 9.5]$.

   a. Graph the probability distribution of T, and state its density function.

   b. Determine $P(3.2 \leq T \leq 6.7)$ and $P(5.6 \leq T \leq 9.1)$.

   c. What is the density at $T = 4$, and what is $P(T = 4)$?

   d. What is the density at $T = 1.8$, and what is $P(T = 1.8)$?

[7]The shaded triangle is similar to the large triangle of the original probability distribution. Since the shaded triangle's base is one-half the base of the large triangle, its height must also be one-half the height of the large triangle.

*   4. The following diagram shows the probability distribution of the random variable, *G*, which is another kind of continuous probability distribution called a **step distribution**—the union of two uniform probability distributions, one over the interval [4, 8) and the other over the interval [8, 12]. From the diagram it is assumed that the density of this distribution over the interval [8, 12] is twice the density over the interval [4, 8).

a. If the total area under this probability distribution must be equal to 1, find the density at *G* = 5 and at *G* = 10.

   Hint: Start with the equation x + 2x = 1.

b. Find $P(4.2 \le G \le 10.4)$.

*   5. Give several examples of continuous random variables that would have step-distribution models.

GROUP   6. The train *City of New Orleans* is scheduled to arrive at Union Station in Chicago at 8:00 P.M., but in the past it has arrived as late as 10:30 P.M. If *R* is the continuous random variable with a value equal to the arrival time of the train, the following diagram could show the distribution of *R* over the interval [8, 10.5].

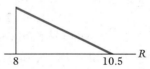

a. Find the density at *R* = 8.

b. Find the density at *R* = 9.

   Hint: The similar triangles in the following diagram imply that $\frac{a}{b} = \frac{c}{d}$.

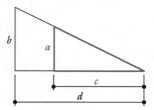

c. Find the probability that the train arrives after 9:00 P.M.

d. Find the probability that the train arrives between 8:00 and 9:00 P.M.

e. According to this probability density function, what is the most likely 30-minute time interval for the arrival of this train, and what is that probability?

6.6   THE NORMAL PROBABILITY DISTRIBUTION

The normal probability distribution will be extremely important later in the text.

The uniform probability distribution is a straight horizontal line, signifying that probabilities are spread out evenly over the interval on which it is defined. Although the uniform distribution has many applications, one of the most important and widely applied continuous probability distributions is the normal distribution; Figure 6-11 shows its model. This model was derived from a bell- or mound-shaped histogram that may have looked somewhat like Figure 6-12.

Thousands of random variables have mound-shaped histograms, which lead to normal probability distributions. A continuous random variable in which values tend to cluster around an expected mean and then tail off symmetrically in both directions most often has the properties of a normal probability distribution. Just a few examples are human life expectancies, heights, weights, IQs, and salaries; weather records; and a great variety of manufacturing, social science, and business data.

Just as in the case of the uniform probability distribution, the total area under this distribution (or any continuous probability distribution) must equal 1. However, instead of the probability's being distributed "uniformly," the intervals with greatest probability in this distribution are clustered about the theoretical mean value of the random variable, μ. Probabilities then decline symmetrically on both sides of μ. The probability density function of the normal distribution is very

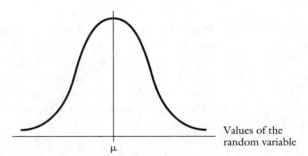

Figure 6-11 The model for the normal distribution.

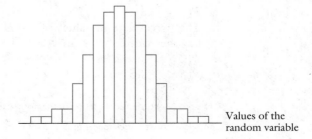

Figure 6-12

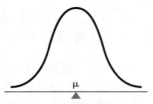

Figure 6-13 The mean is the balance point of the normal distribution.

complicated but need not be dealt with in order to find the desired areas.[8] What must be known is the variable's mean value, μ, and its standard deviation, σ.

Definition ▶ The **mean value, μ,** of a continuous random variable is the balance point of the probability distribution (Figure 6-13).

Sometimes μ is estimated from historical data, and sometimes it is determined theoretically.

Definition ▶ A **standard normal distribution** is a normal distribution with a mean value of 0 and a standard deviation of 1.

However, most practical applications of the normal distribution are **nonstandard;** that is, their means are not 0, nor are their standard deviations 1. The same table (to be introduced shortly) is used to find areas of intervals under both standard normal and nonstandard normal distributions. The following example shows how to use that table for a standard normal model, and further examples will show how to adapt it to nonstandard normal models.

E x a m p l e
9

A continuous random variable, S, has the standard normal distribution; that is, its means is zero ($\mu = 0$) and its standard deviation is one ($\sigma = 1$). Find the following:

a. $P(S = 1.3)$
b. The percentage of values of S that are at least $-.56$ but no more than 1.05; in other words, $P(-.56 \le S \le 1.05)$
c. $P(.78 \le S \le 2.08)$
d. $P(S \le -.45)$

[8]The probability density function of the normal distribution is $Y = \dfrac{1}{\sigma\sqrt{2\pi}}\, e^{-\frac{1}{2}\left(\frac{x-\mu}{\sigma}\right)^2}$. More theoretical approaches to statistics might examine some of the properties of this function.

Solution   This example implies that the frequency curve of all possible values of S constructed from a frequency histogram would be identical (or "close enough") to the standard normal probability distribution. The relationship between historical data and the likelihood of future events justifies the use of the standard normal probability distribution as a model.

a. $P(S = 1.3) = 0$. The probability of any continuous random variable's achieving one specific value is always zero.

b. $P(-.56 \le S \le 1.05)$. This probability is the sum of the two shaded areas in the following diagram.

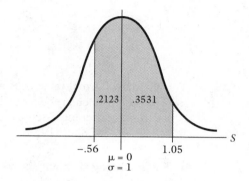

Use Appendix A-2, "Standard Normal Curve Areas," to find the area of the shaded region. First find the area between the mean, 0, and 1.05. A score from a standard normal distribution is referred to as a **Z-score,** or simply as Z in Appendix A-2.[9] According to the diagram at the top of the table, each table entry gives the area under the probability density function between 0 and the desired Z-score.

Find the row 1.0 in the Z column. Now scan the top of the table and find the column headed by .05. The intersection of the 1.0 row and the .05 column lists the area between $z = 0$ and $z = 1.05$, which is .3531. Column and row intersections provide a quick way of finding areas associated with values of Z measured to the nearest hundredth.

To find the area between 0 and $-.56$, you must make a slight adjustment. This particular table is constructed without negative values of Z, because the normal curve is symmetrical and negative values are therefore not needed.[10] Simply find the area between 0 and $+.56$, which is the same as the one between 0 and $-.56$. This area is the intersection of the .5 row and the .06 column, or .2123. Add the two probabilities together: $.3531 + .2123 = .5654$. In other words, 56.54% of all values of S are at least $-.56$ but no more than 1.05.

---

[9]The reason for naming this variable Z, for "standardize," will become clear after the next example.

[10]There are many different formats for the standard normal table. Some formats do show negative values of Z.

c. The next diagram shows the area corresponding to $P(.78 \leq S \leq 2.08)$.

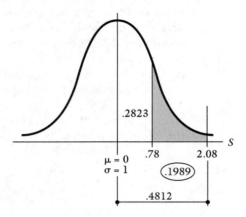

The standard normal table of Appendix A-2 is constructed so that areas are given from 0 to a particular Z-score. Thus, subtraction must be used to find the area between two different Z-scores on the same side of the mean: $P(.78 \leq S \leq 2.08)$ $= .4812 - .2823 = .1989$.

d. The following diagram shows the area corresponding to $P(S \leq -.45)$.

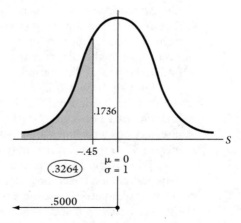

Even though there are, theoretically, no "ends" to a normal probability distribution, the total area is still equal to 1 and the distribution is symmetrical, implying that the area to the left of 0 is .5000. Once again, with some subtraction the desired probability, .3264, is obtained. •

The next example shows how to use the standard normal table with a normal distribution in which the mean is not zero, nor the standard deviation one—the usual case in practical applications.

E x a m p l e
10

In a certain school district where students have taken SAT tests for many years, the results on the mathematics portion of the test are normally distributed with a mean of 610 and a standard deviation of 28.5. What percentage of the students in this district scored between 650 and 700?

Solution   The shaded area in the following diagram represents the required percentage.

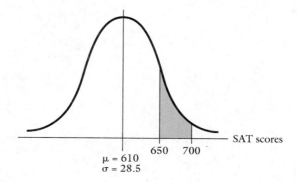

The first step is to standardize the scores 650 and 700. To **standardize** a score is to relate it to a standard normal distribution, which has a mean of 0 and a standard deviation of 1. This process is also referred to as finding Z-scores. If x is a score from a normal distribution with mean μ and standard deviation σ, then the Z-score of x, denoted by $Z(x)$, is found with the following formula.

Standardization Formula

$$Z(x) = \frac{x - \mu}{\sigma}$$

(6-7)

In a *standard* normal distribution with $\mu = 0$ and $\sigma = 1$, every score is a Z-score and reflects how many standard deviations away from the mean that score is. The above standardization formula accomplishes the same thing by first finding how far the score is away from the mean and then dividing the standard deviation into that difference. In this case, the Z-score of 650 is found to be

$$Z(650) = \frac{650 - 610}{28.5}$$

$$\approx 1.40.$$

This means that 650 is (approximately) 1.40 standard deviations bigger than the mean. Likewise, $Z(700) \approx 3.16$.

The first diagram can now be translated into a new, more familiar one that relates it to the standard normal model used earlier.

The scores of 650 and 700 have been *standardized* to 1.40 and 3.16, respectively.

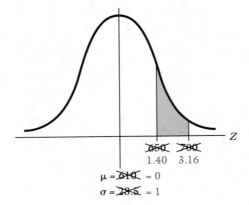

The corresponding areas are as follows:

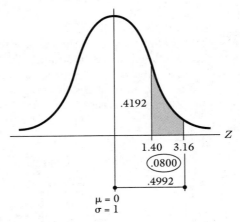

As the above diagram implies, .0800 = 8.00% of the students scored between 650 and 700. ●

Referring to the last example, note that the area from 1.40 to 3.16 under the *standard* normal distribution is the same as the area from 650 to 700 under a normal distribution with $\mu = 610$ and $\sigma = 28.5$. This example contained several additional points to notice.

- Z-scores were rounded to the hundredths place because that is the limit of accuracy for the standard normal table in Appendix A-2.
- The inclusion (or exclusion) of the end points of an interval would not have affected any percentages or probabilities.
- To say that 8.00% of the students scored between 650 and 700 is considered equivalent to saying that the probability that a score is between 650 and 700 is 8.00%.

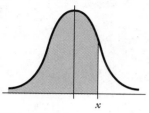

Figure 6-14

Finding Areas Under a Normal Curve Without Using a Table

Most statistical software programs will give areas under a normal probability density function, and some calculators may be programmed to do the same. Minitab, for instance, allows the user to specify x, μ, and σ and then calculates and prints the area to the left of x, the shaded area in Figure 6-14.

There is no need to compute a Z-score; the program does this, although it does not print the result. The area obtained in this way is often more accurate than one obtained through the use of Appendix A-2, because the calculated Z-score is rounded past the hundredths place and a more sophisticated method is used. Remember that, whereas Appendix A-2 gives the area between 0 and the desired score, Minitab calculates the cumulative area to the left of the desired score.

The command that is used to find the area is the same CDF command that was used for the binomial and Poisson random variables. The next example illustrates the use of this command with a normal distribution.

E x a m p l e 11

The results of a certain form of IQ test are normally distributed with a mean of 100 and a standard deviation of 10. A person who took the test is chosen at random. Let the random variable X designate his or her score. Compute the following.

a. $P(X \leq 94.5)$
b. $P(X \leq 115)$
c. $P(96 \leq X \leq 108)$
d. $P(88 \leq X \leq 98.5)$

Solution   It is often helpful, as a first step, to sketch the areas in question.

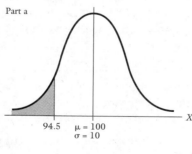

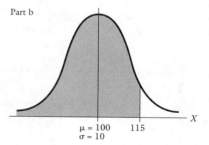

There is no need to standardize scores when using Minitab.

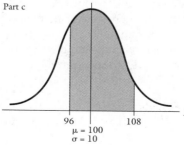

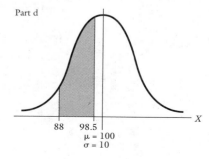

Now, using Minitab:

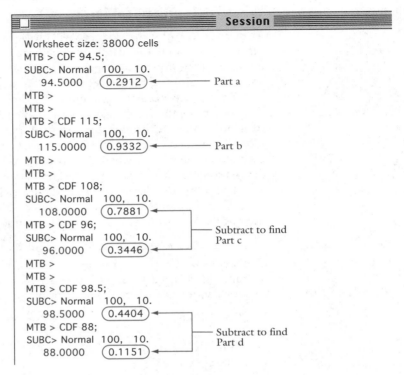

The probabilities for Parts a and b are the numbers circled: .2912 and .9332, respectively. The probabilities for Parts c and d are found by subtraction. The answer for Part c is .7881 − .3446 = .4435, and the answer for Part d is .4404 − .1151 = .3253. Each answer is also the percentage of people who had those scores.   ●

E x a m p l e 12   A certain machine produces ball bearings. Experience has shown that the mean diameter of these ball bearings is $\mu = 1.25$ cm and that $\sigma = .09$ cm. Use the standard normal curve areas in Appendix A-2 to answer the following questions.

a. From a box of 5000 of these ball bearings, how many would probably be returned by a company that places them in machine parts and can use only ball bearings with diameters smaller than 1.375 cm?

b. In an effort to examine the quality of the larger-sized ball bearings being produced, the manufacturing staff will subject those ball bearings with diameters in the 80th percentile to special tests. What diameter marks the smallest of these larger-sized ball bearings?

Solution   a. The shaded area in Figure 6-15 represents the percentage of ball bearings with diameters greater than 1.375, which would be returned by the company.

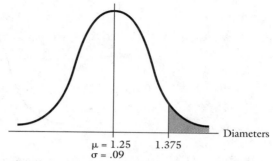

Figure 6-15 The percentage of ball bearings that would be returned.

The first step in using Appendix A-2 is to standardize this normal distribution by computing the Z-score of a diameter of 1.375 cm. Using the standardization formula,

$$Z(1.375) = \frac{1.375 - 1.25}{.09}$$

$$\approx 1.39.$$

Next, consulting Appendix A-2, it can be seen that an area of .4177 lies between the mean and a Z-score of 1.39. Figure 6-16 shows that 8.23% of the ball bearings have diameters greater than 1.375 cm. This corresponds to $.0823 \cdot 5000$, or about 412 ball bearings.

b. The diameter that marks the 80th percentile is greater than 80% of all the ball bearings produced. Since 50% of the ball bearings produced have diameters less than the mean, an additional 30% are needed to reach the 80th percentile. Figure 6-17 illustrates this relationship.

Find the Z-score that determines this additional 30% by using Appendix A-2 "backwards." First find the four-digit area in the body of Appendix A-2 that is closest to an area of .3000. This is .2995, which corresponds to a Z-score of 0.84. Thus, $z = 0.84$ in Figure 6-17.

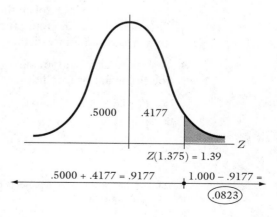

Figure 6-16

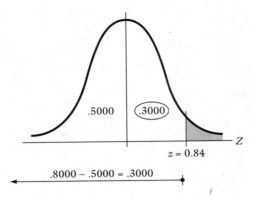

Figure 6-17

Next, using the standardization formula[11]

$$Z(x) = \frac{x - \mu}{\sigma},$$

substitute all of the known quantities and solve for x:

$$0.84 = \frac{x - 1.25}{.09}$$

$$.0756 = x - 1.25$$

$$x \approx 1.30.$$

Thus, a diameter of 1.30 marks the 80th percentile. All ball bearings with diameters of at least 1.30 cm will be tested.   ●

Exercises 6.6

1. Explain why it is not possible to develop a more efficient table of values for areas under a normal curve so as to make it unnecessary to standardize a score before using the table.

2. In your own words, explain the similarities and differences between probability distributions for discrete random variables and probability distributions for continuous random variables.

3. Assume that a random variable, Z, has a standard normal distribution. Find the following probabilities.

   a. $P(-.87 \leq Z \leq 1.25)$

   b. $P(1.38 \leq Z \leq 2.78)$

   c. $P(-.67 \leq Z)$

---

[11]Recall that the variable $Z(x)$ represents the Z-score for a value of the variable X. $Z(x)$ may also be referred to less formally as z.

*   4. Assume that a random variable, Z, has a standard normal distribution. Find the following probabilities.

a. $P(|Z| \leq .35)$

b. $P(|Z| \geq .45)$

5. In Chapter 4, the Empirical Rule generalized about mound-shaped histograms or frequency curves.

- About 68% of the scores fall within one standard deviation of the mean.
- About 95% of the scores fall within two standard deviations of the mean.
- Practically all of the scores fall within three standard deviations of the mean.

Explain how this rule is justified if a normal probability distribution serves as the model for a mound-shaped histogram or frequency curve.

*   6. In Chapter 2, possible outliers were identified as lying 1.5 IQRs beyond the 75th percentile, and probable outliers, as lying 3 IQRs beyond the 75th percentile. Using the standard normal distribution as a model, determine how many standard deviations away from the mean the possible and probable outliers would lie.

Hint: Start by finding the Z-score of the 75th percentile, and let the IQR be twice that Z-score.

*   7. Refer to Exercise 6. What percentage of scores from a random variable with a normal distribution would be expected to lie between the inner and outer fences? What percentage would be beyond the outer fences?

*   8. Refer to Exercise 7 and the percentage of scores beyond the outer fences. Does this mean that probable outliers never occur? Explain.

9. The continuous random variable X has a normal distribution with the stated mean and standard deviation. Determine areas under the normal distribution that correspond to the following probabilities:

	μ	σ	Probability
a.	90.0	9.00	$P(100 \leq X \leq 115)$
b.	3.2	.67	$P(3 \leq X \leq 4)$
c.	25.0	3.80	$P(X \geq 30)$; $P(X \leq 21.2)$
d.	100.0	10.00	$P(91 \leq X \leq 113)$
e.	490.0	30.00	$P(410 \leq X \leq 475)$; $P(X \leq 505)$

10. Arrival times of trans-Atlantic mail are normally distributed with a mean of 7.5 days and standard deviation of 1.25 days.

a. Determine the first quartile of arrival times.

b. Determine the third quartile of arrival times.

c. What is the interquartile range?

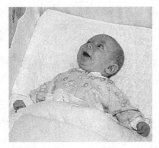

What is the probability that this newborn boy fits into the weight class 6.00 to 7.00 lb? (*Joy Small*)

11. The shape of a frequency curve showing the length of drive, D, for a particular golf pro is close enough to normal to be treated as such. The mean drive is 337 yards with a standard deviation of 45 yards. If the pro hits 8000 drives in the course of a year, how many of them are between 356 and 390 yards?

12. The average birth weight for babies at a certain hospital is 6.78 pounds with a standard deviation of .45 pounds. Treat as normal the frequency curve that shows the distribution of these weights.

   a. What is the probability that a baby born at this hospital has a weight between 6.00 and 7.00 pounds?

   b. What percentage of babies born weigh at least 8.00 pounds?

   c. If 12,728 babies were born at this hospital, how many boys weighing at least 7.75 pounds were born? (Assume that boy and girl births are equally likely.)

   d. Unusually small babies are given special attention. If the hospital defines "unusually small" as having a birth weight in the first quartile, then what is the maximum birth weight that merits special attention?

13. Refer to Exercise 12. Determine the percentile at that hospital for a newborn weighing 7 pounds 9 ounces.

14. The revenues from fines or traffic citations in a certain community are normally distributed with a yearly mean of $45,000 and a standard deviation of $3500. These revenues go toward maintaining the fleet of patrol cars, which amounts to an annual expense of $38,000.

   a. What is the probability that next year's revenues will cover the fleet's maintenance expense?

   b. Assuming that the revenue collected in one year is independent of the revenue collected in any other year, what is the probability that fleet maintenance will be covered for the next 2 years? The next 4 years?

   c. Determine the revenues that mark the 10th and 90th percentiles.

15. The mean score on a certain mathematics aptitude test is 72, with a standard deviation of 8.5. Assume a normal distribution of scores.

   a. What score, S, would one have to earn so that no more than 15% of the scores would be higher than S? No more than 5%?

   b. If a student's Z-score on the test is known to have been $-.67$, what was the student's actual score?

16. In 1993 students averaged 20.7 on the ACT test and 902 on the SAT test. Take the word "average" to imply the mean, with standard deviations of 4.5 on the ACT and 108 on the SAT. Determine the minimum score on each test that would place the student in the upper quintile. Assume scores are normally distributed on both tests.

GROUP   17. Two companies are both bidding on a contract to supply tires for a fleet of rental cars. The first company claims that the mean life of its tires is 24,000 miles with a standard deviation of 1000 miles. The second company claims that the mean life of its tires is 28,000 miles with a standard deviation of 2500 miles. Assuming that tire life is normally distributed, which company should be chosen if a tire should last:

a. At least 20,000 miles?

b. At least 22,000 miles?

c. At least 25,000 miles?

GROUP

18. Refer to the results of Exercise 17. Use trial and error to determine the maximum required tire life (mileage), X, that would warrant the choice of the company with the claimed mean of 24,000 miles and standard deviation of 1000 miles.

* 19. A vending machine sells soda pop in 10-ounce paper cups. Assume that the actual amount dispensed is normally distributed, with a standard deviation of .40 ounces. What amount should the machine be set up to dispense if only 3% of the cups are to overflow?

Stat Fact

The leading career money winner among professional women bowlers as of March 1993 was Lisa Wagner, with total earnings of $508,019.

20. For all pro bowlers on tour for at least 10 years, the standard deviation of the scores of their games is 28 pins. However, the mean score of a right-handed bowler is 191, and the mean score of a left-handed bowler is 218. Assume a normal distribution of scores.

a. What percentage of left-handed bowlers have an average score of at least 230?

b. What percentage of right-handed bowlers have an average score of at least 230?

* c. If 15% of all bowlers are left-handed, what percentage of those who average at least 230 are left-handed?

Case Study *Fat Tails*

Wealth is a fat-tailed distribution because rich people tend to get richer than do most of the population. (*Philadelphia Stock Exchange*)

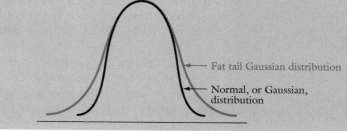

The normal distribution is also called the Gaussian distribution in honor of Carl Friedrich Gauss (1777–1855), who was responsible for much of its development. A **fat-tail** Gaussian distribution is one in which unlikely events, those in the distribution's tails, occur slightly more often than they theoretically should.

← Fat tail Gaussian distribution

← Normal, or Gaussian, distribution

Many events have fat-tail Gaussian distributions. For instance, some economic conditions such as wealth are fat-tailed because there are disproportionately more very rich people than, say, very short people. (Wealth reinforces itself; wealthy people tend to get richer.)

Some believe that the theory of fat-tailed distributions can be applied to make stock-market predictions. In addition, many mathematicians believe that stock-market behavior is made up of a series of random processes, which can be studied through a new branch of mathematics called fractal theory. The October 9, 1993, issue of the London *Economist* contains an article (pp. 10–13) that describes a recent effort to use fat-tail distributions and fractal theory to predict market trends. Obtain a copy of this article from your library, and use it to answer the following questions.

I. The **Hurst exponent** is the probability that one event will be followed by a similar event. If the Hurst exponent for an event is .5, then the experiment has a Gaussian distribution as a model. If the Hurst exponent is greater than .5, then outcomes of the experiment tend to occur in runs. What is the name given to random processes with Hurst exponents greater than .5? What is the name given to random processes with Hurst exponents less than .5?

II. In your own words, explain what is meant by a **leptokurtotic** process.

III. In your own words, explain what is meant by the **degrees of freedom** of an event.

IV. Does the article maintain that weather systems are completely random, leptokurtotic, or contrarian processes? Cite the sentence in the article that supports your answer.

V. Give three examples of contrarian processes.

CHAPTER SUMMARY

Random variables are variables whose values depend on some type of probability function. The two kinds of random variables studied in this chapter were discrete random variables and continuous random variables. The binomial random variable is one type of discrete random variable and is often used to "keep score," or indicate the number of successes achieved in a certain number of repetitions of an experiment. The Poisson random variable is another special type of discrete random variable and is often used in situations in which an event is occurring at random over time or space.

A continuous random variable is one that can assume any value in an interval of values. The chapter included several examples.

The terms "probability density function" and "probability distribution" were used interchangeably, and both often take the form of graphs that show the concentrations of probabilities. A probability distribution can be modeled on a theoretical basis or constructed from a frequency curve.

The normal distribution is a special type of probability distribution of a continuous random variable. It is applied widely in nearly all fields of study and will be used often throughout the rest of this text.

Key Concepts, Terms, and Formulas

Videotape Suggestions

Program 17 from the series *Against All Odds: Inside Statistics* shows additional applications of the binomial distribution, along with the application of a quincunx to illustrate how a normal distribution may be used to approximate a binomial distribution. Also, *Program 4* (suggested earlier) and *Program 5* from the same series include more applications of the normal distribution along with further applications of the standardization formula.

(Picture Perfect™)

A randomly selected
sample can be used to
represent an entire
population.

Chapter

7 Sampling and Estimating Population Parameters

It is usually not practical or even possible to compute the mean and standard deviation using all the scores in a set of data. In an effort to obtain the mean gasoline mileage for all 1994 minivans, for instance, it would be nearly impossible to trace all of the minivans' owners and obtain an accurate gasoline mileage for each vehicle.

This chapter shows how data obtained from a randomly selected sample can be used in place of all the data from the entire population. Such a sample must be selected carefully, however. The **case study** at the end of this chapter examines some of the difficulties that occurred when proper steps were not taken in a random selection for the military draft.

7.1 SOME BASIC DEFINITIONS

Many of the examples and exercises presented so far were based on data from populations.

Definitions ▶

A **population** is all the members of a set (no matter how many), which have a given characteristic.

Population data are data associated with a certain population.

A **population parameter,** or simply **parameter,** is a measure (such as the mean or standard deviation) that is computed using population data.

The word "population" has been used informally up to this point.

283

Refer to Example 10 in Chapter 6 (on page 271). The mean of 610 and the standard deviation of 28.5 are population parameters because they were based on the results of all the students in the district. Less explicitly, Example 11 also involves population parameters. The mean of 100 and the standard deviation of 10 are computed from all the results of the IQ test under consideration.

Definitions ▶

A **sample** of a population is a subset of the population; that is, every member of the sample is also a member of the population.

A **sample statistic,** or simply **statistic,** is an estimated population parameter computed from the data in the sample.

An uppercase N will denote the size[1] of the population, and a lowercase n will denote the size of the sample.

One of the main purposes of this chapter is to explain how to compute sample statistics from a specific sample. First, however, it is necessary to outline a method of choosing members randomly from the population to form a sample, called assembling a sample.

7.2 HOW TO ASSEMBLE A SAMPLE

Consider the sketch in Figure 7-1. There are usually many possible samples that can be chosen, or assembled, from any given population in order to compute estimated values of the population parameters. If a population has a size of, say, $N = 100$, and if statistics are to be computed using a sample of size 40,

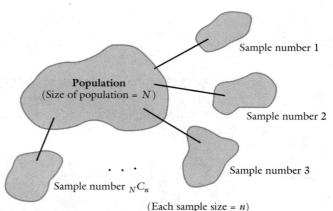

Figure 7-1 The number of samples in the population.

---

[1]In this context, "size" means the number of members in a set and not the set's physical dimensions.

then an almost unbelievable number of samples of size 40 can be formed from the 100 members of the population. In fact, there are $_{100}C_{40}$, or 13,746,234,145,800,000,000,000,000,000,000, possible samples of size 40.

The number of possible samples, $_NC_n$, is the same binomial coefficient that was studied in the last chapter but interpreted differently, as the number of possible subsets of size n taken from a population of size N.

Regardless of how many possible samples can be assembled to compute the desired statistics, it is very important that the actual sample used be a random one.

Definition ▶ A **random sample** is a sample of which every member of the population has an equal chance of being a part.[2]

S t a t   F a c t

The EPA's Environmental Monitoring & Assessment Program (EMAP), launched in 1988, originally used probabilistic data sampling to monitor environmental changes on a regional scale. By 1994 scientific reservations about this procedure, along with budgetary and political considerations, led to a proposed EMAP smaller in size and more hypothesis driven and research oriented.

If a sample is not random, then the sample statistics computed will reflect the special factors that formed the sample and may not be typical of the population.

Imagine that you would like to know the mean age of all the students at a certain college. Since there are probably many thousands of students, it is not reasonable to compute μ (or σ) directly from the population data (assuming that such data are even obtainable). You decide to use a sample of size 50 instead and estimate the mean from it. How do you pick the 50 students? How do you assemble (form) the sample? If you stand at a specific point on campus and poll the first 50 students who pass by, you may not get a random sample, for several reasons. First, depending on the time of day, you will either miss older students who attend classes in the evenings or younger students who attend during the day. Second, you may be polling many students who happen to be going to the same class, and that class may not reflect the make-up of the college as a whole. Picking 50 students at random from the student directory and then finding out their ages would be more work but would be a better method of forming a sample that is truly random.

If, for example, 50 student names are to be randomly chosen from a student directory for the purpose of determining sample statistics, then the method followed must give any name in the directory an equal chance of being part of the sample. The 50 names could be selected by using slips of paper in a box. If the slips were thoroughly mixed and 50 names withdrawn, they would constitute a valid random sample. Thorough mixing is often much more difficult than it would seem, however. The 1970 draft lottery was conducted by "mixing" capsules that contained birth dates. But the capsules were not mixed thoroughly enough (even though a revolving drum was used), and the draft order ended up very strongly resembling the order in which the capsules had been put in the drum. Many lawsuits were filed, and the 1971 draft lottery was modified as a result.[3]

---

[2]In more sophisticated applications of statistics, stratified random sampling and sequential sampling are sometimes preferred over the truly random methods that will be described here.

[3]See the case study at the end of this chapter for more information on draft lotteries and other randomization methods.

```
Appendix A-3
Random Numbers

71274    84346    75444    85690
64017    01373    14665    31891
43747    17686    11045    15549
59688    48689    41591    47042
95016    73467    11447    59500

65207    30591    65947    58339
34510    78657    08883    49489
56299    60624    91572    31734
02113    12579    86172    03819
00884    87979    45184    61572

89367    53577    97412    19603
99781    56740    42659    46617
92024    12100    76013    12587
82861    09215    87342    72789
11286    13011    67982    74101
```

Figure 7-2 Portion of the random number table shown in Appendix A-3.

Alternatively, one of the best methods for choosing members from the population to make up a random sample is to use a **random number table,** which is a table of the digits 0 through 9 repeated many times in random order. For ease of reading and manipulation, the numbers are usually arranged in blocks of five. Figure 7-2 shows part of a random number table. Appendix A-3 is a whole page of random numbers taken from a book full of pages of random numbers.

To use a random number table, first select a part of the page. Figure 7-2, for instance, is the upper left-hand corner of Appendix A-3. Next, select digits from the table in any pattern, such as vertically, horizontally, diagonally, or every other one going across. The pattern does not matter as long as digits are not selected purposefully.

**E x a m p l e
1**

Outline a method of using the random number table to choose a sample of 35 days of the year that can be used to estimate the mean yearly noontime temperature.

Solution   Since the year has 365 days in it, each day can be identified with a number from 1 through 365. Thirty-five of these 365 numbers must be picked at random. The following figure shows one way to proceed. Arbitrarily choose the first three digits of each five-digit block and work down one column, then another, until 35 numbers have been chosen. For example, if the first three digits are 345, that means that the 345th day of the year, or December 11, is to be included in the sample. Numbers that do not apply, such as the first six (712, 640, 437, 596, 950, and 652), can be ignored. Also, if a number comes up more than once, its repetitions are ignored. Follow this procedure until 35 dates are obtained to include in the sample. It would also have been just as valid to use, say, the middle three digits and proceed across the random number page rather than down it. Any arbitrary pattern is legitimate.

```
Appendix A-3
Random Numbers

71274    84346       75444    85690
64017    01373       14665    31891
43747    17686       11045    15549
59688    48689       41591    47042
95016    73467       11447    59500

65207    30591       65947    58339
34510    78657       08883    49489
56299    60624       91572    31734
02113    12579       86172    03819
00884    87979       45184    61572

89367    53577       97412    19603
99781    56740       42659    46617
92024    12100       76013    12587
82861    09215       87342    72789
11286    13011       67982    74101
```

Dec.11 — 34510
Jan. 21 — 02113
Jan. 8 — 00884
April 22 — 11286

Most scientific and graphics calculators generate random numbers; the procedures vary. On a TI-82 graphics calculator, the keystroke sequence

MATH, right arrow key (▶) three times, 1

produces the following screen.

```
rand
```

When the ENTER key is pressed, a nine-digit random number between 0 and 1 is printed. Which digits to use may be decided arbitrarily, and this procedure repeated any number of times, to obtain a list of random numbers.

Example 2

The average 1992 salaries of public secondary school teachers, by state, follow. A two-digit state identification number for each state is also included.

01	AL	27,000	11	HI	34,500	21	MA	37,300	31	NM	27,000	41	SD	23,700
02	AK	45,400	12	ID	26,800	22	MI	41,300	32	NY	44,500	42	TN	29,800
03	AZ	31,800	13	IL	40,000	23	MN	34,600	33	NC	29,400	43	TX	29,700
04	AR	27,300	14	IN	36,000	24	MS	24,900	34	ND	24,600	44	UT	27,200
05	CA	41,700	15	IA	30,200	25	MO	29,800	35	OH	33,800	45	VT	34,400
06	CO	33,800	16	KS	30,700	26	MT	28,800	36	OK	26,400	46	VA	33,600
07	CT	48,400	17	KY	32,500	27	NE	27,200	37	OR	35,100	47	WA	35,700
08	DE	35,500	18	LA	27,100	28	NV	35,100	38	PA	39,400	48	WV	27,800
09	FL	31,100	19	ME	31,500	29	NH	33,200	39	RI	36,100	49	WI	37,000
10	GA	29,500	20	MD	41,000	30	NJ	42,500	40	SC	29,500	50	WY	30,800

Source: *Statistical Abstract of the United States: 1993,* p. 161.

Estimate the population mean state salary[4] based on a sample of five states.

Solution   Once again, use the random number table. An arbitrary decision has been made to use only the last four digits in the five-digit blocks, which are to be read as two two-digit numbers. Furthermore, movement across the page will be diagonal, across the five-digit blocks.

```
Appendix A-3
Random Numbers

7(12)(74)   84346      75444      85690
64017    0(13)(73)   14665      31891
43747    17686      1(10)(45)   15549
59688    48689      41591      4(70)(42)
95016    73467      11447      59500

65207    30591      65947      58339
34510    78657      08883      49489
56299    60624      91572      31734
02113    12579      86172      03819
00884    87979      45184      61572

89367    53577      97412      19603
99781    56740      42659      46617
92024    12100      76013      12587
82861    09215      87342      72789
11286    13011      67982      74101
```

This procedure yields states with identification numbers 12, 74, 13, 73, 10, 45, 70, and 42. Of these, the numbers 12, 13, 10, 45, and 42 can be used. They correspond to the states of Idaho, Illinois, Georgia, Vermont, and Tennessee, respectively. The mean of the salaries for these five states, \$32,100, is the estimated population mean state salary.[5]   ●

In the preceding example, the statistic \$32,100 can also be referred to as a **point estimate** of the population mean, since it is a single number (point) that attempts to estimate an unknown parameter. Chapter 8 will be concerned with a different type of estimate, an **interval estimate,** which places the unknown parameter between two numbers with a certain degree of confidence.

The size of a sample is very critical in determining whether the sample statistic is a good estimate of the population parameter. Obviously, the larger the sample,

[4]This is not the same as the national mean salary, which would take into account the number of teachers in each of the states and result in a weighted mean. In other words, New York's mean salary, \$44,500, would count much more heavily in the calculation of the national mean salary than would South Dakota's mean salary, \$23,700.

[5]It turns out that this estimated mean state salary is within 3% of the actual population mean state salary of \$33,040. Three percent is very good agreement for such a small sample.

the more likely it is to yield a good estimate. Later chapters will examine the implications of sample size more closely.

Exercises 7.1–7.2

Some exercises in this chapter specify a method of using the random number table so that answers may be checked. However, one method of using the random number table is not necessarily better than another.

1. Refer to the discussion of forming a random sample of 50 students at a college on page 285. Outline a method for choosing a random sample when a student directory is not available.

2. Is it possible for a random sample to appear as though its members were not randomly chosen? Explain.

3. If a population has 45 members, then how many ten-member samples are possible?

4. The value of $_NC_n$ increases until n becomes close to what number?

5. Refer to Example 2 on page 287. Find a point estimate of the population mean state salary by using a random sample of size $n = 8$. Generate the random sample from the random number table in Appendix A-3 by working down the last column. Use the first two digits of the five-digit blocks. Then find a second point estimate by choosing eight random numbers in a pattern of your own, or use a calculator to generate them. How do your two point estimates compare?

GROUP

6. The following array shows the 1993 unemployment rate for each of the 50 states.

AL	7.5	AK	7.6	AZ	6.2	AR	6.2
CA	9.2	CO	5.2	CT	6.2	DE	5.3
FL	7.0	GA	5.8	HI	4.2	ID	6.1
IL	7.4	IN	5.3	IA	4.0	KS	5.0
KY	6.2	LA	7.4	ME	7.9	MD	6.2
MA	6.9	MI	7.0	MN	5.1	MS	6.3
MO	6.4	MT	6.0	NE	2.6	NV	7.2
NH	6.6	NJ	7.4	NM	7.5	NY	7.7
NC	4.9	ND	4.3	OH	6.5	OK	6.0
OR	7.2	PA	7.0	RI	7.7	SC	7.5
SD	3.5	TN	5.7	TX	7.0	UT	3.9
VT	5.4	VA	5.0	WA	7.5	WV	10.8
WI	4.7	WY	5.4				

Source: *Statistical Abstract of the United States, 1994,* p. 399.

a. Find a point estimate of the population mean state unemployment rate by using a random sample of size $n = 10$, chosen as follows. Assign the same state identification numbers, 01 through 50, to the states as in Example 2 on page 287. Then use the random number table to select ten scores by starting at the top of the seventh column and moving down, using the third and fourth digits.

b. Repeat Part a, but use the third column instead of the seventh column.

c. Find the difference in the estimates for Parts a and b.

d. Find the actual mean state unemployment rate, and then determine what percentage of the actual population mean state unemployment rate is represented by the difference from Part c.

7. The following array shows the population projections, by state, for the numbers of people under 18 years of age in the year 2000. All projections are in thousands.

Stat Fact

70% of all births in the United States are to mothers under the age of 30. Only 1% of births are to women at least 40 years old.

AL	1093	AK	174	AZ	1188
AR	618	CA	8664	CO	813
CT	780	DE	199	FL	3520
GA	2087	HI	310	ID	276
IL	2981	IN	1427	IA	608
KS	628	KY	887	LA	1105
ME	333	MD	1373	MA	1393
MI	2365	MN	1148	MS	756
MO	1352	MT	179	NE	385
NV	324	NH	351	NJ	2003
NM	485	NY	4200	NC	1809
ND	145	OH	2694	OK	683
OR	689	PA	2777	RI	236
SC	987	SD	190	TN	1256
TX	4671	UT	651	VT	156
VA	1734	WA	1225	WV	375
WI	1215	WY	109		

Source: *State and Metropolitan Area Data Book 1991*, 4th ed., p. 207.

Define four regions as follows.

Region	States
Northeast	ME, NH, VT, MA, RI, CT, NY, NJ, PA
Midwest	OH, IN, IL, MI, WI, MN, IA, MO, ND, SD, NE, KS
South	DE, MD, VA, WV, NC, SC, GA, FL, KY, TN, AL, MS, AR, LA, OK, TX
West	MT, ID, WY, CO, NM, AZ, UT, NV, WA, OR, CA, AK, HI

Randomly choose projections by region to find a point estimate of the mean projection for all 50 states. Use the following procedure.

Assign the same state identification numbers, 01 through 50, to the states as in Example 2 on page 287. Use the random number table to assemble a sample of size $n = 12$ with the following characteristics. Two scores are from the Northeast, three scores are from the Midwest, four scores are from the South, and three scores are from the West. (Not all scores for each region need be randomly chosen before a score is chosen for another region, and once the required number of scores have been chosen for a given region, ignore all other scores for that region.) Use the mean of these 12 scores to estimate the mean of all 50 projections.

  Use the last two digits of the five-digit random number block. Start at the top of the first column and work down.

8. Exercise 7 used **proportional allocation sampling.**

 a. What appears to be the reason for sampling in this way?

 b. What factor or factors have still not been considered?

GROUP 9. A certain factory manufactures light bulbs. One hundred light bulbs are put in each box, and ten boxes are put in each case. At the end of the day, 50 cases are full. Although the manufacturing process is highly reliable, the factory wants to randomly select ten bulbs per day and test them. Describe a way of choosing ten light bulbs randomly (from different cases) that uses the position of each bulb in the box and the fact that the boxes and cases are numbered.

10. Identify each of the following as true or false.

 a. Statistics obtained from relatively large samples must be closer to the population parameters they estimate than statistics obtained from smaller samples.

 b. The mean of the sample statistics obtained from two random samples must be closer to the population parameter than either of the two separate sample statistics.

 c. For a given value of n, the mean of all possible sample means of size n is exactly the same as the population mean.

 d. It is permissible to use any of the digits in the five-digit blocks and any pattern of blocks from a random number table.

 e. When using the random number table, it is impossible to obtain a sample that is highly uncharacteristic of the population.

 f. Two different samples must yield two different estimates of the population parameter, μ.

7.3 CONSIDERATIONS WHEN ESTIMATING μ AND σ

As mentioned at the start of this chapter, in many practical questions the population parameters μ and σ will never be known, or even obtainable. Therefore, reliable estimates of these parameters are usually necessary. This section will explore the relationship between all sample means of a given size from a population and introduce a new sample statistic, which will be used to estimate the population standard deviation.

Using $\overline{X}$ to Estimate μ

Consider the data in the following array.

**Life Expectancies
(in years) for Males**

Finland	70.9
France	72.0
Canada	73.0
Netherlands	73.8
Iceland	75.7
Japan	75.9

Source: *Statistical Bulletin* 73:3, July–Sept. 1992, p. 14.

The mean life expectancy, μ, for this population of six countries could easily be computed directly from the population scores and would not need to be estimated from a sample.[6] For the sake of illustration, however, assume that someone desires to estimate μ from a sample of size $n = 4$. Since $N = 6$ is very small, all possible samples of size $n = 4$ can be listed, and there are $_6C_4$, or 15, of them. The notation $\overline{X}$ (read "X-bar") will denote the mean of an individual sample, which in each case is a different point estimate of the population parameter μ.

Sample No.	Scores in Sample	Mean of Sample, $\overline{X}$
1	{70.9, 72.0, 73.0, 73.8}	72.425
2	{70.9, 72.0, 73.0, 75.7}	72.9
3	{70.9, 72.0, 73.0, 75.9}	72.95
4	{70.9, 72.0, 73.8, 75.7}	73.1
5	{70.9, 72.0, 73.8, 75.9}	73.15
6	{70.9, 72.0, 75.7, 75.9}	73.625
7	{70.9, 73.0, 73.8, 75.7}	73.35
8	{70.9, 73.0, 73.8, 75.9}	73.4
9	{70.9, 73.0, 75.7, 75.9}	73.875
10	{70.9, 73.8, 75.7, 75.9}	74.075
11	{72.0, 73.0, 73.8, 75.7}	73.625
12	{72.0, 73.0, 73.8, 75.9}	73.675
13	{72.0, 73.0, 75.7, 75.9}	74.15
14	{72.0, 73.8, 75.7, 75.9}	74.35
15	{73.0, 73.8, 75.7, 75.9}	74.6

Since this population has a very small value of N, the population parameter, μ, can be computed easily; it is equal to 73.55.

The preceding table shows that a point estimate for μ found by using the mean, $\overline{X}$, of one of these samples could range from 72.425 to 74.6. Of course,

[6]A sample that includes every member of a population is called a **census**.

with a little luck, a sample could be chosen with $\overline{X}$ very close to μ. However, N is usually so large that the actual proximity of the estimate $\overline{X}$ to μ is never known.

It is no accident that $\mu_{\overline{X}}$, the mean of all the $\overline{X}$'s, is exactly equal to μ (73.55). In other words, in this particular instance,

$$\mu_{\overline{X}} = \frac{72.425 + 72.9 + 72.95 + \cdots + 74.6}{15}$$

$$= 73.55.$$

And, in general, $\mu_{\overline{X}} = \mu$.

In general, for a given value of n, all the possible values of $\overline{X}$ cluster tightly around μ.

The fact that the mean of all the possible sample means of a given size is equal to the mean of the population suggests that the best possible point estimate that can be made of the population mean, μ, is the mean of a random sample, $\overline{X}$.

Using s to Estimate σ

The letter s will stand for the estimate of σ that is made using a random sample. Although it is tempting to think in analogy with the estimation of μ with $\overline{X}$ and to compute s as the standard deviation of the sample, a refinement is necessary. The following formula is used to find s.

Formula to determine s, the estimate of the population standard deviation, σ:

$$S = \sqrt{\frac{\sum_{i=1}^{n} (x_i - \overline{X})^2}{n - 1}} \qquad (7\text{-}1)$$

Shortcut formula to determine s:

$$S = \sqrt{\frac{\sum_{i=1}^{n} x_i^2 - \frac{\left(\sum_{i=1}^{n} x_i\right)^2}{n}}{n - 1}} \qquad (7\text{-}2)$$

To reiterate, application of the above formula computes the sample statistic s, which is a point estimate for the population standard deviation, σ.

A comparison of Formula 7-1 with the definition formula for σ in Chapter 3 reveals some differences. First, the summation takes place over n scores instead of N scores, since the lowercase n denotes the number of scores in a sample and the uppercase N signifies the number of scores in a population, and s, being a sample statistic, is based on n scores. Second, $\overline{X}$ replaces μ. This, too, should seem reasonable. If the population parameter μ is unknown, then $\overline{X}$ is its best estimate. Finally, the denominator to determine s is $n - 1$. If this formula were truly analogous to that for σ, one would expect the denominator to be n. A demonstration of *why*

Most calculators will print only a lowercase n (instead of N) and the symbol $\overline{X}$ (instead of μ) for the mean.

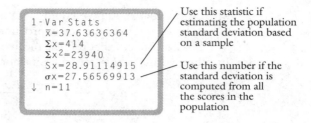

```
1-Var Stats
  x̄=37.63636364
  Σx=414
  Σx²=23940
  Sx=28.91114915
  σx=27.56569913
↓ n=11
```

Use this statistic if estimating the population standard deviation based on a sample

Use this number if the standard deviation is computed from all the scores in the population

Figure 7-3

the denominator must be $n - 1$ instead of n to provide the best possible estimate of the population standard deviation is well beyond the level of a first course in statistics. The denominator must be $n - 1$ for this estimate, nevertheless.[7]

It is important to note that in practical situations it is usually not necessary to apply Formula (7-1) or the shortcut Formula (7-2) directly to determine s. Any calculator with a statistics mode will have a key labeled s (or σ_{n-1} or σ'), and s can be found easily by using that function. Figure 7-3 shows the final screen on the TI-82 graphics calculator when it is used to work Example 4 in Chapter 3 (on page 91).

Notice from Figure 7-3 that the data set contained 11 scores. Also, the values of s (Sx) and σ (σx) are very close; one is arrived at by using a denominator of 11, and the other by using a denominator of $11 - 1 = 10$. When interpreting output such as this, one must know whether calculations have used all the scores of a given population. If so, then the standard deviation can be determined, and it is the number labeled σx. If not all the scores in the population have been used, then the standard deviation of the population can only be estimated from a sample, and the estimate is the sample statistic labeled Sx.

The following chart should also help to clarify matters.

**Comparison of Calculations for Population Parameters
and Sample Statistics**

Actual Population Parameter	Estimated Population Parameter (sample statistic)
Size of population $= N$	Size of sample $= n$
$\mu = \dfrac{\sum\limits_{i=1}^{N} x_i}{N}$	$\overline{X} = \dfrac{\sum\limits_{i=1}^{n} x_i}{n}$
$\sigma = \sqrt{\dfrac{\sum\limits_{i=1}^{N} (x_i - \mu)^2}{N}}$	$s = \sqrt{\dfrac{\sum\limits_{i=1}^{n} (x_i - \overline{X})^2}{n - 1}}$

[7]Later courses in statistics will show that using n as the denominator would result in a slight underestimate of the standard deviation; $n - 1$ corrects this slight negative bias.

E x a m p l e
3

The mortality rates from sudden infant death syndrome (SIDS) seem to vary significantly among states. Twenty-nine states have "higher" mortality rates, defined in this context as being more than 150 deaths per 100,000 live births. The following is a random sample of size $n = 12$ from these states.

TN	186.7	MT	188.2	DE	153.8	WA	245.5
WY	279.3	GA	159.5	IL	175.3	NM	151.8
NV	249.9	WI	182.2	AZ	178.3	IA	167.9

Source: *Statistical Bulletin* 74:1, Jan.–March 1993, p. 12.

Based on this sample, determine point estimates of the mean mortality rate and the standard deviation of those rates for the population of all 29 states with higher mortality rates.

Solution   The population consists of the 29 states that have mortality rates in excess of 150. A sample of size $n = 12$ from the population is shown. The mean, μ, of this population is best estimated by $\overline{X}$, which is the mean of this sample. According to the formula

$$\overline{X} = \frac{\sum_{i=1}^{n} x_i}{n},$$

$$\overline{X} = \frac{186.7 + 279.3 + 249.9 + \cdots + 167.9}{12}.$$

Finishing these calculations, $\overline{X} = 193.2$, which is the best point estimate that can be made of μ, the actual population mean.

   The best point estimate that can be made of the population standard deviation, σ, is

$$s = \sqrt{\frac{\sum_{i=1}^{n} (x_i - \overline{X})^2}{n - 1}}.$$

Applying this formula,

$$s = \sqrt{\frac{(186.7 - 193.2)^2 + (279.3 - 193.2)^2 + \cdots + (167.9 - 193.2)^2}{12 - 1}}.$$

A value of $s \approx 41.71$ is obtained.   ●

E x a m p l e
4

Refer to Example 3. If the higher mortality rates are believed to be normally distributed, then what is the best estimate that can be made of the percentage of higher-mortality states that have rates over 220?

Solution   Assume a normal distribution with mean 193.2 and standard deviation 41.71, since both of these sample statistics, which were computed in Example 3, are the best available estimates of the population parameters.

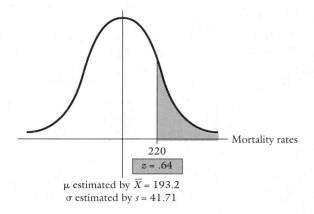

Figure 7-4

The shaded area in Figure 7-4 represents the percentage of all states in the higher-mortality-rate population that have rates greater than 220. The Z-score of 220 is .64. This Z-score is computed by using the standardization formula,

$$Z(x) = \frac{x - \mu}{\sigma},$$

with $\overline{X}$ substituted for μ and s substituted for σ. These two statistics are the best available estimates for μ and σ and will yield an estimated Z-score. In other words,

$$\text{Estimated } Z(x) = \frac{x - \overline{X}}{s}$$

$$Z(220) = \frac{220 - 193.2}{41.71}$$

$$\approx .64.$$

Now, according to Appendix A-2, the area from the estimated mean to the estimated value of $z = .64$ is .2389. Since the area to the right of the mean is .5000, subtraction is called for. It results in an area of

$$.5000 - .2389 = .2611.$$

Thus, an estimated 26.11% of all states in the higher-mortality population have rates greater than 220.[8] ●

If the random sample in Example 4 were formed with extreme, unrepresentative scores (which is always possible), then the results would not be valid. Increasing the sample size, n, can help to avoid such a possibility but does not eliminate it entirely.

---

[8]Actually, seven out of 29, or 24.1%, of the states in the higher-mortality-rate population have rates in excess of 220.

7.4 TOOLS

A statistical software program could have performed much of the work in the preceding examples quickly. The next example illustrates the use of Minitab.

**E x a m p l e
5**

The life of a new type of automobile tire is believed to be normally distributed. Fifty customers were chosen at random. The mileages they obtained from this tire in thousands of miles, follow.

44.5	38.9	40.8	44.0	38.9
43.1	37.4	22.7	38.5	28.0
43.9	44.5	45.2	35.9	32.5
44.9	42.1	35.6	50.0	47.8
35.8	22.2	28.9	33.6	40.2
44.9	29.4	30.0	31.6	43.6
42.8	46.8	40.1	43.8	39.9
34.9	24.8	24.5	43.4	44.5
47.3	41.4	31.4	33.3	40.9
44.5	38.9	45.6	45.7	39.5

a. Estimate the mean and standard deviation of the mileages obtained from all tires of this type.

b. If a dealer sells 500 of the tires and they are guaranteed to have a life of at least 25,000 miles, how many tires can the dealer expect to have returned for replacement?

c. If the tire company wishes to guarantee these tires for a certain number of miles so that only 2% of them are returned, what should the guaranteed mileage be?

Solution   a. First enter all the data into one column of a Minitab worksheet. For convenience, all scores are entered as given, in thousands.

	C1	C2	C3	C4	C5
↓					
1	44.5				
2	43.1				
3	43.9				
4	44.9				

Untitled Worksheet

Then use the **Descriptive Statistics** option from the **Stat** menu.

Minitab: Descriptive Statistics Calculations

	N	MEAN	MEDIAN	TRMEAN	STDEV	SEMEAN
C1	50	38.668	40.150	39.066	7.047	0.997

	MIN	MAX	Q1	Q3
C1	22.200	50.000	33.525	44.500

The circled statistics are the answers to Part a.

- The mean of all the entered scores, which in this case is a point estimate of the population mean, is $\overline{X}$ = 38.668, or 38,688 miles.

- The STDEV statistic that Minitab computes is the estimated standard deviation of the population, based on a sample. A denominator of $(n - 1)$ is used in its computation, and that is exactly what is wanted. Thus, the estimated population standard deviation, s, is 7.047, or 7047 miles.

b. Use Minitab's Cumulative Density Function (CDF) command (explained in Chapter 6) for this calculation. Recall that the CDF command calculates the probability that the random variable is less than or equal to a given value of r, which in this case is 25 (25,000 miles). Note that the estimated population parameters calculated for Part a are used in the subcommand shown in Figure 7-5. The circled output indicates that approximately 2.62% of all scores in a normal distribution with mean 38,668 and standard deviation 7047 will be less than 25,000. Thus, the dealer can expect to have (.0262)·500 = 13.1, or about 13, tires returned to him.

c. Determine this part with Minitab's INVCDF command. For a given probability or percentage, p, and specified distribution with parameters, INVCDF finds a value x such that p = CDF(x). Figure 7-6 illustrates this command and the result. The display in Figure 7-6 says that, in a normal distribution with mean 38,668 and

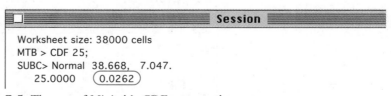

Figure 7-5 The use of Minitab's CDF command.

Figure 7-6 The use of Minitab's INVCDF command.

standard deviation 7047, 2% of the scores will be less than 24,195 miles. Therefore, if the company wants only about 2% of its tires returned, it should guarantee them for about 24,000 miles.   •

Exercises 7.3–7.4

GROUP

1. A small population of size $N = $ five, $\{2, 6, 10, 11, 21\}$, is given.

   a. How many samples of size $n = 3$ are possible?

   b. Compute $\overline{X}$ and s for each possible sample of size $n = 3$. Then use those results, along with the calculated values of μ and σ, to help answer Parts c and d.

   c. For this population, must a sample of size 4 yield a value of $\overline{X}$ that is closer to μ than a value of $\overline{X}$ based on a sample of size 3?

   d. Consider two samples, both of size $n = 3$. If $\overline{X}$ from sample number 1 is closer to μ than $\overline{X}$ from sample number 2, then does it necessarily have to follow that s from sample 1 is closer to σ than s from sample 2?

2. If Minitab's STDEV function were used to compute the standard deviation of a population from all the scores of that population, would the resulting value be slightly too large or slightly too small?

* 3. Consider a sample of size n, with σ the population standard deviation and s the estimate of σ. Then σ and s are related by the equation

$$\sigma = s \cdot \sqrt{\frac{n-1}{n}}.$$

True or false?

4. The following list shows 76 American magazines with leading circulations as of December 31, 1993.

01	*American Health*	820,087
02	*Better Homes & Gardens*	7,600,960
03	*Bon Appetit*	1,294,945
04	*Business Week*	833,718
05	*Car and Driver*	1,055,403
06	*Chatelaine*	903,286
07	*Conde Nast Traveler*	866,557
08	*Cosmopolitan*	2,627,491
09	*Country Home*	1,061,226
10	*Country Living*	1,977,214
11	*Discover*	1,031,496
12	*Ebony*	1,939,500
13	*Elle*	924,429
14	*Essence*	954,351
15	*Family Circle*	5,114,030
16	*The Family Handyman*	1,040,372
17	*Field & Stream*	2,007,901
18	*Glamour*	2,304,769
19	*Globe*	1,011,117
20	*Golf Digest*	1,461,566
21	*Golf Magazine*	1,221,554
22	*Good Housekeeping*	5,162,597
23	*Gourmet*	906,299
24	*Health*	961,113
25	*Home*	1,043,964
26	*Home Mechanix*	1,025,071
27	*Hot Rod*	785,259
28	*House Beautiful*	1,009,446
29	*Jet*	974,764
30	*Ladies' Home Journal*	5,153,565
31	*Life*	1,625,096
32	*McCall's*	4,605,441
33	*Mademoiselle*	1,218,985
34	*Money*	2,100,039
35	*Motor Trend*	932,881
36	*Nation's Business*	861,402
37	*National Enquirer*	3,403,330
38	*National Geographic*	9,390,787
39	*New Woman*	1,314,294
40	*Newsweek*	3,156,192
41	*Outdoor Life*	1,502,676
42	*Parents Magazine*	1,776,470
43	*Penthouse*	1,201,692
44	*People Weekly*	3,446,569
45	*Playboy*	3,402,617
46	*Popular Mechanics*	1,656,951
47	*Prevention*	3,220,763

48	Reader's Digest	16,261,968
49	Reader's Digest (Canadian ed.)	1,230,888
50	Redbook	3,345,451
51	Rolling Stone	1,236,525
52	Self	1,314,315
53	Sesame Street Magazine	1,187,862
54	Seventeen	1,940,601
55	Smithsonian	2,212,418
56	Soap Opera Digest	1,437,758
57	Southern Living	2,368,678
58	Sport	792,093
59	Sports Illustrated	3,356,729
60	Star	2,957,915
61	Sunset	1,441,506
62	Teen	1,170,842
63	Time	4,103,772
64	Travel & Leisure	1,010,939
65	True Story	825,060
66	TV Guide	14,122,915
67	TV Guide (Canada ed.)	817,877
68	U.S. News & World Report	2,281,369
69	US	1,110,056
70	Vogue	1,250,008
71	Weight Watchers	1,035,322
72	Woman's Day	4,858,625
73	The Workbasket	752,020
74	Workbench	807,810
75	Working Woman	805,157
76	YM	1,701,615

Source: *Information Please Almanac, 1995,* p. 311.

Assemble a random sample of size $n = 25$ by using the random number table in the following way. Start at the bottom right-hand corner, with the five-digit block 39074; using the last two digits, work up the column.

a. Compute the sample statistics $\overline{X}$ and s, the point estimates of the parameters μ and σ for this population of 76 magazines.

b. Treat the population as normally distributed and find the magazine title that comes closest to marking the third quartile.

5. Use the following data and the similar data for males on page 292 to determine, on the basis of a random sample of size $n = 3$, whether life expectancy is more variable for males or for females in the population composed of these six countries. Assign each country an identification number from 1 through 6, in alphabetical order. Choose three countries using the random number table by starting at the top of the third column and working down, using the last digit of each five-digit block. Use the same countries for both males and females. Use the estimated coefficient of variation, $\frac{s}{\overline{X}}$, as the measure of variability.

**Life Expectancies
(in years) for
Females**

Finland	78.9
France	80.3
Canada	79.7
Netherlands	80.1
Iceland	80.3
Japan	81.8

Source: *Statistical Bulletin*
73:3, July–Sept. 1992, p.
14.

6. Refer to Exercise 5. The statistic $\frac{s}{X}$ was used to estimate the parameter $\frac{\sigma}{\mu}$.
   Compute the actual value of this parameter for both the male and female populations. By what percentage do the estimates differ from the actual values?

7. Refer to Example 5 on page 297. Why does it seem reasonable to believe that tire life is normally distributed, or at least mound-shaped?

8. The mortality rates from SIDS in all the states with lower rates (rates less than 150) are as follows. Consider this set of 21 states as a population.

NH	149.7	VT	123.3	MA	89.6	RI	42.2
CT	37.4	NY	70.9	NJ	74.7	PA	125.6
OH	127.1	MI	142.4	MN	148.3	KS	118.6
MD	132.0	VA	123.5	WV	114.4	SC	121.6
FL	139.6	AL	131.7	LA	119.1	TX	103.8
HI	63.0						

Source: *Statistical Bulletin* 74:1, Jan.–March 1993, p. 12.

Use the third column (MA, NJ, MN, WV, and LA) as a random sample of size $n = 5$ of this population.

a. Estimate the population mean and standard deviation on the basis of this sample.

b. If the distribution of these rates is mound-shaped, use the estimates from Part a to determine an interval in which about 68% (or 14) of these lower rates would lie.

c. What is the actual percentage of the 21 states that have mortality rates in the interval found in Part b?

9. The retail prices for a pound of coffee (in cents) for the years 1921 through 1970 are as follows.

'21	36.3	'22	36.1	'23	36.9	'24	42.6
'25	50.4	'26	50.2	'27	47.4	'28	48.2
'29	47.9	'30	39.5	'31	32.8	'32	29.4
'33	26.4	'34	26.9	'35	25.7	'36	24.3
'37	25.5	'38	23.2	'39	22.4	'40	21.2
'41	23.6	'42	28.3	'43	30.0	'44	30.1

'45	30.5	'46	34.4	'47	46.9	'48	51.4
'49	55.4	'50	79.4	'51	86.8	'52	86.8
'53	89.2	'54	110.8	'55	93.0	'56	103.4
'57	101.7	'58	90.7	'59	78.0	'60	75.3
'61	73.6	'62	70.8	'63	69.4	'64	81.6
'65	83.3	'66	82.3	'67	76.9	'68	76.4
'69	76.5	'70	91.1				

Source: *Historical Statistics of the United States, Colonial Times to 1970*, Bicentennial Edition, Part 1, p. 213.

Use a random sample of size $n = 12$ to estimate the mean and standard deviation of this population of 50 prices. Assemble the sample as follows. Start at the bottom of the third column of the random number table and work up. Use the last two digits of the five-digit blocks, with each such two-digit number identifying a year.

10. Refer to Exercise 9. Use the computed sample statistics to define an interval with end points $\overline{X} - 2s$ and $\overline{X} + 2s$. Based on the percentage of the population's scores contained in this interval, would Chebyshev's Theorem or the Empirical Rule more closely apply to the distribution of scores in this population?

Case Study   *Random Selection and the Draft Lottery*

The 1970 draft lottery for service in the Vietnam War violated some important rules of random selection. (*Arthur R. Hill / Visuals Unlimited*)

Human societies have long recognized random selection techniques as acceptable methods of making decisions that affect many people. The abundant historical and present-day examples range from biblical accounts of drawing lots to selection procedures for court juries. Draft lotteries, another example of the use of random selection, developed in the United States during the First and Second World Wars, but it was difficult to conduct them fairly. Furthermore, the designers of the 1970 draft lottery evidently did not learn from earlier mistakes.

Obtain a copy of the January 22, 1971, issue of *Science.* Read the article by Stephen E. Fienberg (pp. 255–261), titled *Randomization and Social Affairs: The 1970 Draft Lottery.* Use the information in that article to answer the following questions.

I. Give three examples of social dilemmas that have been solved through the use of chance or random selection.
II. In your own words, describe the famous 1842 court case that involved the random selection of passengers to board a lifeboat.
III. Briefly describe the randomization procedures used in the 1917 and 1940 draft lotteries.
IV. Name three randomization difficulties, or irregularities, that were encountered in the 1940 draft lottery.
V. What lesson from the 1940 draft lottery seems not to have been learned?
VI. Several mistakes were made in mixing the capsules for the 1970 draft lottery. What were they?
VII. In your own words, explain how the popular meaning of "randomness" is not consistent with its statistical meaning.
VIII. If the birth-date capsules for the 1970 draft lottery had indeed been randomly mixed, what would have been the probability of obtaining the 1970 result?

CHAPTER SUMMARY

Computing the mean and standard deviation for most large data sets, or populations, is often neither practical nor cost-effective. The mean and standard deviation of a population can be effectively estimated from a random sample of the population. However, when one estimates the population parameters (μ and σ) from a sample, it is very important that the sample be truly random. A random sample is a subset of the population of which every member of the population has an equal chance of being a part. A random number table is an effective tool for assembling a random sample.

Once a random sample of size n is formed from a population of size N, two sample statistics can then be used as the best point estimates of the population parameters: $\overline{X}$ to estimate the population mean and s to estimate the population standard deviation. $\overline{X}$ is, quite logically, the mean of the n scores in the sample. The formula for s, however, is slightly different from what one would expect:

$$s = \sqrt{\frac{\sum_{i=1}^{n}(x_i - \overline{X})^2}{n-1}}.$$

Key Concepts, Terms, and Formulas

Use of s as an estimate of the population standard deviation:

$$s = \sqrt{\frac{\sum_{i=1}^{n}(x_i - \overline{X})^2}{n-1}} \quad (7\text{-}1)$$

Videotape Suggestions

Program 8 from the series *Against All Odds: Inside Statistics* provides more information about the 1970 draft lottery. In addition, *Program 12* from the same series shows a method of constructing a random number table, and *Program 14* explains the differences between a simple random sample and stratification.

CUMULATIVE
REVIEW
CHAPTERS 5A–7

▶ O v e r v i e w

The four chapters in the second unit, Chapters 5A through 7, provided the foundation of probability theory as well as illustrated ways in which probability is used in the study of statistics. A knowledge and understanding of probability theory is very important for inferential statistics, which is the next unit of the text.

Chapter 5A examined many ideas in probability through games of chance, which is how early studies of probability began over a hundred years ago. Some of the games featured were roulette, dice, cards, slot machines, lotteries, and horse racing. Two important fundamentals necessary for understanding and using probability are the classic definition of the probability of an event and the Fundamental Counting Principle. Counting the number of successful outcomes of an event is often the key step for solving many probability questions. Chapter 5B presented some optional material in probability, including more sophisticated counting procedures, a more in-depth look at conditional probability, and further work with probability trees.

The work in Chapter 6 extended the treatment of probability to discrete and continuous random variables, and examined the graphs associated with both types of these random variables. The binomial random variable, which keeps track of the number of successes in a Bernoulli process, was shown to have many applications, ranging from determining probabilities of having three boys in a family of six children, to how many people may be affected by a new strain of the flu. Continuous random variables described by the normal probability distribution also have widespread applications, and will form the theoretical foundation for much of the work ahead in inferential statistics.

Finally, Chapter 7 showed how to use a random number table to choose a random sample from a population to determine sample statistics. These sample statistics, $\overline{X}$ and s, are point estimates of the population parameters, μ and σ, respectively.

Procedure Index

Procedures in parentheses indicate optional material.

EXERCISES ▶ *Exercises in parentheses indicate optional material.*

1. In a five-horse race, you think horse number 5 will win and horse number 4 will finish second.

   a. Use a tree diagram to determine the number of possible trifecta bets under these two conditions.

   b. What is the probability that a trifecta bet in which horse number 5 wins and horse number 4 places is a winning bet in this five-horse race? Assume that all trifecta bets are equally likely.

2. Special dice are constructed that have 6, 8, 10, and 12 sides. On each die the faces are numbered, as expected, from 1 through the number of sides. One die of each type is tossed.

   a. What is the probability that the sum of the numbers tossed is 4?

      Hint: Use the Fundamental Counting Principle to determine how many outcomes (not sums) are possible.

   b. What is the probability that the sum of the numbers tossed is 6?

      Hint: One of the combinations that add to 6 is 2 + 2 + 1 + 1. Consider all the possible permutations of these and other numbers.

3. Use the following Venn diagram to answer all parts of this exercise. Assume that $U = \{$All playing cards in a standard deck$\}$, $C = \{$Clubs$\}$, $D = \{$Diamonds$\}$, $H = \{$Hearts$\}$, $S = \{$Spades$\}$, $T = W \cap S$, $Y = W \cap D$, $X = W \cap C$, and $Q = W \cap H$.

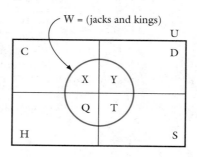

a. Find $P(C)$, $P(D)$, $P(H)$, $P(S)$, $P(W)$, $P(X)$, $P(Y)$, $P(Q)$, and $P(T)$.

b. Find $P(H \cup W)$.

c. If $Z = \{$Red cards$\}$, then find $P(Z \cup W)$ and $P(Z \cup C)$.

d. If $K = \{$All cards except spades$\}$, then find $P(K \cap W)$, $P(K \cup W)$, $P(K \cup H)$, and $P(K \cup S)$.

4. A hotel is equipped with two types of smoke detectors. One type has a reliability of .90, which means that, if smoke is present, the probability that the alarm will sound is .90. If the probability that at least one alarm will sound is .95 and the probability that both will sound is .93, what is the reliability of the second type of smoke detector?

5. What is the probability of rolling a sum of 7 at least once on four rolls of two dice?

6. Draw Venn diagrams to help determine which of the following pairs of events could be independent.

	Event A	**Event B**
a.	Being a single parent	Being widowed
b.	Being married	Being single
c.	Being single	Being a divorced parent

(7.) Refer to the Birthday Problem on page 214 in Chapter 5B. Use trial and error to determine the minimum number of people who must be chosen at random so that the probability that at least two of them have the same birthday is at least $\frac{3}{4}$.

(8.) In your own words explain the difference between the odds for an event and the probability of an event.

(9.) Find the number of different possible positive integers that can be formed by using all the digits in each of the following numerals.

a. 314159   b. 1223223   c. 66675555997

(10.) The following cross-tabs table shows 1993 employment status (in thousands) for people of different age groups.

Employment Status

	Employed	Unemployed
16–19 years old	5530	1296
20–24 years old	12,136	1421
25–54 years old	86,921	5349
55 years and older	14,720	667

Based on data from *Statistical Abstract of the United States, 1994*, p. 403.

Compute the necessary totals, and determine the following probabilities in relation to this table.

a. $P(\text{Unemployed})$

b. $P(20\text{–}24 \text{ years old})$

c. $P(\text{Employed} \mid 55 \text{ years and older})$

d. $P(20\text{–}24 \text{ years old} \mid \text{Unemployed})$

e. $P(25\text{–}54 \text{ years old} \cap \text{Employed})$

(11.) In a certain town, 30% of the voters are Republicans and 70% are Democrats. Seventy-five percent of the Republicans are in support of an issue, but only 50% of the Democrats support it. If a person is chosen at random and she supports the issue, what is the probability that she is a Republican?

12. You play a card game with a standard deck of cards by the following rules. A card is drawn at random from the deck. If it is a heart (♥) honor card ("honor" means jack, queen, king, or ace), you win $5. If it is a red (♥ or ♦) nonhonor card, you win $2.75. If any other card is drawn, you lose $4, unless it is a black (♣ or ♥) queen, in which case you lose $15.

   a. Find the average amount of money you could expect to win (or lose) in this game.

      Hint: *Use negative numbers to represent money lost.*

   b. A game is **fair** if its mean amount of winnings is zero. Is this game fair? If not, whom does it favor?

13. Suppose advocates of gun control believe that 60% of all Americans favor some type of new gun control legislation, and a polling organization interviews 500 people selected at random. Assume that the advocates of gun control are correct, and use the model of a binomial random variable to answer the following questions.

   a. About how many people in that survey should express a desire for new legislation?

   b. What is the probability that exactly eight out of ten people surveyed at random express a desire for new legislation?

   c. What is the probability that at least seven out of nine people surveyed at random express a desire for new legislation?

   d. What is the probability that at most three people out of eight surveyed at random do *not* express a desire for new legislation?

(14.) Five hundred catfish were released in a 10,000-square-foot pond. If you covered 1000 square feet of the pond in a rowboat, determine the probability that the part of the pond over which you rowed contained 55 fish.

15. A continuous random variable, C, has a uniform probability distribution over the interval $[2.3, 18)$.

    a. Graph the probability distribution of C, and state its density function.

    b. Determine $P(02.56 \leq C \leq 10.03)$ and $P(1.10 \leq C \leq 6.50)$.

    c. What is $P(C = 4)$, and what is the density at $C = 4$?

16. The life span, L, of a certain type of tiger in captivity is 19 years with a standard deviation of 2.3 years. Assume a normal distribution model.

    a. What percentage of these tigers live no more than 20 years?

    b. What percentage of these tigers live no more than 15 years?

    c. What percentage of these tigers live at least 21 years?

    d. What percentage of these tigers live at least 17 years but no more than 23 years?

17. Briefly explain how the 1970 draft lottery could have been conducted using a random number table. What do you think are some reasons why it was not conducted in that way?

18. The following table shows the number of nonfatal-injury accidents on urban interstate highways in each state in 1992.

01	AL	624		26	MT	81
02	AK	320		27	NE	366
03	AZ	907		28	NV	967
04	AR	456		29	NH	96
05	CA	15,064		30	NJ	3117
06	CO	2117		31	NM	757
07	CT	2381		32	NY	8032
08	DE	293		33	NC	1266
09	FL	4427		34	ND	58
10	GA	4530		35	OH	7056
11	HI	807		36	OK	1896
12	ID	163		37	OR	1169
13	IL	6510		38	PA	2501
14	IN	734		39	RI	570
15	IA	426		40	SC	651
16	KS	767		41	SD	126
17	KY	1112		42	TN	2040
18	LA	1971		43	TX	18,862
19	ME	235		44	UT	1168
20	MD	2682		45	VT	13
21	MA	1524		46	VA	2667
22	MI	4905		47	WA	4218
23	MN	1584		48	WV	474
24	MS	369		49	WI	1018
25	MO	3633		50	WY	67

Source: *Highway Statistics 1992*, Federal Highway Administration, p. 218.

a. Use the random number table in Appendix A-3 and the given two-digit state identification numbers to select a random sample of $n = 15$ states. Use the second and third digits of the five-digit random number block. Start at the top of the second column and work down.

b. Use the random sample from Part a to determine a point estimate of the mean number of nonfatal-injury accidents for all 50 states.

Important decisions should be made based on knowledge of inferential statistics. (©1993, Comstock, Inc.)

3 Inferential Statistics

People use inferential statistics to make decisions in uncertain situations. This unit brings together the main ideas from probability, random variables, and random sampling to form reliable confidence intervals and to test opposing claims.

Chapter 8 uses mathematical modeling to build the theoretical foundation for constructing confidence intervals; Chapter 9 uses the models presented in Chapter 8 to examine hypothesis testing. Together, these two concepts—confidence intervals and hypothesis testing—are the basis for the rest of the text.

Chapter 10 applies confidence intervals and hypothesis testing to two populations, and Chapter 11 introduces another mathematical model, the chi-square distribution, with additional applications.

The final three chapters of the text, Chapters 12–14, apply the methods of inferential statistics to nonparametric situations, multiple populations, and correlation.

(Charles D. Winters)

Confidence intervals are
formed using random
samples of the
population.

Chapter

8 Confidence Intervals

T he last chapter dealt with estimates for the population parameters μ and σ. As already mentioned, these estimates are point estimates. Recall that a **point estimate** of a population parameter is a single number that estimates the unknown parameter. The point estimate for μ is $\overline{X}$, and the point estimate for σ is s.

The point estimates $\overline{X}$ and s depend on the chance selection of scores for a sample and are usually not equal to the population parameters they are meant to estimate. Increasing the sample size may improve the accuracy of these statistics but is not always practical. The **case study** at the end of this chapter examines this dilemma more closely, from the perspective of a research biologist.

Recall from Chapter 3 that $[a, b]$ denotes all real numbers greater than or equal to a and less than or equal to b.

8.1 INTERVAL ESTIMATES

A more sophisticated alternative to a point estimate is an interval estimate. An **interval estimate** for the population mean, μ, specifies two numbers between which that population parameter is likely to be found with a certain degree of confidence, or probability. If the degree of confidence is, say, 80%, then once the interval has been found, it can be said that μ would lie in 80% of such intervals if the intervals were formed from repeated samples of the same size, using the same procedure. This is more simply referred to as an 80% **confidence interval.**

For example, if someone is interested in determining the mean selling price for a home in a certain neighborhood, one way to estimate that parameter, μ, is to find the mean, $\overline{X}$, of a random sample of size n and use that as a point estimate of μ. Since $\overline{X}$ is very unlikely to have exactly the same value as μ, however, a confidence interval, $[a, b]$, in which μ would lie a certain percentage of the time is often preferred. The contrast between a point estimate, $\overline{X}$, and a confidence interval estimate, $[a, b]$, is significant. Although the determination of $[a, b]$ still depends on the random sample, and μ may not lie in it at all, the degree of risk is made clear. Also, a range of likely values for μ is very helpful.

315

8.2 CHARACTERISTICS OF A SAMPLING DISTRIBUTION

To construct confidence intervals, two characteristics that describe the relationship between a population and a sample from that population must be understood. Consider a population of size N and all the possible samples of size n that can be formed from it. Recall from Chapter 7 that there are $_NC_n$ such samples, and reexamine Figure 8-1.

Definition ▶ The **sampling distribution of the $\overline{X}$'s**, referred to more simply as the **sampling distribution**, is a frequency curve, or histogram, constructed from all the $_NC_n$ possible values of $\overline{X}$.

Every sampling distribution has two characteristics that apply to all the possible values of $\overline{X}$ for the given sample size, and the characteristics hold true for any population, whether it is large or small and whether it has a normal or nonnormal distribution. When studying these characteristics, keep in mind that N denotes the population size and n denotes the sample size.

Characteristic 1. The mean of all the $_NC_n$ possible values of $\overline{X}$ is equal to the population mean, μ:

$$\mu_{\overline{X}} = \mu. \tag{8-1}$$

Characteristic 2. The standard deviation of the sampling distribution, which is the standard deviation of all the $_NC_n$ values of $\overline{X}$, is equal to the standard deviation of the population divided by the square root of the sample size:

$$\sigma_{\overline{X}} = \frac{\sigma}{\sqrt{n}}. \tag{8-2}$$

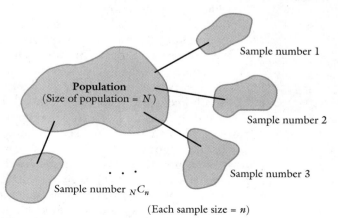

Figure 8-1 The number of samples in the population.

The standard deviation of the sampling distribution is known as the **standard error of the sample mean** or, more simply, the **standard error** (*SE*). The standard error is very important in this chapter and much of the rest of the text. Characteristic 2 indicates the relationship between *SE* and the standard deviation of the population, σ:

$$SE = \sigma_{\overline{X}}$$

$$= \frac{\sigma}{\sqrt{n}}.$$

If the point estimate s is used in place of σ, then the estimated value of the standard error, denoted by $S_{\overline{X}}$, becomes

It is this value, $\frac{s}{\sqrt{n}}$, that Minitab computes and lists under the SEMEAN (standard error of the mean) heading when the **Describe** command is executed for a given set of data.

$$SE \approx S_{\overline{X}}$$

$$= \frac{s}{\sqrt{n}}.$$

It is very important not to confuse *SE* with s. The sample statistic, s, is an estimated value of the standard deviation of the population, σ, whereas the standard error, *SE*, is the standard deviation of all the $_{N}C_{n}$ values of $\overline{X}$ obtained from all the possible samples of size n. When σ is unknown, *SE* is estimated[1] by the fraction $\frac{s}{\sqrt{n}}$.

The following example, which involves an artificially small population, illustrates these two characteristics of a sampling distribution.

E x a m p l e
1

From the population

$$\{1, 6, 9, 10, 11, 13, 20\}$$

of size $N = 7$, form all possible samples of size $n = 5$. Then perform the following tasks.

a. Illustrate the fact that the mean of all the sample means is equal to the mean of the population; that is, $\mu_{\overline{X}} = \mu$. (This is characteristic 1.)
b. Verify that, for this population, *SE* can be found with the formula

$$SE = \sigma_{\overline{X}}$$

$$= \frac{\sigma}{\sqrt{n}} \cdot \sqrt{\frac{N-n}{N-1}}.$$

(This is characteristic 2.)

---

[1]If the population size, N, is known, a finite correction factor, $\sqrt{\frac{N-n}{N-1}}$, can be used to refine *SE* further. In this case, $SE = \frac{\sigma}{\sqrt{n}} \cdot \sqrt{\frac{N-n}{N-1}}$. If σ is unknown, *SE* can be approximated by $\frac{s}{\sqrt{n}} \cdot \sqrt{\frac{N-n}{N-1}}$. In many practical situations, however, N is unknown; at other times, when N is known, the value of the finite correction factor is so close to 1 as to make no significant change in the calculated value of *SE*. For these two reasons, and to simplify future calculations, the finite correction factor will not be used after this section.

Solution   All possible samples of size $n = 5$ are formed, and $\overline{X}$ is computed for each of the $_NC_n = {_7}C_5$, or 21, possible samples.

Sample No.	Scores in Sample	$\overline{X}$
1	9, 10, 11, 13, 20	12.6
2	6, 10, 11, 13, 20	12.0
3	6, 9, 11, 13, 20	11.8
4	6, 9, 10, 13, 20	11.6
5	6, 9, 10, 11, 20	11.2
6	6, 9, 10, 11, 13	9.8
7	1, 10, 11, 13, 20	11.0
8	1, 9, 11, 13, 20	10.8
9	1, 9, 10, 13, 20	10.6
10	1, 9, 10, 11, 20	10.2
11	1, 9, 10, 11, 13	8.8
12	1, 6, 11, 13, 20	10.2
13	1, 6, 10, 13, 20	10.0
14	1, 6, 10, 11, 20	9.6
15	1, 6, 10, 11, 13	8.2
16	1, 6, 9, 13, 20	9.8
17	1, 6, 9, 11, 20	9.4
18	1, 6, 9, 11, 13	8.0
19	1, 6, 9, 10, 20	9.2
20	1, 6, 9, 10, 13	7.8
21	1, 6, 9, 10, 11	7.4

a. The mean of this small population can be computed directly as $\mu = 10$—the same value that would be obtained as the mean of the 21 values of $\overline{X}$. That is to say,

$$\frac{1 + 6 + 9 + 10 + 11 + 13 + 20}{7} = 10, \quad \text{and}$$

$$\frac{12.6 + 12.0 + 11.8 + \cdots + 7.4}{21} = 10.$$

Thus, for this population and this sample, $\mu_{\overline{X}} = \mu$.

b. The standard deviation of the seven scores in the population, rounded to the nearest hundredth, is $\sigma = 5.45$. Since the population size is known ($N = 7$) and is small, the finite correction factor will be used. Substituting the values $\sigma = 5.45$, $N = 7$, and $n = 5$ into the formula for *SE* gives

$$SE = \sigma_{\overline{X}}$$

$$= \frac{\sigma}{\sqrt{n}} \cdot \sqrt{\frac{N - n}{N - 1}}$$

$$= \frac{5.45}{\sqrt{5}} \cdot \sqrt{\frac{7 - 5}{7 - 1}}$$

$$\approx 1.41.$$

Characteristic 2 states that the standard deviation of the 21 values in the $\overline{X}$ column (the standard error, *SE*) must be 1.41. You can verify for yourself that this is true.   ●

The next example further demonstrates characteristic 2 of a sampling distribution. In this case, however, the finite correction factor is not used.

**E x a m p l e
2**

Forty combined SAT scores were chosen at random from the records of seniors at a certain high school over the last 20 years.

770	680	510	520	660	680	730	460
660	500	570	680	800	660	520	440
560	690	700	720	500	460	770	780
600	620	660	800	420	380	560	680
740	770	680	560	500	660	540	710

Use these scores to compute the estimated standard deviation of the sampling distribution, *SE*. Compare *SE* with the estimated standard deviation of the population, *s*.

Solution   The standard deviation of the sampling distribution, *SE*, can be found using characteristic 2.

$$SE = \sigma_{\overline{X}}$$
$$= \frac{\sigma}{\sqrt{n}}.$$

Since the population standard deviation, σ, is unknown, one may use *s* to estimate σ. Therefore,

$$SE \approx S_{\overline{X}}$$
$$= \frac{s}{\sqrt{n}}.$$

The sample statistic, *s*, is found to be approximately 115.04.

S t a t   F a c t

In 1992 the mean verbal SAT score of all students whose parents had bachelor's degrees was 444—higher by 50 points than the mean score for those students whose parents had only high-school diplomas, and more than 100 points higher than the mean score for those students whose parents had no high-school diplomas.

See Example 4 in Chapter 3 (page 91) to review the use of the TI-82 graphics calculator to find these statistics.

```
1 - V a r   S t a t s
  x̄ = 6 2 2 . 5
  Σx = 2 4 9 0 0
  Σx² = 1 6 0 1 6 4 0 0
s→Sx = 1 1 5 . 0 4 1 7 9 8 4
  σx = 1 1 3 . 5 9 4 6 7 4 2
↓ n = 4 0
```

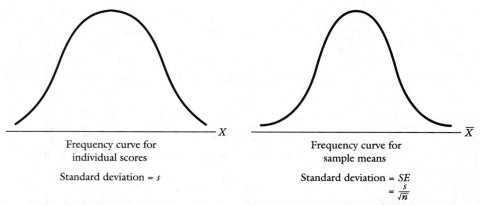

Figure 8-2 A comparison of frequency curves for individual scores and for sample means.

Thus,

$$SE \approx \frac{s}{\sqrt{n}}$$

$$= \frac{115.04}{\sqrt{40}}$$

$$\approx 18.19.$$

Compare SE and s. The value of s, 115.04 (about six times greater than SE), is the estimated standard deviation of the population of individual scores, with each senior in the last 20 years contributing one score. The value of SE (only 18.19) is the estimated standard deviation of the means of all possible groups of size 40 taken from the population on which s is based.

The fact that SE is much less than s indicates that the distribution of the $\overline{X}$'s is much more compact and less spread out than the distribution of the individual values of X. This implies that there will always be less variation in a set of means than in a set of individual scores. Figure 8-2 shows how the fact that $s > SE$ affects the frequency curves when the population of all SAT scores over the last 20 years is normally distributed.   ●

Alternative Solution to Example 2   Once the data have been entered onto a Minitab worksheet, the `Descriptive  Statistics` command, executed from the `Basic Stats` option in the `Stat` menu, can be used to arrive at a value for SE. This can take place in Minitab's interactive mode.

Minitab: Descriptive Statistics Calculations

	N	MEAN	MEDIAN	TRMEAN	STDEV	SEMEAN
C1	40	622.5	660.0	625.0	115.0	18.2

	MIN	MAX	Q1	Q3
C1	380.0	800.0	520.0	707.5

Note that the sample statistic, s, also appears, labeled STDEV, with a rounded value of 115.0.   ●

Exercises 8.1 – 8.2

1. Use examples to show that for a fixed n, as the population size, N, increases, the finite correction factor gets very close to 1.

2. Will use of the finite correction factor make the value of SE slightly larger or slightly smaller? Explain.

3. What differences are there, if any, among SE, $\sigma_{\overline{X}}$, and $S_{\overline{X}}$?

4. Consider the population $\{30, 40, 60, 80, 85, 100\}$. Assume that a random sample of size 3 is chosen from this population.

   a. State the values of N and n.

   b. State a range of possible values for $\overline{X}$.

   c. Which sample of three scores do you believe would give a value of s closest to σ?

GROUP

5. Refer to Exercise 4. In a manner similar to Example 1 on page 317, form all $_NC_n$ possible samples and then use them to perform the following tasks.

   a. Illustrate the two characteristics of a sampling distribution.

   b. Using the calculations of each of the $_NC_n$ possible values of s from Part a, verify (or correct) your answer to Part c of Exercise 4. Does this sample also yield a value of $\overline{X}$ that is closest to the population mean, μ?

8.3   CONFIDENCE INTERVALS AND THE CENTRAL LIMIT THEOREM

The information about interval estimates at the start of this chapter can be formalized into the following definition.

Definition ▶ A **C% confidence interval** for a population mean, μ, is an interval $[a, b]$ such that μ would lie within $C\%$ of such intervals if repeated samples of the same size were formed and interval estimates made.

Later examples will demonstrate the procedure for determining such a confidence interval. Its construction depends on one of the most important theorems in all of mathematics, called the Central Limit Theorem.

Theorem ▶ **The Central Limit Theorem**

Under certain conditions, the sampling distribution of the $\overline{X}$'s results in a normal distribution.

Under these "certain" conditions—which will be specified in the course of this chapter—the confidence interval is obtained by the use of a normal distribution as a model. (When a certain distribution serves as a **model,** the ideal shape and characteristics of that distribution function as a pattern for other data.) Whenever the Central Limit Theorem does not apply, other models are used for the construction of confidence intervals.

Figure 8-3 illustrates the Central Limit Theorem with three populations, each having $N = 6$ scores. The three are distributed quite differently, but notice that

Distribution of the population

Distribution of all possible sample means of size $n = 4$

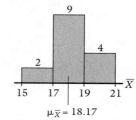

$\mu_{\bar{X}} = 18.17$

All possible sample means =
{20.25, 20, 19.75, 18.75, 18.25, 19, 18.75, 17.75, 17.25, 18.5, 17.5, 17, 17.25, 16.75, 15.75}

$\mu = 18.17$

Population =
{12, 16, 17, 18, 22, 24}

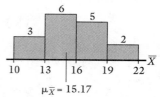

$\mu_{\bar{X}} = 15.17$

All possible sample means =
{19.75, 19, 17.75, 16.25, 15.25, 17.5, 16.25, 14.75, 13.75, 15.5, 14, 13, 12.75, 11.75, 10.25}

$\mu = 15.17$

Population =
{3, 9, 12, 17, 23, 27}

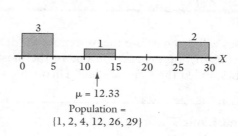

$\mu = 12.33$

Population =
{1, 2, 4, 12, 26, 29}

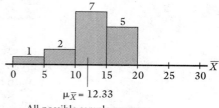

$\mu_{\bar{X}} = 12.33$

All possible sample means =
{17.75, 17.25, 15.25, 11.75, 11, 17, 15, 11.5, 10.75, 14.5, 11, 10.25, 9, 8.25, 4.75}

Figure 8-3

in each case the distribution of all possible sample means of size $n = 4$ (the sampling distribution) is generally more concentrated about the mean and is mound-shaped regardless of the shape of the population distribution.

As the population and sample sizes increase, the Central Limit Theorem assures that the histograms of all possible sample means will come closer and closer to fitting into a normal distribution. The histograms in Figure 8-4, for a population of size $N = 10$ and all the possible sample means of sample size $n = 8$, further illustrates this relationship. (There are $_{10}C_8 = 45$ means in the sampling distribution.)

Smaller interval widths of some of the sampling distribution histograms in Figures 8-3 and 8-4 reflect the fact that the standard deviation of the sampling distribution is always less than the standard deviation of the population.

The flow chart in Figure 8-5 indicates two possible mound- or bell-shaped models that can be used for the construction of confidence intervals, depending on which conditions are met. The path leading to the correct model has much to do with sample size. As can be seen by examining the flow chart and following some of the examples, several different paths can lead to the same model.

For any given model, the two characteristics of a sampling distribution, along with the Central Limit Theorem, form the theoretical foundation for the construction of the confidence interval. For instance, consider a 95% confidence interval for the population mean, using the normal distribution model (Figure 8-6). Appendix A-2 assures that, if the values of $\overline{X}$ are normally distributed, then 95% of the possible values for $\overline{X}$ will lie within 1.96 standard errors on either side of the mean of the $\overline{X}$'s. Since the mean of the $\overline{X}$'s is also the population mean, however, this implies that 95% of all such intervals centered about an observed value of $\overline{X}$ will contain the population mean. The probability is 95% that any interval so constructed will contain the population mean.

Distribution of the population

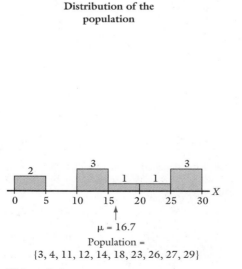

$\mu = 16.7$

Population =
$\{3, 4, 11, 12, 14, 18, 23, 26, 27, 29\}$

Distribution of all possible sample means of size $n = 8$

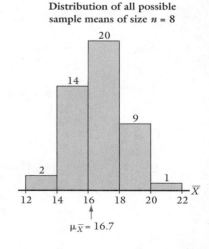

$\mu_{\overline{X}} = 16.7$

Figure 8-4

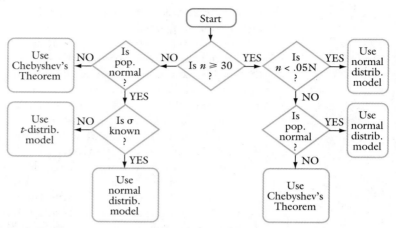

Figure 8-5 A flow chart for the construction of confidence intervals.

Consider Figure 8-7; it illustrates the fact that any value of $\overline{X}$ within 1.96 standard errors of the population mean (such as $\overline{X}_1$ and $\overline{X}_2$) must determine an interval that contains the population mean. However, any value of $\overline{X}$ that is more than 1.96 standard errors away from the population mean (such as $\overline{X}_3$) will not determine an interval that contains the population mean.

The following explanations and examples illustrate the procedure for constructing a confidence interval and are basically of two types: those involving large sample sizes, $n \geq 30$, and those involving small sample sizes, $n < 30$. (This definition of a "large" sample is a rule of thumb that will be used throughout this text.) The sample size determines the two main branches of the flow chart.

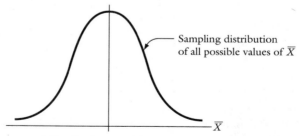

Figure 8-6 The normal distribution model.

Mean of all $\overline{X}$'s = Mean of population
$$(\mu_{\overline{X}} = \mu)$$
Standard deviation of all $\overline{X}$'s = Standard error
$$(\sigma_{\overline{X}} = SE)$$

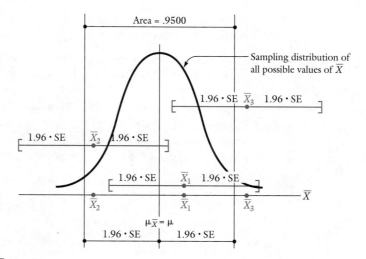

Figure 8-7

8.4 FLOW CHART BRANCHES TO THE RIGHT: LARGE SAMPLES ($n \geq 30$)

The right side of the flow chart shows that there are two conditions under which the normal distribution model can be used for the construction of a confidence interval with a large sample.

When employing the normal distribution model for the construction of a C% confidence interval, use the following steps (which will be illustrated and explained in Example 3 on page 326).

Steps for the Construction of a C% Confidence Interval Using the Normal Distribution Model

$Z(\overline{X})$ is functional notation and represents the *Z*-score of $\overline{X}$—that is, the number of standard errors that $\overline{X}$ is away from the mean of all $\overline{X}$'s.

Step 1   *Visualize which area of the normal distribution model the confidence interval occupies (a sketch is helpful).*

Step 2   *Use Appendix A-2 to find the relevant Z-scores that enclose the middle C% of the standard normal curve—the value of $Z(\overline{X})$ in Step 4.*

Step 3   *Calculate $\overline{X}$ and SE (SE $= \dfrac{\sigma}{\sqrt{n}}$ or its estimate, $\dfrac{s}{\sqrt{n}}$).*

Step 4   *Substitute the calculated values from Steps 2 and 3 into the formula*

$$Z(\overline{X}) = \frac{\overline{X} - \mu_{\overline{X}}}{SE}.$$

Step 5   *Solve the equation from Step 4 for $\mu_{\overline{X}}$ to find the end points of the confidence interval where $\mu_{\overline{X}}$ would lie C% of the time.*

Since the first characteristic of any sampling distribution is $\mu_{\overline{X}} = \mu$, however, the confidence interval found for $\mu_{\overline{X}}$ is a confidence interval for μ as well.

With a little experience, it will be possible to efficiently combine some of these steps, and later examples will illustrate how.

E x a m p l e 3

Example 2 dealt with 40 SAT scores from a random sample taken over the last 20 years. Find a 95% confidence interval for μ, the population mean for those 20 years. For ease of reference, the scores in the random sample are repeated here.

770	680	510	520	660	680	730	460
660	500	570	680	800	660	520	440
560	690	700	720	500	460	770	780
600	620	660	800	420	380	560	680
740	770	680	560	500	660	540	710

Solution The following set of conditions is one under which the Central Limit Theorem applies, stating that the sampling distribution of the $\overline{X}$'s is normal.

The sample size is large (i.e., $n \geq 30$).

The population size is large relative to the sample size.

The latter condition is satisfied if the sample size is less than 5% of the population size—in other words, if $n < .05\,N$. These two conditions make up the first branch to the right on the flow chart.

The given $n = 40$ implies a large sample. The population is the set of all seniors who have taken the SAT test in the last 20 years at this school. If this population's size, N, is greater than 800, then $.05\,N$ exceeds n, and the population is considered large relative to the sample. Assume that $N > 800$ and that the Central Limit Theorem will be applied.[2] Thus, the sampling distribution of all the values of $\overline{X}$ forms a normal distribution.

Steps 1 and 2

Consider Figure 8-8. Since a 95% confidence interval is desired for μ, find the Z-scores that define the middle 95% of the standard normal model by using Appendix A-2 as follows.

- The area from $\mu_{\overline{X}}$ to each Z-score is $\dfrac{95\%}{2} = .4750$.

- Find .4750 (or the number closest to it) in the body of Appendix A-2. This yields Z-scores of $z = +1.96$ and $z = -1.96$, which enclose the middle 95% of the area under the normal distribution model.

---

[2]See Exercise 4 (in Exercises 8.6) on page 339 to explore the implications of *not* assuming $N > 800$, meaning that $.05\,N$ is therefore not greater than n.

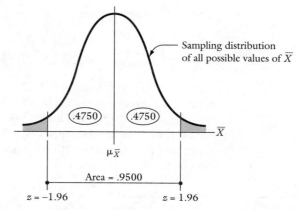

Figure 8-8 A normal distribution model for the construction of a confidence interval.

Chapter 6 presented the formula for computing the Z-score for a particular score, x, from the population:

$$Z(x) = \frac{x - \mu}{\sigma}.$$

An analogous formula can be used to find the Z-score for a particular value of $\overline{X}$ in the sampling distribution of the $\overline{X}$'s:

$$Z(\overline{X}) = \frac{\overline{X} - \mu_{\overline{X}}}{SE}.$$

This formula enables one to relate $\mu_{\overline{X}}$ to the sample mean, $\overline{X}$, as well as to the Z-score that encloses the middle 95% of the normal distribution model.

Step 3

The sample statistic $\overline{X}$ can be computed as 622.50. In Example 2 on page 320, SE was estimated at 18.19.

Step 4

Substitute all of these calculated values into the formula for $Z(\overline{X})$,

$$\pm 1.96 = \frac{622.50 - \mu_{\overline{X}}}{18.19}.$$

Step 5

Solve for $\mu_{\overline{X}}$.

$$\pm 1.96 \cdot (18.19) = 622.50 - \mu_{\overline{X}},$$
$$\pm 35.6524 = 622.50 - \mu_{\overline{X}},$$
$$\mu_{\overline{X}} = 622.50 \pm 35.6524.$$

The two possible values

$$622.50 - 35.6524 = \mathbf{586.85} \quad \text{and}$$
$$622.50 + 35.6524 = \mathbf{658.15}$$

determine the end points of a 95% confidence interval for the mean of the sample means, $\mu_{\overline{X}}$. Since $\mu_{\overline{X}} = \mu$, $[586.85, 658.15]$ is also a 95% confidence interval for the population mean, μ. That is, one can be 95% certain that the mean SAT score for this population is between 586.85 and 658.15, inclusive. However, since SAT scores are not reported with such accuracy, a rounded value of $[590, 660]$ might also be used.   ●

E x a m p l e
4

Construct an 80% confidence interval for the population mean, μ, in Example 3 (page 326).

Solution   The values of $\overline{X}$ and *SE* are the same as in Example 3. However, one must determine new *Z*-scores that enclose the middle 80% (instead of 95%) of the normal model. Figure 8-9 illustrates this change. Note that $\dfrac{80\%}{2} = .4000$, so the area from $\mu_{\overline{X}}$ to each *Z*-score is .4000. The number in the body of Appendix A-2 closest to .4000, namely .3997, yields *Z*-scores of $z = +1.28$ and $z = -1.28$.
Substituting these new values into the formula,

$$Z(\overline{X}) = \frac{\overline{X} - \mu_{\overline{X}}}{SE},$$

$$\pm 1.28 = \frac{622.50 - \mu_{\overline{X}}}{18.19}.$$

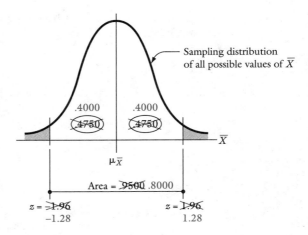

Figure 8-9

This locates the end points of an 80% confidence interval for $\mu_{\overline{X}}$ (and therefore also for μ) at

$$622.50 - 1.28 \cdot (18.19) = \textbf{599.22} \qquad \text{and}$$
$$622.50 + 1.28 \cdot (18.19) = \textbf{645.78}.$$

Thus, an 80% confidence interval for μ is [599.22, 645.78].   ●

Notice that the 95% confidence interval in Example 3 (page 326) is wider than the 80% confidence interval just found. This difference should seem reasonable, since one should feel more confident that μ lies in a wider interval.

The steps for the construction of a confidence interval were not numbered in Example 4 as they were in Example 3. This process may be streamlined even more. The key step is to supply the values for the equation

$$Z(\overline{X}) = \frac{\overline{X} - \mu_{\overline{X}}}{SE},$$

then solve it for $\mu_{\overline{X}}$, which is equal to μ. If one solves this equation for $\mu_{\overline{X}}$ in general terms, one obtains a formula that may be used to determine the end points of a confidence interval.

> The greater the percentage of confidence, the wider the confidence interval.

Location of the end points of a confidence interval for μ, using a normal distribution model:

$$\text{Left end point} = \overline{X} - Z(\overline{X}) \cdot SE,$$
$$\text{Right end point} = \overline{X} + Z(\overline{X}) \cdot SE.$$
$$\text{(Place end points at } \overline{X} \pm Z(\overline{X}) \cdot SE) \tag{8-3}$$

The next example illustrates the application of this formula and shorter method.

E x a m p l e 5

Fifty electric bills incurred by apartments of a certain size within a city during the month of August are chosen at random. The mean electric bill of this sample is $\overline{X} = \$109.50$, with $s = \$21.75$. It is believed that the population from which the sample was taken (electric bills for all the apartments of that size within the city) has a normal distribution. Construct a 98% confidence interval for μ, the population mean.

Solution   This is a large sample ($n \geq 30$) taken from a normal population. Thus, the second branch on the right side of the flow chart indicates a normal distribution model. (Note that, since the population is normally distributed and the sample is large, the normal distribution model is used regardless of whether $n < .05\,N$.)

The following figure shows that Z-scores of $z = \pm 2.33$ enclose the middle 98% of the standard normal model.

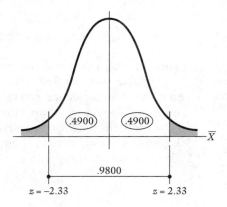

It is given that $\overline{X} = \$109.50$. Also,

$$SE \approx \frac{s}{\sqrt{n}}$$

$$= \frac{21.75}{\sqrt{50}}$$

$$\approx 3.08.$$

Therefore, the end points of a 98% confidence interval for μ are

$$\overline{X} \pm Z(\overline{X}) \cdot SE = 109.50 \pm 2.33 \cdot (3.08)$$
$$= [\$102.32, \$116.68].$$

Thus, one can be 98% certain that the mean electric bill for all apartments of this size within the city was at least \$102.32 but no more than \$116.68.   •

8.5 *Technology* TOOLS

Many statistical software programs can be used to construct confidence intervals. Minitab has two macro commands for that purpose; the choice depends on whether a normal distribution or a t-distribution model is used.[3] The ZINTERVAL macro uses the normal distribution model and requires that a standard deviation of the population σ (or its estimate, s) be supplied. The following example demonstrates the use of the ZINTERVAL command.

---

[3]The TINTERVAL macro uses the t-distribution model, which is discussed later in this chapter.

E x a m p l e
6

The gasoline mileage obtained from 48 rental cars, randomly chosen from a large fleet of cars of the same size, follows.

31.2	30.3	21.3	22.7
27.6	19.7	26.9	20.9
30.3	31.2	24.3	30.1
22.1	24.3	27.9	15.8
28.3	17.5	30.0	25.6
20.7	24.9	19.4	24.7
33.0	22.7	31.6	24.9
28.4	23.6	16.2	23.7
32.6	21.4	27.6	26.9
21.1	22.0	17.1	26.7
34.2	25.0	27.6	27.4
20.7	31.7	27.3	25.9

Construct 99%, 95%, 90%, 80%, and 50% confidence intervals for the mean gas mileage of the entire fleet of all cars of this size.

Solution Although it is not specifically stated, assume that the fleet (population) is large relative to this sample. Such an assumption, along with the fact that the sample itself is a large one ($n \geq 30$), justifies the use of the normal distribution model for the construction of the required confidence interval. Use Minitab's ZINTERVAL macro command, which requires that the user furnish the standard deviation of the population, σ, or its estimate, s. The value of s can be obtained by hand or with Minitab's **Descriptive Statistics** command.

First enter the 48 scores in the Minitab worksheet in column C1:

	C1	C2
↓		
1	31.2	
2	27.6	
3	30.3	
4	22.1	
5	28.3	
6	20.7	
7	33.0	
8	28.4	
9	32.6	

Then, in the interactive mode, execute the **Descriptive Statistics** command from the **Basic Statistics** option in the **Stat** menu, with the following result.

Minitab: Descriptive Statistics Calculations

	N	MEAN	MEDIAN	TRMEAN	STDEV	SEMEAN
C1	48	25.354	25.300	25.405	4.647	0.671

s

	MIN	MAX	Q1	Q3
C1	15.800	34.200	21.550	28.375

Use the rounded value of *s*, **4.65** miles per gallon, as the estimated standard deviation of the entire fleet for the construction of confidence intervals based on the normal distribution model. Accomplish this with the macro command ZINTERVAL, as follows. Obtain a Session window. After the **MTB>** prompt, type ZINTERVAL. Follow this with the desired percent of confidence, the value of *s*, and finally the location of the data, C1. The completed command line should appear as follows:

```
                                                          Session
Worksheet size: 38000 cells
MTB > ZINTERVAL  99  4.65   C1
```

When you press the [RETURN] key, a **99%** confidence interval is printed. Then repeat the ZINTERVAL macro command four times for 95%, 90%, 80%, and 50% confidence intervals. Study the illustrations in Figure 8-10 carefully. (Each time, Minitab prints **THE ASSUMED SIGMA = 4.65** as part of the output.)

The entire command line does not need to be retyped. Minitab Session windows allow for copy and paste editing.

As the degree of confidence decreases, the width of the confidence interval diminishes. (One is less confident that μ lies in a smaller interval.) •

E x a m p l e 7

Refer to Example 6 on page 331. Use Minitab to find a 99% confidence interval when long experience has shown that $\sigma = 5.00$ and use of the estimate *s* is therefore unnecessary.

Solution   Once again, use of the normal distribution model is justified. Accordingly, use the macro command ZINTERVAL, but instead of specifying the estimated standard deviation of 4.65, enter the known value, 5.00.

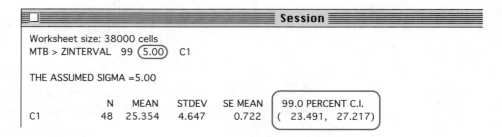

```
                                                          Session
Worksheet size: 38000 cells
MTB > ZINTERVAL  99  5.00   C1

THE ASSUMED SIGMA =5.00

        N    MEAN    STDEV    SE MEAN    99.0 PERCENT C.I.
C1     48   25.354   4.647     0.722    ( 23.491,  27.217)
```

```
MTB > ZINTERVAL    99   4.65   C1

THE ASSUMED SIGMA =4.65
```

	N	MEAN	STDEV	SE MEAN	99.0 PERCENT C.I.
C1	48	25.354	4.647	0.671	(23.622, 27.087)

```
MTB > ZINTERVAL    95   4.65   C1

THE ASSUMED SIGMA =4.65
```

	N	MEAN	STDEV	SE MEAN	95.0 PERCENT C.I.
C1	48	25.354	4.647	0.671	(24.037, 26.672)

```
MTB > ZINTERVAL    90   4.65   C1

THE ASSUMED SIGMA =4.65
```

	N	MEAN	STDEV	SE MEAN	90.0 PERCENT C.I.
C1	48	25.354	4.647	0.671	(24.249, 26.459)

```
MTB > ZINTERVAL    80   4.65   C1

THE ASSUMED SIGMA =4.65
```

	N	MEAN	STDEV	SE MEAN	80.0 PERCENT C.I.
C1	48	25.354	4.647	0.671	(24.493, 26.215)

```
MTB > ZINTERVAL    50   4.65   C1

THE ASSUMED SIGMA =4.65
```

	N	MEAN	STDEV	SE MEAN	50.0 PERCENT C.I.
C1	48	25.354	4.647	0.671	(24.902, 25.807)

Figure 8-10 Confidence intervals computed by Minitab for five different percents of confidence.

Notice that, even though Minitab computes the value $s = 4.647$ as before, it uses the supplied value, 5.00, for the construction of the confidence interval. Also notice that this confidence interval is wider than the one found in Example 6. This result should be expected, because a larger value for the standard deviation results in a larger value for SE; here the standard error (labeled **SE MEAN**) is 0.722, whereas in Example 6 it was 0.671. ●

Exercises 8.3 – 8.5

1. Exactly what is meant by a "large" sample and by a "relatively large" population?

2. Loosely speaking, what is meant by the statement that a population has a normal distribution?

3. Refer to Example 6 on page 331. Why do you think the cars were specified as being the same size?

4. What are the two ways to narrow a confidence interval?

5. Refer to Example 5 on page 329. Using all the statistics presented, recompute the confidence interval for the following sample sizes.

   a. $n = 75$

   b. $n = 100$

   c. $n = 250$

6. Conventionally, confidence intervals enclosed in brackets, such as those worked out in Examples 3, 4, and 5 (pages 326, 328, and 329) include their end points. The confidence intervals in Example 6 on page 331, enclosed in parentheses, do not include their end points. Thus, the inclusion of the end points of a confidence interval is arbitrary. Try to explain why.

7. Identify each of the following statements as true or false.

   a. Constructed from the same random sample of size n, an 80% confidence interval for μ must be wider than a 65% confidence interval.

   b. As the sample size increases, the confidence interval narrows.

   c. The population size is large relative to the sample size if the inequality $n < .05 N$ is satisfied.

   d. The standard error is always more than the standard deviation of the population, σ, or its estimate, s.

MTB   e. When one uses Minitab's ZINTERVAL command, there is no need to specify the standard deviation or its estimate. Minitab computes this value and uses it to construct the confidence interval.

8. Refer to Example 6 on page 331. What is the minimum fleet size that would justify the application of the Central Limit Theorem and a normal distribution model? Does it seem reasonable to have assumed this minimum size?

9. For the gasoline mileage problem of Examples 6 and 7, suppose that the standard deviation of the entire fleet is actually 5.5 mpg. If a 99% confidence interval were constructed on the basis of this new information, how would its width compare to the 99% confidence intervals already found?

GROUP   10. It is believed that life span in a certain country has a standard deviation of 7.75 years. A random sample of 100 death certificates is examined, and the mean age is found to be 73.56 years. Construct 95%, 90%, and 75% confidence intervals for the mean life span of all people in the country.

11. Ninety students' scores on a mathematics achievement test are considered to be normally distributed. A random sample of 30 of the scores follows.

790	660	500	680	760	660
580	780	670	660	750	700
600	620	580	590	700	640
650	700	480	760	700	620
660	600	640	710	680	660

Construct a 98% confidence interval for the mean score, μ, of the 90 scores.

12. A major metropolitan telephone company asserts that the standard deviation of the length of a telephone call is 3.55 minutes. Fifty calls are sampled at random, with a mean calling time of 8.75 minutes. Construct 75% and 90% confidence intervals for the mean length of all telephone calls.

GROUP 13. Refer to Exercise 12. Assume that the standard deviation is not 3.55 but is to be estimated by s instead. Construct a 90% confidence interval for the following combinations of n and s.

a. $n = 50, s = 5.00$

b. $n = 100, s = 5.00$

c. $n = 75, s = 8.00$

d. $n = 75, s = 4.00$

e. $n = 300, s = 5.00$

f. $n = 75, s = 9.50$

14. Base answers to the following questions on the confidence intervals obtained in Exercise 13.

a. Does a change in the sample size or a change in the standard deviation seem to have a greater effect on the width of a confidence interval?

* b. Mathematically justify your answer to Part a of this exercise.

15. The following table presents, by state, the 1992 rates of abortion per 1000 women in a random sample of 30 states.

OH	19.5	AL	18.2	WI	13.6
GA	24.0	WY	4.3	KY	11.4
CA	42.1	UT	9.3	TX	23.1
CO	23.6	HI	46.0	PA	18.6
IL	25.4	NV	44.2	OK	12.5
AZ	24.1	VT	21.2	FL	30.0
SD	6.8	NJ	31.0	MA	28.4
NY	46.2	NM	17.7	RI	30.0
IN	12.0	ME	14.7	MS	12.4
CT	26.2	WA	27.7	MT	18.2

a. What is the population, and what does μ represent?

b. Assume that the population is normally distributed, and construct an 85% confidence interval for the population mean.

c. Does the assumption of a normal distribution for the population in Part b seem reasonable for these data? Why or why not?

16. The following array is a random sample of 30 states and their 1991 numbers of full-time equivalent employees per 100 average daily patients in community hospitals.

DE	674	OH	666	IL	614
LA	602	KY	539	HI	525
ME	573	NJ	478	WY	480
GA	517	MA	636	VA	559
TX	630	CO	614	NY	512
ID	548	KS	580	WI	537
FL	589	NV	556	TN	569
MS	487	WA	694	NH	604
ND	401	CA	644	WV	547
MN	434	NC	580	NM	671

Source: *Health, United States, 1993,* U.S. Department of Health and Human Services, p. 217.

a. Describe the population, and what does μ represent?

b. Determine the point estiamtes $\overline{X}$ and s.

c. Assuming that the population of 50 states is normally distributed, construct 85% and 95% confidence intervals for the population mean.

17. A commuter railroad claims that the standard deviation of arrival times of one of its commuter trains at Union Station is 1.00 minute. A random sample of 50 arrival times follows.

3:30	3:33	3:32	3:30	3:35	3:39	3:28	3:30	3:45	3:36
3:38	3:35	3:31	3:30	3:26	3:33	3:32	3:38	3:31	3:30
3:20	3:30	3:52	3:37	3:30	3:34	3:33	3:33	3:33	3:40
3:29	3:32	3:36	3:30	3:37	3:38	3:41	3:39	3:30	3:33
3:29	3:35	3:30	3:40	3:27	3:30	3:33	3:28	3:27	3:35

Accepting the railroad's value of σ, determine 80%, 90%, and 98% confidence intervals for the mean arrival time of this commuter train at Union Station. Assume that $n < .05 N$.

Hint: To simplify calculations, drop the hours from the above times; for example, the time 3:30 becomes 30. The resulting confidence interval is then in terms of minutes past 3 o'clock.

GROUP

18. Refer to Exercise 17. Do not use the railroad's value of σ; instead, use the statistic s obtained from samples of the 50 times given.

a. Use four random samples of size 30 from the times listed to determine four 90% confidence intervals for the mean arrival time of this train at Union Station. Form the four random samples by using columns 1 through 6, 2 through 7, 4 through 9, and 5 through 10.

b. Should the widths of all four of these confidence intervals be the same? Explain.

c. What possible values of μ are in all four confidence intervals?

d. If a confidence interval were formed using all 50 times, would the confidence interval necessarily contain all of the confidence intervals from Part a?

19. Following is a random sample of annual starting salaries offered to bachelor's degree candidates in March 1993.

Advertising	$23,753	Sociology	$19,058
Journalism	17,814	Real estate	27,716
Civil engineering	29,385	Elementary education	20,434
Computer science	31,118	Nursing	31,851
Pharmacy	43,570	Mathematics	24,698
Accounting	27,716	Chemical engineering	39,793
Criminal justice	22,836	Architectural engineering	28,667
Special education	21,390	Actuarial science	34,450
Electrical engineering	34,035	Agribusiness	29,983
Writing/editing	20,650	Law enforcement	23,268
Bioengineering	37,000	Public relations	21,204
Insurance (underwriting)	23,130	Accounting (public)	28,228
Animal sciences	21,246	Nuclear engineering	33,833
Petroleum engineering	39,157	Communications	21,964
Textiles and clothing	20,913	Hotel/restaurant management	21,809
Psychology	20,538	Visual and performing arts	20,429

Source: *College Placement Council, Inc. CPC Salary Survey*, March 1993, pp. 2–4.

Assume that all salaries offered to bachelor's degree candidates are normally distributed.

a. Construct a 90% confidence interval for the salary offer a bachelor's degree candidate could expect to receive in March 1993.

b. Remove the highest salary and the lowest salary, then recompute the confidence interval from Part a.

20. Refer to Exercise 19. Was the confidence interval in Part b shifted to the left (lower values) or to the right (higher values)? Why?

8.6 USE OF CHEBYSHEV'S THEOREM TO DETERMINE CONFIDENCE INTERVALS

A sample comes from a relatively small population if $n \geq .05\,N$.

The only time the normal distribution is not used as a model for the construction of a $C\%$ confidence interval involving a large sample is when the sample comes from a relatively small population that is not normally distributed. In such a situation, Chebyshev's Theorem (stated in Chapter 4) implies that at least $(1 - \frac{1}{k^2})$ of the sample means must be within k standard errors of the mean of all the samples—that is, within $k \cdot SE$ of μ. Once the equation

$$1 - \frac{1}{k^2} = C\%$$

is solved for k and SE is determined, the end points of the desired confidence interval for μ can be located at $\overline{X} \pm k \cdot SE$.

E x a m p l e 8

About 250 artists are believed to live and work in the immediate vicinity of Santa Fe, New Mexico. A random sample of $n = 50$ of these artists yielded a mean annual income of $\overline{X} = \$22,800$ and an estimated standard deviation of $s = \$8500$. Use this information to determine an 80% confidence interval for the mean annual income of the population of all 250 artists.

Solution   Although this is a large sample, it comes from a relatively small population because n is not less than $.05\,N$; that is, $50 \geqslant .05 \cdot (250)$. Also, because of the unusual nature of art sales, an assumption that the sales of all 250 artists are normally distributed does not seem reasonable. Thus, the flow chart in Figure 8-5 suggests that Chebyshev's Theorem be used.

The first step in applying Chebyshev's Theorem to construct an 80% confidence interval is to solve

$$1 - \frac{1}{k^2} = 80\%$$

for k.

$$1 - \frac{1}{k^2} = .80,$$
$$\frac{1}{k^2} = .20,$$
$$k^2 = 5,$$
$$k = \sqrt{5},$$
$$k \approx 2.24.$$

Also,

$$SE = \frac{s}{\sqrt{n}}$$
$$= \frac{8500}{\sqrt{50}}$$
$$\approx \$1202.08.$$

Therefore, the end points of the confidence interval are at $\overline{X} \pm k \cdot SE$, or $\$22,800 \pm 2.24 \cdot \$1202.08$. This yields an 80% confidence interval of approximately [\$20,107.34, \$25,492.66]. •

Exercises 8.6

1. Refer to Example 8 on page 338 and offer further explanation of why an assumption of normal distribution does not seem reasonable.

2. Refer to Exercise 12 (in Exercises 8.3–8.5) on page 335. Imagine that the telephone company is very small instead of large, and all call times are not normally distributed. Use Chebyshev's Theorem to construct the required 75% and 90% confidence intervals.

* 3. Compare the results of the preceding exercise and Exercise 12 (in Exercises 8.3–8.5) on page 335.

    a. How much wider, in percentage, is the 90% confidence interval found using Chebyshev's Theorem than the confidence interval found using the normal distribution model?

    b. Is it always the case that confidence intervals found using Chebyshev's Theorem are wider than those found using the normal distribution model?

4. Refer to Example 3 on page 326. Determine 70% confidence intervals for the population mean, with $N \leq 800$, in each of the following cases.

    a. The population is normally distributed.

    b. The population is not normally distributed.

5. Refer to Exercise 15 (in Exercises 8.3–8.5) on page 335. Determine an 85% confidence interval for the population mean if the population is not normally distributed.

8.7 CONFIDENCE INTERVALS FOR POPULATION PROPORTIONS

An important and familiar application of confidence intervals is in the construction of interval estimates of a population proportion.

Definition ▶ A **population proportion** is the percentage of the population that has a given characteristic. This proportion is determined by yes/no data.

For example, the percentage of people who prefer a new brand of cereal, the percentage of buses that arrive on time, and the percentage of voters who favor a certain political candidate are all examples of population proportions.

The actual population proportion is denoted by π, and the percentage of a sample that has the desired characteristic is denoted by P. Thus, P is a point estimate of π. If the sample size, n, is great enough, the Central Limit Theorem also applies to the distribution of the sample proportions, P, and an interval estimate for π

S t a t   F a c t

Nearly 75% of professional pollsters believe that polls will be accurate enough to replace elections by the year 2000, and nearly half of them believe that polling techniques have already changed the function of voting to confirmation of the polls' results. More than one-third of these pollsters think elections will be necessary only about once in every 25 years, to recalibrate the polling systems.

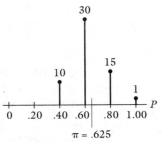

Figure 8-11 The sampling distribution of sample proportion *P*.

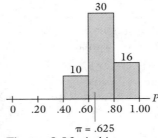

Figure 8-12 A histogram of possible values of sample proportion *P*.

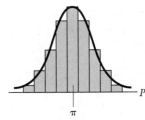

Figure 8-13

using *P* is formed in a manner analogous to the formation of an interval estimate for μ using $\overline{X}$.

As an illustration of a sampling distribution of *P*, consider a population of size $N = 8$. If five members of this population have a certain characteristic, then $\pi = \frac{5}{8} = .625$ denotes the percentage of the population with that characteristic. If a sample of size $n = 5$ is chosen from the population, then the percentage of the sample with that characteristic, denoted by *P*, can be one of four values: $\frac{2}{5}, \frac{3}{5}, \frac{4}{5}$, or $\frac{5}{5}$. If all $_8C_5 = 56$ possible samples are examined, the sampling distribution of *P* looks like Figure 8-11.[4] Figure 8-11 may also take the form of a left-ended histogram as shown in Figure 8-12. (The last interval contains both end points.)

If the sample is large enough in relation to *P*, the mound-shaped histogram of possible values of *P* fits a normal distribution model, centered about the population proportion, π, as shown in Figure 8-13.

The sample is considered large enough in relation to *P* to justify a normal distribution model if **Cochran's Rules,** stated next, are satisfied. Notice that Cochran's Rules imply that the more "extreme" the proportion, the larger the sample must be.

Cochran's Rules for Normality of the Sampling Distribution of the Sample Proportion, *P*

P	Minimum Sample Size, *n*, Needed to Use a Normal Distribution Model to Construct a Confidence Interval for π
.50	30
.40 or .60	50
.30 or .70	80
.20 or .80	200
.10 or .90	600
.05 or .95	1400

[4]These calculations may be verified. See Exercise 3 in Exercises 8.7, on page 344 for details.

For instance, if $P \approx .20$, then n should be at least 200. Once Cochran's Rules are satisfied, then:

The standard deviation of all possible values of P is also called the standard deviation of the sampling distribution or the standard error of the sample proportion.

a. The mean of all possible values of P, for a given sample size, n, is equal to π.
b. The standard deviation of all possible values of P, SE, is found as follows:

$$SE = \sqrt{\frac{\pi(1 - \pi)}{n}}$$

$$\approx \sqrt{\frac{P(1 - P)}{n}}.$$

c. Appendix A-2 may be used to find the relevant Z-scores, $Z(P)$, that enclose the middle $C\%$ of the area under the normal distribution model.

Determine the end points of the desired confidence interval for π with the following formula.

Location of the end points of a confidence interval for π

$$\text{Left end point} = P - Z(P) \cdot SE,$$
$$\text{Right end point} = P + Z(P) \cdot SE.$$

(Place end points at $P \pm Z(P) \cdot SE$.)   **(8-4)**

**E x a m p l e
9**

A national political candidate polled 500 voters in Illinois and found that 270 of them preferred her. In Michigan, she was preferred by 430 of 900 polled. Construct 95% confidence intervals for the proportions of all voters in each of these two states who prefer her as a candidate. Round all calculations to the nearest thousandth.

Solution   In Illinois,

$$P_I = \frac{270}{500}$$

$$= .540.$$

In Michigan,

$$P_M = \frac{430}{900}$$

$$= .478.$$

To simplify the use of Cochran's Rules, the minimum sample size for an unlisted value of P will always be the larger value of n between which P falls. For example, if $P = .34$, then $n = 80$, and if $P = .17$, then $n = 600$.

In accordance with Cochran's Rules, use of the normal distribution model requires a minimum sample size of about 50. Since the given sample sizes are several hundred each, Cochran's Rules are easily satisfied and the normal distribution model will be the basis for the construction of the desired confidence intervals.

Stat Fact

In a recent special election to fill a state senate vacancy in Philadelphia, Republican Bruce Marks lost a close contest to Democrat William Stinson because Stinson received a very large majority of the absentee ballot votes. A court-appointed expert in probability, using nonabsentee vote totals as an indicator of absentee voting and employing other statistical methods, concluded that there was a 94% chance that irregularities in the absentee ballots, and not chance alone, had swung the election to the Democrat. The decision whether to declare Marks the winner or order a new election was left to a court.

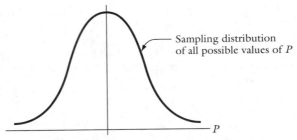

Sampling distribution of all possible values of P

Mean of all possible values of $P = \pi$

The next step is to find the standard deviation of each sampling distribution, SE. Since the candidate is trying to determine the actual percentage of voters in each state who prefer her, these parameters, π_I and π_M, are unknown. Therefore, approximate SE for each of the states are as follows:

$$SE_I \approx \sqrt{\frac{P_I(1 - P_I)}{n}}$$
$$= \sqrt{\frac{.540(1 - .540)}{500}}$$
$$\approx .022.$$
$$SE_M \approx \sqrt{\frac{P_M(1 - P_M)}{n}}$$
$$= \sqrt{\frac{.478(1 - .478)}{900}}$$
$$\approx .017.$$

Earlier examples demonstrated that Z-scores of $z = \pm 1.96$ enclose 95% of the area under the normal distribution model. Accordingly, the end points of a 95% confidence interval for voter preferences in Illinois, denoted by π_I, are

$$P_I \pm Z(P_I) \cdot SE.$$

Substituting, the confidence interval is determined as

$$.540 \pm 1.96 \cdot (.022) = [.497, .583].$$

For Michigan, a 95% confidence interval for π_M is

$$P_M \pm Z(P_M) \cdot SE, \quad \text{or}$$
$$.478 \pm 1.96 \cdot (.017) = [.445, .511]. \quad \bullet$$

The Margin of Error and the Sample Size for Population Proportions

It is often useful to specify a width for a confidence interval of a population proportion. For example, a political candidate may want to know her voter appeal within 3 percentage points. This implies that the confidence interval for the population proportion, P, has a width of 6%, with P in the center of it. This enables predetermination of the accuracy and easier comparison of different estimates.

Definition ▶

The margin of error is often specified arbitrarily.

The **margin of error** of a confidence interval, denoted by ME, is the maximum error the statistician will allow. It can be found with the formula

$$ME = Z(P) \cdot SE. \tag{8-5}$$

In Example 9 on page 341, the end points of the confidence interval for π_I were found by evaluating

$$.540 \pm Z(P_I) \cdot SE.$$

This implies that ME is the amount added to and subtracted from the population proportion estimate, P, and that twice the margin of error is the width of the confidence interval.

ME is often written as a percentage. In the Illinois example,

$$\begin{aligned} ME &= Z(P_I) \cdot SE \\ &= 1.96 \cdot (.022) \\ &\approx 4.31\%. \end{aligned}$$

The minimum sample size, n, required for a specified margin of error can be found with the formula

$$n \geq \left(\frac{Z(P) \cdot .5}{ME} \right)^2. \tag{8-6}$$

If the candidate in Example 9 wants a 95% confidence interval for her voter appeal, with a margin of error of $\pm 3\%$, then the minimum number of potential voters who must be surveyed is

$$n \geq \left(\frac{1.96 \cdot .5}{.03} \right)^2$$

$$\approx 1068.$$

On the other hand, if an 80% confidence interval is desired, with a margin of error of $\pm 5\%$, then $Z(P) = 1.28$ instead of 1.96, and the minimum number of voters in the sample need be only

$$n \geq \left(\frac{1.28 \cdot .5}{.05} \right)^2$$

$$\approx 164.$$

In the latter case, however, Cochran's Rules may not be satisfied, depending on the value of P, and a larger sample may be required.

Exercises 8.7

1. Usually, the actual value of the population proportion, π, is never known with complete certainty. Why?

2. Give three examples of situations in which the population proportion, π, could be known with complete certainty.

GROUP * 3. Refer to footnote 4 on page 340, which cites this exercise. Determine the frequency of each possible value of the sample proportion, P, by executing these steps:

  1. Let the numbers 1 through 8 represent the population.

  2. Randomly choose any five of the numbers from the population.

  3. Make an organized table of all $_8C_5 = 56$ subsets of the population.

  4. Choose any subset from those in Step 3, and determine what percentage of the numbers chosen in Step 2 are in that subset; that is a value of P.

  5. Repeat Step 4 for each of the 56 subsets, and check your answers against those presented earlier.

4. Refer to Example 9 on page 341. What is the approximate minimum number of voters who would have to be surveyed in Illinois to justify the use of the normal distribution as a model?

5. A sales manager for a new laundry detergent is interested in knowing what percentage of customers prefer the new product to their old ones. A random sample of 350 customers reveals that 195 prefer the new product.

  a. Describe what the population proportion π would represent.

  b. Determine a 90% confidence interval for π.

6. The Gallup poll surveys random samples of 1500 voters in an effort to predict the results of presidential elections. The following table presents the results of such polls for the major-party candidates in the years 1960, 1964, 1968, and 1972.

Year	Democrat	P	Republican	P
1960	Kennedy	.51	Nixon	.49
1964	Johnson	.64	Goldwater	.36
1968	Humphrey	.50	Nixon	.50
1972	McGovern	.38	Nixon	.62

The actual outcomes were: Kennedy, $\pi_K = .501$; Johnson, $\pi_J = .613$; Humphrey, $\pi_H = .497$; and McGovern, $\pi_M = .382$.

Construct 95% confidence intervals for the voter preferences for the Democratic candidates. Then determine which, if any, fail to include the actual outcomes.

7. Refer to Exercise 6.

   a. Determine the margin of error (*ME*) for each of the four Democratic candidates' confidence intervals.

   b. What three factors affect the margin of error?

8. A 90% confidence interval is to be constructed for a professional basketball team's field-goal shooting percentage. Assuming that Cochran's Rules are satisfied, how many shots must be sampled so that the margin of error, *ME*, is

   a. 2%?

   b. 3%?

   c. 5%?

   d. 10%?

9. Many polls relating to political candidates and issues are based on an implied 95% confidence interval with a margin of error of no more than 3%. If a pollster wanted to reduce the number of people to survey, would the percentage of confidence, margin of error, or both have to change? Explain.

* 10. The formula

$$n \geq \left(\frac{Z(P) \cdot .5}{ME} \right)^2$$

   was derived from the facts that $ME = Z(P) \cdot SE$ and that SE is a maximum when $p = .5$. Show why SE is a maximum under this condition.

8.8   FLOW CHART BRANCHES TO THE LEFT: SMALL SAMPLES ($n < 30$)

The left side of the flow chart in Figure 8-5 on page 324 shows that there is another set of conditions under which the normal distribution model can be used for the construction of a confidence interval.

Small-Sample Case 1

If the sample is small ($n < 30$), and

• the population is normally distributed, and
• the population standard deviation is known,

then the sampling distribution of the $\overline{X}$'s can be considered normal, regardless of the size of the population.

The vertical branch on the left side of the flow chart illustrates these conditions.

E x a m p l e
10

In the Chicago area, the price of a new, high-quality automobile tire is believed to be normally distributed with a standard deviation of $\sigma = \$11.50$. A random sample of the selling prices of five of these tires indicated a mean selling price of $\overline{X} = \$98.70$. Construct an 85% confidence interval for the mean selling price, μ, of this new tire in the Chicago area.

Solution   This is a small-sample case taken from a normally distributed population with known (i.e., not estimated) standard deviation. Therefore, the confidence interval can be constructed with a normal distribution model as it was in the large-sample case. First, the distribution model can be depicted as follows.

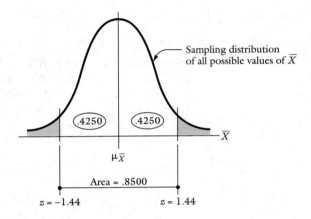

The required Z-scores of $z = \pm 1.44$ can be obtained by studying this model and using Appendix A-2.

Next, the standard error is

$$SE = \frac{\sigma}{\sqrt{n}}$$

$$= \frac{11.50}{\sqrt{5}}$$

$$\approx 5.143.$$

The end points of an 85% confidence interval for μ are at

$$\overline{X} \pm Z(\overline{X}) \cdot SE = 98.70 \pm 1.44 \cdot (5.143)$$
$$= [\$91.29, \$106.11]. \quad \bullet$$

Returning to the left side of the flow chart, it can be seen that if σ is not known, the t-distribution model must be used (instead of the normal distribution) to construct the confidence interval.

Small-Sample Case 2

If the sample is small ($n < 30$), and

* the population is normally distributed, and
* the population standard deviation is *not* known,

then the *t*-distribution model is used to determine the confidence interval.

The *t*-Distribution

Often, in a small-sample case, all that is known about the population is the fact that it is normally distributed. Example 10 was a bit unusual in the sense that the standard deviation, σ, was somehow known. When one is constructing confidence intervals with small samples, if the only fact known about the population is that it is normally distributed, then a different but similar *t*-distribution model replaces the normal model for the sampling distribution of the $\overline{X}$'s.

Student's *t*-distribution, or simply *t*-**distribution,** is actually a whole family of mound-shaped curves. They all resemble the normal curve but have slightly different shapes, depending on the value of *n*. If *s* is a small-sample estimate of σ, it may not be a very accurate estimate. The *t*-distribution curves were developed to compensate for this potential error. As the sample size, *n*, increases, *s* becomes a more reliable estimate of σ, and the *t*-distribution model and the normal distribution model become more similar. For sample sizes of at least 30, the *t*-distribution model is very close to the normal distribution model, in which case this text treats it identically.

Figure 8-14 shows how some *t*-distribution models (for $n = 3$ and $n = 15$) approximate the normal distribution model. Notice that as *n* increases, the mound-shaped *t*-distribution curve nears the normal distribution curve.

As in the case of the normal distribution, a table of areas is available as a tool for the construction of confidence intervals with the *t*-distribution model. This table is arranged differently, however, because it must account for different values of *n*.

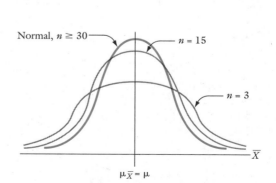

Figure 8-14 The *t*-distribution model for different sample sizes.

The general steps for determining a C% confidence interval with the *t*-distribution model are similar to those for the normal distribution model.

a. Visualize the area under the appropriate *t*-distribution model that would be occupied by the confidence interval.
b. Use Appendix A-4 to find the relevant *t*-scores, $t(\overline{X})$, that enclose the middle C% of the area under the *t*-distribution model.
c. Calculate $\overline{X}$ and $SE \approx \dfrac{s}{\sqrt{n}}$.

Determine the end points of the confidence interval with the following formula.

Location of the end points of a confidence interval for μ using a *t*-distribution model:

$$\text{Left end point} = \overline{X} - t(\overline{X}) \cdot SE,$$
$$\text{Right end point} = \overline{X} + t(\overline{X}) \cdot SE.$$
$$(\text{Place end points at } \overline{X} \pm t(\overline{X}) \cdot SE.)$$

(8-7)

The following example illustrates this process.

E x a m p l e 11

The ages of all full-time professors at a certain university are normally distributed. A random sample of ten professors revealed the following ages:

45.5   52.0   48.5   46.5   50.5   44.8   51.0   45.5   55.5   54.2

Find a 95% confidence interval for the mean age of all full-time professors at that university. Round all answers to the nearest hundredth.

Solution   Since this is a small-sample case from a normal population with unknown standard deviation, use the *t*-distribution model. To find the relevant *t*-scores, $t(\overline{X})$, that enclose the middle 95% of the *t*-distribution model, examine Appendix A-4.

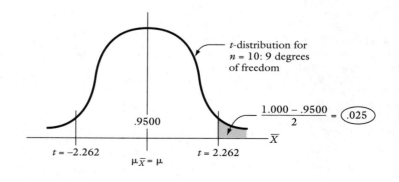

Although this appendix is constructed very differently from the one used for the standard normal table, the area under a *t*-distribution model is also 1.

The column labeled "Degrees of Freedom" is a function of the sample size, *n*. In the construction of confidence intervals, the **degrees of freedom** are always the sample size minus 1, or $n - 1$. Thus, in this example there are $10 - 1 = 9$ degrees of freedom.[5]

Unlike the areas provided in Appendix A-2 for the standard normal distribution, areas in Appendix A-4 for the *t*-distribution are tail areas, so a little extra arithmetic may be needed. Since a 95% confidence interval is being constructed, an area of

$$(1.000 - .9500) = .0500$$

is left over for the two tails. Because the *t*-distribution is symmetrical, the right tail, labeled α in Appendix A-4, is

$$\frac{.0500}{2} = .025.$$

The intersection of the $(n - 1) = 9$ degrees of freedom row with the $\alpha = .025$ column yields a *t*-score of $t = \pm 2.262$.[6] Thus, for this confidence interval, $t(\overline{X}) = \pm 2.262$ in Formula (8-7).

Using the ten scores provided, calculate the statistics $\overline{X} = 49.40$ and $s = 3.82$. Therefore,

$$SE = \frac{3.82}{\sqrt{10}}$$

$$\approx 1.21.$$

The end points of the desired confidence interval are at

$$\overline{X} \pm t(\overline{X}) \cdot SE = 49.4 \pm 2.262 \cdot (1.21).$$

Using these numbers, a 95% confidence interval for μ is [46.66, 52.14]—wider than would have been obtained with the normal distribution model.   •

E x a m p l e
12

Use Minitab to find the confidence interval for Example 11.

Solution   Use the macro command TINTERVAL. Like the ZINTERVAL command, it computes the estimated mean, estimated standard deviation (*s*), and *SE* automatically from the worksheet. Unlike the ZINTERVAL command, however, it does not require that the population standard deviation be entered. Under the TINTERVAL command, Minitab automatically uses *s* for the construction of the confidence interval and uses the *t*-distribution model with the correct degrees of freedom.

---

[5]The degrees of freedom (d.f.) in a statistical calculation do not necessarily always equal $n - 1$; they do, however, when confidence intervals are constructed with the *t*-distribution and in some other situations.

[6]The notion of *t*-score for $\overline{X}$ is completely analogous to that of *Z*-score for $\overline{X}$ except that it is computed with a small sample size and *s*, the approximation of σ.

The first step, as usual, is to enter the data into a Minitab worksheet:

	C1	C2
↓		
1	45.5	
2	52.0	
3	48.5	
4	46.5	
5	50.5	
6	44.8	
7	51.0	
8	45.5	
9	55.5	
10	54.2	

Next obtain a Session window and execute the TINTERVAL macro, with the following results.

```
                                                    Session
Worksheet size: 38000 cells
MTB > TINTERVAL  95  C1

             N    MEAN    STDEV   SE MEAN    95.0 PERCENT C.I.
C1          10    49.40    3.82     1.21    (  46.66,   52.14)
```

Notice that the percentage of confidence for the interval and the column designation for the data follow the TINTERVAL macro. The standard deviation is not supplied. Also, one can construct *t*-distribution confidence intervals for a greater variety of percentages of confidence with Minitab than would be possible using Appendix A-4. ●

Exercises 8.8

1. Examine Appendix A-4. As the degrees of freedom increase, do corresponding *t*-scores increase or decrease?

2. As the degrees of freedom increase, a confidence interval obtained with the *t*-distribution model increasingly resembles one obtained with the normal distribution model. Explain why.

3. When either the normal distribution or the *t*-distribution is used as a model for the construction of a confidence interval, what happens to the interval width as the sample size increases? Explain.

4. Of the three models that can be used for the construction of confidence intervals—normal distribution, *t*-distribution, and Chebyshev's Theorem—which produces the narrowest confidence interval? Which the widest?

MTB

5. Solve Example 3 on page 326 with Minitab's TINTERVAL command. Besides possible rounding differences, why are the intervals not identical but fairly close?

GROUP

6. The following table lists male car-related deaths per 100,000 population in a random sample of countries.

Country	Deaths	Country	Deaths
Bahamas	26	Japan	15
Kuwait	43	Romania	27
United States	25	Ecuador	38
Spain	27	Poland	34
Britain	13	Greece	28
Venezuela	44	Panama	31
Portugal	41	France	24
Hungary	35	Belgium	26
Switzerland	17	Australia	25
New Zealand	30	Puerto Rico	26

Source: *Time,* April 19, 1993, p. 23.

a. Construct 50%, 80%, and 98% confidence intervals for mean male car-related deaths per 100,000 of the population for all countries, assuming that deaths are normally distributed.

b. Construct a 75% confidence interval for the population mean if the deaths in all countries are not normally distributed.

Hint: *Use Chebyshev's Theorem.*

7. A 1993 survey determined the percentages of people in various cities who thought their branches of the U.S. Postal Service were "good" or "improved." The approval percentages of a random sample of $n = 11$ of those cities follow.

City	Percentage of People Who Approve
Albany, NY	94
Boston, MA	90
Chicago, IL	72
Detroit, MI	87
Houston, TX	87
Miami, FL	86
New Orleans, LA	89
New York, NY	74
Omaha, NE	91
Salt Lake City, UT	93
San Francisco, CA	85

Stat **F**act

About 4 million tons of unwanted mail are deposited in landfills each year.

Assume that the approval percentages for all cities in the survey are normally distributed, and construct a 90% confidence interval for the population mean approval percentage.

8. Assume that in 1992 the average (mean) stays in community hospitals in large cities were normally distributed. A sample of ten large cities, chosen at random, had the following stays.

City	Mean Stay (days)
St. Louis, MO	7.6
Oakland, CA	6.0
San Jose, CA	5.2
Toledo, OH	6.8
New York, NY	10.1
El Paso, TX	5.9
Atlanta, GA	7.2
Milwaukee, WI	6.5
Honolulu, HI	8.4
New Orleans, LA	7.4

Source: *American Hospital Association Hospital Statistics,* 1993–94 ed., Table 7.

a. Describe the population mean.

b. Determine a 90% confidence interval for the population mean.

9. The following array shows a random sample of the number of births in the United States per 1000 of the population for a random selection of 15 years since 1910.

Year	Birth Rate
1964	21.0
1920	27.7
1954	25.3
1993	15.7
1959	24.3
1980	15.9
1973	14.9
1987	15.7
1955	25.0
1956	25.2
1976	14.8
1940	19.4
1990	16.7
1971	17.2
1992	16.0

a. Describe the population.

b. What are the necessary conditions to use the *t*-distribution model to construct a confidence interval for μ?

   c. Assume the conditions from Part b are satisfied and construct an 80% confidence interval for the population mean.

10. The family incomes within a certain zip code in the city are believed to be normally distributed. A random sample of 41 of the incomes, in dollars, follows.

65,500	54,750	60,700	72,010	66,600
57,800	44,800	50,000	50,000	56,400
45,900	32,800	40,300	43,000	52,780
37,250	28,560	88,900	56,890	66,000
66,780	56,347	58,900	44,275	34,678
65,780	64,000	45,500	55,980	67,000
29,500	57,600	66,675	49,500	50,500
33,750	34,750	45,500	45,000	55,600
42,650				

   a. Construct a 90% confidence interval for the mean family income, μ, within this zip code, using the normal distribution model, as the flow chart in Figure 8-5 suggests.

   b. Construct a 90% confidence interval for the mean family income, μ, within this zip code if the t-distribution model is used with $(41 - 1) = 40$ degrees of freedom.

   c. How much wider, in percentage, is the confidence interval found in Part b than the confidence interval found in Part a?

8.9 ONE-SIDED CONFIDENCE INTERVALS

Definition ▶ A **one-sided confidence interval** is a confidence interval that establishes either a likely minimum or a likely maximum value for μ, but not both.

Consider Figure 8-15. A one-sided $C\%$ confidence interval to identify the maximum value of μ is usually written in the form

$$(-\infty, \overline{X} + Z(\overline{X}) \cdot SE],$$

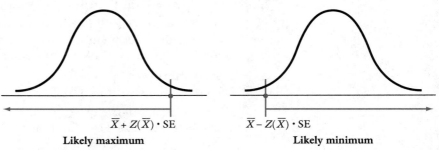

$\overline{X} + Z(\overline{X}) \cdot SE$	$\overline{X} - Z(\overline{X}) \cdot SE$
Likely maximum	**Likely minimum**

Figure 8-15 One-sided confidence intervals.

and a one-sided confidence interval to identify the minimum value of μ is usually written in the form

$$[\overline{X} - Z(\overline{X}) \cdot SE, \infty).$$

In each case, $Z(\overline{X})$ is the Z-score that encloses the upper (lower) $C\%$ of the area under the normal distribution model. One-sided confidence intervals may also be constructed using the t-distribution model.

Exercises 8.9

1. The following table shows the number of persons injured in motor vehicle crashes in Illinois during the six major holiday periods in 1991.

Holiday	Number of Persons Injured
Memorial Day	1693
Fourth of July	1969
Labor Day	1385
Thanksgiving	1535
Christmas	643
New Year's Day	540

Source: *Illinois Crash Facts and Statistics: 1991*, Illinois Department of Transportation, Division of Traffic Safety, p. 9.

Assume the following: (1) the number of injuries in 1991 was typical of most years; (2) the distribution of injuries for these six holiday periods over many years is normally distributed.

a. Construct a 90% confidence interval for the mean number of injuries that can be expected over a holiday period.

b. Find a 90% one-sided confidence interval that would establish the minimum number of injuries to be expected.

c. Explain why assumptions 1 and 2 are specified. If these assumptions were not made, how would the confidence interval in Part a have to be constructed?

2. Refer to Exercise 7 (in Exercises 8.8) on page 351.

a. Construct a one-sided 90% confidence interval for the maximum value of the population mean, and construct a one-sided 90% confidence interval for the minimum value of the population mean.

b. Explain why the end points for Part a of this exercise are different from the end points found for a 90% confidence interval in Exercise 7 on page 351.

c. Give your best estimate of the percentage of confidence for the two-sided confidence interval formed from the upper and lower estimates of μ from Part a.

3. Suppose that the waiting times in a dentist's office are normally distributed. Twenty-three people, chosen at random, waited the following lengths of time, in minutes:

5.0	3.5	8.5	10.5	6.0	7.5	15.0	4.5	3.5	19.5
9.5	5.5	6.5	4.5	6.5	9.0	17.5	6.5	2.5	22.5
7.5	11.5	10.5							

Find a 90% confidence interval for the minimum of the mean waiting time of all people who visit that dentist's office.

4. Solve Exercise 3 under the condition that the waiting times are normally distributed and it is known that $\sigma = 5.00$ minutes.

5. Solve Exercise 3 under the condition that the waiting times are not normally distributed and the standard deviation is unknown. Assume that the distributions in the two tails are approximately the same.

   Hint: *This problem must be modeled on Chebyshev's Theorem. Start by solving the equation*

   $$1 - \frac{1}{k^2} = .80$$

   for k.

GROUP

6. Following are motor vehicle deaths by month in months chosen at random from the years 1987 through 1991.

S t a t   F a c t

Single-car crashes account for 45% of all car occupant fatalities. The typical fatal crash most likely involves an intoxicated lone male driver who loses control of his speeding car on a deserted road in the wee hours of a Sunday morning.

Month	Year	Fatalities
January	1990	3460
February	1987	2997
March	1990	3590
April	1989	3788
May	1991	3780
June	1988	4319
July	1990	4470
August	1987	4815
September	1990	4230
October	1988	4567
November	1991	3660
December	1989	4150

Source: *National Safety Council, Accident Facts,* 1992 edition, p. 61.

   a. Describe the population.

   b. What assumptions must be made in order to use the *t*-distribution model to construct confidence intervals for the population mean? Use these assumptions to complete the last parts of this exercise.

   c. Construct 50%, 80%, and 99% confidence intervals for the population mean.

d. Construct 60%, 95%, and 99% one-sided confidence intervals for the minimum value of the population mean.

e. Construct 75%, 99.5%, and 99.9% one-sided confidence intervals for the maximum value of the population mean.

Case Study

Proper sampling procedures are critical to integrity of the biological disciplines. (*New Jersey Department of Fish, Games, and Wildlife*)

A Tool for the Biologist

Many biological studies use samples to estimate a population mean. For instance, biologists may want to determine the mean number of a species in a certain location over time, to judge the effects of pollution or predators. However, taking too many samples may not be affordable, and taking too few samples is likely to produce misleading results. The biologist's choice of how many samples to take is extremely important and directly affects the probability that the estimated mean will accurately describe the population mean.

The May 1991 issue of *BioScience* (pp. 346–348) examines the problem of sample size and provides several equations for determining sample size to achieve a given level of accuracy. Obtain a copy of this article from your library, and use it to answer the following questions.

I. What is a practical consideration, besides cost, that may influence sample size?

II. What theorem allows the formulas given for sample size to be used with populations that are not necessarily normally distributed?

III. Show algebraically how the first formula given for sample size can be transformed into its alternative version, which uses the coefficient of variation.

IV. Refer to "A simple example" in the article. In your own words, explain what the sample size 112 represents.

V. Again refer to "A simple example." Determine the sample size if the desired accuracy is ±15% at a 0.10 level of significance.

CHAPTER SUMMARY

This chapter described methods of forming an interval estimate of the population parameters μ and π. An interval estimate is a more refined, and often much more useful, way to estimate a population parameter than a point estimate. An interval estimate for μ is an interval in which μ will lie with a certain degree of confidence, C%; that is, μ will lie in that interval C% of the time.

The Central Limit Theorem, one of the most important theorems in all of mathematics, is the foundation for establishing the algorithm to construct confidence intervals.

Several similar algorithms can be followed to construct a confidence interval, depending on the sample size and whether the population parameter σ is known or only estimated. Statistical software such as Minitab can help greatly with this procedure.

This chapter provided a flow chart that pointed the way to the appropriate model for constructing confidence intervals. Chebyshev's Theorem may be used when neither the normal distribution nor the t-distribution model applies. The end points of a confidence interval for a population parameter are found by evaluating one of several formulas presented in the chapter, all of which involve the standard error of the sample means, SE. The standard error of the sample means is the standard deviation of the sampling distribution, and SE is equal to $\frac{\sigma}{\sqrt{n}}$ if σ is known; otherwise, SE is estimated by $\frac{s}{\sqrt{n}}$. Under certain circumstances, the finite correction factor can be used to refine the value of SE.

One may also construct one-sided confidence intervals to establish likely minimum or maximum values of the population mean.

Key Concepts, Terms, and Formulas

Videotape Suggestions

Program 18 from the series *Against All Odds: Inside Statistics* shows some applications of the Central Limit Theorem in modern gambling casinos, and *Program 19* has additional information about confidence intervals. Also *Program 21* contains some historical background about the development of the t-distribution.

A legal trial exemplifies
the application of
hypothesis testing.

9 Hypothesis Testing

The theory behind the construction of confidence intervals is the basis for **hypothesis testing**—an important formal procedure that determines whether a claim about a population parameter (μ, for example) seems unreasonable and should be rejected or judgment should be reserved. The **case study** at the end of this chapter presents an alternative explanation of the hypothesis testing procedure.

9.1 THE NULL AND ALTERNATIVE HYPOTHESES

An unproven claim that is tentatively assumed to provide a basis for statistical investigation, and that may eventually be rejected, is called the **null hypothesis** and denoted by H_0. In the hypothesis testing procedure, the null hypothesis is always opposed by an **alternative hypothesis,** denoted by H_A. Together the null and alternative hypotheses exhaust all possibilities about the population parameter, and they have nothing in common.

Definitions ▶

The **null hypothesis, H_0,** is a claim that is tentatively assumed to provide a basis for statistical investigation.

The **alternative hypothesis, H_A,** opposes the null hypothesis. Together H_0 and H_A exhaust all possibilities, and they have nothing in common.

The alternative hypothesis is the only hypothesis that may be supported on the basis of a hypothesis test. This support is provided by rejecting, or "nullifying" the null hypothesis.

If the null hypothesis is rejected, then the alternative hypothesis is accepted. However, because hypothesis testing uses sample statistics, a rejection decision regarding H_0 can never be made with complete certainty, and the conditions that warrant rejection of H_0 will always be stated in terms of probabilities at the start of the test.

It is important to understand that, if the null hypothesis is not rejected, it is also not proved true. Statistical evidence can support only the alternative hypothesis, H_A.

In some ways, hypothesis testing is analogous to a legal trial.

	Legal Trial	**Hypothesis Test**
H_0	The defendant is not guilty.	Claim about a population parameter.
H_A	The defendant is guilty.	Opposing claim about a population parameter.
Result	The evidence convinces the jury to reject the assumption of innocence. The verdict is "guilty."	The statistics indicate a rejection of H_0, and the alternative hypothesis is accepted.
	or	*or*
	The evidence does not convince the jury of the defendant's guilt. The defendant is acquitted.	The sample statistics do not indicate a rejection of H_0. Judgment is reserved.

If the defendant is acquitted, she is not proved innocent; rather, the assumption of innocence was always present, and the evidence failed to cause a rejection of that assumption. Similarly, in a hypothesis test, if the sample statistics fail to indicate a rejection of the null hypothesis that was assumed tentatively throughout the test, that failure does not prove the null hypothesis. Like the work of the unsuccessful prosecutor, the statistical evidence supporting the alternative hypothesis has not been convincing. In this situation, the prudent course is to reserve judgment about the actual validity of the null hypothesis.

How to Choose the Null and Alternative Hypotheses

Rejection of the null hypothesis and the resulting acceptance of the alternative hypothesis are the only possible actions that may be taken as the result of a hypothesis test. Therefore, for a claim to be supported on the basis of statistical evidence, it is often necessary to define the null hypothesis as the opposite of what is to be shown. For instance, suppose a medical researcher wants to demonstrate a difference in the reaction times of two different types of blood pressure medication. The null and alternative hypotheses should be formulated as follows.

H_0: There is no difference in reaction times for the two types.

H_A: There is a difference in reaction times for the two types.

Now, if the null hypothesis is rejected, statistical evidence will support the acceptance of the alternative hypothesis (different reaction times).

In other cases, a claim about the value of a population parameter may be tested by tentatively using the null hypothesis to assign that value to the population parameter. In turn, the alternative hypothesis states that the population parameter is not that value. If eventually the null hypothesis is rejected, then statistical evidence will exist to support the alternative-hypothesis challenge. For example:

Debate occurs among mangement and union officials about the mean number of work hours lost annually, per automobile assembly-line worker, due to the common cold. Management believes the mean is 32 hours per year (with some unknown standard deviation), but the union believes this estimate is not accurate. The union offers the following hypotheses.

$$H_0: \quad \mu = 32.$$

$$H_A: \quad \mu \neq 32.$$

Examination of a sample of $n = 50$ records of workers who were chosen at random from the population of all assembly-line workers reveals that the mean number of hours mixed due to the common cold, $\overline{X}$, is 29.4, with $s = 7.6$. The hypothesis test (to be described in Example 1 on page 363) will determine whether these sample statistics ($\overline{X} = 29.4$, $s = 7.6$) are sufficient evidence to reject the null hypothesis and support the union's claim that $\mu \neq 32$.

In this illustration, the fact that $\overline{X}$ is not exactly the same as the value of μ assumed under the null hypothesis should not be surprising. After all, probably many millions of samples of size $n = 50$ are possible, and the chances of choosing one with $\overline{X}$ exactly equal to 32 is very small. The question that needs to be examined is:

Is the value of $\overline{X}$ so far away from the assumed value of μ under the null hypothesis that its very occurrence invalidates the null hypothesis that $\mu = 32$?

The degree of certainty that is desired in the conclusion, or the level of significance of the hypothesis test, will determine the answer to this question.

9.2   THE SIGNIFICANCE LEVEL OF A HYPOTHESIS TEST

Among all the $\overline{X}$'s that are possible when μ is indeed 32, the maximum allowable percentage of them that are considered unusual enough to invalidate the null hypothesis is called the **significance level** or **size of the hypothesis test** and is denoted by the Greek letter alpha, α.

For the question about automobile workers now under consideration, the sampling distribution of all the $\overline{X}$'s has a normal model because the sample is large, and it seems safe to assume that the population size, N, is large relative to the sample size. Furthermore, suppose it is decided that the significance level, or size, of the hypothesis test to be performed is $\alpha = 10\%$. This is equivalent to stating that the calculated value of $Z(\overline{X})$ must fall somewhere within the middle

$$(100 - \alpha)\% = 100\% - 10\%$$
$$= 90\%$$

of a normal distribution model with a mean of 32 in order for the null hypothesis, that $\mu = 32$, to escape rejection.

If the calculated value of $Z(\overline{X})$ falls within the shaded region of Figure 9-1, one must reject the null hypothesis that $\mu = 32$, because such a value of $Z(\overline{X})$,

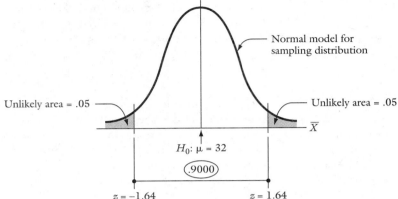

Figure 9-1   The normal distribution model for rejection of the null hypothesis.

while possible, is just too unlikely if μ is indeed 32. The shaded area in Figure 9-1 is called the rejection region of the hypothesis test.

Definition ▶   The **rejection region** of a hypothesis test is the part of the sampling distribution that is so far away from the assumed value of μ that it causes rejection of the null hypothesis.

The area of the rejection region represents the significance level (size) of the hypothesis test, α. Whether or not the null hypothesis value of μ is rejected depends upon the size of the hypothesis test, α, specified at the start of the test procedure.

If the calculated value of $Z(\overline{X})$ falls anywhere in the rejection region, it is termed a **statistically significant result,** and a **statistically significant difference** exists between $\overline{X}$ and the value of μ under the null hypothesis. Such a difference warrants rejection of the null hypothesis. If $Z(\overline{X})$ does not fall in the rejection region, then the difference in values between $\overline{X}$ and μ can be attributed completely to chance and is not statistically significant; it does not warrant rejection of the null hypothesis at the given significance level, α.[1] Judgment about the actual validity of the null hypothesis is reserved.

The next set of examples shows how all of the ideas examined so far in this chapter are brought together to conduct a formal hypothesis test. In such a test, the value of μ under the null hypothesis is used to calculate a **test statistic,** $Z(\overline{X})$ or $t(\overline{X})$, and the test statistic's relation to the rejection region determines whether the claimed value of μ is to be rejected.

The student is encouraged to use the five-step method for hypothesis testing, soon to be described, to work the exercises that follow. Also, while reading the

---

[1] For a different value of α, however, the conclusions could differ.

examples in this chapter, consult the flow chart on page 324 in Chapter 8 for a better understanding of how the distribution models for $\overline{X}$ are determined.

9.3 TWO-TAIL HYPOTHESIS TESTS

In the examples and exercises of this section, the rejection region is split equally into two parts. This configuration is called a **two-tail** (or **two-sided**) **hypothesis test.** A later section will treat a **one-sided hypothesis test,** in which the rejection region is concentrated in a single tail. The nature of the claim to be tested determines whether or not the rejection region is to be split in half.

E x a m p l e
1

Consider the hypothetical information about auto workers on pages 360–361. For convenience, it is summarized here.

In a large population, management claims that $\mu = 32$, with an unknown standard deviation. The union disputes this claim, and a random sample of $n = 50$ from this population yields $\overline{X} = 29.4$ and $s = 7.6$.

Conduct a hypothesis test at the 10% ($\alpha = .10$) level of significance to determine whether statistical evidence exists to support the union's position that $\mu \neq 32$.

Solution   Conduct the hypothesis test in five steps.

Step 1   *State the hypothesis being tested, H_0, and an alternative hypothesis, H_A, in such a way that exactly one of them is true.*

$$H_0: \quad \mu = 32.$$
$$H_A: \quad \mu \neq 32.$$

The goal of Steps 2–5 will be to decide whether to reject H_0 and accept H_A or reserve judgment.

Step 2   *Decide on a model for the sampling distribution.*

Because this situation involves a large sample taken from a relatively large population, the sampling distribution model of the $\overline{X}$'s will be the normal distribution.

Step 3   *Determine the end points of the rejection region, and state the rule by which the decision to reject H_0 will be made.*

The determination of the rejected region end points rests on the fact that this is a 10%-level-of-significance hypothesis test based on a normal distribution model. It is also a two-tail hypothesis test, in which half the rejection region is in the outer 5% of the left tail, and half is in the outer 5% of the right tail. Together these sections constitute 10% of the area that is farthest from the assumed value of μ under the null hypothesis and make up the rejection region for the test. Figure 9-2, which matches Figure 9-1, illustrates the rejection region.

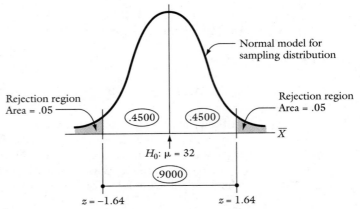

Figure 9-2

The end points of the rejection region are $z = \pm 1.64$. Obtain them by finding the number closest to .4500, which is .4495, in the body of Appendix A-2. The corresponding Z-score that defines this area is 1.64.[2] Allowing for symmetry, the other Z-score is -1.64. The decision rule is[3]

Reject H_0 and accept H_A if $Z(\overline{X}) > 1.64$ or if $Z(\overline{X}) < -1.64$. Otherwise, reserve judgment.

Step 4   Compute the test statistic, $Z(\overline{X})$, using the parameters of H_0. Recall that

$$Z(\overline{X}) = \frac{\overline{X} - \mu_{\overline{X}}}{SE},$$

with $\mu_{\overline{X}} = \mu$. By the assumption of the null hypothesis, $\mu = 32$. Since σ is unknown, the value of SE will have to be estimated by s:

$$SE \approx \frac{s}{\sqrt{n}}$$

$$= \frac{7.6}{\sqrt{50}}$$

$$\approx 1.07.$$

Substituting these values into the formula for the test statistic yields

$$Z(\overline{X}) = \frac{29.4 - 32}{1.07}$$

$$\approx -2.43.$$

[2]Actually, Z-scores of 1.64 and 1.65 both correspond to areas that are equally close to .4500. To simplify cases such as this, the smaller of the two Z-scores is always used in this text.

[3]Note that the distribution of the $\overline{X}$'s, not the distribution of the population, is under consideration even though the validity of a claim about the population mean, μ, is being tested.

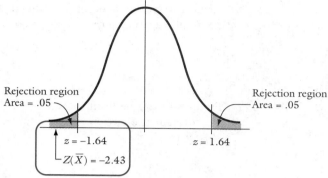

Figure 9-3 Location of the test statistic in the rejection region.

Step 5   State the conclusion.

The test statistic $Z(\overline{X}) = -2.43$ lies in the left side of the rejection region shown in Figure 9-3, marked by end point $z = -1.64$. This implies that the difference between $\overline{X}$ and the value of $\mu = 32$ under H_0 is statistically significant and cannot be attributed to chance.

Therefore, **reject H_0 and accept H_A.** Statistical evidence exists (at the 10% level of significance) to support the union's claim that the population mean is not equal to 32.   •

E x a m p l e 2

Repeat the hypothesis test of Example 1 at the 0.5% ($\alpha = .005$) level of significance.

Solution   Several steps are the same as in Example 1.

As before, the only possible outcomes of this hypothesis test are acceptance of the union's claim that $\mu \neq 32$ and reservation of judgment.

Step 1   State the hypotheses.

$$H_0: \quad \mu = 32.$$
$$H_A: \quad \mu \neq 32.$$

Step 2   Decide on a model for the sampling distribution.

As in Example 1, the sampling distribution of the $\overline{X}$'s will be based on the normal distribution model because the sample is large and is taken from a relatively large population.

This example and Example 1 will now be seen to differ at the following points.

• The determination of the Z-scores for the end points of the rejection region in Step 3
• The decision rule in Step 3
• Possibly the conclusion in Step 5

Step 3   Determine the end points of the rejection region, and state the decision rule.

The phrase "0.5% level of significance" means that H_0 will be rejected if $Z(\overline{X})$ falls in one of the outer two tails, which together compose 0.5% (.005) of the area farthest from the value of μ assumed under the null hypothesis. This means that each tail has an area of

$$\frac{\alpha}{2} = \frac{.005}{2}$$

$$= .0025.$$

Figure 9-4 can be used to determine the end points of the rejection region.

Obtain the end points of the new rejection region by finding the number closest to

$$.5000 - .0025 = .4975$$

in the body of Appendix A-2. The corresponding Z-scores that mark the end points of the rejection region in Figure 9-4 are $z = \pm2.81$. The decision rule is:

Reject H_0 and accept H_A if $Z(\overline{X}) > 2.81$ or if $Z(\overline{X}) < -2.81$. Otherwise, reserve judgment.

Step 4   Compute the test statistic.
From Example 1, the value of the test statistic $Z(\overline{X}) = -2.43$.

Step 5   State the conclusion.
The value of the test statistic $Z(\overline{X}) = -2.43$ does not lie in the new rejection region, shown in Figure 9-5. Therefore, **reserve judgment.** At the .005 level of significance there is not enough statistical evidence to support the union's alternative hypothesis that $\mu \neq 32$. This fact does not in any way support the null hypothesis that $\mu = 32$. The difference between $\overline{X}$ and μ should be attributed completely to chance.

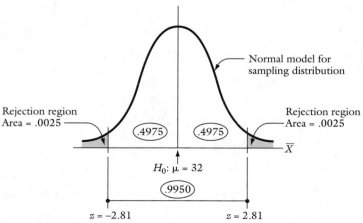

Figure 9-4

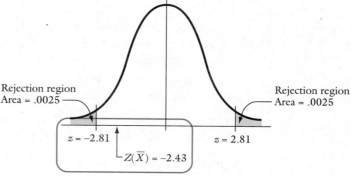

Rejection region
Area = .0025

Rejection region
Area = .0025

$z = -2.81$

$z = 2.81$

$Z(\overline{X}) = -2.43$

Figure 9-5 ●

Exercises 9.1 – 9.3

1. A hypothesis test has been compared to a legal trial. What is the statistical analogy to a "reasonable doubt?"

2. In what ways is the null hypothesis temporarily accepted?

3. Refer to Example 1 on page 363. What other factor, besides the value of $\overline{X}$ and the level of significance, could have affected the outcome of the test?

4. For each of the following situations, define the null and alternative hypotheses.

   a. Is the mean height of NBA players 6 feet 7 inches?

   b. Is the mean salary for major league baseball players $875,000?

   c. Is the mean salary for professional football players different from the mean salary for professional baseball players?

   d. Is the mean price for a minivan in the Chicago area really $16,775?

   e. Does a particular gasoline additive change gasoline mileage?

   f. Do women score differently from men on the SAT?

5. Refer to Exercise 4. Using your choices of null and alternative hypotheses, which statements could be supported by statistical evidence?

6. If the difference between $\overline{X}$ and the value of μ under the null hypothesis is attributed to chance, explain why judgment about the null hypothesis should be reserved.

9.4 HYPOTHESIS TESTING USING THE t-DISTRIBUTION MODEL

The next example shows that the normal distribution does not necessarily have to be the model for the sampling distribution of the $\overline{X}$'s.

The attorneys in a certain firm claim that their incomes are normally distributed. The question arises whether their mean income is $120,000.00. A sample of ten attorneys from this firm, chosen at random, has the following incomes.

$109,575	$ 89,500	$145,500	$200,575	$180,000
64,800	158,600	175,800	225,000	190,680

Perform a hypothesis test at the 5% ($\alpha = .05$) level of significance to determine whether statistical evidence exists that the mean income is not $120,000.

Solution   Follow the five-step method.

Step 1   *State the hypotheses.*

$$H_0: \quad \mu = \$120{,}000.$$
$$H_A: \quad \mu \neq \$120{,}000.$$

The only possible outcomes of this hypothesis test are the conclusion that their mean income is not $120,000 and reservation of judgment.

Step 2   *Decide on a model for the sampling distribution.*

The sampling distribution model for the $\overline{X}$'s is the *t*-distribution. It is chosen because the sample is small ($n < 30$) and is taken from a population that is claimed to be normally distributed with unknown standard deviation.

Step 3   *Determine the end points of the rejection region, and state the decision rule.*

Find the *t*-scores that determine the end points of the rejection region, as follows.

- A 5% level of significance means that the rejection region for H_0 is in the outer

$$\frac{\alpha}{2} = \frac{5\%}{2}$$
$$= .025$$

    of each tail.
- According to Appendix A-4, the intersection of the .025 column with the $n - 1$, or 9, degrees of freedom row establishes *t*-scores that will mark the end points of the rejection region: $t = \pm 2.262.$[4]

The decision rule is:

Reject H_0 and accept H_A if $t(\overline{X}) > 2.262$ or if $t(\overline{X}) < -2.262$. Otherwise, reserve judgment.

Step 4   *Compute the test statistic.*

The test statistic is $t(\overline{X})$ and is found with the formula

$$t(\overline{X}) = \frac{\overline{X} - \mu_{\overline{X}}}{SE},$$

where $\mu_{\overline{X}} = \mu$, and with the assumption of the null hypothesis that $\mu = 120{,}000$.

---

[4]Recall that this particular *t* table is constructed so that it assumes negative values of *t*, just as does the standard normal table. There is an advantage to this type of table construction when a one-tail test is conducted.

Since σ is unknown, one must use the estimate s for the computation of SE. The values of s and $\overline{X}$ can be found in a number of ways. If a TI-82 graphics calculator is used, the final screen display appears as follows:

```
1-Var Stats
→ x̄=154003
   Σx=1540030
   Σx²=2.6106E 11
s→ Sx=51522.13501
   σx=48878.18896
↓ n=10
```

This display shows that $\overline{X} = 154{,}003$ and $s \approx 51{,}522.14$. One can now determine SE using this rounded value of s.

$$SE \approx \frac{s}{\sqrt{n}}$$

$$\approx \frac{51{,}522.14}{\sqrt{10}}$$

$$\approx \$16{,}292.73.$$

Compute the test statistic $t(\overline{X})$ as follows:

$$t(\overline{X}) = \frac{\overline{X} - \mu_{\overline{X}}}{SE}$$

$$= \frac{154{,}003 - 120{,}000}{16{,}292.73}$$

$$\approx 2.087.$$

Step 5   *State the conclusion.*

This value of $t(\overline{X})$ did not quite make it into the rejection region.

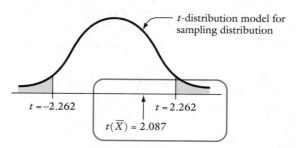

t-distribution model for sampling distribution

$t = -2.262$ $t = 2.262$

$t(\overline{X}) = 2.087$

Therefore, **reserve judgment.** At the 5% level of significance there is not enough statistical evidence to support the alternative hypothesis that the mean salary of all attorneys in the firm is not \$120,000. The difference between $\overline{X}$ and the value of μ under the null hypothesis should be attributed to the chance selection of scores in the sample.   ●

Statistical software programs can be used to conduct hypothesis tests. The macro command ZTEST or TTEST is needed to conduct hypothesis tests using Minitab, depending on whether the normal model or the *t*-distribution model, respectively, is used for the sampling distribution. The next two examples illustrate the use of these commands.

E x a m p l e

4

S t a t   F a c t

Connecticut and Tennessee have the greatest number of automobiles per capita of all the states in the union: .74. Arkansas' per-capita registration, .40, is the lowest.

A leading auto manufacturer claimed that a popular version of its minivan was available in the Midwest for an average price of $16,000 with a standard deviation of $800. A consumer group doubted that report and surveyed 50 recent purchasers of the minivan to dispute the manufacturer's claim. The purchase prices, in dollars, follow.

16,700	16,500	16,775	15,750	15,400
16,200	15,750	15,895	16,750	15,890
16,000	17,000	17,350	16,350	16,000
15,375	15,875	16,750	16,750	17,450
17,350	16,750	17,250	18,075	16,000
14,900	16,750	17,225	16,500	14,250
15,450	14,725	16,500	14,950	14,775
16,900	15,800	15,500	17,090	16,474
15,750	15,750	17,750	16,000	16,665
16,000	16,875	16,600	16,000	16,750

Assume that these prices reflect minivans that are as identical as possible. Conduct a hypothesis test at the 1.0% ($\alpha = .01$) level of significance to determine whether statistical evidence exists to support the consumer group's claim that $\mu \neq 16,000$.

Solution   The manufacturer is claiming that $\sigma = 800$ and that $\mu = 16,000$. You must use both of these claimed parameters in the test. The steps follow.

Step 1   State the hypotheses.

$$H_0: \quad \mu = 16,000.$$
$$H_A: \quad \mu \neq 16,000.$$

Step 2   Decide on a model for the sampling distribution.

This is a large sample. Also, since the company is a leading auto manufacturer, it seems reasonable to assume that the population (all minivans of this particular type available in the Midwest) is large relative to the sample. Therefore, use the normal distribution model with a mean of $\mu = 16,000$ and a standard deviation of $\sigma = 800$.

With Minitab, Steps 3 and 4 can be combined as follows.

Steps 3 and 4

• Enter the data into column C1 of a Minitab worksheet.
• With the normal distribution model, use the macro command ZTEST (which is used with a known or assumed standard deviation). Obtain a Session window. Then type ZTEST, followed by the values of the assumed mean and standard deviation under the null hypothesis and the location of the data, C1.

Figure 9-6 shows the result once the [RETURN] key is pressed. (The diagram and circles have been added for clarification and are not part of the output.)

Figure 9-6 reveals that there are 50 scores with a mean of 16,277 and a standard deviation of 816. The value of *SE*, 113, was calculated with an assumed standard deviation of 800. The value of *Z*, listed as 2.45, is Minitab's calculation of the test statistic, $Z(\overline{X})$.

The **P value**, or **P**, labeled .014, represents the probability of obtaining a test statistic of $Z(\overline{X}) = \pm 2.45$ or farther from the claimed mean. It can be interpreted geometrically as the sum of the areas of the two shaded regions.

Step 5   State the conclusion.

The *P* value of .014 represents a larger rejection region than that specified ($\alpha = .01$). Because the test statistic 2.45 is the end point of this larger rejection region, it cannot be in the smaller, specified rejection region. Therefore, **reserve judgment.** At the 1% level of significance there is not enough statistical evidence to support the consumer group's alternative hypothesis that the mean selling price is not $16,000.

Note that $2.45 \approx \dfrac{16277 - 16000}{113}$.

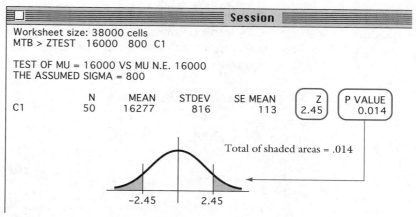

Figure 9-6 Minitab's ZTEST command and output.   ●

9.6 *P* VALUE AND ATTAINED SIGNIFICANCE LEVEL OF A HYPOTHESIS TEST

The following definitions and guidelines summarize the information in Example 4.

Definitions ▶
> The ***P* value**, or ***P***, is the probability that the test statistic is as far or farther from μ if the null hypothesis is true.
>
> The **attained significance level** of a hypothesis test is the *P* value of its test statistic.

The *P* value of the test statistic may be used to make a decision about the outcome of a hypothesis test for any α specified, according to the following guidelines.

> Guidelines for rejecting H_0 based on the *P* value:
>
> If $P < \alpha$, then reject H_0 and accept H_A.
>
> If $P \geq \alpha$, then reserve judgment about H_0.

E x a m p l e
5

Refer to Example 4 on page 370. Of the 50 purchase prices given there, ten are chosen at random:

16,200	15,895	15,875	16,750	17,450
18,075	17,225	16,900	16,875	16,750

With only these ten purchase prices, conduct the same hypothesis test as in Example 4. Instead of using the manufacturer's claimed standard deviation of $\sigma = \$800$, however, use the value of s computed from the smaller sample. Assume the population is normally distributed.

Solution   Proceed as follows.

Step 1   State the hypotheses.
$$H_0:\quad \mu = \$16,000.$$
$$H_A:\quad \mu \neq \$16,000.$$

Step 2   Decide on a model for the sampling distribution.
Use the *t*-distribution model, because this is a small sample taken from a normally distributed population with unknown standard deviation.

In Figure 9-7, T represents the value of the test statistic $t(\overline{X})$.

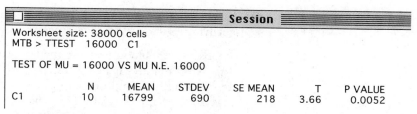

```
┌─────────────────────────────────────────────────────────────┐
│ ▣ ══════════════════════ Session ═══════════════════════════ │
├─────────────────────────────────────────────────────────────┤
│ Worksheet size: 38000 cells                                  │
│ MTB > TTEST   16000   C1                                     │
│                                                              │
│ TEST OF MU = 16000 VS MU N.E. 16000                         │
│                                                              │
│              N     MEAN    STDEV    SE MEAN     T    P VALUE │
│ C1          10    16799      690        218   3.66   0.0052  │
```

Figure 9-7 Minitab's TTEST command and output.

Steps 3 and 4

- Enter the data into column C1 of a Minitab worksheet.
- With the *t*-distribution model, use the macro command TTEST. The standard deviation is not specified. Minitab will automatically compute *s* and use it in calculating *SE, T,* and the *P* value.

Figure 9-7 shows the result once the RETURN key is pressed.

Step 5   *State the conclusion.*

Note that $P < \alpha$—that is, $.0052 < .01$—implying that the test statistic, $t(\overline{X})$, is in the rejection region. Therefore, **reject H_0 and accept H_A.** Under these new conditions, there is enough statistical evidence at the 1.0% level of significance to support the alternative hypothesis that the population mean is not $16,000.   ●

Thus, a smaller sample and a different distribution model turned a reserved judgment about the null hypothesis in Example 4 (page 370) into rejection of the same null hypothesis in Example 5.

All of the prior examples and the exercises that follow involve two-tail hypothesis tests. Each tail of the distribution model contains a rejection region, and if the test statistic is either too large or too small, the null hypothesis is rejected. Section 9.7 will examine a one-tail, or one-condition, test.

Exercises 9.4–9.6

1. At the beginning of this chapter it was stated that the validity of a claim about a population parameter depends on the degree of certainty about the decision. Briefly explain how luck plays a part in this process.

2. What is the test statistic?

3. What is meant by the rejection region of a hypothesis test?

4. Explain why an examination of the distribution of the sample means is justified in a test of the validity of a claim about the population parameter μ.

5. One possible origin of the term "*null* hypothesis" is the word "nullify." Explain the connection.

6. As the level of significance increases, what happens to the size of the rejection region? Explain.

7. Which hypothesis test is more stringent (that is, more likely to reject H_0): one specifying a 10% level of significance or one specifying a 5% level of significance?

8. Suppose a new blood pressure medication that claims a mean reaction time of $\mu = 90$ minutes is being tested. For the well-being of patients who will use the drug, it is important to test this claim very stringently. Should α be set relatively large or relatively small?

9. Refer to Exercise 8. Would a relatively large P value or a relatively small P value imply that the mean reaction time was not 90 minutes?

* 10. Explain how it is possible that nearly any null hypothesis can be rejected by adjusting the significance level of the hypothesis test. Under what conditions (if any) would the significance level not affect the probability of rejection of the null hypothesis?

11. Refer to Example 1 on page 363. Estimate the maximum level of significance that would result in reserved judgment about the null hypothesis.

12. Refer to Example 3 on page 368. Estimate the minimum level of significance that would result in rejection of the null hypothesis.

13. Minitab's macro commands ZTEST and TTEST are similar, but one of them requires the user to input more information. Which command is it, and what is that information?

In Exercises 14–22, use the normal model for the sampling distribution in Step 2 of the hypothesis tests.

14. A manufacturer of gas pipe claims that the mean weight of his pipe is 28.50 pounds per yard with a standard deviation of $\sigma = 6.45$ pounds. A random sample of $n = 40$ pieces of pipe reveals that $\overline{X} = 30.45$ pounds per yard. Conduct a hypothesis test to determine whether statistical evidence exists to support a claim that $\mu \neq 28.50$ at the 5% level of significance. Assume that $n < .05N$.

15. A certain company produces hundreds of thousands of ball bearings each day. It is important that they have, within limits, diameters of .25 inches; bearings that are either too small or too large will cause problems in the machines in which they are installed. Long experience has demonstrated that the diameters of the ball bearings are normally distributed with $\sigma = .05$ inches.

    A random sample of 25 ball bearings reveals a mean diameter of .2670 inches. Conduct a hypothesis test at the 10% level of significance to determine whether there is statistical evidence that the manufacturing process is not running correctly, that is, that $\mu \neq .25$.

16. Automobile ownership (cars per 1000 people) is normally distributed for the top 50 countries, with a standard deviation of $\sigma = 155.63$. Following are the automobile ownership statistics for a random sample of $n = 30$ of these countries.

Country	Cars per 1000 People	Country	Cars per 1000 People
Canada	457	Mexico	71
Kuwait	244	United States	573
Fiji	39	Japan	265
Israel	116	Hungary	169
Lebanon	174	Uruguay	96
Costa Rica	53	Iceland	489
Poland	128	Cyprus	238
France	410	Portugal	155
Austria	381	Chile	53
Greece	127	Italy	424
Spain	295	Jordan	45
Netherlands	362	Tunisia	40
Saudi Arabia	173	Finland	382
Luxembourg	470	Czechoslovakia	200
Former USSR	51	Ireland	263

Source: *The 1993 Information Please Environmental Almanac*, p. 336.

Conduct a hypothesis test at the 4% level of significance to determine whether statistical evidence exists to support the claim that automobile ownership for the top 50 countries is not 150 cars per 1000 people (that is, not 15%).

17. Assume that mean retail sales per household, by state, are normally distributed in a given year. Following is a random sample of sales for $n = 20$ states in 1991. In that year the standard deviation of retail sales for all states was $\sigma = \$2718.43$.

State	Per Household Retail Sales	State	Per Household Retail Sales
WV	$14,777	OH	$18,534
MT	17,810	MA	23,084
TX	20,275	SD	18,663
IL	20,058	OR	20,575
NY	18,337	FL	20,446
UT	18,958	TN	17,677
ME	21,309	KS	17,931
VA	19,485	CA	20,563
MI	19,626	NE	17,649
NM	17,217	AZ	18,797

Source: *Sales & Marketing Management*, August 24, 1992, p. B-4.

a. Explain why the normal distribution model should be used for a hypothesis test even though this sample is small.

b. Conduct a hypothesis test at the 5% level of significance to examine the null hypothesis that the mean retail sales for all 50 states in 1991 was $19,750.00.

GROUP

18. Assume that the number of AIDS cases per 100,000 of the population, by state, is normally distributed with a standard deviation of $\sigma = 9.13$. A random sample of $n = 20$ states for the year 1991 provides the following information.

State	AIDS Cases per 100,000	State	AIDS Cases per 100,000
TX	19.6	MA	16.6
VA	11.1	NC	7.3
VT	4.2	NM	7.2
RI	9.1	WA	11.4
OH	5.6	KY	4.7
CA	26.3	KS	4.1
IL	13.6	TN	7.0
PA	9.5	SD	0.4
NE	3.7	OR	10.2
AK	8.3	MD	19.3

Source: *Statistical Bulletin* 73:2, April–June 1992, p. 23.

a. Conduct hypothesis tests at the 3% and 5% levels of significance to examine the null hypothesis that the state mean for this statistic is 14.0.

b. Use a statistical software program, such as Minitab, to determine the attained significance level of the test in Part a. If software is not available, estimate the P value using Appendix A-2.

19. Recently the *Journal of the American Veterinary Medical Association* reported that 44% of dogs and cats kept as pets are overweight. A veterinarian randomly sampled 50 pets in her care and concluded that 26 of them were overweight. Assuming that this veterinarian treats a typical collection of pets, conduct a hypothesis test at the 10% level of significance to determine whether this sample provides statistical evidence to challenge the journal's statement.

Hint: Assume that Cochran's Rules are satisfied, and use the value of π under H_0 to compute SE as shown in Section 8.7 of Chapter 8.

* 20. Refer to Exercise 19. Determine the minimum and maximum numbers of overweight pets in the veterinarian's sample that would have caused her to reject the journal's statement.

21. Some people claim that gender is a factor in fear of flying. A survey was conducted of 2575 people who were afraid to fly, 1486 of whom were male. Does statistical evidence support the claim at the 5% level of significance?

Hint: Let π = the male percentage of those who are afraid to fly, then H_0: $\pi = .50$ is the null hypothesis.

* 22. A politician believed she had a 70% approval rating in the Midwest. She commissioned a public opinion poll of 3500 voters in that area, and 2405 voters approved of the job she was doing.

a. Does this poll contradict her belief at the 3% level of significance?

b. What is the maximum level of significance that would cause her to reserve judgment?

In Exercises 23–30, use the t-distribution as the model for the sampling distribution in Step 2 of the hypothesis tests.

23. Refer to Exercise 15. Conduct the same hypothesis test, at the same level of significance, under the condition that the diameters are normally distributed, but the standard deviation is unknown and is only estimated by $s = .05$ inches.

24. The average weekly unemployment benefits, by state, are normally distributed. A random sample of ten states reveals the following benefits for 1992.

PA	$201	IL	$183
WY	163	AL	121
ME	167	SD	128
MO	146	OH	180
NE	133	KY	144

Conduct a hypothesis test at the 5% level of significance to determine whether statistical evidence exists to support the claim that the average state benefit is not $170.

25. It is believed that the numbers of persons injured in automobile accidents in cities with populations of 25,000 to 50,000 in large states are normally distributed. A random sample of numbers of persons injured in some Illinois cities of this size in 1992 follows.

City	Number of Persons Injured
Lansing	359
Quincy	491
Berwyn	445
Urbana	299
Addison	373
Wilmette	236
Elmhurst	641
Chicago Heights	519
Niles	274
Pekin	500
Park Ridge	298
Harvey	564
Glenview	476
Granite City	453

Source: *Illinois Crash Facts and Statistics for 1992,* Illinois Department of Public Transportation, pp. 32–33.

a. Describe the population for this set of data.

b. Conduct a hypothesis test at the 5% level of significance to determine whether there is statistical evidence that the population mean is not equal to 500.

c. Use a statistical software program, such as Minitab, to determine the attained significance level for the test in Part b. If software is not available, estimate the *P* value using Appendix A-4.

26. The 50 states' average (mean) costs to community hospitals per stay are normally distributed. A random sample of 12 states in 1989 yielded the following data.

State	Average Cost to Hospital per Stay
NM	$4064.42
IN	4091.41
TX	4263.90
MA	5558.43
AL	3839.06
CT	5821.92
OK	3969.86
NE	4282.79
NC	4032.32
VA	4054.11
GA	3919.97
MS	2881.73

Source: *Source Book of Health Insurance Data, 1991,*
Health Insurance Association of America.

a. Describe the population for this set of data.

b. Determine whether statistical evidence exists at the 10% level of significance to support the claim that the population mean is not $4500.00.

GROUP

27. Assume that the numbers of days per year on which metropolitan areas fail to meet air quality standards for ozone are normally distributed. Following is a random sample of average (mean) numbers of days on which some metropolitan areas failed to meet ozone standards.

Stat Fact

Of all metropolitan areas reporting, the south coast of Los Angeles had the most days—more than 100—that exceeded air quality standards for ozone.

Metropolitan Area	1990–92 Yearly Average
Grand Rapids, MI	3.4
El Paso, TX	3.7
Louisville, KY (IN)	1.8
Atlanta, GA	4.4
York, PA	0.3
Nashville, TN	2.3
Columbus, OH	0.7
Memphis, TN (AR, MS)	0.0
Knox and Lincoln counties, ME	2.8
Canton, OH	0.3
Birmingham, AL	2.1
Raleigh–Durham, NC	0.0
Indianapolis, IN	0.3
Evansville, IN (KY)	0.0
St. Louis, MO (IL)	1.4
Sheboygan, WI	3.2
Portland, ME	4.5
Ventura County, CA	17.6
Reno, NV	1.4
San Joaquin Valley, CA	22.6

Conduct hypothesis tests at the .1%, .5%, 1%, 5%, and 10% levels of significance to determine whether or not to reject the claim that the average number of days for all metropolitan areas in this time period (1990–92) is 6.0 days.

28. A mathematics achievement test was administered to gifted students. The following graph represents the raw scores earned by a random sample of $n = 10$ of those students.

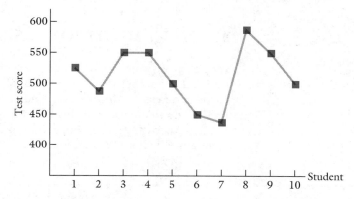

Assume that the population of all scores is normally distributed. Reading the graph as accurately as possible, decide whether statistical evidence exists at the 5% level of significance to support the claim that μ is not 560.

29. A church historian has made a claim that the average reign of the popes of the Roman Catholic Church is 12 years. A random sample of $n = 15$ popes revealed the following data.

Name	Reigned	
	From	*To*
St. Lucius I	253	254
Boniface IX	1389	1404
Benedict XV	1914	1922
Julius II	1503	1513
Clement X	1670	1676
Anastasius	496	498
Paul VI	1963	1978
Alexander VI	1492	1503
Paul V	1605	1621
John IX	898	900
Liberius	352	366
Alexander III	1159	1181
St. Urban I	222	230
Gregory XVI	1831	1846
Eugene II	824	827

Source: *The 1995 Information Please Almanac,* pp. 418–420.

Should this historian's claim be rejected at the 5% level of significance? Assume that the lengths of all reigns are normally distributed.

30. Refer to Exercise 29.

    a. Do the results of the hypothesis test verify the church historian's claim? Explain.

    b. Somehow it is known that the standard deviation of all reigns is 5 years. Conduct the hypothesis test with this new information.

9.7 ONE-TAIL HYPOTHESIS TESTS

A **one-tail hypothesis test** is one in which the rejection region is concentrated in one end, or tail, of the sampling distribution model. A one-tail hypothesis test is used to determine if statistical evidence supports a "more-than" or "less-than" claim. The theory behind this type of hypothesis test is similar to that for a two-tail hypothesis test. The concentration of the rejection region in one tail follows from a null hypothesis that includes the possibility of inequality.

E x a m p l e 6

Exercise 14 in Exercises 9.4–9.6 on page 374 said the following:

A manufacturer of gas pipe claims that the mean weight of his pipe is 28.50 pounds per yard with a standard deviation of $\sigma = 6.45$ pounds. A random sample of $n = 40$ pieces of pipe reveals that $\overline{X} = 30.45$ pounds per yard. Determine whether or not the manufacturer's claim should be rejected at the 5% level of significance. Assume that $n < .05\,N$.

Suppose the manufacturer now claims that the mean weight of his pipe is *more than* 28.50 pounds per yard. If the other statistics and conditions remain the same, determine whether this new claim can be supported by statistical evidence.

Solution This is a one-tail hypothesis test because the claim in question is one of inequality ("more than"), not equality. Also, since statistical evidence can support only the alternative hypothesis, this hypothesis must be H_A: $\mu > 28.50$.

Step 1 State the hypotheses.

$$H_0: \quad \mu \leq 28.50.$$
$$H_A: \quad \mu > 28.50.$$

Notice that, just as in the two-tail case, exactly one of these two hypotheses can be true. Likewise, the goal of this one-tail test is to either reject the null hypothesis and accept the alternative hypothesis or reserve judgment.

Step 2 Decide on a model for the sampling distribution.

This is a large sample ($n = 40$) taken from a relatively large population. Therefore, use the normal distribution model.

Step 3   *Determine the end points of the rejection region and state the decision rule.*
In this case, the null hypothesis will be rejected only if the Z-score for $\overline{X}$ is too large, because the alternative hypothesis, H_A, indicates a "greater than" relationship. As a result, all of the rejection region is concentrated on the right side of the distribution model.

The direction of the inequality for the alternative hypothesis, H_A, will always "point" to the rejection region in a one-tail test.

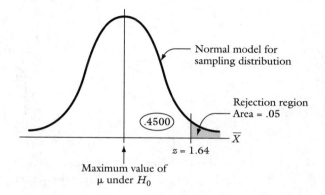

Normal model for sampling distribution

Rejection region
Area = .05

.4500

$z = 1.64$

$\overline{X}$

Maximum value of μ under H_0

Since the right side of the standard normal distribution model has an area of .5000 and all of the rejection region is in the right tail, the area from the assumed maximum value of μ under H_0 to the rejection region is

$$.5000 - .05 = .4500.$$

Using Appendix A-2, the Z-score closest to .4500 is $z = 1.64$, which is the left end point of the rejection region. The decision rule is:

Reject H_0 and accept H_A if $Z(\overline{X}) > 1.64$. Otherwise, reserve judgment.

Step 4   *Compute the test statistic.*
The standard error can be calculated as

$$SE = \frac{\sigma}{\sqrt{n}}$$

$$= \frac{6.45}{\sqrt{40}}$$

$$\approx 1.02.$$

Use this value for SE. Enter the value of $\overline{X}$ and the assumed maximum value of μ for $\mu_{\overline{X}}$ to obtain the test statistic $Z(\overline{X})$:

$$Z(\overline{X}) = \frac{\overline{X} - \mu_{\overline{X}}}{SE}$$

$$= \frac{30.45 - 28.50}{1.02}$$

$$= 1.91.$$

Step 5   State the conclusion.

$Z(\overline{X})$ falls in the rejection region.

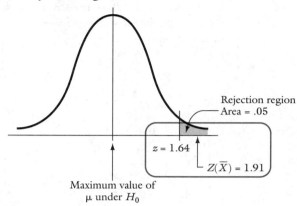

Therefore, **reject H_0 and accept H_A.** Statistical evidence exists at the 5% level of significance to support the alternative hypothesis of the manufacturer, that $\mu > 28.50$. •

Judgment about the claim $\mu = 28.50$ should have been reserved when Exercise 14 on page 374 was completed. Because a one-tail hypothesis test concentrates the rejection region all in one tail, the end point of that one rejection region is closer to the assumed center of the sampling distribution than is either end point in a two-tail test. Consequently, it is easier for the test statistic to fall in the rejection region of a one-tail test.

**E x a m p l e
7**

A manufacturing company claims that one of its new steel alloys will withstand a pressure of at least 1449 pounds per square inch (psi) with a standard deviation of $\sigma = 90$ psi. A random sample of $n = 20$ of these alloys was tested, with the following results.

1520	1400	1375	1475	1400
1400	1525	1455	1520	1325
1375	1475	1400	1250	1450
1450	1540	1500	1250	1375

Assume a normally distributed population, and use Minitab to determine whether the manufacturer's claim about μ and σ should be rejected at the 9% level of significance.

Solution   Because the purpose of this hypothesis test is to make a decision about rejection, the claim in question, H_0: $\mu \geq 1449$, is the null hypothesis.

Step 1   State the hypotheses.

$$H_0: \quad \mu \geq 1449.$$
$$H_A: \quad \mu < 1449.$$

The only possible outcomes from this hypothesis test are the conclusions that the new alloy will withstand pressures of less than 1449 psi (reject manufacturer's claim) and res-ervation of judgment.

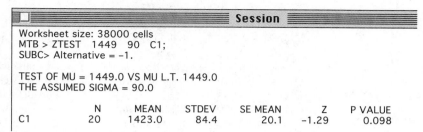

```
                                                      Session
 Worksheet size: 38000 cells
 MTB > ZTEST  1449  90  C1;
 SUBC> Alternative = –1.

 TEST OF MU = 1449.0 VS MU L.T. 1449.0
 THE ASSUMED SIGMA = 90.0

              N      MEAN      STDEV    SE MEAN       Z    P VALUE
 C1          20    1423.0       84.4      20.1    –1.29      0.098
```

Figure 9-8 Minitab's ZTEST command with "Alternative" subcommand.

Step 2   Decide on a model for the sampling distribution.

This is a small sample with given (that is, assumed) standard deviation, σ. Since the population is also assumed normally distributed, use a normal distribution as the model for the sampling distribution of the $\overline{X}$'s. Therefore, utilize Minitab's ZTEST command with the appropriate subcommand for a one-tail test.

Steps 3 and 4

- Enter the data into column C1 of a Minitab worksheet.
- The ZTEST macro command, along with the subcommand

$$\text{Alternative} = -1.$$

defines a one-tail hypothesis test in which the alternative hypothesis, H_A, is that μ is less than claimed. The subcommand "Alternative = 1" defines the alternative hypothesis so that μ is greater than claimed. The use of one of these two subcommands results in a calculated P value for a one-tail test.

   Figure 9-8 shows the use of the correct subcommand and the results. Note that the ZTEST command line ends with a semicolon, which activates the subcommand (**SUBC** >) prompt when the [RETURN] key is pressed.

Step 5   State the conclusion.

When using a statistical software package, such as Minitab, it is possible to conduct hypothesis tests at any level of significance.

Recall that this test was to be conducted at the 9% level of significance. Thus, $\alpha = .09$. Since $P > \alpha$ (that is, $.098 > .09$), **reserve judgment.** There is not enough statistical evidence at the 9% level of significance to reject the manufacturer's claim and support the alternative hypothesis that $\mu < 1449$. This result does not in any way support the manufacturer's claim. •

E x a m p l e
8

Someone makes the claim that the mean batting average of National League batting champions during the years 1876 through 1994 is more than .350. Assume that the batting averages of each year's champions are normally distributed, and use the following random sample of size $n = 20$ to determine whether statistical evidence exists to support this claim at the 5% level of significance.

Year	Name	Team	Batting Average
1975	Bill Madlock	Chicago Cubs	.354
1885	Roger Connor	N.Y. Giants	.371
1968	Pete Rose	Cincinnati Reds	.335
1920	Rogers Hornsby	St. Louis Cardinals	.370
1955	Richie Ashburn	Philadelphia Phillies	.338
1989	Tony Gwynn	San Diego Padres	.336
1914	Jake Daubert	Brooklyn Dodgers	.329
1936	Paul Waner	Pittsburgh Pirates	.373
1887	Cap Anson	Chicago Cubs	.421
1988	Tony Gwynn	San Diego Padres	.313
1963	Tommy Davis	L.A. Dodgers	.326
1946	Stan Musial	St. Louis Cardinals	.365
1907	Honus Wagner	Pittsburgh Pirates	.350
1924	Rogers Hornsby	St. Louis Cardinals	.424
1952	Stan Musial	St. Louis Cardinals	.336
1981	Bill Madlock	Pittsburgh Pirates	.341
1900	Honus Wagner	Pittsburgh Pirates	.380
1959	Hank Aaron	Milwaukee Braves	.355
1930	Bill Terry	N.Y. Giants	.401
1967	Roberto Clemente	Pittsburgh Pirates	.357

Solution   Since statistical evidence can support only the alternative hypothesis, H_A, this hypothesis must be H_A: $\mu > .350$.

Step 1   State the hypotheses.

$$H_0: \quad \mu \leq .350.$$
$$H_A: \quad \mu > .350.$$

Step 2   Decide on a model for the sampling distribution.

This is a small sample taken from a normally distributed population with unknown standard deviation. Therefore, use the *t*-distribution model for the sampling distribution of the $\overline{X}$'s.

Step 3   Determine the end points of the rejection region and state the decision rule.
The *t*-score that determines the end point of the rejection region can be found as follows.

- A 5% level of significance means that the rejection region has an area of .05 and is in the right tail.

The only possible outcomes of this hypothesis test are the conclusion that the mean batting average is greater than .350 and reservation of judgment.

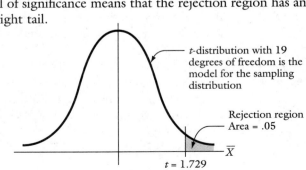

t-distribution with 19 degrees of freedom is the model for the sampling distribution

Rejection region
Area = .05

$\overline{X}$

$t = 1.729$

- According to Appendix A-4, the intersection of the .05 column with the $n - 1 = 19$ degrees of freedom row establishes a t-score of $t = 1.729$ that marks the left end point of the rejection region.

The decision rule is:

Reject H_0 and accept H_A if $t(\overline{X}) > 1.729$. Otherwise, reserve judgment.

Step 4   Compute the test statistic.
The test statistic is $t(\overline{X})$:

$$t(\overline{X}) = \frac{\overline{X} - \mu_{\overline{X}}}{SE},$$

where $\mu_{\overline{X}} = \mu$, which is assumed to be .350 under the null hypothesis.

To supply $\overline{X}$ and SE in the formula for the test statistic, it will be necessary to compute $\overline{X}$ and s from the random sample. A TI-82 graphics calculator can be used to determine these statistics.

```
      1-Var Stats
   →  x̄=.35875
      Σx=7.175
      Σx²=2.591431
s  →  Sx=.0302617963
      σx=.0294955505
    ↓ n=20
```

Using a rounded value of $s \approx .03$,

$$SE \approx \frac{s}{\sqrt{n}}$$

$$\approx \frac{.03}{\sqrt{20}}$$

$$\approx .0067.$$

Thus,

$$t(\overline{X}) = \frac{\overline{X} - \mu_{\overline{X}}}{SE}$$

$$= \frac{.35875 - .350}{.0067}$$

$$\approx 1.31.$$

Step 5   State the conclusion.
The value of $t(\overline{X})$ does not make it into the rejection region. Therefore, **reserve judgment.** At the 5% level of significance, there is not enough statistical evidence to support the alternative hypothesis that the mean batting average is greater than .350. ●

E x a m p l e
9

Use Minitab to solve Example 8.

Solution   As usual, the first step is to enter the data into column C1 of a Minitab worksheet. Then obtain a Session window and employ the macro command TTEST. Use the subcommand "Alternative = 1" to indicate a "greater than" alternative hypothesis, H_A. The completed command lines appear in Figure 9-9. Notice that the test statistic has a value of 1.29 rather than 1.31 when it is computed by hand, a difference that can be attributed to rounding. Since the *P* value of .11 is greater than α (.05), judgment about H_0 is reserved.

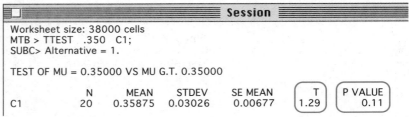

		Session				
Worksheet size: 38000 cells
MTB > TTEST .350 C1;
SUBC> Alternative = 1.

TEST OF MU = 0.35000 VS MU G.T. 0.35000

	N	MEAN	STDEV	SE MEAN	T	P VALUE
C1	20	0.35875	0.03026	0.00677	1.29	0.11

Figure 9-9 ●

Exercises **9.7**

1. Explain why a one-tail hypothesis test can result in the rejection of H_0, whereas a two-tail test using the same data may not.

MTB

2. Minitab's ZTEST and TTEST macro commands can be used for two-tail hypothesis testing. With the appropriate "Alternative" subcommand, they can be used for one-tail hypothesis testing as well. What part of the output does the subcommand change?

Exercises 3–7 refer to Exercises 9.4–9.6 on pages 373–380.

3. Refer to Exercise 17 on page 375. Does statistical evidence exist to support the claim that the mean retail sales for all 50 states in 1991 were less than $19,750.00? Perform a hypothesis test at the 3% level of significance.

* 4. Refer to Exercise 22 on page 376. Determine whether a claim that the politician's approval rating is more than 65% can be supported at the 5% level of significance.

   Hint: Use π = *.65 to calculate SE.*

5. Refer to Exercise 24 on page 377. Can a claim that the average state benefit is less than $170 be supported at the 10% level of significance?

6. Refer to Exercise 26 on page 378. Does statistical evidence exist to support the claim that the mean cost per stay for all states is greater than $4000.00? Perform a hypothesis test at the 10% level of significance.

7. Refer to Exercise 27 on page 378. Does statistical evidence exist to support the claim that the mean number of days for all metropolitan areas in the given time period is less than 3.75 days? Perform a hypothesis test at the 5% level of significance.

GROUP

8. A new drug has been developed to treat severe cases of high blood pressure. Its developers claim that it reduces blood pressure to an acceptable level in less than 75 minutes. The drug was administered to a random sample of 15 patients. The times at which the drug was administered and at which it produced favorable reactions follow.

Patient Number	Time Administered	Reaction Time
1	8:15	9:40
2	8:00	9:12
3	7:55	9:06
4	7:15	8:05
5	10:00	10:58
6	6:15	7:52
7	9:25	10:02
8	8:30	9:45
9	6:25	8:08
10	5:10	5:55
11	7:00	8:22
12	11:50	12:27
13	8:10	9:45
14	9:15	9:49
15	6:30	7:23

How stringently tested should a drug be before it is administered to patients? (*Courtesy of Smithkline Beecham*)

Assuming that all patient reaction times are normally distributed, determine whether statistical evidence exists at the 10% and 5% levels of significance to support the drug's claimed reaction time.

GROUP

9. Refer to Exercise 8. Imagine that the manufacturer also claims a standard deviation of reaction times of $\sigma = 15$ minutes. Conduct the required hypothesis tests using this new information.

   Hint: This standard deviation of reaction times will also change the distribution model.

10. During a 1994 House of Representatives election campaign, an incumbent congressman came under heavy criticism for making too much use of his free mailing privileges. The congressman, however, claimed that his use of this privilege was "less than average."

    A random sample of $n = 22$ members of the House and the amounts of money they spent on mailings follow.

For possible reasons members of Congress might spend so much on mailing when they have franking privileges, see "Post Office Inquiry Leads to Scrutiny Of Franking Privileges, Stamp Purchases" in the 1992 *Congressional Quarterly Almanac,* pp. 48–50.

Representative	Postage Cost	Representative	Postage Cost
A	$168,806.92	L	$115,041.85
B	28,837.42	M	163,146.65
C	73,919.15	N	121,433.47
D	16,297.33	O	176,499.07
E	95,250.18	P	45,500.65
F	167,607.98	Q	64,498.30
G	147,881.83	R	92,276.57
H	79,578.10	S	111,218.12
I	139,309.14	T	146,421.73
J	127,528.19	U	109,800.21
K	155,786.96	V	106,404.85

If the incumbent congressman is Representative A above, perform a hypothesis test at the 5% level of significance to determine whether the congressman's claim can be supported.

Hint: The implied alternative hypothesis is H_A: $\mu >$ $168,806.92.

* 11. Refer to Exercise 10. Determine the maximum amount of money Representative A could have spent on postage in order for his claim to be supported at the 5% level of significance. (Assume that the value of SE remains the same.)

12. Refer to Exercise 10. If the actual value of μ equals $99,912 in the House, then what is the P value of Representative A's postal spending? (Assume that the value of SE remains the same.)

GROUP   13. Refer to Example 4 on page 370. Determine whether statistical evidence exists to support the claim that the mean price for that minivan is less than $16,500. Perform hypothesis tests at the 1%, 5%, and 10% levels of significance. Do not use the manufacturer's claimed standard deviation; use s based on the sample instead.

14. Assume that the numbers of full-time-equivalent licensed practical nurses (LPNs) employed in large cities are normally distributed. A random sample of 12 of these cities in 1993 follows.

City	Number of LPNs Employed
Denver, CO	288
Jacksonville, FL	327
Tulsa, OK	282
San Diego, CA	332
Buffalo, NY	537
Philadelphia, PA	1171
Kansas City, MO	395
Detroit, MI	827
Virginia Beach, VA	78
Honolulu, HI	293
St. Louis, MO	944
Portland, OR	108

Source: *American Hospital Association Hospital Statistics,* 1993–94 edition, Table 7.

Determine whether statistical evidence exists to support the claim that the mean number of full-time-equivalent LPNs in large cities is more than 400. Perform a hypothesis test at the 10% level of significance.

* 15. Refer to Exercise 14. Determine the greatest value of N for which the data would support the claim that the mean number of full-time-equivalent LPNs in large cities is more than N. Use the 10% level of significance.

MTB *Although Exercises 16–19 may be worked by hand, they are especially suitable for Minitab's TTEST command with the "Alternative" subcommand.*

16. As of January 31, 1994, 112 chief and associate justices have served in the United States Supreme Court. A random sample of size $n = 20$ reveals the following data.

Name	Term	Name	Term
Henry Baldwin	1830–1844	Earl Warren	1953–1969
William Rehnquist	1972–1986	Pierce Butler	1923–1939
William Howard Taft	1921–1930	Samuel Blatchford	1882–1893
Joseph McKenna	1898–1925	Samuel Chase	1796–1811
Alfred Moore	1800–1804	Horace Gray	1882–1902
Abe Fortas	1965–1969	Potter Stewart	1958–1981
John Marshall	1801–1835	Joseph Story	1812–1845
David Davis	1862–1877	Salmon P. Chase	1864–1873
Charles Evans Hughes	1901–1916	Thomas Todd	1807–1826
Henry Brown	1890–1906	Wiley Rutledge	1943–1949

Assuming that the lengths of the 112 terms are normally distributed, is there statistical evidence that the mean length of term is less than 19 years? Perform a hypothesis test at the 10% level of significance.

17. Assume that the sales of America's 100 fastest growing companies in the last four quarters are normally distributed. A random sample of the sales of $n = 25$ of these companies follows.

Company	Sales (millions)	Company	Sales (millions)
Mesa Airlines	$ 340.1	QVC Network	$1110.7
Starbucks	122.9	Future Now	428.6
CompUSA	1273.9	Qual-Med	499.2
Mirage Resorts	831.6	Rally's	132.5
Blockbuster Entertainment	1337.6	Grancare	342.8
		Checker's Drive-Ins	105.7
Office Depot	1881.8	Samuel Goldwyn	107.8
Novacare	495.0	Intuit	103.2
Dell Computer	1273.2	Value Health	317.1
Jenny Craig	480.5	Cisco Systems	554.5
In Home Health	90.6	California Energy	132.5
Video Lottery Technologies	139.9	President Riverboat Casinos	124.9
Read-Rite	517.0	Outback Steakhouse	149.2

Source: *Fortune,* August 9, 1993, pp. 40–54.

a. Describe the population mean, μ.

b. Does statistical evidence exist to support the claim that the population mean is less than \$600 million at the 2% level of significance?

c. What is the attained significance level of the test in Part b?

18. Assume that the percentages of caesarean births each year, by state, are normally distributed. Following is a random sample of $n = 20$ states and the percentages of total births that were caesarean in 1990.

State	Percent	State	Percent
TN	29.5	IA	20.6
FL	34.4	IL	26.4
AZ	28.2	MA	25.9
MO	25.0	HI	18.0
NM	25.8	VA	25.0
MT	24.4	RI	20.0
WI	19.7	SD	12.7
OH	28.5	WA	24.7
NE	22.0	NY	43.7
CO	28.2	LA	39.2

Source: *Statistical Bulletin* 73:1, January–March 1992, p. 14.

Conduct a hypothesis test at the 20% level of significance to determine whether evidence exists to support the claim that the mean percentage of caesarean births for all 50 states is more than 25%.

19. A random sample of $n = 10$ prices of women's walking shoes follows.

Brand and Model	Price
Soft Spot Wings	\$72
Reebok Comfort I Ultra	75
Payless 9628	15
Nike Air Essential	60
Avia ArchRocker 382	65
Etonic Trans Am Walker	50
MacGregor Walkers for Her	20
Easy Spirit Mach I	74
NaturalSport Challenger	65
Keds True Lite Walker	57

Source: *Consumer Reports,* July 1993, p. 424.

Assume that the prices of all women's walking shoes are normally distributed.

a. Conduct a hypothesis test at the 10% level of significance to determine whether evidence exists to support the claim that the population mean is more than \$50.

b. Explain why a different null hypothesis, that the population mean is at least \$50, would not be rejected at any level of significance.

20. Assume that the numbers of calories in different brands and types of hot dogs are normally distributed. A random sample of the numbers of calories in $n = 24$ brands and types of hot dogs follows.

Brand and Type	No. of Calories
Jennie-O Turkey	70
Armour Premium Beef	149
Oscar Mayer Pork and Turkey	150
Hygrade's Chicken and Pork	181
Healthy Choice Minimal-Fat	50
Ball Park Beef	167
Farmer John	130
Lady Lee Chicken	79
Hormel Wranglers	180
Louis Rich Bun-Length Turkey	130
Oscar Mayer Light	130
Armour Premium Lean Beef Jumbo	80
Eckrich Bunsize (pork, turkey, and beef)	180
Hebrew National Kosher	149
Lyke's Meaty Jumbo	180
Dubuque (chicken and pork)	108
Butterball Bun Size Turkey	130
Nathan's Famous 8 Skinless	176
Ball Park Lite (beef and pork, chicken)	140
Bar S Jumbo (chicken, pork, and beef)	216
Oscar Mayer Beef	140
Shoprite (chicken and pork)	120
Best's Kosher (lower salt and fat)	100
Bryan Juicy Jumbos	180

a. Perform a hypothesis test at the 5% level of significance to determine whether evidence exists to support the claim that the mean number of calories in a hot dog (chosen at random by brand and type) is less than 150.

b. Round the following calculations to the nearest percent.

1. What is the highest level of significance for which the hypothesis test in Part a would not end in rejection?

2. What is the lowest level of significance for which the hypothesis test in Part a would end in rejection?

9.8 TYPES OF ERRORS RELATED TO HYPOTHESIS TESTING

Due to the random nature of $\overline{X}$, the final conclusion of a hypothesis test may be incorrect. Many times, the statistician has no way of knowing about this until long after he or she has decided whether or not to reject H_0 and has seen the consequences.

For example, suppose an automobile manufacturer requires a certain type of engine to run a specified number of miles per gallon. Because it is impossible to test every engine, a random sample of engines must be tested. The resulting $\overline{X}$ may indicate that the engine does not meet the required specifications, but it is also possible that the random sample is (unluckily) not typical of the population, in which case it causes a rejection error. If this is to be the only qualifying test, the engine will not be approved when it should be, and the resulting slowdown in production could increase the eventual price of the automobile.

Two types of errors are possible in a hypothesis test.

Type I error: rejecting H_0 when H_0 is, in fact, true

Type II error: failing to reject H_0 when H_0 is, in fact, false

---

The engine example relates to the probability of a Type I error, which could be reduced by decreasing α, the area of the rejection region of the test. However, this adjustment would only increase the probability of a Type II error. Any effort to reduce the chances of one type of error must increase the chances of making the other type.

At times, one type of error is much less acceptable than the other. Consider, for example, a new drug that is being tested with the following hypotheses.

$$H_0: \quad \text{The drug is unsafe.}$$
$$H_A: \quad \text{The drug is safe.}$$

In the interest of public safety, the rejection region, α, should be very small so that a Type I error and rejection of the null hypothesis (possibly causing the drug to be put on the market when it is not safe) are very unlikely.

In general, α, the level of significance, is determined by the willingness to make one type of error or the other. Further courses in statistics will examine this dilemma more carefully.

Case Study   *Another Explanation of Hypothesis Testing*

The December 4, 1993, issue of *New Scientist* contains a supplement (following page 40) that presents another explanation of how to conduct a hypothesis test. Several of the examples in the article apply the binomial distribution and illustrate on an intuitive level the critical role sample size can play in hypothesis testing. Obtain a copy of this article from your library, read it, and answer the following questions.

I. The author claims that, unlike arithmetic, statistics is not predictable. Explain what you think he means.

II. Explain why the choice of the null hypothesis is not always clear, and cite the example the author uses to illustrate this fact.

III. How does the author define the **acceptance level** of a hypothesis test?

IV. In your own words, define the **critical cure rate** as used by the author in his drug example.

V. In your own words, define the **megalithic yard** and explain why the hypothesis test for its existence is flawed.

CHAPTER SUMMARY

Hypothesis testing is a formal mathematical procedure that determines the likelihood of a claim about an unknown population parameter. It is accomplished by stating two diametrically opposed hypotheses: the null hypothesis (H_0) and the alternative hypothesis (H_A). The alternative hypothesis must be logically accepted if the hypothesis test ends in rejection of the null hypothesis. A hypothesis test can provide only statistical evidence for the alternative hypothesis or reserved judgment.

In this chapter, the normal distribution and the t-distribution served as models with which to judge the sample statistic to arrive at a decision regarding the rejection of H_0.

The exact wording of the claim, along with the use of a given population parameter, σ, or its estimated value, s, can make the difference between rejection and nonrejection of the null hypothesis. Careful reading of the claim and instructions and the correct choice of distribution model are critical.

Key Concepts, Terms, and Formulas

Videotape Suggestions

Program 20 from the series *Against All Odds: Inside Statistics* compares hypothesis testing with a paternity test and discusses P values. *Program 21,* suggested earlier, presents additional examples of the use of the t-distribution in hypothesis testing.

(Picture Perfect™)

Inferences about the
difference of population
means might concern
age, ethnicity, and other
factors.

10 Inferences About the Difference of Population Means

Confidence intervals and hypothesis testing are especially useful to those drawing inferences about the difference of two population means. Two drugs, for example, may both be effective in treating a certain disease, but it is important to know which, on the average, works faster, and how much faster. Or one may want to know which of two tires produced by different manufacturing processes can withstand more use.

In social science applications, inferences about the difference of population means could be important in the interpretation of test results that indicate differences between men and women or between ethnic groups.

The **case study** at the end of this chapter examines the "before and after" effects of a massive government public information campaign.

10.1 METHODS OF SAMPLING

Ways of drawing inferences about the difference of population means depend on whether the samples were derived through independent sampling or through matched-pair sampling.

Definition ▶   **Independent sampling** is performed by picking two random samples, one from each of two populations, in such a way that the make-up of one sample does not affect the make-up of the other. It is not necessary that the two samples be the same size.

Exercises in this chapter will examine some of the differences in salaries for RNs.

Independent sampling is often used to obtain a single type of information about two populations in which individual members may not share many characteristics. One might use independent sampling to determine whether there is a difference between full-time hourly pay for RNs in the Southwest and in New England; to conduct a marketing study of sales of a new breakfast cereal to adults and children; or to determine the effect on children's grades of watching more than 20 hours of TV per week.

Matched-pair sampling is an alternative to independent sampling.

Definition ▶ | **Matched-pair sampling** is the matching of each member of one sample with a corresponding "twin" in the other sample on the basis of factors other than the one being measured.

Matched-pair sampling is used when members from two populations can be matched in all characteristics but the one being studied. For example, suppose researchers want to compare the mileage obtained from tires manufactured by an experimental process to that obtained from tires manufactured by the traditional process. Each set of experimental tires can be matched to a set of traditional tires under the conditions of identical automobile type, driver profile, and weather.

This chapter makes two important assumptions about the populations from which independent samples and matched-pair samples are drawn.[1]

> Two necessary assumptions about populations:
>
>   a. The two populations are normally distributed.
>   b. The two populations have approximately equal standard deviations.

10.2 INDEPENDENT SAMPLING

First it is necessary to establish some conventional notation. If A and B are two populations, then μ_A and μ_B denote their respective means. N_A denotes the size of population A, and N_B, the size of population B. If samples are taken from populations A and B, then n_A and n_B, respectively, denote the sizes of those samples. The mean of the sample taken from population A is denoted by $\overline{X}_A$, and the mean of the sample from population B, by $\overline{X}_B$.

One way to form a point estimate of the difference of the two population means, $\mu_A - \mu_B$, is to choose a simple random sample from population A and

---

[1]Assuming that populations are normally distributed can be quite reasonable, as the examples and exercises of the last two chapters have shown. However, when either or both of these assumptions are not justified, other approaches may be more appropriate. One, the Mann-Whitney technique for constructing a confidence interval for the difference of two populations' medians, is treated later in this chapter. Also, Excursion 10.7A at the end of the chapter treats a special case: drawing inferences from small samples taken from normally distributed populations in which the standard deviations are not necessarily equal.

another simple random sample from population B. The difference of the means of the samples, $\overline{X}_A - \overline{X}_B$, can then be used as a point estimate for the difference of the population means. However, for reasons explained in Chapter 8, an interval estimate for $\mu_A - \mu_B$ is often preferable to the point estimate, $\overline{X}_A - \overline{X}_B$.

The construction of a confidence interval for $\mu_A - \mu_B$ is similar to the construction of a confidence interval for the mean of a single population, μ. Likewise, the characteristics of a sampling distribution of $\overline{X}_A - \overline{X}_B$ are analogous to those of a sampling distribution of a single population.

Characteristics of the Sampling Distribution of $\overline{X}_A - \overline{X}_B$

These two characteristics of the sampling distribution of $\overline{X}_A - \overline{X}_B$ should be compared to those of a single population, discussed in Chapter 8, Section 8.2 (page 316).

Characteristic 1   The mean of the sampling distribution of all the possible values of $\overline{X}_A - \overline{X}_B$ is $\mu_A - \mu_B$. In symbols,

$$\mu_{\overline{X}_A - \overline{X}_B} = (\mu_A - \mu_B). \qquad (10\text{-}1)$$

That is, the mean of all the possible differences is equal to the difference of the population means.

Characteristic 2   The standard deviation of the sampling distribution of all the possible values of $\overline{X}_A - \overline{X}_B$, called the standard error, *SE*, can be found with the formula

$$SE = \sigma_{\overline{X}_A - \overline{X}_B}$$

$$= \sqrt{\frac{\sigma_A^2}{n_A} + \frac{\sigma_B^2}{n_B}}, \qquad (10\text{-}2)$$

where σ_A^2 and σ_B^2 are the variances of populations A and B, respectively.[2] If the population standard deviations, σ_A and σ_B, are unknown, the standard error can be approximately by

$$SE \approx s_{\overline{X}_A - \overline{X}_B}$$

$$= \sqrt{\frac{s_A^2}{n_A} + \frac{s_B^2}{n_B}}. \qquad (10\text{-}2a)$$

In this formula, s_A^2 and s_B^2 represent the estimated values of each population's variance, based on s_A and s_B, respectively.

Just as in the construction of confidence intervals, the sizes of the samples affect the choice of sampling distribution model.[3] In this chapter, either the normal

---

[2] Recall that the variance is the square of the standard deviation. If two standard deviations are approximately equal, then so are their variances.

[3] In the construction of a confidence interval for a single population parameter, μ, the distribution of the population also affects the choice of sampling distribution model. The assumption in this chapter, however, is that all populations are normally distributed. Thus, the choice of a normal distribution or t-distribution model for the construction of the confidence interval of $\mu_A - \mu_B$ must depend on other factors.

distribution or the t-distribution model is used for the construction of confidence intervals for $\mu_A - \mu_B$. Follow the rules of thumb laid out in the next section to choose the appropriate model in a specific situation. Remember that these two sampling distribution models require the assumption that both populations are normally distributed with approximately equal standard deviations.

10.3 CHOOSING THE SAMPLING DISTRIBUTION MODEL OF $\overline{X}_A - \overline{X}_B$

The Normal Distribution Model

Use the normal distribution model for the construction of a confidence interval for $\mu_A - \mu_B$ if either (or both) of the following are true.

- $n_A \geq 30$ and $n_B \geq 30$, and both are from relatively large populations.

OR

- The two samples (large or small) are taken from populations with known standard deviations.

If at least one of these two conditions is satisfied, then the sampling distribution of all the possible values of $\overline{X}_A - \overline{X}_B$ is considered normally distributed.

The t-Distribution Model

If both samples are small and are taken from populations with unknown standard deviations, then the t-distribution model should be used for the sampling distribution of all the possible values of $\overline{X}_A - \overline{X}_B$. The degrees of freedom are $(n_A + n_B - 2)$.

Recall the assumption underlying this part of the chapter: that the two populations under consideration are normally distributed and have approximately equal standard deviations. Because statistics obtained from small samples may be less reliable than those obtained from large samples and the population standard deviations are assumed approximately equal, it is recommended that the two sample variances, s_A^2 and s_B^2, be "pooled" to arrive at a weighted average of the common (unknown) variance. The **pooled variance,** s_P^2, is the weighted mean of s_A^2 and s_B^2 and is found with the following formula.

Formula for the **Pooled Variance**

$$s_P^2 = \frac{(n_A - 1) \cdot s_A^2 + (n_B - 1) \cdot s_B^2}{n_A + n_B - 2} \qquad \textbf{(10-3)}$$

The pooled variance, s_P^2, is used in place of both s_A^2 and s_B^2 in the approximation formula for the standard error, SE (Formula 10-2a), and is always used when the t-distribution model is used for a difference of population means.

The following examples demonstrate the use of the normal distribution model and the *t*-distribution model to construct confidence intervals for the difference of two population means. They also show how to conduct hypothesis tests for the difference of two population means.

**E x a m p l e
1**

A test was devised to judge the reflex times of candidates applying to operate high-speed machinery. Thousands of applicants took the test, and their results were divided into two groups: the times of those over 35 years of age (population A) and the times of those 35 and younger (population B).

A random sample of 40 times was chosen from each group. The sample statistics from the older group were a mean reflex time of $\overline{X}_A = 13.75$ seconds and an estimated standard deviation of $s_A = 2.10$ seconds. The sample statistics from the younger group were a mean reflex time of $\overline{X}_B = 12.50$ seconds and an estimated standard deviation of $s_B = 2.30$ seconds. Construct an 85% confidence interval for the difference in mean response times of the two populations.

Solution   The following chart summarizes the information presented so far.

Older	Younger
$n_A = 40$	$n_B = 40$
$\overline{X}_A = 13.75$	$\overline{X}_B = 12.50$
$s_A = 2.10$	$s_B = 2.30$

This is a case of independent sampling in which each sample is large: $n_A = n_B = 40$. Also, since thousands of people were tested, it seems safe to assume that $n_A < .05\,N_A$ and $n_B < .05\,N_B$. Therefore, the populations are large relative to the samples, and conditions are satisfied for the use of the normal distribution model to construct the required confidence interval (Figure 10-1).

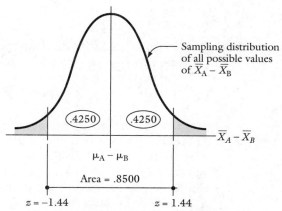

Figure 10-1   The normal distribution model for the construction of a confidence interval for the difference of population means.

Because the standard deviations of the two populations are unknown (although they are assumed approximately equal), the standard deviation of the sampling distribution just shown, denoted by SE, must be approximated using s_A and s_B:

$$SE \approx s_{\overline{X}_A - \overline{X}_B}$$

$$= \sqrt{\frac{s_A^2}{n_A} + \frac{s_B^2}{n_B}}$$

$$= \sqrt{\frac{(2.10)^2}{40} + \frac{(2.30)^2}{40}}$$

$$= \sqrt{.11025 + .13225}$$

$$\approx .49.$$

(Since the normal distribution model is used, it is not necessary to first compute a pooled variance.)

The Z-scores of ± 1.44 that are used to determine the end points of the confidence interval are found as they were in Chapter 8.

- $\dfrac{85\%}{2} = 42.50\% = .4250.$

- The Z-score in the body of Appendix A-2 that comes closest to enclosing an area of .4250 from the mean is $z = 1.44$.

The Z-scores of ± 1.44 are denoted by $Z(\overline{X}_A - \overline{X}_B)$. In a manner analogous to that in Chapter 8, the end points of the required 85% confidence interval for $\mu_A - \mu_B$ can be found by evaluating

$$\overline{X}_A - \overline{X}_B \pm Z(\overline{X}_A - \overline{X}_B) \cdot SE.$$

Substituting the known values,

$$85\% \text{ confidence interval for } \mu_A - \mu_B = (13.75 - 12.50) \pm 1.44 \cdot .49$$

$$= 1.25 \pm .7056$$

$$= [.54, 1.96].$$

Thus it can be said with 85% confidence that the older population takes between .54 seconds and 1.96 seconds longer to react than the younger population. •

The next example illustrates how to formally apply the method of hypothesis testing to the difference of population means.

**E x a m p l e
2**

The claim is made that the mean price for a certain drug at privately owned drugstores is greater than the mean price for the same drug at drugstores that are parts of large chains. Let A denote the population of privately owned drugstores and B the population of chain drugstores. A random sample of $n_A = 10$ private stores revealed a mean price for the drug of $\overline{X}_A = \$8.75$ with a standard deviation of $s_A = \$1.15$. A random sample of $n_B = 14$ chain stores had a mean price of $\overline{X}_B = \$7.97$ with $s_B = \$0.95$.

Perform a hypothesis test at the 2.5% ($\alpha = .025$) level of significance to determine whether statistical evidence exists to support this claim.

Solution   The statistics presented can be summarized as follows:

Private (A)	Chain (B)
$n_A = 10$	$n_B = 14$
$\overline{X}_B = 8.75$	$\overline{X}_B = 7.97$
$s_A = 1.15$	$s_B = .95$

Step 1   State the hypotheses.

In a hypothesis test, it is possible for statistical evidence to support only the alternative hypothesis, H_A, which occurs when the null hypothesis is rejected. Thus, the null hypothesis must be formulated as the opposite of the claim to be tested. Using the notations μ_A and μ_B for the means of populations A and B, respectively,

H_0: $\mu_A - \mu_B \leq 0$.   (This is equivalent to stating that $\mu_A \leq \mu_B$ or that μ_A is "no more than" μ_B.)

H_A: $\mu_A - \mu_B > 0$.   ($\mu_A > \mu_B$)

Remember that one of the necessary assumptions in this chapter is that the populations are normally distributed.

Step 2   Decide on a model for the sampling distribution.

The t-distribution model is appropriate for the sampling distribution of all possible values of $\overline{X}_A - \overline{X}_B$ because both samples are small and are taken from populations with unknown (but approximately equal) standard deviations.

Step 3   Determine the end points of the rejection region, and state the decision rule.

The alternative hypothesis points to the rejection region, and Figure 10-2 should be helpful in determining its end point, $t = 2.074$.

The degrees-of-freedom formula,

$$\text{d.f.} = n_A + n_B - 2,$$

gives 22 degrees of freedom for this model. According to Appendix A-4, the intersection of the 22 degrees of freedom row and the $\alpha = .025$ column indicates

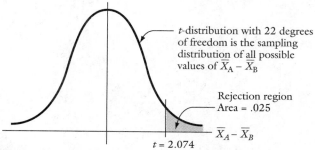

t-distribution with 22 degrees of freedom is the sampling distribution of all possible values of $\overline{X}_A - \overline{X}_B$

Rejection region
Area = .025

$\overline{X}_A - \overline{X}_B$

$t = 2.074$

Figure 10-2 The t-distribution model for the construction of a confidence interval for the difference of population means.

a *t*-score of 2.074 to mark the end point of this rejection region. Accordingly, the decision rule is:

Reject H_0 and accept H_A if $t(\overline{X}_A - \overline{X}_B) > 2.074$. Otherwise, reserve judgment.

Step 4   *Compute the test statistic.*

The test statistic is $t(\overline{X}_A - \overline{X}_B)$. It can be calculated with the following formula, which is analogous to the formula for computing the *t*-score for a sample mean from a single population, discussed in Chapter 9.

$$t(\overline{X}_A - \overline{X}_B) = \frac{(\overline{X}_A - \overline{X}_B) - (\mu_A - \mu_B)}{SE}.$$

The two terms in the numerator can be determined in a straightforward manner.

- $(\overline{X}_A - \overline{X}_B)$ can be calculated from the given statistics:

$$(\overline{X}_A - \overline{X}_B) = 8.75 - 7.97$$
$$= .78.$$

- The value of $\mu_A - \mu_B$ is the maximum allowed under the null hypothesis, which is 0.

To determine SE, first find the pooled variance, s_P^2. Remember that this must be done whenever the *t*-distribution model is used for a difference of population means.

Start by substituting known quantities into the formula:

$$s_P^2 = \frac{(n_A - 1) \cdot s_A^2 + (n_B - 1) \cdot s_B^2}{n_A + n_B - 2}$$

$$= \frac{(10 - 1) \cdot (1.15)^2 + (14 - 1) \cdot (.95)^2}{10 + 14 - 2}$$

$$= \frac{(9) \cdot 1.3225 + (13) \cdot .9025}{22}$$

$$\approx 1.07.$$

Next, substitute this value of s_P^2 for s_A^2 and s_B^2 in the formula for the standard error, SE:

$$SE \approx \sqrt{\frac{s_A^2}{n_A} + \frac{s_B^2}{n_B}}$$

$$= \sqrt{\frac{1.07}{10} + \frac{1.07}{14}}$$

$$\approx .43.$$

Everything needed to compute the test statistic $t(\overline{X}_A - \overline{X}_B)$ is now known.

$$t(\overline{X}_A - \overline{X}_B) = \frac{(\overline{X}_A - \overline{X}_B) - (\mu_A - \mu_B)}{SE}$$

$$= \frac{(.78) - (0)}{.43}$$

$$\approx 1.81.$$

Step 5   State the conclusion.

$t(\overline{X}_A - \overline{X}_B)$ is not quite in the rejection region.

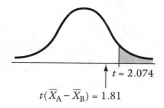

$$t(\overline{X}_A - \overline{X}_B) = 1.81$$

Therefore, **reserve judgment.** Although the mean price for this drug is higher at the sampled private stores than at the sampled chain stores, it is not enough higher to support the alternative hypothesis (at the 2.5% level of significance) that it is higher in the population of private stores, as well.   •

Many statistical software programs can be used to construct confidence intervals for differences of population means or to conduct hypothesis tests of differences of population means. The next two examples illustrate Minitab's TWOSAMPLE macro command.

E x a m p l e
3

Researchers set out to compare the amounts of money spent at Christmastime at two competing malls, A and B. Fourteen shoppers at mall A are chosen at random. These shoppers say they spent the following amounts of money.

Mall A

$90.81	90.00	30.00	20.56	78.50	16.00	79.13
494.00	88.13	149.00	6.50	90.79	72.75	250.00

Sixteen shoppers at mall B are chosen at random. They spent the following amounts of money.

Mall B

$468.00	32.56	4.19	97.25	11.25	12.56	6.00	3.00
14.50	11.13	21.19	11.00	69.94	75.69	31.02	22.13

Determine a 90% confidence interval for the difference in mean spending at the two malls.

Solution   Because the samples are small ($n_A = 14$ and $n_B = 16$) and taken from populations with unknown standard deviations, the model for the sampling distribution of all the possible values of $\overline{X}_A - \overline{X}_B$ is the *t*-distribution. Once again, one must pool the two estimated variances, s_A^2 and s_B^2, for use in the calculation of *SE*. It will not be necessary to do this by hand as in the last example. Use Minitab's "Pooled" subcommand instead.

As usual, the first step is to enter the data into columns C1 and C2 of a Minitab worksheet.

	C1	C2	C3	C4	C5
↓	Mall A	Mall B			
1	90.81	468.00			
2	494.00	14.50			
3	90.00	32.56			

Untitled Worksheet

Minitab's TWOSAMPLE command always uses the *t*-distribution model, regardless of the sample sizes.

Next, obtain a Session window. After the **MTB** > prompt, type the command TWOSAMPLE followed by the percentage of confidence for the interval and the columns where the data are stored. This line ends with a semicolon, which activates the **SUBC** > prompt when the RETURN key is pressed. The subcommand "Pooled" is then entered on the second line and ended with a period. When the RETURN key is pressed, the output appears on the screen (Figure 10-3).

Figure 10-3 reveals that a 90% confidence interval for the difference in mean spending at the two malls is

$$[-\$19.00, \$130.00].$$

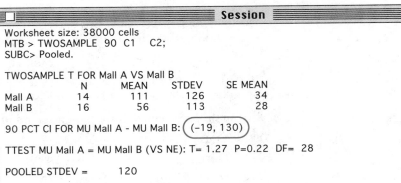

```
Worksheet size: 38000 cells
MTB > TWOSAMPLE  90  C1    C2;
SUBC> Pooled.

TWOSAMPLE T FOR Mall A VS Mall B
               N      MEAN      STDEV      SE MEAN
Mall A        14       111       126         34
Mall B        16        56       113         28

90 PCT CI FOR MU Mall A - MU Mall B: (-19, 130)

TTEST MU Mall A = MU Mall B (VS NE): T= 1.27  P=0.22  DF= 28

POOLED STDEV =        120
```
Session

Figure 10-3 Minitab's TWOSAMPLE command with "Pooled" subcommand.

This is a 90% confidence interval for

$$\text{Mean of mall A} - \text{mean of mall B}.$$

The negative left end point, $-\$19.00$, indicates the possibility that the spending at mall B is greater than that at mall A. Because the end points have different signs, zero must be in the confidence interval. Therefore, the researchers cannot conclude (at the 90% confidence level) that the mean spending at the two malls is different.

Also, notice that the TWOSAMPLE command automatically includes the necessary information to conduct a *t*-distribution model hypothesis test for

H_0: $\mu_A - \mu_B = 0$.   (There is no difference in mean spending at the two malls.)

H_A: $\mu_A - \mu_B \neq 0$.   (There is a difference in mean spending at the two malls.)

Assume, for example, a 10% level of significance. Since $P = .22$, then $P > \alpha$, and judgment of the hypothesis that there is no difference in mean spending at the two malls must be reserved. ●

E x a m p l e
4

Refer to the cholesterol screening device on page 42 in Chapter 2. The data from the two groups are repeated here for convenience.

Group H (identified by screening device)				Group R (chosen at random)			
139	157	160	201	160	198	140	130
122	188	194	240	138	85	98	108
200	180	230	179	126	208	178	170
194	190	180	188	175	208	161	195
179	194	140	136	130	120	110	128
180	193	159	152	223	98	163	165
176	157	178	206	150	186	248	83
160	120	176	192	186	161	165	188
203	210	176	188	163	145	126	195
194	189	163	196	210	123	116	164

Determine whether statistical evidence exists that the screening device identifies people with higher cholesterol counts. Use Minitab to perform a hypothesis test at the 5% level of significance.

If the screening device does its intended job, the mean cholesterol count of the population it identifies should be greater than the mean cholesterol count of a population chosen at random. That is, it should be true that $\mu_H > \mu_R$.

Solution   Treat the two groups as random samples, H and R, chosen from large populations through the device and in random fashion, respectively. Since statistical evidence can support only the alternative hypothesis, H_A—that the device does its intended job—the null hypothesis, H_0, must be the opposite of that.

H_0: $\mu_H - \mu_R \leq 0$.   ($\mu_H \leq \mu_R$)

H_A: $\mu_H - \mu_R > 0$.   ($\mu_H > \mu_R$)

```
┌──────────────────────────────════ Session ════──────────────────┐
│ ▤ ▢ ══════════════════════════════════════════════════════════  │
│ Worksheet size: 38000 cells                                      │
│ MTB > TWOSAMPLE   C1   C2;                                        │
│ SUBC> Alternative = 1.                                           │
│                                                                  │
│ TWOSAMPLE T FOR C1 VS C2                                         │
│              N      MEAN     STDEV     SE MEAN                    │
│ C1          40      179.0    26.0       4.1                      │
│ C2          40      155.6    39.4       6.2                      │
│                                                                  │
│ 95 PCT CI FOR MU C1 - MU C2:  (8.5, 38.3)                        │
│                                                                  │
│ TTEST MU C1 = MU C2 (VS GT): (T= 3.13)(P=0.0013) DF=  67         │
└──────────────────────────────────────────────────────────────────┘
```

Figure 10-4 Minitab's TWOSAMPLE command with "Alternative" subcommand.

Because this test involves large samples taken from (it is assumed) relatively large populations, the normal distribution model for the sampling distribution of all possible values of $\overline{X}_H - \overline{X}_R$ is the appropriate choice. Minitab's TWOSAMPLE command, however, always uses the t-distribution as a model for the sampling distribution and a complex formula for the degrees of freedom. Since both samples are large, it is safe to assume that the degrees of freedom will be large and the resulting t-distribution model will be close enough to a normal model to be treated as normal.

Accordingly, the first step is to enter the data into columns C1 and C2 of a Minitab worksheet. Then obtain a Session window and type the TWOSAMPLE macro command followed by the location of the data. End the line with a semicolon to allow for a subcommand. Use the "Alternative = 1" subcommand to specify the "greater than" relationship of the alternative hypothesis, H_A.[4] The output in Figure 10-4 should appear on the screen when the (RETURN) key is pressed. Since $P < \alpha$ (that is, since .0013 < .05), **reject H_0 and accept H_A.** At the 5% level of significance there is statistical evidence that this device identifies a population with higher cholesterol counts. ●

Notice that the TWOSAMPLE command printed a 95% confidence interval for the difference of the population means. This happens by default whenever no percentage is specified. Also, Minitab's degrees of freedom calculation is different from that which would be obtained by hand.

Exercises 10.1–10.4

1. What two important assumptions are made about populations when confidence intervals are constructed or hypothesis tests are performed for the difference of those populations' means?

2. Is it possible for both end points of a confidence interval for the difference of two population means to be negative? If so, what is the meaning of the two negatives?

---

[4]Since a normal distribution model would have been used by hand, the "Pooled" subcommand is not used.

3. Specify the conditions under which the t-distribution model is used to construct confidence intervals for the difference of two population means.

4. When the t-distribution model is used for hypothesis testing of the difference of two population means without a statistical software program such as Minitab, how are the degrees of freedom determined?

5. Must standard deviations be pooled when the normal distribution model is used?

6. Explain the meaning of a confidence interval for a difference of population means that has both positive and negative end points.

7. Is the following statement true or false? " 'Before and after' types of comparisons are usually not made using independent sampling."

8. Use the data in Example 1 on page 399 to determine whether statistical evidence exists at the 10% level of significance to support the claim that the older population has a greater mean reflex time than the younger population.

9. Refer to Example 2 on page 400. Estimate the minimum level of significance that would result in the rejection of H_0.

MTB   10. What two pieces of information are always part of the output of Minitab's TWOSAMPLE command?

11. Refer to Example 3 on page 403. Use trial and error to find the maximum percentage confidence interval for which both end points are positive.

12. Work Example 4 on page 405 without using Minitab.

For Exercises 13–23, use the normal distribution model.

GROUP   13. A random sample of 60 young men from a university had a mean height of 66.15 inches with $s = 1.05$ inches. A sample of 35 young women from the same university had a mean height of 59.55 inches with $s = .85$ inches. Determine 50%, 75%, and 95% confidence intervals for the difference in the mean heights of the men and women.

14. Long experience has demonstrated that the standard deviation of hourly wages in a certain type of industry is the same in Chicago as in Los Angeles and that $\sigma = \$2.75$ per hour. A random sample of 30 workers in Chicago revealed a mean wage of $22.75 per hour, and a random sample of 40 workers in Los Angeles revealed a mean wage of $18.55 per hour. Determine an 80% confidence interval for the difference in the mean hourly wages of the two cities' workers in this type of industry.

15. A shipping manager believes that the mean time required for materials to be sent from loading dock A differs from the mean time required for materials to be sent from dock B. A random sample of 35 orders from dock A revealed a mean shipping time of $\overline{X}_A = 5.05$ days with $s_A = 1.00$ day. A random sample of 40 orders from dock B revealed a mean shipping time of $\overline{X}_B = 3.78$ days with $s_B = .85$ day. Perform a hypothesis test to determine whether statistical evidence exists at the 10% level of significance to support the manager's belief.

16. Refer to Exercise 15. Perform a hypothesis test at the 7.5% level of significance to determine whether statistical evidence exists to support the claim that the mean shipping time is longer from dock A than from dock B.

GROUP 17. The owner of a lumber yard is about to purchase one of two freight cars of framing lumber. A random sample of size $n_A = n_B = 30$ boards from each car is examined. The mean number of knots per board in the sample taken from the first car is $\overline{X}_A = 6.50$ with $s_A = 2.55$. The mean number of knots per board in the sample from the other car is $\overline{X}_B = 5.00$ with $s_B = 3.00$.

   a. Construct a 90% confidence interval for the difference of the mean numbers of knots per board in the lumber from these two freight cars.

   b. Perform a hypothesis test at the 5% level of significance to determine whether statistical evidence exists that the mean number of knots per board from these two freight cars differ.

   c. Perform a hypothesis test at the 1% level of significance to determine whether statistical evidence exists that the mean number of knots per board in boards from car A is greater than that in boards from car B.

The sample means given in Exercises 18–23 were obtained from an earnings survey in the October 1993 issue of RN *magazine (pages 54–61), titled "Double-Digit Raises Disappear." Sample sizes and standard deviations were not published and are this author's best estimates.*

Nurses' salaries are related to region, size of hospital, years of experience, day shift vs. night shift, profit vs. nonprofit hospital, among other factors. (*Anne Sherman*)

18. The mean hourly pay for full-time RNs in New England was $19.20; that for RNs in the Southwest was $17.00. Perform a hypothesis test to determine whether these figures provide statistical evidence at the 5% level of significance that RNs earned more in New England than in the Southwest. Use equal sample sizes of $n = 350$ and equal standard deviations of $s = \$3.00$.

19. The mean hourly pay of full-time RNs with 11 to 15 years of experience was $18.80, while that of those with 3 to 5 years of experience was $16.30. Determine a 90% confidence interval for the difference in mean pay for RNs in these two populations. Use $n_{11-15} = 800$, $n_{3-5} = 850$, $s_{11-15} = \$2.50$, and $s_{3-5} = \$2.70$.

20. The mean hourly pay of full-time RNs working during the day was $18.10, and that of those working in the evening was $17.60. Determine a 95%

confidence interval for the difference in mean pay for RNs in these two populations. Use equal sample sizes of $n = 1000$ and equal standard deviations of $s = \$2.75$.

21. The mean hourly pay of part-time RNs working in hospitals with fewer than 100 beds was $16.10, and that of those working in hospitals with 500 or more beds was $19.10. Construct an 85% confidence interval for the difference in mean pay for RNs in these two populations. Use $n_{100} = 500$, $n_{500} = 1000$, $s_{100} = \$2.85$, and $s_{500} = \$2.65$.

22. The mean hourly pay of full-time RNs working in private nonprofit hospitals was $18.10, and that of those working in for-profit hospitals was $17.30. Perform a hypothesis test at the 10% level of significance to determine whether these data provide statistical evidence that full-time RNs earned less in for-profit hospitals. Use equal sample sizes of $n = 800$ and equal standard deviations of $s = \$3.00$.

23. The mean hourly pay of part-time RNs working in suburban hospitals was $19.30, and that of those working in urban hospitals was $18.70. Determine an 80% confidence interval for the difference in mean pay for RNs in these two populations. Use equal sample sizes of $n = 1200$ and equal sample standard deviations of $s = \$3.00$.

24. Refer to Exercises 18–23. If the actual sample standard deviations were much smaller, would that fact alter the results of the hypothesis tests that were performed?

25. Refer to Exercises 18–23. If the actual samples were much larger, how would that fact affect the confidence intervals?

For Exercises 26–30, use the t-distribution model with pooled standard deviation.[5]

26. An insurance investigator wants to determine whether the mean numbers of moving violations received by young men and young women are the same. A random sample of 17 men's records yielded a mean number of moving violations in the last 5 years of 3.7, with $s_M = 2.05$. A random sample of 15 women's records yielded a mean number of moving violations in the last 5 years of 2.1, with $s_W = 1.55$.

    Perform a hypothesis test at the 10% level of significance to determine whether statistical evidence exists that the mean numbers of violations differ.

27. In a trial of a new manufacturing process for tires, a random sample of 16 tires lasted the following numbers of miles.

34,500	40,500	39,000	41,500
43,000	38,500	39,000	42,500
42,000	38,500	37,500	39,000
38,500	46,500	38,000	37,500

---

[5] Use of the degrees of freedom formula will occasionally necessitate consulting the *t*-distribution table for 30 or more degrees of freedom. In that case, round the degrees of freedom to the nearest entry in the table.

A random sample of 16 tires manufactured with the old process lasted the following numbers of miles.

29,700	36,500	40,000	29,500
33,600	45,000	40,000	31,500
38,900	29,700	36,000	28,500
29,000	28,700	30,500	31,000

Determine whether support exists for the claim that this new process changes expected tire life. Perform a hypothesis test at the 10% level of significance.

28. A consumer group randomly sampled 20 of an expensive brand of light bulbs and 25 of a generic brand of light bulbs to determine which brand has a longer life. The results follow.

Expensive Brand: Number of Hours Burned				**Generic Brand:** Number of Hours Burned			
1500	1450	1376	1510	1400	1460	1510	1360
1550	1500	1450	1490	1510	1475	1475	1480
1490	1560	1500	1495	1610	1440	1420	1520
1480	1520	1500	1420	1475	1560	1480	1420
1550	1500	1430	1480	1400	1350	1450	1380
				1475	1500	1400	1320
				1430			

Determine whether there is statistical evidence for the following claims.

a. The expensive brand and the generic brand have different mean life expectancies. Use $\alpha = .05$.

b. The expensive brand burns longer than the generic brand. Use $\alpha = .01$.

c. Use Appendix A-4 to find the greatest approximate value of α that would reverse the decision in Part a.

29. Some educators claim that SAT scores have been changing over the years. A random sample of $n_A = 25$ SAT scores from the class of 1990 and a random sample of $n_B = 30$ SAT scores from the class of 1993 were chosen from a large pool of scores. They follow.

Class of 1990			Class of 1993		
660	800	770	740	510	480
680	560	680	520	540	660
780	740	600	780	640	580
610	800	800	760	690	770
640	480	360	780	500	480
680	550	760	580	600	490
800	660	720	780	690	560
610	760	760	560	680	780
	660		740	640	660
			600	770	550

Construct 50%, 75%, 80%, 90%, 95%, and 99% confidence intervals for the difference in the mean SAT scores of the classes of 1990 and 1993. Interpret the intervals obtained.

30. Example 8 of Chapter 9 (page 383) presented a random sample of $n_{NL} = 20$ championship batting averages in the National League in the years 1876 through 1992. They are repeated here.

Year	Player	Team	Batting Average
1975	Bill Madlock	Chicago Cubs	.354
1885	Roger Connor	N.Y. Giants	.371
1968	Pete Rose	Cincinnati Reds	.335
1920	Rogers Hornsby	St. Louis Cardinals	.370
1955	Richie Ashburn	Philadelphia Phillies	.338
1989	Tony Gwynn	San Diego Padres	.336
1914	Jake Daubert	Brooklyn Dodgers	.329
1936	Paul Waner	Pittsburgh Pirates	.373
1887	Cap Anson	Chicago Cubs	.421
1988	Tony Gwynn	San Diego Padres	.313
1963	Tommy Davis	L.A. Dodgers	.326
1946	Stan Musial	St. Louis Cardinals	.365
1907	Honus Wagner	Pittsburgh Pirates	.350
1924	Rogers Hornsby	St. Louis Cardinals	.424
1952	Stan Musial	St. Louis Cardinals	.336
1981	Bill Madlock	Pittsburgh Pirates	.341
1900	Honus Wagner	Pittsburgh Pirates	.380
1959	Hank Aaron	Milwaukee Braves	.355
1930	Bill Terry	N.Y. Giants	.401
1967	Roberto Clemente	Pittsburgh Pirates	.357

A random sample of $n_{AL} = 15$ championship batting averages in the American League during this same time period follows.

Year	Player	Team	Batting Average
1930	Al Simmons	Philadelphia A's	.381
1985	Wade Boggs	Boston Red Sox	.368
1943	Luke Appling	Chicago White Sox	.328
1911	Ty Cobb	Detroit Tigers	.420
1973	Rod Carew	Minnesota Twins	.350
1961	Norm Cash	Detroit Tigers	.361
1964	Tony Oliva	Minnesota Twins	.323
1907	Ty Cobb	Detroit Tigers	.350
1990	George Brett	Kansas City Royals	.329
1980	George Brett	Kansas City Royals	.390
1933	Jimmie Foxx	Philadelphia A's	.356
1955	Al Kaline	Detroit Tigers	.340
1944	Lou Boudreau	Cleveland Indians	.327
1969	Rod Carew	Minnesota Twins	.332
1958	Ted Williams	Boston Red Sox	.328

Determine whether these data represent statistical evidence, at the 10% level of significance, that the mean batting average of National League champions is greater than the mean batting average of American League champions.

10.5 DIFFERENCES BETWEEN POPULATION PROPORTIONS

It is often important to know whether the difference between two sample proportions is due to chance or can be attributed to the fact that the corresponding population proportions, which the samples estimate, are actually different. For instance, a politician might want to know whether the apparent difference between her voter appeal and a rival's is real or due only to the chance selection of those who were polled. Public health officials might want to know whether the percentage incidence of flu among those who received flu shots matches or differs from the incidence among those who did not obtain shots. An advertising executive might want to know whether the percentages of people who remember a particular ad during certain types of television programs differ.

It is possible to construct a confidence interval for the difference of two population proportions. If both end points of such a confidence interval have the same sign, then the two population proportions are considered to be different at that level of confidence. This assumption should seem reasonable, because if 0 is not in the confidence interval, there must be some (nonzero) difference in the population proportions. However, if the end points have different signs, then 0 is in the confidence interval and one may conclude that the two population proportions are not different at that level of confidence.

Two assumptions will be made when forming confidence intervals for the difference of population proportions.

Two necessary assumptions for the construction of a confidence interval for the difference of two population proportions:

a. The data obtained are simple random samples.
b. The normal distribution model is used for the sampling distribution of the difference of the sample proportions.

The next example illustrates the method used to form such an interval.

E x a m p l e
5

An ad agency is interested in knowing whether people remember certain types of ads more during football games or during movies. After running a test ad during a football game, the agency randomly contacted 900 people who had watched the game and asked whether they remembered the ad; 507 responded yes. The agency ran the same ad on a channel where a movie was playing. It randomly contacted 800 people who had watched the movie, and 405 of those people remembered the ad.

Form a 95% confidence interval for the difference in population proportions.

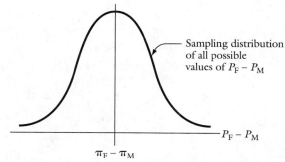

Figure 10-5 The normal distribution model for the construction of a confidence interval for the difference of population proportions.

Solution Consider Figure 10-5. Define these population proportions:

π_F = Actual percentage of population who will remember the ad when it is shown during a football game.

π_M = Actual percentage of population who will remember the ad when it is shown during a movie.

Also, let

n_F = Number surveyed who watched the ad during a football game
= 900

and

n_M = Number surveyed who watched the ad during a movie
= 800.

Define the sample proportions:

P_F = Percentage in random sample who remembered the ad when it was shown during a football game.

P_M = Percentage in the other random sample who remembered the ad when it was shown during a movie.

The sample statistics P_F and P_N can be computed as follows:

$$P_F = \frac{507}{900} \qquad P_M = \frac{405}{800}$$
$$\approx .563. \qquad\quad \approx .506.$$

In a manner analogous to previous examples, the end points of the required confidence interval can be found by evaluating

$$(P_F - P_N) \pm Z(P_F - P_M) \cdot SE.$$

In this expression, $Z(P_F - P_M)$ represents the Z-scores enclosing the middle 95% of the normal distribution model, which is $z = \pm 1.96$. The formula to obtain the standard error is

$$SE = \sqrt{\frac{P_F \cdot (1 - P_F)}{n_F - 1} + \frac{P_M \cdot (1 - P_M)}{n_M - 1}}$$

$$= \sqrt{\frac{.563 \cdot (1 - .563)}{900 - 1} + \frac{.506 \cdot (1 - .506)}{800 - 1}}$$

$$= \sqrt{\frac{.563 \cdot (.437)}{899} + \frac{.506 \cdot (.494)}{799}}$$

$$\approx .024.$$

Everything needed to obtain the end points of the confidence interval for $\pi_F - \pi_N$ is now known.

$$\pi_F - \pi_N = (P_F - P_N) \pm Z(P_F - P_M) \cdot SE$$
$$= (.563 - .506) \pm 1.96 \cdot (.024)$$
$$= .057 \pm .04704$$
$$\approx [.010, .1040].$$

Because 0 is not in this interval, one may be 95% confident that the two population proportions differ and that the difference ranges from 1.0% to 10.4%.   ●

Exercises 10.5

1. A new insurance benefit package was proposed to the employees of a large company. A 90% confidence interval for the difference in the percentages of employees who prefer the new package and those who prefer the old one, $(\pi_{new} - \pi_{old})$, is $[-.08, .34]$.

   a. What conclusions, if any, should management draw from this confidence interval?

   b. In order for both end points of this confidence interval to be positive, must the percentage of confidence be raised or lowered?

   c. Could increasing the sample size sufficiently produce a 90% confidence interval in which both end points were positive? Explain.

2. A politician surveyed 800 city voters and found that 650 approved of the job he was doing. In the suburbs, however, only 570 of 950 people surveyed approved of him. Find a 90% confidence interval for the difference in approval ratings between cities and suburbs.

GROUP
3. Following is a random sample of numbers of promotions given to male and female workers in a 5-year period at a certain company.

	Male	Female
Promotions	345	150
No promotions	420	210

Construct 60%, 80%, and 95% confidence intervals for the difference in percentages of male and female promotions at this company.

GROUP

4. Researchers compared an experimental drug with the standard drug for the treatment of a certain disease. Of 300 patients who were given the experimental drug, 225 recovered. Of 500 patients given the standard drug, 371 recovered.

   a. Define each population proportion.

   b. Construct a 75% confidence interval for the difference between the population proportions. At this level of confidence, could one conclude that the experimental drug is more effective or less effective than the standard drug?

* 5. Refer to Exercise 4. Find the maximum percentage for a confidence interval that does not contain 0. Interpret the meaning of such a confidence interval.

6. Researchers tested a new computer chip in a guided missile. A random sample of 150 missiles with the new chip hit the target 137 times, and a random sample of 200 missiles with the old chip hit the target 169 times. Determine whether these data constitute statistical evidence that the new chip makes a difference in the accuracy of the missile. Perform a hypothesis test at the 10% level of significance.

10.6 MATCHED-PAIR SAMPLING

The aim of sampling by matched pairs is to transform two-population data into data on which one-population methods may be used.

As an alternative to independent sampling of two populations, matched-pair sampling is appropriate when "twins" from the two populations can be found. The set of matched-pair differences obtained from the twins can then be treated as a one sample problem by methods examined in the last two chapters.

**E x a m p l e
6**

Pharmaceutical researchers are testing a new type of acne medicine to see whether it has a shorter reaction time than the old medicine. Twenty pairs of patients who are the same age and have similar skin conditions and lifestyles were formed. One member of each pair was given the new medicine, and the other member was given the old medicine. The following array shows the results. Reaction times are in hours.

S t a t   F a c t

A recent study found that skin disorders affect 50% of Americans between the ages of 25 and 49 but only 25% of teenagers.

Reaction Times to Acne Medicine

Pair No.	Old Drug	New Drug	Pair No.	Old Drug	New Drug
1	20	18	11	25	30
2	22	22	12	27	22
3	19	27	13	25	25
4	25	19	14	30	22
5	26	17	15	15	17
6	21	20	16	28	23
7	18	22	17	30	28
8	24	30	18	28	22
9	24	16	19	26	28
10	22	19	20	25	14

Use the method of matched-pair differences to construct a 95% confidence interval for the difference in reaction times between the old and new drugs.

Solution   A few calculations will be necessary. First combine the two columns of old and new reaction times, by subtraction, into a single set of scores.

Matched-Pair Differences for Reaction Times to Acne Medicine

	Reaction Time		Matched-Pair Difference:
Pair No.	*Old Drug*	*New Drug*	Old − New
1	20	18	2
2	22	22	0
3	19	27	−8
4	25	19	6
5	26	17	9
6	21	20	1
7	18	22	−4
8	24	30	−6
9	24	16	8
10	22	19	3
11	25	30	−5
12	27	22	5
13	25	25	0
14	30	22	8
15	15	17	−2
16	28	23	5
17	30	28	2
18	28	22	6
19	26	28	−2
20	25	14	11

Then treat this single set of scores (circled) as a simple random sample taken from a single population of all possible matched-pair differences. You may construct confidence intervals and perform hypothesis tests on these differences by methods analogous to those already described in Chapters 8 and 9.

Location of the End Points of a Confidence Interval for the Mean of All Possible Matched-Pair Differences, μ_D

$$\text{Left end point} = \overline{D} - Z(\overline{D}) \cdot SE,$$
$$\text{Right end point} = \overline{D} + Z(\overline{D}) \cdot SE. \qquad \textbf{(10-4)}$$

(Place end points at $\overline{D} \pm Z(\overline{D}) \cdot SE$ or $\overline{D} \pm t(\overline{D}) \cdot SE$.)

In Formula 10-4, $\overline{D}$ represents the mean of this sample of matched-pair differences, $Z(\overline{D})$ represents the Z-scores that enclose the required percentage of the normal distribution model, and SE is the standard error of the mean matched-pair differences. If the *t*-distribution model is used, $t(\overline{D})$ represents the *t*-scores that enclose the required percentage of that model.

In this case, the number of matched-pair differences is small ($n < 30$), and it will be assumed that the population of all possible matched-pair differences is normally distributed. Because the standard deviation of the population is unknown, use the *t*-distribution as the model for the sampling distribution.[6]

The next step is to determine $\overline{D}$, $t(\overline{D})$, and the standard error, *SE*, for the data in the "Matched-Pair Differences" column. The 20 circled scores may be entered in a TI-82 graphics calculator. The final screen display appears as follows:

```
        1-Var Stats
 D ────►x̄=1.95
        Σx=39
        Σx²=619
 s ────►Sx=5.345681976
        σx=5.210326285
      ↓ n=20
```

From these statistics the standard error, *SE*, can be approximated as

$$SE \approx \frac{s}{\sqrt{n}}$$

$$\approx \frac{5.35}{\sqrt{20}}$$

$$\approx 1.20.$$

The degrees of freedom for this example are equal to the number of matched pairs minus one, or $n - 1$. Thus, the value of $t(\overline{D})$ that encloses the middle 95% of the *t*-distribution model is found at the intersection of the $20 - 1 = 19$ degrees of freedom row and the $\alpha = \dfrac{1 - .95}{2} = .025$ column of Appendix A-4. The corresponding *t*-scores are $t = \pm 2.093$ (Figure 10-6).

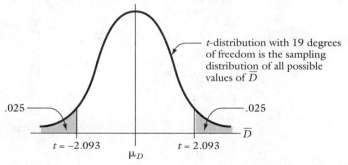

Figure 10-6 The *t*-distribution model for the construction of a confidence interval for the mean of all possible matched-pair differences, μ_D.

[6]For 30 or more matched pair differences, use the normal distribution model.

All of the quantities needed for construction of the 95% confidence interval are now known.

$$\text{End points of confidence interval} = \overline{D} \pm t(\overline{D}) \cdot SE$$
$$= 1.95 \pm 2.093 \cdot (1.20)$$
$$= 1.95 \pm 2.5116.$$

The confidence interval determined in this way is $[-.56, 4.46]$. Because 0 is in this interval, one may conclude that there is no difference in reaction times between the old and new drugs at this confidence level. •

Alternative Solution to Example 6

Minitab may be used to carry out all of the calculations in the preceding example. Enter the old and new reaction times in columns C1 and C2, respectively, of a Minitab worksheet. Next, obtain a Session window and use the macro command SUBTRACT C2 C1 C3. This command instructs Minitab to subtract the value of each cell in column C2 from the corresponding one in column C1, then place the difference in column C3. Figure 10-7 illustrates the command line and the new worksheet that appears when the [RETURN] key is pressed.

Now the data Minitab has placed in column C3 can be treated as a simple random sample taken from a single population. The macro command TINTERVAL, followed by the percentage of confidence and the location of the data should be typed in the active Session window. Figure 10-8 shows the command line and the required confidence interval. Except for very slight rounding differences, these results match the ones obtained by hand.

```
                                      Session
Worksheet size: 38000 cells
MTB > SUBTRACT   C2   C1   C3
```

	C1	C2	C3	C4	C
↓					
1	20	18	2		
2	22	22	0		
3	19	27	-8		
4	25	19	6		

Figure 10-7 Minitab's SUBTRACT command.

```
                                      Session
Worksheet size: 38000 cells
MTB > TINTERVAL   95   C3

           N      MEAN    STDEV    SE MEAN     95.0 PERCENT C.I.
C3         20     1.95    5.35     1.20      (  -0.55,    4.45)

MTB >
```

Figure 10-8 •

E x a m p l e
7

For the data in Example 6, perform a hypothesis test at the 10% level of significance to determine whether statistical evidence exists to support the claim that the difference in reaction times is less than 4 hours. Assume that the population of all matched-pair differences is normally distributed.

Solution   Let μ stand for the population mean of all possible matched-pair differences. Since statistical evidence can support only the alternative hypothesis, H_A, this must be the hypothesis that $\mu < 4$.

Step 1   State the hypotheses.

$$H_0: \quad \mu \geq 4.$$
$$H_A: \quad \mu < 4.$$

Step 2   Decide on a model for the sampling distribution.

For the reasons stated in Example 6 on page 415, use the t-distribution with 19 degrees of freedom as a model for the sampling distribution of all the possible values of $\overline{D}$.

Step 3   Determine the end point of the rejection region and state the decision rule.

According to Appendix A-4, the intersection of the .1 column and the 19 degrees of freedom row indicates the following location for the rejection region.

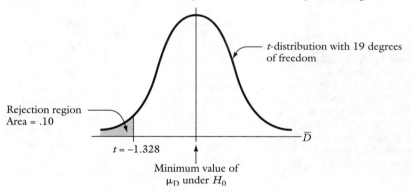

Accordingly, the decision rule is:

Reject H_0 and accept H_A if $t(\overline{D}) < -1.328$. Otherwise, reserve judgment.

Step 4   Compute the test statistic.

The test statistic is $t(\overline{D})$ and is found with this familiar formula:

$$t(\overline{D}) = \frac{\overline{D} - \mu}{SE}.$$

Using the values from Example 6, and the minimum value of μ under H_0,

$$t(\overline{D}) = \frac{1.95 - 4}{1.20}$$

$$\approx -1.71.$$

Step 5   State the conclusion.

Since $-1.71 < -1.328$, **reject H_0 and accept H_A.** At this level of significance, statistical evidence exists to support the claim that the difference in reaction times is less than 4 hours.   •

Exercises 10.6

1. Name some situations in which matched-pair sampling would be preferred over independent sampling.

2. Explain what is meant by the notation μ_D. What do you think is the meaning of $\mu_{\overline{D}}$?

3. Explain why Minitab does not need a separate macro command for use with matched-pair sampling.

GROUP   4. Construct 50%, 75%, and 99% confidence intervals for Example 6. By trial and error, try to find the greatest percentage for the confidence interval so that both end points are positive.

In Exercises 5–10 continue to assume that all populations are normally distributed.

5. Annual tuitions for residents and nonresidents at 19 public colleges and universities in 1993, chosen at random, follow.

S t a t   F a c t

Between 1990 and 1993, the mean cost to attend a private college in the United States rose 22%, while the cost of attending a 4-year public institution increased by 37%. About 80% of higher-education students attend public colleges and universities.

		Tuition	
Institution	Location	*Nonresident*	*Resident*
Auburn Univ.	Auburn, AL	$5850	$1950
Fort Lewis College	Durango, CO	6059	1705
State Univ. of N.Y.	Albany, NY	6783	2883
South Dakota Univ.	Vermillion, SD	4027	2190
Norfolk State Univ.	Norfolk, VA	5760	2530
Univ. of Illinois	Champaign, IL	7580	3328
Grambling State Univ.	Grambling, LA	3588	2038
Kansas State Univ.	Manhattan, KS	6493	1957
Rutgers Univ.	Camden, NJ	7227	3857
Salisbury State Univ.	Salisbury, MD	5694	3026
Univ. of Missouri	Kansas City, MO	8582	3152
Mesa State College	Grand Junction, CO	4442	1620
Indiana State Univ.	Terre Haute, IN	5960	2452
New Mexico State Univ.	Las Cruces, NM	5686	1756
Eastern Michigan Univ.	Ypsilanti, MI	6114	2529
Eastern Kentucky Univ.	Richmond, KY	4220	1540
Keene State College	Keene, NH	7345	2865
Kent State Univ.	Kent, OH	7192	3596
Univ. of Maine	Fort Kent, ME	3690	2460

Use the method of matched-pair differences to determine 80% and 90% confidence intervals for the mean difference in tuitions for nonresidents and residents.

6. Refer to Exercise 5. Determine whether statistical evidence exists to support

the claim that the mean difference between nonresident and resident tuitions is greater than $2900. Perform a hypothesis test at the 5% level of significance.

7. Twelve states were chosen at random, and in each the annual salaries of the governor and the chief justice of the state supreme court were obtained.

	Governor's Salary	Chief Justice's Salary
MA	$ 75,000	$ 90,450
IL	100,681	100,681
NY	130,000	120,000
CA	114,000	121,207
WY	70,000	72,000
OH	105,000	101,150
VA	110,000	102,700
CO	70,000	84,000
FL	103,909	100,444
DE	95,000	106,700
TN	85,000	93,540
WI	92,283	91,252

Source: *The 1994 Information Please Almanac*, pp. 789–790.

Use the method of matched-pair differences to determine whether statistical evidence exists to support the claim that the mean salary of all 50 governors is different from the mean salary of all 50 chief justices. Perform a hypothesis test at the 5% level of significance.

GROUP

8. A new gasoline additive was tested. Forty randomly chosen drivers recorded their gasoline mileages for 1 week with the additive and for 1 week without it. The results follow.

Driver Number	With Additive (mpg)	Without Additive (mpg)	Driver Number	With Additive (mpg)	Without Additive (mpg)
1	28.7	26.2	21	22.2	20.9
2	22.6	20.0	22	29.6	24.5
3	18.5	18.0	23	30.0	27.9
4	27.9	21.6	24	32.5	31.6
5	32.1	29.0	25	19.8	21.5
6	25.5	25.0	26	24.5	22.3
7	29.9	19.9	27	29.9	28.3
8	30.2	26.6	28	26.5	25.7
9	29.5	26.8	29	31.5	29.5
10	19.8	17.0	30	32.9	29.8
11	18.8	18.8	31	27.9	27.0
12	21.2	19.6	32	25.5	22.4
13	28.9	28.5	33	17.9	17.9
14	27.6	25.0	34	27.8	24.9
15	28.9	27.6	35	27.8	28.0
16	33.0	29.5	36	24.6	22.7
17	14.5	13.0	37	28.7	26.9
18	22.2	20.0	38	19.9	18.0
19	26.7	26.5	39	21.9	21.2
20	30.0	28.7	40	28.9	27.1

Use the method of matched-pair differences to form 50%, 75%, 90%, 95%, and 99% confidence intervals for the difference in mean gas mileage to be expected from the use of this gasoline additive.

9. Exercise 7 (in Exercises 8.8) on page 351 showed a random sample of $n = 11$ cities and the percentages of people in those cities who approved of the performance of the U.S. Postal Service. Added to that information here are the percentages of overnight first-class mail delivered on time in those cities.

City	Percentage of People Who Approve	Percentage of Overnight First-Class Mail Delivered on Time
Albany, NY	94	82
Boston, MA	90	80
Chicago, IL	72	77
Detroit, MI	87	77
Houston, TX	87	82
Miami, FL	86	75
New Orleans, LA	89	85
New York, NY	74	73
Omaha, NE	91	88
Salt Lake City, UT	93	89
San Francisco, CA	85	81

Use the method of matched-pair differences to determine whether statistical evidence exists of a difference between approval ratings and percentages of mail delivered on time. Perform a hypothesis test at the 20% level of significance.

10. Refer to Exercise 9. If statistical evidence exists to support the alternative hypothesis, explain what that might mean.

10.7 THE MANN-WHITNEY CONFIDENCE INTERVAL

The following example shows how to construct a confidence interval for the difference of two population's *medians,* utilizing a nonparametric approach.

Definition ▶   A **nonparametric approach** is one that does not assume (or require) a normally distributed population.

A nonparametric approach is often useful for dealing with distributions that contain outliers.

The Mann-Whitney confidence interval is appropriate when one or both of the two basic assumptions made at the start of this chapter do not seem justified—that is, when the underlying populations are not normally distributed or when the two populations do not have approximately equal standard deviations.

E x a m p l e 8

A financial speculator wants to compare prices of rare coins to those of rare stamps. Ten prices paid for rare coins and ten prices paid for rare stamps are selected at random from recent auction catalogs.

Coins

$ 55	125	210	75	15
1375	175	20	550	865

Stamps

$30	10	210	90	100
12	25	100	75	40

Form a 95% confidence interval for the difference in the average prices paid for rare coins and for stamps.

Solution   The word "average" often implies the mean. However, a confidence interval for the difference of two population means requires that the two populations (1) be normally distributed and (2) have approximately equal standard deviations.

The large range of coin prices ($1360) and the much smaller range of stamp prices ($200) imply that the second condition may not be satisfied. This conclusion is reinforced if the two samples are used to estimate the standard deviations of prices paid for coins and stamps.

Coins	Stamps
1-Var Stats	1-Var Stats
$\bar{x}=346.5$	$\bar{x}=69.2$
$\Sigma x=3465$	$\Sigma x=692$
$\Sigma x^2=3040975$	$\Sigma x^2=81194$
(Sx=452.1986412)	(Sx=60.8345662)
$\sigma x=428.9932983$	$\sigma x=57.7127369$
↓ n=10	↓ n=10

These statistics suggest that the population standard deviations are very different. In such a situation, the Mann-Whitney nonparametric approach is appropriate, and it will determine a 95% confidence interval for the difference in population medians instead of population means.[7]

[7]For another alternative, see Example 9 on page 427.

Step 1   Arrange the data in a square array,[8] from least to greatest.

COINS

		15	20	55	75	125	175	210	550	865	1,375
S	10										
T	12										
A	25										
M	30										
P	40										
S	75										
	90										
	100										
	100										
	210										

Step 2   In each cell, enter the difference between the number at the top of the column and the number at the left side of the row.

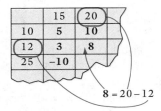

$$8 = 20 - 12$$

Such differences may be calculated by hand or software may be used to speed up the process. In some popular spreadsheet programs, one can accomplish all of these calculations by executing a single array formula as shown in Figure 10-9.

Array formula

$\{=B1:K1-A2:A11\}$

Worksheet 1

	A	B	C	D	E	F	G	H	I	J	K
1		15	20	55	75	125	175	210	550	865	1375
2	10	5	10	45	65	115	165	200	540	855	1365
3	12	3	8	43	63	113	163	198	538	853	1363
4	25	−10	−5	30	50	100	150	185	525	840	1350
5	30	−15	−10	25	45	95	145	180	520	835	1345
6	40	−25	−20	15	35	85	135	170	510	825	1335
7	75	−60	−55	−20	0	50	100	135	475	790	1300
8	90	−75	−70	−35	−15	35	85	120	460	775	1285
9	100	−85	−80	−45	−25	25	75	110	450	765	1275
10	100	−85	−80	−45	−25	25	75	110	450	765	1275
11	210	−195	−190	−155	−135	−85	−35	0	340	655	1165

In other spreadsheet programs, the array formula may differ slightly from that shown.

Figure 10-9 Use of a spreadsheet program to determine cell values.[9]

[8]Although a Mann-Whitney confidence interval can be constructed when the two sample sizes are different and with different percentages of confidence, such applications will not be part of this text.

[9]Microsoft Excel 4.0 for the Macintosh is illustrated here. Some other spreadsheet programs for Macintosh computers and PCs have similar capabilities.

Step 3   Consult Table 1.[10]

Table 1   Determination of End Points of a 95% Confidence Interval for the Difference of Two Populations' Medians

Sample Size of Group	Rank of Largest and Smallest Differences That Determine End Points
5	3
6	6
7	9
8	14
9	18
10	24
11	31
12	38

Source: Brightman and Schneider, *Statistics for Business Problem Solving,* Southwestern, 1992, p. 415.

The sample size of each group is 10. Therefore, Table 1 indicates that the 24th largest and the 24th smallest differences mark the end points of the confidence interval.

Step 4   Carefully study the differences shown in the spreadsheet.

Notice that -195 is the smallest difference, -190 is the next smallest, -150 the next smallest, and so on. With care and patience, -10 can be identified as the 24th smallest difference. In the same manner, one can determine that 520 is the 24th largest difference. The required confidence interval is, accordingly, $[-10, 520]$. Because 0 is in this interval, there is no statistically significant difference (at the 95% confidence level) in the median prices paid for rare coins and rare stamps at that auction house.   •

This example contains 100 differences, and it is somewhat difficult to find the rank of the 24th largest and 24th smallest. A stem-and-leaf display for the differences or the use of the sorting capabilities of a graphics calculator makes the task easier.

Exercises 10.7

1. Under what conditions may the use of the Mann-Whitney confidence interval be appropriate for describing the difference in population averages?
2. Refer to Example 8 on page 423. If both end points of the confidence interval are to be made positive, does the percentage of confidence need to be higher or lower?
3. Find a 95% confidence interval for the difference in median career winnings of the top male and female golfers. A random sample of those winnings follows.

Male Golfer	Winnings	Female Golfer	Winnings
Tom Kite	$8,131,482	Sally Little	$1,524,903
Greg Norman	5,745,496	Amy Alcott	2,863,558
Fred Couples	5,859,040	Jane Geddes	1,871,697
Tom Watson	6,163,802	Rosie Jones	1,914,752
Jack Nicklaus	5,352,513	Betsy King	4,164,000

Source: *The Universal Almanac, 1994,* pp. 666–668.

[10]For sample sizes greater than 12 or different percentages of confidence, consult tables in an applied nonparametric statistics textbook.

4. Consider the following set of sample data.

Sample 1	2	4	0	3	5	5
Sample 2	50	79	29	35	38	48

Construct a 95% confidence interval for the difference between the population medians represented by these samples.

5. An automobile manufacturer is testing a new method of attaching steering wheels on the assembly line. A random sample of seven installation times for each method follows.

	Minutes Required						
Old method	6	9	9	10	19	9	12
New method	5	7	5	7	5	5	5

Construct a 95% confidence interval for the difference between the population median times. Interpret your answer.

6. A commuter group wants to determine whether public transportation by rail or by bus is more efficient. Twelve people who use rail transportation are randomly selected and asked to report how many minutes late their trains were on a given day.

Rail

1	0	3	2	0	2
0	5	2	0	1	1

In a similar manner, 12 people who take the bus were surveyed.

Bus

8	10	1	3	29	6
3	9	4	19	20	2

a. Describe the populations represented by these samples.

b. Form a 95% confidence interval for the difference of the population medians from Part a.

c. Is there a statistically significant difference (at the 95% confidence level) between the population medians? Why or why not?

Excursion 10.7A • A SPECIAL CASE OF SMALL-SAMPLE CONFIDENCE INTERVALS

The material and exercises for the special case of small-sample confidence intervals are optional. They may be covered at any time or omitted.

The Mann-Whitney confidence interval provides a way of using small samples to determine a confidence interval for two population medians in cases where the two populations may neither be normally distributed nor have approximately equal standard deviations. However, if the two populations are normally distributed, then it is possible to use small samples to construct a confidence interval for the difference of the two population means, regardless of whether or not the populations have approximately equal standard deviations. The following example illustrates the procedure.

E x a m p l e   9

Refer to the data in Example 8 on page 423—prices paid for coins and stamps. Assume that the two populations (coin prices and stamp prices) are normally distributed. Determine a 95% confidence interval for the difference of the two population means.

Solution   This is a special case in which small samples are taken from normally distributed populations that appear to have unequal standard deviations. It is still possible to draw inferences about the difference of the population means if two adjustments are made to the methods given earlier for small samples.

First, the two sample standard deviations should not be pooled to compute the standard error, *SE*. Instead, find *SE* with the formula

$$SE = \sqrt{\frac{s_A^2}{n_A} + \frac{s_B^2}{n_B}}.$$

[This is the same as Formula (10-2a) given earlier.]

Second, use the following formula to determine the degrees of freedom.

$$\text{d.f.} = \frac{\left(\dfrac{s_A^2}{n_A} + \dfrac{s_B^2}{n_B}\right)^2}{\dfrac{\left(\dfrac{s_A^2}{n_A}\right)^2}{n_A - 1} + \dfrac{\left(\dfrac{s_B^2}{n_B}\right)^2}{n_B - 1}}.$$

Then round this calculation down to the nearest integer.

Now let the subscript A represent coin prices and the subscript B stamp prices. Figure 10-10 repeats the sample statistics found earlier. Substituting into

Coins	Stamps
1-Var Stats	1-Var Stats
$\bar{x}=346.5$	$\bar{x}=69.2$
$\Sigma x=3465$	$\Sigma x=692$
$\Sigma x^2=3040975$	$\Sigma x^2=81194$
$Sx=452.1986412$	$Sx=60.8345662$
$\sigma x=428.9932983$	$\sigma x=57.7127369$
↓ n=10	↓ n=10

Figure 10-10

the formula just given for *SE*,

$$SE = \sqrt{\frac{s_A^2}{n_A} + \frac{s_B^2}{n_B}}$$

$$= \sqrt{\frac{452.199^2}{10} + \frac{60.835^2}{10}}$$

$$\approx \sqrt{20448.394 + 370.090}$$

$$\approx 144.286.$$

The degrees of freedom are found by evaluating

$$\text{d.f.} = \frac{\left(\frac{s_A^2}{n_A} + \frac{s_B^2}{n_B}\right)^2}{\frac{\left(\frac{s_A^2}{n_A}\right)^2}{n_A - 1} + \frac{\left(\frac{s_B^2}{n_B}\right)^2}{n_B - 1}}$$

$$= \frac{(20448.394 + 370.090)^2}{\frac{20448.394^2}{9} + \frac{370.090^2}{9}}$$

$$\approx \frac{433409276.1}{46459646.35 + 15218.512}$$

$$\approx 9.33$$

$$\approx 9.$$

Next, consult Appendix A-4 for the intersection of the 9 degrees of freedom row with the $\alpha = \dfrac{1 - .95}{2} = .025$ column. This indicates *t* scores of $t = \pm 2.262$ to determine the end points of the 95% confidence interval.

Finally, evaluating

$$(\overline{X}_A - \overline{X}_B) \pm 2.262 \cdot (144.285)$$

provides $[-49.07, 603.67]$ as the required confidence interval. Note that this interval is wider than that found for the difference in population medians using the Mann-Whitney technique. However—just as in the case of the confidence interval for the median—because 0 is in this interval, one may conclude that there is no statistical difference (at the 95% confidence level) in the mean prices paid for rare coins and rare stamps at the auction house.   ●

The *F*-Test

Sometimes it is not obvious if two population standard deviations are the same. A hypothesis test called an ***F*-test** determines whether statistical evidence exists that two population standard deviations are not approximately equal. It consists of the following steps.

Step 1   *The hypotheses are*

$$H_0: \quad \sigma_1 = \sigma_2.$$
$$H_A: \quad \sigma_1 \neq \sigma_2.$$

Step 2   *Form the **F-statistic**, which is* $F = \left(\dfrac{s_1}{s_2}\right)^2.$

(Choose sample number 1 and sample number 2 so that $s_1 > s_2$.)

Step 3   *Consult Appendix A-6, the model for the sampling distribution of the F-statistic.*

Use the table that contains correct α values.

- Numerator degrees of freedom $= n_1 - 1$.
- Denominator degrees of freedom $= n_2 - 1$.

(The notations n_1 and n_2 are the sizes of samples number 1 and number 2, respectively.)

Step 4   *Reject H_0 and accept H_A if the F-statistic is greater than the entry shown in Appendix A-6 at the intersection of the degrees-of-freedom row and degrees-of-freedom column obtained from Step 3. Otherwise, reserve judgment.*

Step 5   *State the conclusion.*

Exercises 10.7A

1. Briefly explain the difference between the assumptions made about two populations in the use of the Mann-Whitney technique and the assumptions made in the special case of small-sample confidence intervals.

2. What calculation seems to be the most tedious in the special case of small-sample confidence intervals?

* 3. Are confidence intervals formed by application of the special case of small samples wider or narrower than they would be if the population standard deviations were assumed approximately equal? Explain why.

4. Refer to Exercise 4 (in Exercises 10.7) on page 426. Assume that the two populations are normally distributed, and construct a 95% confidence interval for the difference in the two population means.

5. Refer to Exercise 5 (in Exercises 10.7) on page 426. Assume that the two populations are normally distributed, and construct a 95% confidence interval for the difference in the two population means.

* 6. Compare the results of Exercises 4 and 5 in Exercises 10.7 (page 426) and Exercises 4 and 5 in this section. Does it seem possible to make a general observation about the relative widths of confidence intervals found using the methods in these two sections? Explain.

7. Assume that reading times for children and young adults are normally distributed. Ten sampled children read a 25-page story in the following times, in minutes.

55	40	109	30	120
74	35	120	75	110

Ten young adults read a story of comparable length in these times.

30	42	39	50	41
35	35	40	45	55

Find a 90% confidence interval for the difference in mean reading times of the two populations.

8. Study the following pairs of sample histograms, which represent samples taken from different populations. Assuming that sample histograms reflect population distributions, indicate which pairs would call for the Mann-Whitney technique and which would call for the method described in this section.

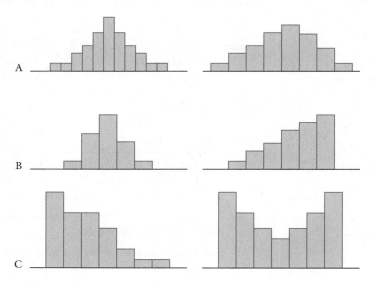

9. Refer to Example 8, page 423. Determine whether statistical evidence exists that the two population standard deviations are not approximately the same. Perform a hypothesis test at the 5% level of significance.

10. Refer to Exercise 7, and determine whether statistical evidence exists that the two population standard deviations are not approximately the same. Perform a hypothesis test at the 5% level of significance.

* 11. Refer to the procedure for the *F*-test outlined on pp. 428–29. Why does the rejection region seem to have only one tail?

GROUP 12. Researchers examined differences in SAT verbal scores for several ethnic groups, cross-tabulated by number of years of academic study. The following table shows typical scores, chosen at random.

	Typical SAT Verbal Score	
	20 or More Years of Academic Study	**Under 15 Years of Academic Study**
American Indian	440	347
Asian American	467	328
Black	400	306
Mexican American	422	324
Puerto Rican	406	314
Other Hispanic	430	318
White	480	380
Other	466	344

Based on data from a College Board press release, August 27, 1992, p. 10.

a. Perform an *F*-test for these data at the 5% level of significance.

b. Using the results of Part a, construct a 95% confidence interval for either the difference in population means or the difference in population medians. Justify your choice.

Case Study

Public information campaigns with a clear objective can still yield contradictory results. (*Courtesy of the American Red Cross*)

A Public Information Campaign

In 1987 the government launched a massive public information campaign to warn the nation about the danger of AIDS and provide information about its transmission and prevention. Two Gallup surveys were conducted, one just before the campaign began in 1987 and another several months after its completion in 1988, to examine changes in the public's knowledge of AIDS. Those surveyed were identified in several categories: age, race, education level, marital status, and gender. The 1987 survey addressed a sample of 1569 people, and the 1988 survey, 1027 people.

Some of the results were surprising. For instance, consider the respondents, grouped by educational level, who were asked whether donating blood is a way to catch AIDS from someone who has it.

	High School or Less	**More Than High School**
1987	Yes, 38.7%	Yes, 25.5%
1988	Yes, 44.6%	Yes, 22.4%

These seemingly contradictory results contributed to the argument that the effects of the campaign had been minimal.

Volume 55 (Summer 1991) of *Public Opinion Quarterly* contains an article (pp. 164–179) with more information about this campaign as well as the full results of the two Gallup polls. Obtain a copy of that article from your library, and use it to answer the following questions.

I. Refer to the responses of participants with high-school educations or less to the question about donating blood. Construct a 90% confidence interval for the change in the population proportion of those responding yes.

II. Although the article does not address this question, do you think independent sampling or matched-pair sampling was used to gather the information in the Gallup surveys? Explain why.

III. Another question that seemed to produce contradictory results was the right of an employer to fire an employee because that person has AIDS. Construct 90% confidence intervals for the changes in opinion of married and unmarried people on this issue, and then interpret the results.

IV. The two Gallup polls indicated that all groups increased their levels of belief that AIDS eventually would become an epidemic for the population at large. Find a 90% confidence interval for this change, for those 18 through 44 years old.

V. What are some reasons why it is difficult to evaluate the effect of a public information campaign such as this one?

CHAPTER SUMMARY

In many practical situations, an interval estimate of the difference between two population means is extremely important. This chapter examined two methods of constructing confidence intervals or performing hypothesis tests: independent sampling and matched-pair sampling. Independent sampling of two populations is very flexible and can be conducted even when the samples have different numbers of scores. Matched-pair sampling, however, requires the identification of sets of "twins" drawn from the two populations and reduces the two-sample problem to a one-sample problem.

Either of these methods assumes that the two populations under consideration are normally distributed and have approximately equal standard deviations. When this is not the case, nonparametric methods, such as the Mann-Whitney technique for the difference between two population medians, may be preferred. Excursion 10.7A described an additional method for such situations.

Computer software can be especially helpful for determining confidence intervals and conducting hypothesis tests about the difference of population means or medians.

Key Concepts, Terms, and Formulas

Independent sampling   *Page 395*

Characteristics of the sampling distribution of the difference of sample means:

$\mu_{\overline{X}_A - \overline{X}_B} = (\mu_A - \mu_B)$ **(10-1)**   *Page 397*

$\sigma_{\overline{X}_A - \overline{X}_B} = \sqrt{\dfrac{\sigma_A^2}{n_A} + \dfrac{\sigma_B^2}{n_B}}$ **(10-2)**   *Page 397*

Pooled variance: $s_P^2 = \dfrac{(n_A - 1) \cdot s_A^2 + (n_B - 1) \cdot s_B^2}{n_A + n_B - 2}$ **(10-3)**   *Page 398*

Standard error for the difference of population proportions:

$$SE = \sqrt{\frac{P_1(1 - P_1)}{n_1 - 1} + \frac{P_2(1 - P_2)}{n_2 - 1}}$$

Matched-pair sampling

Location of end points for a confidence interval when using the method of matched-pair sampling:

$$\overline{D} \pm Z(\overline{D}) \cdot SE, \text{ or } \overline{D} \pm t(\overline{D}) \cdot SE \quad \textbf{(10-4)}$$

Nonparametric approach

F-statistic, $F = \left(\dfrac{s_1}{s_2}\right)^2$

Videotape Suggestions

Program 21 from the series *Against All Odds: Inside Statistics,* suggested earlier, shows a good application of matched-pair sampling. *Programs 22 and 23* from the same series contain additional examples and applications of two-sample problems as well as constructions of confidence intervals for population proportions based on two independent samples.

CUMULATIVE REVIEW CHAPTERS 8–10

Overview

The material in the previous three chapters form the foundation of inferential statistics, which is concerned with drawing conclusions about a population from sample data. Probability is used to make clear the reliability of these conclusions.

The normal, t-distribution, and Chebyshev's Theorem were used as models for constructing confidence intervals in Chapter 8. The choice of the correct model was determined in large measure by the sample size, although other factors, such as the shape of the distribution of the population, also played a role. The models used in Chapter 8 were also important parts of the formal five-step hypothesis testing procedure introduced in Chapter 9 and used throughout the remainder of the text. The flow chart in Figure 8-5, page 324, is used for choosing the correct model when constructing confidence intervals or performing hypothesis tests.

Chapter 10 applied the concepts of confidence intervals and hypothesis testing to two populations. Although formulas for standard deviation and degrees of freedom were different from the single population applications in Chapters 8 and 9, the procedures were analogous. The Mann–Whitney confidence interval in Chapter 10 was a preview of work with nonparametric statistics in Chapter 12.

Procedure Index

EXERCISES ▶

1. In your own words, explain the similarities and differences between:

   a. A population mean and a population proportion.

   b. The standard deviation and the standard error.

   c. The normal distribution and the *t*-distribution.

2. Following is a random sample of suburbs in the Chicago metropolitan area, chosen from the population of suburbs that have relatively high home values, and their average home values in thousands of dollars.

LaGrange	213	Hinsdale	330
Glencoe	466	Darien	181
Evanston	260	Wheaton	188
East Dundee	178	Winnetka	687
Flossmoor	195	Deerfield	274
Wayne	202	Bull Valley	289
Naperville	201	Northfield	421
Palatine	189	Oak Brook	464
Glenview	257	River Forest	300
Wilmette	349	Huntley	179
Glen Ellyn	188	Lincolnwood	261
Medinah	203	Buffalo Grove	194
Bannockburn	720	Sleepy Hollow	179
Lake Forest	651	Palos Park	247
Deer Park	314	Northbrook	276
Clarendon Hills	192	St. Charles	188
Inverness	392	Park Ridge	210
Green Oaks	233	Willowbrook	228

   Source: "Living in Greater Chicago, 1993," *Chicago Sun-Times,* p. 276.

   a. Explain what this population mean, μ, represents.

   b. Assume that the population is normally distributed. Construct an 80% confidence interval for the population mean, μ.

   c. If the population is mound-shaped but skewed to the left, then under what conditions may the normal model still be used for the sampling distribution?

3. The family incomes within a certain zip code in the city are believed to be normally distributed. A random sample of 20 of the incomes follows.

$65,500	$54,750	$60,700	$72,010	$66,600
57,800	44,800	50,000	50,000	56,400
45,900	32,800	40,300	43,000	52,780
37,250	28,560	88,900	56,890	66,000

   a. Construct a 90% confidence interval for the mean family income, μ, within this zip code, assuming that the population is normally distributed.

   b. Construct a 90% confidence interval for the minimum value of the population's mean family income, assuming that the population is normally distributed.

4. One thousand potential purchasers of a new model car were surveyed about three possible color schemes that could be used to paint the car. The following array shows the results.

Option 1: Black and White Two-Tone	Option 2: Black with White Trim	Option 3: Basic Black	No Opinion
100	125	475	300

   a. Find 98% confidence intervals for the percentages of all future purchasers who prefer each of the three options.

   b. Using the confidence intervals found in Part a, and assuming that 20,000 cars will be manufactured initially, what is the lowest number that should be painted basic black?

   c. What result of the survey makes the company's decision about how many cars to paint in each color scheme less critical?

   d. Determine the margin of error (ME) for each confidence interval found in Part a.

5. Suppose one wants to show that statistical evidence supports a statement, S. Explain why the null hypothesis must be the opposite of S.

6. The size of a hypothesis test, α, can also be thought of as the probability of rejecting the null hypothesis when the null hypothesis is, in fact, true. Explain.

7. If the null hypothesis is not rejected, then is the null hypothesis proved true?

8. A basketball fan believes that the mean height of a player in the NBA is 7 feet 0 inches. A friend disagrees and finds the heights of $n = 12$ players, randomly selected from his collection of basketball cards:

Player	Height
Lucious H. Harris, Jr.	6 feet 5 inches
Warren Kidd	6 feet 9 inches
Pete Myers	6 feet 6 inches
Moses Eugene Malone	6 feet 10 inches
Franklin Leonard Johnson	6 feet 1 inch
Derrick Allen Mahorn	6 feet 10 inches
Collier Brown, Jr.	6 feet 11 inches
Gregory Lawrence Graham	6 feet 4 inches
Donald Hodge	7 feet 0 inches
Richard Thomas King	7 feet 2 inches
Rex Andrew Walters	6 feet 4 inches
George Muresan	7 feet 7 inches

Conduct a hypothesis test to determine whether these cards provide statistical evidence that the mean height is not 7 feet 0 inches. Answer this question at the 5% level of significance.

9. Refer to Exercise 8. At the 5% level of significance, do these cards provide statistical evidence that the mean height is more than 6 feet 6 inches?

10. Assume that the amount of gasohol used, as a percentage of the gasoline used on the highways per state, is normally distributed with a standard deviation of $\sigma = 13.18\%$. A random sample of $n = 10$ states reveals the following percentages.

	% Use of Gasohol		% Use of Gasohol
TN	7.91	OH	27.73
ND	17.43	VA	3.43
IL	34.95	MO	9.19
CA	0.45	NE	50.98
IN	24.64	MI	12.16

Source: *Highway Statistics 1992*, Federal Highway Administration, p. 14.

a. Determine whether the claim that the mean percentage consumption for all states is 20% should be rejected at the 10% level of significance.

b. Determine whether there is statistical evidence for the claim that the mean percentage consumption for all states is more than 15% at the 10% level of significance.

11. A political candidate claims that she is more popular with women than with men. A survey identified 8973 people who supported the candidate, 4610 of whom were women. Test the candidate's claim at the 5% level of significance.

12. Give some instances when it is important to test null hypotheses very stringently.

13. What two assumptions are usually made when inferences are drawn about the difference of two population means? What nonparametric approach may be used if one or both of these assumptions do not seem valid?

14. An agent for a sawmill must choose between one of two forests. Thirty-five trees sampled from one forest have a mean diameter of 30.5 inches with $s_1 = 9.45$ inches. Fifty trees from the other forest have a mean diameter of 28.2 inches with $s_2 = 10.6$ inches. Assume that the diameters of the trees in both forests are normally distributed with approximately equal standard deviations.

a. Perform a hypothesis test at the 10% level of significance to determine whether evidence exists to support the claim that the two forests contain trees of different sizes.

b. Perform a hypothesis test at the 8% level of significance to determine whether statistical evidence exists to support the claim that the mean diameter of trees from the first forest is more than 1 inch greater than that of trees from the second forest.

15. A random sample of 14 people from a new suburb has the following incomes.

$90,500	$67,675	$44,500	$88,456	$39,000
66,789	78,900	65,122	44,565	47,901
46,989	65,900	70,000	68,750	

A random sample of ten people from an older section of the city has the following incomes.

$50,890	$29,000	$52,500	$79,890	$47,500
60,000	27,500	71,000	34,250	61,100

Assume that the populations represented by these samples are normally distributed and have approximately equal standard deviations. Form a 95% confidence interval for the difference in the mean incomes of these two populations, and interpret your result.

16. Refer to Exercise 15. Do not assume that the two populations are normally distributed or that their standard deviations are approximately equal. Use the ten middle-ranked suburban scores to form a 95% confidence interval for the difference of the two population medians.

17. The following table shows the numbers of patients treated in a test of two different toothpastes and the numbers in whom at least one cavity occurred within 6 months.

	Toothpaste A	Toothpaste B
Patients treated	750	800
Patients in whom a cavity occurred	310	295

Construct 50%, 75%, 85%, 90%, 95% and 99% confidence intervals for the difference in percentages of people using these two toothpastes who will develop cavities. Interpret the meaning of any difference in sign of the end points of the confidence intervals.

18. The following array shows the numbers of licensed female drivers in two age groups, in a random sample of 10 states.

	19 Years and Under	75–79 Years
NH	20,966	10,307
FL	187,961	202,207
IL	205,517	95,220
AL	92,801	40,568
AK	8,800	1,100
MD	65,049	34,623
SC	63,116	28,516
KS	54,101	30,734
WY	12,187	4,364
IA	59,171	34,883

Source: *Highway Statistics 1992*, U.S. Department of Transportation, p. 33.

Stat Fact

In 1992 only 14 states had a greater percentage of female drivers than male drivers. Georgia had the smallest percentage of male drivers, 47.96%, and Montana had the greatest, 54.82%.

a. Assume that the populations represented by these samples are normally distributed and have approximately equal standard deviations. Determine a 95% confidence interval for the difference in the mean numbers of registrations for these two age groups.

b. Construct a 95% confidence interval for the difference in population medians.

c. (Pertains to Excursion 10.7A.) Is there statistical evidence to support the claim that the assumption of approximately equal standard deviations in Part a is invalid? Perform a hypothesis test at the 5% level of significance to answer this question.

(David Fish)

Policymakers must make decisions based on a welter of data and opinions. Both qualitative and quantitative variables are considered.

11 Some Applications of the Chi-Square Distribution

Stat Fact

The English mathematician Karl Pearson (1857–1936) was largely responsible for the development of the chi-square distribution. He demonstrated that analytical statistics is applicable to a wide variety of problems in all the sciences. For example, using statistical analysis made possible by his chi-square distribution, Pearson showed that tuberculosis is related more to heredity than to environmental factors. This result led to the elimination of widespread use of sanitoria to treat tuberculosis.

Unknown to Pearson, a German mathematician Friedrich Helmert discovered the chi-square distribution in 1875 in a different context.

T he last several chapters have demonstrated the importance of the normal and *t*-distribution models and provided some examples of their use in certain types of situations. In other situations, problems arise that call for different models. One of those models, the chi-square distribution,[1] is often used in situations where information is obtained from *counting* rather than measuring.

The **case study** at the end of this chapter uses the chi-square distribution to examine some congressional election results.

11.1 TESTING THE INDEPENDENCE OF QUALITATIVE VARIABLES

Nearly all of the variables studied so far have been quantitative variables, that is, variables whose values are expressed as numbers. A **qualitative variable** (as mentioned in Chapter 1) is one whose value is not necessarily expressed as a number. The religious conviction of the president of the United States, the color of a car chosen at random, the sex of a newborn, and the type of tree on the front lawn of your house are all examples of qualitative variables.

Definition ▶ Two (or more) qualitative variables are **independent** if the value of one is completely unaffected by the value of the other(s).

[1]The Greek character chi (χ) is pronounced "kī."

For example, the color of car a person drives and the type of tree planted in front of that person's house almost certainly have nothing to do with each other and are therefore statistically independent or, more simply, independent. However, an individual's blood type and the blood type of that person's parents are examples of qualitative variables that are not independent.

Often it is not so clear whether two qualitative variables are independent. This chapter uses a statistic called the chi-square test statistic as part of a hypothesis test to judge the independence of two qualitative variables. The symbol χ^2 will denote chi-square.

As an example, suppose this question arises: Is the gender of a driver independent of the type of traffic ticket that driver receives? A total of 218 traffic citations are chosen at random. The following cross-tabs table summarizes the findings.

Cross-Tabs Table of Traffic Citations

Type of Ticket	Male Drivers	Female Drivers
Speeding	100	30
Improper turn	12	10
Failure to obey signal	34	15
Driving with expired plates	7	10

Data might also have been collected to cross-tabulate the type of ticket with the age of the driver.

This is a question about independence of two qualitative variables: gender of driver and type of ticket. Two values are possible for the variable gender of driver: male and female. Four values are indicated for the variable type of ticket: speeding, improper turn, failure to obey signal, and driving with expired plates.

The numbers in the eight cells of the preceding cross-tabs table represent joint frequencies of possible values for the variables. For instance, of all the recipients of citations examined, 100 males and 30 females received speeding tickets. These are actual frequencies and are denoted by f_o, for **observed frequencies.** If the gender of the driver and the type of ticket received are independent,[2] then it should be possible, for each cell, to determine the **expected number of frequencies,** denoted by f_e, on the basis of the total number of citations, as follows.

First redraw the cross-tabs table in a more convenient format called a **contingency table,** shown in Figure 11-1. The initial step in constructing this contingency table is to place each observed frequency in the upper half of each cell. That allows room for the expected frequencies, which will soon be calculated. Also, each row total (RT), column total (CT), and grand total (GT) should be calculated and placed as shown.

Now consider the expected frequency for the male speeding cell. The first column total, 153, indicates that there are 153 males out of a grand total of 218 people, which implies that the percentage of males in this sample, $P(M)$, is

$$P(M) = \frac{153}{218}.$$

The first row total indicates that there is a total of 130 speeding tickets in this sample. If the two variables are independent, the percentage of speeding tickets

---

[2]You will see that this assumption of independence is, in fact, always the null hypothesis of this type of hypothesis test.

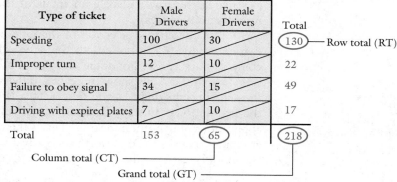

Figure 11-1 A contingency table of traffic citations.

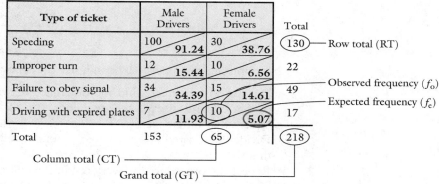

Figure 11-2

received by males should match the percentage of males in the sample. That is, males should receive

$$\frac{153}{218} \cdot 130 \approx 91.24$$

speeding tickets, which is the value of the expected frequency, f_e, of the male speeding cell.[3] This value implies a formula for the computation of f_e for any cell:

Formula for Computing **Expected Frequencies from a Contingency Table**

$$f_e = \frac{(\text{Row total}) \cdot (\text{Column total})}{(\text{Grand total})}$$

$$= \frac{(RT) \cdot (CT)}{GT} \tag{11-1}$$

Figure 11-2 is the completed contingency table. All expected frequencies are in boldface. One can verify that the row and column totals of the expected frequen-

[3]Round all expected frequencies in this chapter to the nearest hundredth.

cies match the row and column totals of the observed frequencies (except for rounding differences).

Given the chance selection of the random sample of 218 citations, the fact that each value of f_e is not exactly the same as its corresponding value of f_o should come as no surprise. For some cells, the observed and expected frequencies are very close; other cells show more of a difference. Do the theoretical results and the actual results differ by too much? That is, do f_o and f_e differ by so much that the assumption of independence used to calculate f_e must be rejected, as in general applications of hypothesis testing? This is the question a chi-square hypothesis test attempts to answer. To perform such a test the next step is to compute the chi-square (χ^2) test statistic.

Formula for the χ^2 Test Statistic

$$\chi^2 = \Sigma\left(\frac{(f_o - f_e)^2}{f_e}\right) \tag{11-2}$$

If computing this statistic by hand, use the following steps.

- Starting with any cell, find the difference between the observed frequency, f_o, of that cell and its expected frequency, f_e.
- Square that difference.
- Divide the squared difference by f_e, rounding all calculations to the nearest hundredth.
- Repeat the procedure for every cell in the contingency table, then add together all the calculations.

Consult Figure 11-2 and use the indicated values of f_o and f_e to make the following computation.

$$\chi^2 = \frac{(100 - 91.24)^2}{91.24} + \frac{(12 - 15.44)^2}{15.44} + \cdots + \frac{(10 - 5.07)^2}{5.07}$$

$$= \frac{(8.76)^2}{91.24} + \frac{(-3.44)^2}{15.44} + \cdots + \frac{(4.93)^2}{5.07}$$

$$= 12.24.$$

An illustration of the use of a graphics calculator to efficiently perform this type of calculation appears on the bottom of page 452.

(Notice that the chi-square statistic can never be negative.)

If the variables gender of driver and type of ticket are independent, then all the possible values of χ^2 should form a frequency curve called the chi-square distribution.

The Chi-Square Distribution

Figure 11-3 suggests that, like the *t*-distribution, the chi-square distribution is a whole family of curves, and its shape depends on its degrees of freedom. In turn, the method of determining the degrees of freedom depends on the nature of the application.

The Chi-Square Distribution

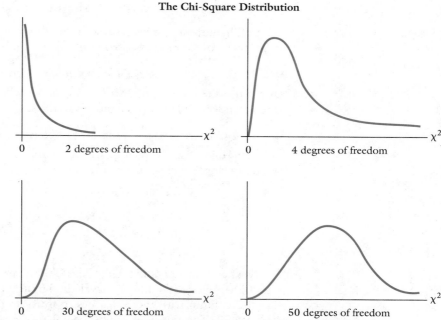

Figure 11-3 The chi-square distribution for various degrees of freedom.

Figure 11-3 indicates that possible values of the χ^2 test statistic are bounded by 0 on the left and skewed toward the higher values of χ^2. As the degrees of freedom increase, the chi-square distribution, like the t-distribution, approaches the normal curve more closely. Should each pair of values of f_o and f_e agree exactly, then the value of χ^2 is zero, which is the left end of the chi-square distribution. Some variation in the values between f_o and f_e is reasonable and is represented by the middle mound-shaped part of the chi-square distribution. The right tail, which represents larger sums and greater differences between f_o and f_e, is the location of the rejection region.

A hypothesis test will now be performed to determine whether the claim of independence of these two variables should be rejected.

E x a m p l e
1

Using the data just shown, determine whether evidence exists that the two qualitative variables, gender of driver and type of ticket, are not independent. Perform a chi-square hypothesis test at the 2% level of significance.

Solution   In this type of situation, the word "evidence" implies statistical evidence, and the following procedure includes the same five-step method of hypothesis testing that appeared in previous chapters.

The only possible outcomes from this hypothesis test are the conclusion that the two qualitative variables are not independent of each other and reservation of judgment.

Step 1   State the hypotheses.

H_0:   The two qualitative variables, gender of driver and type of ticket, are independent of each other.

H_A: The two qualitative variables are not independent of each other.

The null hypothesis always assumes the independence of the variables in order to justify the procedure used to compute f_e on the basis of row and column totals.

Step 2   Decide on a model for the sampling distribution.

Always use the chi-square distribution model when conducting a hypothesis test of independence for two qualitative variables.

Step 3   Determine the end points of the rejection region, and state the decision rule.

First determine the degrees of freedom. In a hypothesis test of two qualitative variables using the chi-square distribution, the degrees of freedom, d.f., are

$$\begin{aligned}
\textbf{d.f.} &= (\textbf{No. of rows} - 1) \cdot (\textbf{No. of columns} - 1) \\
&= (4 - 1) \cdot (2 - 1) \\
&= 3 \cdot 1 \\
&= 3.
\end{aligned}$$

Next consult Appendix A-5, "The Chi-Square Distribution." Because this test is to be conducted at the 2% level of significance, find the intersection of the $\alpha = .02$ column with the 3 degrees of freedom row to obtain a critical value of $\chi^2 = 9.837$. The rejection region begins to the right of this value.

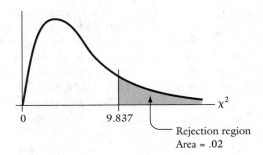

Therefore, the decision rule is:

Reject H_0 and accept H_A if $\chi^2 > 9.837$. Otherwise, reserve judgment.

Step 4   Compute the chi-square test statistic.

Recall that the chi-square test statistic is calculated with the formula

$$\chi^2 = \Sigma \left(\frac{(f_o - f_e)^2}{f_e} \right).$$

Earlier in this chapter on page 444, a value of $\chi^2 = 12.24$ was obtained for these data.

Step 5   State the conclusion.

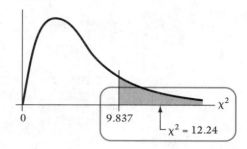

Since $12.24 > 9.837$, **reject H_0 and accept H_A.** There is statistical evidence at the 2% level of significance that the qualitative variables gender of driver and type of ticket are not independent. In other words, there *is* a relationship between the two variables, and evidence supports the claim that these two variables are dependent. ●

Recall that the level of significance, 2%, implies a willingness to erroneously reject the null hypothesis exactly 2% of the time when performing the test over and over again. If the level of significance is lowered to, say, 0.1% ($\alpha = .001$), then the willingness to erroneously reject the null hypothesis is lowered to only 0.1% of the time. Logically, this change should be reflected by moving the rejection region further to the right. If Appendix A-5 is consulted for 3 degrees of freedom with $\alpha = .001$, the rejection region starts at 16.266 instead of 9.837. Thus, at the 0.1% level of significance there is not enough evidence to support the alternative hypothesis and judgment must be reserved.

Chi-square tests for independence of qualitative variables are easier with statistical software such as Minitab. Using Example 1 as an illustration, the first step is to enter the gender-of-driver values into columns C1 and C2 of a Minitab worksheet, as shown in Figure 11-4.[4] Next obtain a Session window and enter the macro command CHISQUARE followed by the column designations of the data. Figure 11-5 shows the result once the [RETURN] key is pressed. The chi-square statistic 12.243 given by Minitab can be compared to the end point of the rejection region found with Appendix A-5.

The chi-square distribution may not always be an appropriate model with which to test the independence of qualitative variables. Its appropriateness becomes more likely as the expected frequencies of the cells increase. The following guidelines constitute a good rule of thumb.

---

[4]The data may also be entered into columns C1 through C4 of a worksheet, in two rows per column.

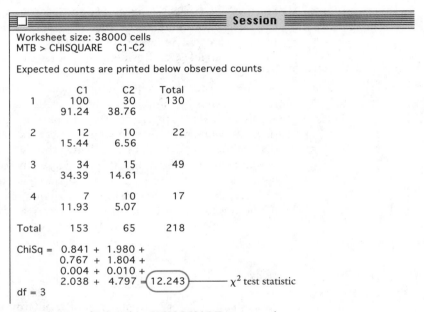

Figure 11-4

Figure 11-5 Use of Minitab's CHISQUARE command.

Rule ▶ Guidelines for using the chi-square distribution model to test the independence of qualitative variables:

a. Not more than 20% of the cells should have an expected frequency of less than 5.

and

b. No cell should have an expected frequency of less than 1.

The next example illustrates a method for dealing with a contingency table in which these guidelines are not satisfied.

E x a m p l e
2

If a survey were conducted of 331 frequent travelers who log more than 2500 miles per year, its data might take the following form.

Cross-Tabs Table of Work Status and Modes of Travel

	White-Collar	Blue-Collar	Retired	Student
Air	65	30	25	20
Train	30	10	30	8
Auto	25	15	40	5
Bus	2	1	5	20

Determine, at the 5% level of significance, whether evidence exists that the two qualitative variables, work status and travel mode, are not independent of each other.

Solution   Form a contingency table that has row and column totals, then use those totals to compute expected frequencies in each cell. Figure 11-6 is the completed table. There are two cells with expected frequencies less than 5: blue-collar bus riders ($f_e = 4.74$) and student bus riders ($f_e = 4.48$). Two cells are acceptable under the rule of thumb mentioned earlier, because they are less than 20% of the total (16 cells).

For illustration purposes, however, cells will be combined to eliminate all those with expected frequencies less than 5. Combine rows or columns so as to cause the minimum change in the total number of cells in the table. In this case, add the bottom two rows.[5] This new contingency table (Figure 11-7) has 12 cells; this is the **adjusted number of cells.** The observed frequencies of the bottom two rows of the original table (Figure 11-6) were simply added to produce the adjusted observed frequencies of the last row in the new table. The adjusted expected frequencies were found in the same way.

	White collar	Blue collar	Retired	Student	Total
Air	65 / 51.60	30 / 23.69	25 / 42.30	20 / 22.42	140
Train	30 / 28.75	10 / 13.20	30 / 23.56	8 / 12.49	78
Auto	25 / 31.33	15 / 14.38	40 / 25.68	5 / 13.61	85
Bus	2 / 10.32	1 / 4.74	5 / 8.46	20 / 4.48	28
Total	122	56	100	53	331

Figure 11-6 A contingency table of work status and mode of travel.

[5] The "Blue-Collar" and "Student" columns could be combined instead. However, the arbitrary choice was made to maintain all work-status distinctions.

	White collar	Blue collar	Retired	Student	Total
Air	65　51.60	30　23.69	25　42.30	20　22.42	140
Train	30　28.75	10　13.20	30　23.56	8　12.49	78
Auto or bus	27　41.65	16　19.12	45　34.14	25　18.09	113
Total	122	56	100	53	331

Figure 11-7 An adjusted contingency table.

Proceed with the hypothesis test.

The only possible outcomes of this hypothesis test are the conclusion that the two variables are not independent of each other and reservation of judgment.

Step 1   State the hypotheses.

> H_0: The two qualitative variables, work status and travel mode, are independent of each other.

> H_A: The two variables are not independent of each other.

Step 2   Decide on a model for the sampling distribution.
Use the chi-square distribution model.

Step 3   Determine the end points of the rejection region, and state the decision rule.
The number of degrees of freedom in the *adjusted* contingency table is

$$\text{d.f.} = (3 - 1) \cdot (4 - 1)$$
$$= 6.$$

According to Appendix A-5, the critical value of χ^2 is located at the intersection of the 6 degrees of freedom row with the $\alpha = .05$ column. This value is 12.592.

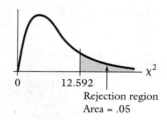

Rejection region
Area = .05

Therefore, the decision rule is:

Reject H_0 and accept H_A if $\chi^2 > 12.592$. Otherwise, reserve judgment.

Step 4   Compute the chi-square test statistic.

$$\chi^2 = \frac{(65 - 51.60)^2}{51.60} + \frac{(30 - 23.69)^2}{23.69} + \cdots + \frac{(25 - 18.09)^2}{18.09}$$

$$= \frac{(13.4)^2}{51.60} + \frac{(6.31)^2}{23.69} + \cdots + \frac{(6.91)^2}{6.91}$$

$$= 28.458.$$

Step 5   *State the conclusion.*

Since $28.458 > 12.592$, **reject H_0 and accept H_A.** There is enough statistical evidence to conclude that the two qualitative variables are not independent. A person's work status and mode of transportation are dependent at the 5% level of significance.   ●

   The follow-up example illustrates how greatly the combining of cells affected the χ^2 test statistic.

E x a m p l e 3

Use Minitab to conduct the hypothesis test of Example 2.

Solution   The first step is to enter the data into columns C1 through C4 of a Minitab worksheet.

		C1	C2	C3	C4	C5
	↓	White	Blue	Retired	Student	
1		65	30	25	20	
2		30	10	30	8	
3		25	15	40	5	
4		2	1	5	20	
5						

Untitled Worksheet

When the CHISQUARE command is executed, the following results appear.

Session

```
Worksheet size: 38000 cells
MTB > CHISQUARE   C1-C4

Expected counts are printed below observed counts

        White    Blue   Retired   Student    Total
  1        65      30       25        20       140
         51.60   23.69    42.30     22.42

  2        30      10       30         8        78
         28.75   13.20    23.56     12.49

  3        25      15       40         5        85
         31.33   14.38    25.68     13.61

  4         2       1        5        20        28
         10.32    4.74     8.46      4.48

Total     122      56      100        53       331

ChiSq = 3.479 + 1.683 + 7.073 + 0.261 +
        0.054 + 0.774 + 1.757 + 1.614 +
        1.279 + 0.027 + 7.986 + 5.447 +
        6.708 + 2.948 + 1.415 + 53.702 = 96.206
df = 9
2 cells with expected counts less than 5.0
```

A more reliable alternative to combining cells is to increase the number of people being surveyed.

Because no more than 20% of the cells had expected frequencies less than 5, and no expected frequency was less than 1, Minitab did not combine any cells. Note that a much larger χ^2 statistic, 96.206, is obtained. Once again, the null hypothesis, H_0, is rejected.   •

Exercises 11.1–11.2

1. Explain why it is impossible to obtain a negative value for the χ^2 test statistic.

2. If the guidelines given in the text are not followed and cells are not combined, then is the calculated value of the χ^2 statistic higher or lower than it would have been if cells were combined?

3. What seems to be the net effect of combining rows or columns in a contingency table?

4. Identify the following statements as true or false. If a statement is false, explain why.

   a. As the degrees of freedom increase, the chi-square distribution resembles a normal distribution.

   b. When the chi-square model is used, the degrees of freedom are always equal to [(No. of rows) · (No. of columns) − 1].

   c. f_e is usually greater than f_o.

   d. Disregarding rounding differences, the sum of the f_e's should equal the sum of the f_o's in any particular row or column of a contingency table.

   e. As the level of significance increases in a chi-square test, one is more likely to reject the null hypothesis.

GROUP

5. Refer to Example 1 on page 445. The chi-square hypothesis test indicated that gender and type of traffic violation are not statistically independent. In your own words, try to explain what that implies. Then try to make a case against the test's conclusion.

* 6. Refer to Example 3 on page 451. Outline a method, using Minitab, of computing a chi-square test statistic on the basis of a contingency table in which no cell has an expected frequency less than 5.

7. Refer to Example 3 on page 451. Examine the observed and expected frequencies of the cells, and identify which three cells contribute most to the value of the chi-square test statistic. Is there one cell in particular in which the difference between the observed and expected frequencies is so great that the test is destined to fail at nearly any level of significance? If so, identify the cell.

For Exercises 8–16, a graphics calculator may be used efficiently to combine many steps in the calculation of χ^2. For instance, the calculation on page 444 may be keyed in as one long expression, started as follows:

```
(100 - 91.24)^2/91
.24+(12 - 15.44)^2
/15.44+(34 - 34.39
```

8. The distribution of gold, silver, and bronze medals among the top six medal-winning countries at the 1992 Summer Olympic Games follows.

	Gold	Silver	Bronze
Unified Team	45	38	29
United States	37	34	37
Germany	33	21	28
China	16	22	16
Cuba	14	6	11
Hungary	11	12	7

Determine whether evidence exists that the type of medal won and the country are not independent. Conduct a hypothesis test at the 20% level of significance.

9. A specialty tea company is interested in determining whether gender plays any role in the public's preferences for certain types of tea that it manufactures. A random sample of 250 customers yielded the following data.

	Number Preferring Type of Tea	
	Male	*Female*
Peppermint	32	32
Chamomile	21	43
Rose	8	19
Spearmint	26	12
Lemon	31	17
Cranberry	6	3

Conduct a hypothesis test at the 30% level of significance to determine whether evidence exists that gender and tea preference are not independent.

10. The following data are based on the results of a survey to examine media preferences.

	TV Fans	Radio Listeners	Newspaper Readers
18 to 34	29	41	31
35 to 49	22	27	29
50 or older	50	32	40

Source: *American Demographics*, July 1992, p. 55.

a. Does evidence support the claim that the type of media and age group are not independent? Perform a hypothesis test at the 10% level of significance.

b. Estimate the attained significance level of this test.

11. The following table lists the numbers and types of automobile accidents in several states, chosen at random.

S t a t   F a c t

In 1992 the state of Rhode Island, of all the states in the union, had the smallest number of fatal-injury accidents per 100 million vehicle-miles of travel. The state of Missouri had the smallest number in the nonfatal category.

	Injury Accidents	
	Fatal	*Nonfatal*
IL	1249	103,246
KS	337	19,675
NV	224	13,263
SD	141	5112
TX	2691	170,513
VT	87	3000
WI	586	40,778

Source: Federal Highway Administration, *Highway Statistics 1992*, p. 214.

a. Does evidence support the claim that the type of accident and state are not independent? Perform a chi-square hypothesis test at the 20% level of significance.

b. What is the practical difficulty of using a chi-square hypothesis test in this type of situation?

12. The following data are based on the results of a survey to examine television viewing preferences.

	CNN	Cable	Pay	Early News
White	272	566	299	147
Black	147	366	247	160
Spanish-speaking	192	421	266	131
Other	114	297	247	110

Source: Harold W. Stanley and Richard G. Niemi, *Vital Statistics on American Politics*, 3d ed., 1992, p. 54.

Determine whether evidence exists to support the claim that television viewing preferences and ethnicity are not independent. Conduct a chi-square hypothesis test at the 20% level of significance.

13. The following table shows billions of miles traveled by air, bus, and rail during the last 5 decades.

	Air	Bus	Rail
1940s	1	1	24
1950s	10	26	32
1960s	34	19	22
1970s	119	25	11
1980s	219	28	12

Conduct a chi-square hypothesis test at the 10% level of significance to determine whether the data in this table support the claim that the mode of transportation is not independent of the decade.

14. The following cross-tabs table provides data from a census report studying the conditions of the elderly. This table specifically shows numbers of men and women over age 65 who are likely to be in certain income categories.

	Male	Female
Less than $10,000	400	1080
$10,000 to $29,999	600	600
$30,000 to $49,999	120	60
$50,000 or more	100	25

Based on data from *American Demographics,* December 1992, p. 47.

   a. Conduct a hypothesis test at the 20% level of significance to determine whether evidence exists to suggest that gender and income category are not independent for those over the age of 65.

   b. What can one conclude about the attained significance level of the test in Part a?

15. The following table indicates levels of career commitment of Chicago-area women who plan to return to school.

Women Planning to Return to School

Career Commitment	High School or Less	More Than High School
Low	38	12
Medium	47	38
High	15	50

Source: *The Sociological Quarterly* 34:2, 1993, p. 265.

Determine whether evidence supports the claim that the level of career commitment and the level of schooling already attained are not independent of each other. Conduct a hypothesis test at the 5% level of significance.

GROUP 16. The following tables indicate preferences for watching professional football and for watching professional baseball. The data are cross-tabulated by four variables: gender, age, household income, and education. Conduct four separate hypothesis tests to determine whether evidence exists that a preference for watching a sport is not independent of any of these four variables. Conduct all four tests at the 10% level of significance.

Gender

	Football	Baseball
Men	72	60
Women	48	45

Age

	Football	Baseball
18–34 years	68	50
35–49 years	59	47
50–64 years	55	54
65 and older	48	63

Household Income

	Football	Baseball
$50,000 and over	65	53
$35,000–$49,999	66	55
$20,000–$34,999	63	54
Under $20,000	50	49

Education

	Football	Baseball
Some college or more	64	48
High-school graduate	62	57
Some high school or less	48	50

Source: *Information Plus: Recreation*, 1993 edition, p. 37.

11.3 CONSTRUCTING A CONFIDENCE INTERVAL FOR σ

The construction of a confidence interval for a population's standard deviation, σ, is another application of the chi-square distribution model, when performance variability as well as mean performance needs to be considered. For instance, suppose a medical equipment manufacturer has invented a machine that automatically injects a certain amount of drug per hour into a patient's bloodstream. Although the mean amount actually injected each hour may be very close to the target amount, the standard deviation of the amounts injected is also very important. If σ is greater than it should be, the patient may get either too little or too much of the drug and have a serious reaction. This section describes how to determine a confidence interval for the population standard deviation on the basis of sample statistics.

Given a random sample of size n, determine the value of χ^2 as follows:

Value of χ^2 for Construction of a Confidence Interval for σ

$$\chi^2 = \frac{(n-1) \cdot s^2}{\sigma^2}$$

(11-3)

Several observations can be made about the possible values of χ^2. First, because all the factors in this ratio are positive, χ^2 is always positive. Second, because $n - 1$ is a fixed number and σ^2 is a fixed but unknown number, the value of χ^2 varies as s^2 varies, which implies that the smallest value of χ^2 is 0 when the estimated population variance, s^2, is 0. Also, because s^2 is skewed toward higher values, χ^2 is also skewed toward higher values. These are important characteristics of the chi-square distribution. In fact, the chi-square distribution can be interpreted as the frequency curve of all possible values of $\dfrac{(n - 1) \cdot s^2}{\sigma^2}$ for a given value of n.

If the value of s is computed on the basis of a simple random sample and if the population is normally distributed, then the chi-square distribution can serve as a model in the construction of a confidence interval for σ.

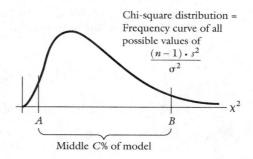

Table 1 contains tail areas to the right of A and B, which will later be used, with the help of Appendix A-5, to find values of χ^2 that enclose the middle $C\%$ of the chi-square distribution.

Table 1 Right-Tail Areas for a $C\%$ Confidence Interval

$C\%$	To the Right of A	To the Right of B
60%	.80	.20
80%	.90	.10
90%	.95	.05
96%	.98	.02
98%	.99	.01

The next example explains how to use Table 1 and Appendix A-5 to construct a confidence interval for σ.

E x a m p l e
4

A simple random sample of size $n = 12$ taken from a normally distributed population yields an estimated population standard deviation of $s = 2.45$. Form a 90% confidence interval for the population standard deviation, σ.

Solution   Because this is a simple random sample taken from a normally distributed population, the chi-square distribution can serve as a model for the construction of the 90% confidence interval.

According to Table 1, the right-tail areas for a 90% confidence interval are .95 to the right of *A* and .05 to the right of *B* (Figure 11-8). In this type of application, the degrees of freedom for the chi-square distribution are

$$\textbf{Degrees of freedom} = \textbf{sample size} - 1$$
$$= 12 - 1$$
$$= 11.$$

According to Appendix A-5, the intersection of the 11 degrees of freedom row with the .95 column is $A = 4.575$, and the intersection of the 11 degrees of freedom row with the .05 column is $B = 19.675$ (Figure 11-9).

Determine the required confidence interval by solving the following inequality for σ.

$$[1] \quad 4.575 \leq \frac{(n-1) \cdot s^2}{\sigma^2} \leq 19.675.$$

This is perhaps easiest accomplished by following these algebraic manipulations.

First rewrite inequality [1] as follows. Note the change in direction of the inequality symbols.

$$[2] \quad \frac{1}{4.575} \geq \frac{\sigma^2}{(n-1) \cdot s^2} \geq \frac{1}{19.675}.$$

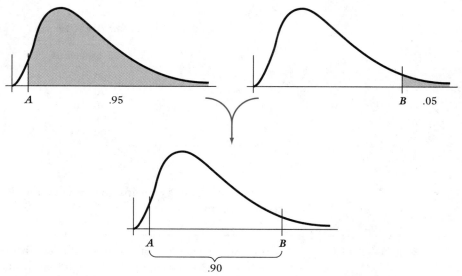

Figure 11-8 Right-tail areas used to determine the end points of a confidence interval for σ.

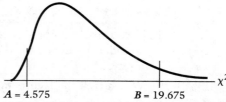

Figure 11-9 χ^2 coordinates of the end points of a confidence interval for σ.

Next, multiply across inequality [2] by $(n - 1) \cdot s^2$ to obtain

[3] $\dfrac{(n - 1) \cdot s^2}{4.575} \geq \sigma^2 \geq \dfrac{(n - 1) \cdot s^2}{19.675}.$

Now just rearrange the terms on different sides:

[4] $\dfrac{(n - 1) \cdot s^2}{19.675} \leq \sigma^2 \leq \dfrac{(n - 1) \cdot s^2}{4.575}.$

Then substitute

$$(n - 1) = 12 - 1 = 11$$

and

$$s^2 = 2.45^2 = 6.0025$$

into inequality [4], and simplify:

$$\dfrac{11 \cdot 6.0025}{19.675} \leq \sigma^2 \leq \dfrac{11 \cdot 6.0025}{4.575},$$

$$3.357 \leq \sigma^2 \leq 14.432.$$

Finally, take the square root:

$$1.83 \leq \sigma \leq 3.80.$$

Therefore, [1.83, 3.80] is a 90% confidence interval for the population standard deviation, σ. ●

The results of this example may be summarized as follows:

Inequality That Determines the End Points of a Confidence Interval for the Standard Deviation, σ

$$\sqrt{\dfrac{(n - 1) \cdot s^2}{\chi^2_{\text{RIGHT}}}} \leq \sigma \leq \sqrt{\dfrac{(n - 1) \cdot s^2}{\chi^2_{\text{LEFT}}}} \qquad \textbf{(11-4)}$$

Exercises 11.3

1. Give three examples of situations (not described in this section) in which a confidence interval for the standard deviation is as important as a confidence interval for the mean.

2. Why are all the factors in the value of χ^2 positive?

3. Explain the consequences of s^2 being skewed toward higher values.

* 4. Algebraically derive Formula 11-4.

* 5. Study Table 1 on page 457. Determine tail areas to the right of A and B for 50% and 70% confidence intervals.

6. Refer to Example 4 on page 456, and construct a 96% confidence interval for the population standard deviation.

7. A postal inspector wants to estimate the population standard deviation for waiting times at a certain window. A random sample of $n = 10$ times (the numbers of minutes people have had to wait) follows.

3.4	4.5	2.0	5.6	6.7
2.1	4.5	3.0	5.5	3.2

a. What assumptions about these waiting times must be made in order to use the method of constructing a confidence interval for σ outlined in Example 4?

b. Assuming that the conditions from Part a are met, construct a 98% confidence interval for the population standard deviation.

c. If the confidence interval from Part b is smaller than the inspector antici- pated, what conclusions might she draw?

8. The following data were randomly selected from a normal population: {45.8, 44.7, 50.5, 44.7, 50.4, 45.6, 49.8, 44.7, 60.0, 52.9}.

a. Calculate point estimates of the mean and estimated standard deviation of this population.

b. Construct an 80% confidence interval for the population mean.

c. Construct an 80% confidence interval for the population standard deviation.

GROUP   9. The medical equipment manufacturer mentioned on page 456 claims that the standard deviation of his machine is less than 0.007 mg. A random sample of 12 injections is used to compute a sample statistic of $s = .00645$ mg.

a. Construct a 96% confidence interval for the standard deviation of the amounts injected.

b. Does this machine seem safe to use? Why or why not?

10. The yearly salaries of eight randomly selected LPN/LVN nurses follow.

$19,794	$20,168	$18,995	$19,800
19,777	18,880	20,200	19,550

a. Assuming that all LPN/LVN salaries are normally distributed, find a 90% confidence interval for σ.

b. Use the result of Part a to find a 90% confidence interval for the population variance.

11.4   TESTING ASSUMPTIONS ABOUT A POPULATION'S DISTRIBUTION

Another important application of the chi-square distribution is in testing the reasonableness of assumptions made about a population's distribution, also called a sample's goodness of fit. A **goodness-of-fit hypothesis test** determines the likelihood that sample data come from, or "fit," a specific type of population distribution.[6] For example, in the last few chapters the shape of the population distribution has at times been assumed normally distributed for the purpose of forming confidence intervals or conducting hypothesis tests. It may be important to test a claim about the shape of a population's distribution if inferences about population parameters are desired. A goodness-of-fit test can reject or fail to reject such a claim at a specified level of significance.

In every goodness-of-fit exercise, the null hypothesis claims that the sample data come from a specific type of population, with or without stated population parameters. As in earlier examples and exercises, this null hypothesis is then the basis for finding expected frequency values, f_e. The alternative hypothesis is that the sample did not come from such a population but from some other type (which need not be specified).

Consider a situation in which 185 people are randomly surveyed about the number of hours each week they spend watching television. The following table presents the survey data, using interval notation.

Recall that $[a, b)$ denotes all the times that are greater than or equal to a but strictly less than b.

Survey of Weekly Hours Spent Watching Television

Hours Spent Watching TV Each Week	Number of People, f_o
$[0, 5)$	6
$[5, 10)$	4
$[10, 15)$	12
$[15, 20)$	15
$[20, 25)$	28
$[25, 30)$	38
$[30, 35)$	34
$[35, 40)$	22
$[40, 45)$	15
$[45, 50)$	6
More than 50	5
	185

[6]The goodness-of-fit test can be conducted with reference to any distribution (such as normal, binomial, uniform, or Poisson). However, the main focus of this section is the application of goodness-of-fit tests to the normal distribution. Later courses in statistics may cover goodness-of-fit tests with other distributions.

The next example outlines a goodness-of-fit hypothesis test conducted to determine whether the claim that this sample comes from a normally distributed population— more specifically, from a normally distributed population with a mean of $\mu = 30$ and a standard deviation of $\sigma = 10$—should be rejected. Before proceeding with the example, it is necessary to examine how to determine the test statistic for a goodness-of-fit test.

The χ^2 test statistic used to test the independence of qualitative variables is also used as the test statistic for goodness-of-fit testing and is computed with the familiar formula

$$\chi^2 = \Sigma\left(\frac{(f_o - f_e)^2}{f_e}\right).$$

The numbers of people who watch TV for the stated numbers of hours are the observed frequencies, that is, values of f_o. Much of the work in a goodness-of-fit hypothesis test involves determining the corresponding values of f_e. The values of f_e depend on the null hypothesis, which indicates the shape of the population's distribution (in this case it is normal) along with any specified parameters ($\mu = 30$, $\sigma = 10$). Figure 11-10 should be helpful in understanding how values of f_e are obtained. Notice that the bands identified under the normal model represent specified times spent watching TV. Times are converted to Z-scores to mark the end points of the bands.

To compute the expected frequency for each of the 11 bands, it is necessary to find the band's area. The shaded area in Figure 11-10, which represents 35 to 40 weekly hours spent watching TV, will illustrate the method.

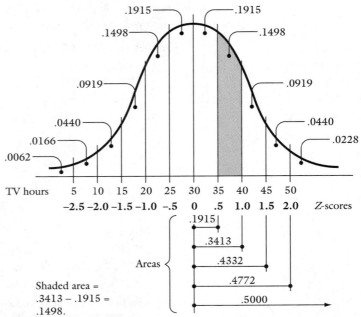

Figure 11-10 A normal distribution model with $\mu = 30$ and $\sigma = 10$.

a. Convert all TV times to Z-scores in the usual way, with the formula

$$Z(x) = \frac{x - \mu}{\sigma},$$

using the claimed values of $\mu = 30$ and $\sigma = 10$. The Z-score end points of the shaded band are

$$Z(35) = \frac{35 - 30}{10} = .5 \quad \text{and}$$

$$Z(40) = \frac{40 - 30}{10} = 1.0.$$

b. Compute the band area between the two Z-scores by using the standard normal table, Appendix A-2. Since Appendix A-2 lists areas from μ to the right, one must subtract to find the band area. The area of the shaded band is

$$.3413 - .1915 = .1498$$
$$= 14.98\%$$
$$= .1498.$$

c. The band area found in Step b represents the percentage of scores that should fall within that band if the population is normally distributed with $\mu = 30$ and $\sigma = 10$. Of a total of 185 scores, 14.98%, or 27.71 scores (rounding to the nearest hundredth), should fall in the band representing viewing times of 35 to 40 hours. Thus, for this band (or time interval), $f_e = 27.71$.

Determine the percentage of viewers in each of the remaining bands in the same way. The following table indicates the expected number of viewers, f_e, in each of the remaining bands.

Table of Observed and Expected Frequencies

Hours Spent Watching TV Each Week	Number of People	
	f_o	f_e
[0, 5)	6	.0062(185) = 1.15
[5, 10)	4	.0166(185) = 3.07
[10, 15)	12	.0440(185) = 8.14
[15, 20)	15	.0919(185) = 17.00
[20, 25)	28	.1498(185) = 27.71
[25, 30)	38	.1915(185) = 35.43
[30, 35)	34	.1915(185) = 35.43
[35, 40)	22	.1498(185) = 27.71
[40, 45)	15	.0919(185) = 17.00
[45, 50)	6	.0440(185) = 8.14
More than 50	5	.0228(185) = 4.22

Example 5 completes this goodness-of-fit hypothesis test.

E x a m p l e
5

Conduct a goodness-of-fit hypothesis test at the 10% level of significance to examine the claim that the sample of 185 TV viewing times came from a normal population with parameters $\mu = 30$ and $\sigma = 10$.

Solution   Use a five-step method as in previous hypothesis tests.

Step 1   *State the hypotheses.*

H_0:   The population is normally distributed with a mean of $\mu = 30$ and a standard deviation of $\sigma = 10$.

H_A: The population is not normally distributed with a mean of 30 and a standard deviation of 10.

Step 2   *Decide on a model for the sampling distribution.*

Use the chi-square distribution model for all goodness-of-fit hypothesis tests.

Step 3   *Define the rejection region, and state the decision rule.*

In a goodness-of-fit test, the degrees of freedom are

> **d.f. = No. of adjusted classes (intervals)**
> **− No. of estimated population parameters − 1.**

As in the first application of the chi-square distribution, it is a rule of thumb that no more than 20% of the intervals should have expected frequencies less than 5. In this case, three intervals have expected frequencies less than 5, and they constitute more than 20% of the total intervals. Therefore, it is necessary to combine the first three intervals and the last two intervals, for a total of eight adjusted intervals. Also, there are no estimated population parameters, since claimed values of μ and σ are part of the null hypothesis. Therefore,

> d.f. = No. of adjusted classes − No. of estimated population parameters − 1
>   = 8 − 0 − 1
>   = 7.

Consulting Appendix A-5, the critical value of χ^2, at the intersection of the 7 degrees of freedom row and the $\alpha = .10$ column, is 12.017. This marks the beginning of the rejection region. The decision rule is:

Reject H_0 and accept H_A if $\chi^2 > 12.017$. Otherwise, reserve judgment.

Step 4   *Calculate the χ^2 test statistic.*

The table of observed and expected frequencies follows, with indications of the intervals that needed to be combined.

The only possible outcomes of this hypothesis test are the conclusion that the population is not normally distributed (with $\mu = 30$ and $\sigma = 10$) and reservation of judgment.

Hours Spent Watching TV Each Week		Number of People	
		f_o	f_e
1	[0, 5) ⎫	6	.0062 · (185) = 1.15
	[5, 10) ⎬ *Combine*	4	.0166 · (185) = 3.07
	[10, 15) ⎭	12	.0440 · (185) = 8.14
		22	12.36
2	[15, 20)	15	.0919 · (185) = 17.00
3	[20, 25)	28	.1498 · (185) = 27.71
4	[25, 30)	38	.1915 · (185) = 35.43
5	[30, 35)	34	.1915 · (185) = 35.43
6	[35, 40)	22	.1498 · (185) = 27.71
7	[40, 45)	15	.0919 · (185) = 17.00
8	[45, 50) ⎫ *Combine*	6	.0440 · (185) = 8.14
	More than 50 ⎭	5	.0228 · (185) = 4.22
		11	12.36

These frequencies are used in the calculation of the χ^2 test statistic

The chi-square statistic is found with the usual formula:

$$\chi^2 = \sum \left(\frac{(f_o - f_e)^2}{f_e} \right)$$

$$= \frac{(22 - 12.36)^2}{12.36} + \frac{(15 - 17.00)^2}{17.00} + \cdots + \frac{(11 - 12.36)^2}{12.36}$$

$$= \frac{(9.64)^2}{12.36} + \frac{(-2.00)^2}{17.00} + \cdots + \frac{(-1.36)^2}{12.36}$$

$$\approx 9.56.$$

Recall that a reservation of judgment in a hypothesis test does not in any way support the null hypothesis.

Step 5   State the conclusion.

Since the computed value of $\chi^2 = 9.56$ is less than 12.017, **reserve judgment.** There is not enough statistical evidence to support the alternative hypothesis that these data did not come from a normal population with $\mu = 30$ and $\sigma = 10$. ●

The next example briefly illustrates how a different conclusion is drawn if the mean and standard deviation represent sample statistics instead of claimed population parameters.

E x a m p l e
6

Refer to Example 5. Conduct the same hypothesis test at the same level of significance, under the condition that the values for the mean and the standard deviation are estimated and are not part of the null hypothesis.[7]

---

[7] For ways in which population parameters may be estimated, review Chapter 4, "Interpreting Histograms and Frequency Curves."

Solution   An important difference between this example and the last one lies in the degrees of freedom used to determine the value of χ^2 that marks the end point of the rejection region. Recall that

d.f. = No. of adjusted intervals − No. of estimated population parameters − 1.

The two population parameters, μ and σ, are now to be treated as estimated values $\overline{X}$ and s. Accordingly, the degrees of freedom are

$$\text{d.f.} = 8 - 2 - 1$$
$$= 5.$$

According to Appendix A-5, the intersection of the 5 degrees of freedom row and the $\alpha = .10$ column establishes the end point of the rejection region at $\chi^2 = 9.236$. The decision rule is:

Reject H_0 and accept H_A if $\chi^2 > 9.236$. Otherwise, reserve judgment.

The computed value of χ^2, 9.56, is greater than 9.236. Therefore, **reject H_0 and accept H_A.** •

Exercises 11.4

1. Refer to Example 5 on page 464. Explain why not all eight interval widths are the same.

2. If the population parameters must be estimated, does that increase or decrease the probability that the null hypothesis of a goodness-of-fit test will be rejected?

3. Example 2 of Chapter 4 (page 127) presented the following school profile.

GPA Interval	Number of Students in Interval
[4.20, 5.00)	32
[3.50, 4.20)	58
[2.20, 3.50)	70
[1.50, 2.20)	12
[0.00, 1.50)	8

Examine the claim that the GPAs for this school are normally distributed with $\mu = 2.85$ and $\sigma = 1$. Apply a goodness-of-fit test at the 10% level of significance.

4. Refer to Exercise 3. Conduct the same test using new estimated values of the population parameters.

    Hint: These estimates were found in Example 2 of Chapter 4 (page 127).

5. Exercise 11 of Chapter 4 (page 138) described a situation in which a car rental company supplied the following mileage information on its fleet of cars.

Accumulated Mileage (thousands)	Tally
50 to under 60	8
40 to under 50	16
20 to under 40	108
15 to under 20	245
10 to under 15	120
5 to under 10	30
1 to under 5	30
0.5 to under 1	22
0.25 to under 0.5	8
under 0.25	10

a. Use the estimated mean and standard deviation found in that exercise to examine the claim that these data fit a normal distribution. Perform a goodness-of-fit test at the 20% level of significance.

b. What can be concluded from Part a about the attained significance level of the test?

6. A new type of blood pressure medicine is tested on 300 volunteers, and the following reaction times are recorded.

Reaction Time	Tally
6 minutes or more	18
5 minutes to under 6 minutes	30
4 minutes to under 5 minutes	59
3 minutes to under 4 minutes	93
2 minutes to under 3 minutes	60
1 minute to under 2 minutes	24
Less than 1 minute	16

Is there evidence that these reaction times are not normally distributed? Conduct a goodness-of-fit test at the 10% level of significance, using an estimated population mean of 3.50 minutes and an estimated standard deviation of 1.25 minutes.

GROUP   7. The claim is made that the average length of career for baseball players in the Hall of Fame is normally distributed, with $\mu = 18$ years and $\sigma = 4$ years. A random sample of 44 Hall of Famers reveals the following career spans.

Roy Campanella (1948–1957)   Roger Connor (1880–1897)
Early Wynn (1939–1963)   Mel Ott (1926–1947)
George Kelly (1915–1932)   Henry Aaron (1954–1976)
William Terry (1923–1936)   Adrian Joss (1902–1910)

Willie McCovey (1959–1980)	Grover Alexander (1911–1930)
Leon Goslin (1921–1938)	Ernie Banks (1953–1971)
Willie Mays (1951–1973)	Juan Marichal (1960–1975)
Harry Hooper (1909–1925)	Carl Yastrzemski (1961–1983)
Tyrus Cobb (1905–1928)	Robert Feller (1936–1956)
Enos Slaughter (1938–1959)	Raymond Schalk (1912–1929)
Mickey Mantle (1951–1968)	Warren Spahn (1942–1965)
John Bench (1967–1983)	Hugh Duffy (1888–1906)
Harmon Killebrew (1954–1975)	Don Drysdale (1956–1969)
Lou Brock (1961–1980)	Jackie Robinson (1947–1956)
John Eyers (1902–1919)	Luis Aparicio (1956–1973)
Lou Boudreau (1938–1952)	Clark Griffith (1891–1914)
Charles Comiskey (1882–1894)	Cy Young (1890–1911)
Sandy Koufax (1955–1966)	James Foxx (1925–1945)
Cap Anson (1876–1897)	Paul Waner (1926–1945)
Ted Lyons (1923–1946)	Frank Frisch (1919–1937)
Travis Jackson (1922–1936)	Charles Hafey (1924–1937)
Waite Hoyt (1918–1938)	Fred Clarke (1894–1915)

a. Compute the length of each player's career.

b. Construct a left-end/minimum-score histogram to illustrate the data from Part a. Use six intervals.

c. Is there evidence that these careers are not normally distributed with $\mu = 18$ and $\sigma = 4$? Perform a goodness-of-fit test at the 5% level of significance.

8. Refer to Exercise 7. Perform the same goodness-of-fit test if $\overline{X} = 18$ and $s = 4$, instead of having claimed population parameters.

Excursion 11.4A • Goodness of Fit and the Binomial Distribution

The material and exercises for goodness of fit and the binomial distribution are optional. They may be covered at any time or omitted.

Goodness-of-fit hypothesis tests may be applied to a wide variety of distribution models. The main difference between applications is the method used to determine the expected frequencies. A careful comparison of the next example with the preceding goodness-of-fit examples should reveal many similarities.

E x a m p l e 7 A building inspector may pass or fail an apartment in any of six segments of the city building code. The following table shows, out of a random sample of 200 apartments, the number of violations each apartment received.

Survey of Building-Code Violations

Number of Segment Violations, X	Number of Apartments with That Many Violations, f_o
0	31
1	51
2	70
3	32
4	9
5	5
6	2
	200

Conduct a goodness-of-fit hypothesis test at the 5% level of significance to examine the claim that the number of violations per apartment is a binomially distributed random variable.

Solution   Recall that a binomial random variable is a discrete random variable. Name the random variable X; it can take any of the values 0 through 6. Consider a violation a "success," and assume that each type of violation is independent of any other type. Each apartment has six chances for success.

Follow the usual five-step method.

The only possible outcomes of this hypothesis test are the conclusion that the number of building-code violations is not binomially distributed and reservation of judgment.

Step 1   State the hypotheses.

H_0:   The number of building-code violations is binomially distributed.

H_A: The number of building-code violations is not binomially distributed.

Step 2   Decide on a model for the sampling distribution.

Use the chi-square distribution model for all goodness-of-fit hypothesis tests.

Step 3   Define the rejection region and state the decision rule.

To identify the rejection region, one must determine the degrees of freedom for the model. Recall that

d.f. = No. of adjusted classes − No. of estimated population parameters − 1.

There are seven classes (0, 1, 2, 3, 4, 5, 6). However, the last two, 5 and 6 violations, have observed frequencies of only 5 and 2, respectively. It is therefore likely that the last two or three classes will have to be combined. For now, make the estimate that there will be five adjusted classes.

Recall from Chapter 6 that the binomial formula gives the probability of r successes in n trials:

$$P(X = r) = {}_nC_r \cdot p^r \cdot (1 - p)^{n - r}.$$

This expression, to be used later in the calculation of f_e for the χ^2 test statistic, contains one population parameter, p. Because p is not part of the null hypothesis, it will have to be estimated. Accordingly there is one estimated population parameter, and, the degrees of freedom are

$$\text{d.f.} = 5 - 1 - 1$$
$$= 3.$$

According to Appendix A-5, the intersection of the 3 degrees of freedom row and the $\alpha = .05$ column establishes the end point of the rejection region at $\chi^2 = 7.815$. Therefore, the decision rule is:

Reject H_0 and accept H_A if $\chi^2 > 7.815$. Otherwise, reserve judgment.

Step 4   Calculate the χ^2 test statistic.

This statistic depends on the estimated value of p, the probability that an apartment will receive a violation. The mean number of violations is

$$\mu_X = 0 \cdot \frac{31}{200} + 1 \cdot \frac{51}{200} + 2 \cdot \frac{70}{200} + \cdots + 6 \cdot \frac{2}{200}$$
$$= 1.8.$$

Recall that Chapter 6 showed that the mean value of a binomial random variable, X, is

$$\mu_X = n \cdot p.$$

Therefore, substituting the calculated value of $\mu_X = 1.8$ and $n = 6$,

$$1.8 = 6 \cdot p,$$
$$.3 = p.$$

Now that an estimate for p is available, one can compute expected frequencies. Consider the first expected frequency, for the number of apartments with 0 violations. The probability of receiving 0 violations out of 6 chances is

$$_6C_0 \cdot (.3)^0 \cdot (1 - .3)^6 = .117649.$$

Multiplying this probability by the number of apartments gives the expected frequency of the 0 violations class:

$$.117649 \cdot 200 \approx 23.53.$$

Find the remaining expected frequencies in a similar manner. The following table shows these expected frequencies, with indications of classes that must be combined.

Number of Segment Violations X	Number of Apartments with That Many Violations	
	f_o	f_e
0	31	23.53
1	51	60.50
2	70	64.83
3	32	37.04
4	9 ⎫	11.91 ⎫
5	5 ⎬ 16 14.10	2.04 ⎬
6	2 ⎭	.15 ⎭
	Combine	*Combine*

5 classes

These frequencies are used in the calculation of the χ^2 test statistic

Now determine the chi-square test statistic.

$$\chi^2 = \sum \left(\frac{(f_o - f_e)^2}{f_e} \right)$$

$$= \frac{(31 - 23.53)^2}{23.53} + \frac{(51 - 60.50)^2}{60.50} + \cdots + \frac{(16 - 14.1)^2}{14.1}$$

$$= \frac{(7.47)^2}{23.53} + \frac{(-9.50)^2}{60.50} + \cdots + \frac{(1.9)^2}{14.1}$$

$$\approx 5.22.$$

Step 5   *State the conclusion.*

Because $5.22 \leq 7.815$, **reserve judgment.** At the 5% level of significance there is not enough statistical evidence to support the alternative hypothesis that the number of violations per apartment is not a binomial random variable.   ●

Exercises 11.4A

1. What assumption(s) must be made when a goodness-of-fit test is conducted for a binomial distribution?

*   2. An auto mechanic believes that the warning lights for low oil pressure in a certain type of automobile come on in a binomially distributed pattern of occurrences for the first 20,000 miles. A random sample of repair records yields the following data.

Number of Times Light Went On in First 20,000 Miles	Observed Frequency (number of automobiles), f_o
0	7
1	12
2	16
3	18
4	25
5	23
6	20
7	15
8	9
9	0
10	4

Perform goodness-of-fit tests at the 1%, 5%, and 20% levels of significance to examine the mechanic's belief.

*   3. Refer to Exercise 2. Do you think that the mechanic's claim must always be rejected? Explain why or why not.

Case Study

Congressional Election Results

Consider the following data.

| | **Representatives, by Age** | | | | | |
	Under 40	**40–49**	**50–59**	**60–69**	**70–79**	**Over 80**
97th Congress (1981)	94	142	132	54	12	1
98th Congress (1983)	86	145	132	57	13	1
99th Congress (1985)	71	154	131	59	17	2
100th Congress (1987)	63	153	137	56	24	2
101st Congress (1989)	41	163	133	74	20	2
102nd Congress (1991)	39	152	134	86	20	4

On the basis of this cross-tabs table, one may argue that the nation is trending toward electing older representatives. *Statistical Abstract of the United States: 1993,* Section 8, presents this and related information. Obtain a copy of the book from your library, and use it to answer the following questions.

I. Perform a chi-square hypothesis test at the 5% level of significance to determine whether statistical evidence exists to support the claim that the nation is trending toward electing older representatives.

II. Construct a cross-tabs table, similar to the preceding one, that contains the ages of elected senators.

III. Perform a chi-square hypothesis test at the 5% level of significance to determine whether statistical evidence exists to support the claim that the nation is trending toward electing older senators.

IV. Do you believe incumbency of candidates is a factor in electing older representatives? Find data that support your opinion in *Statistical Abstract.*

V. Construct a cross-tabs table showing the numbers of blacks, hispanics, and females elected to Congress in 1981 and 1993. Then perform a chi-square hypothesis test at the 5% level of significance to determine whether statistical evidence exists to support the claim that minority representation is not independent of these particular years.

The U.S. Congress became a Republican majority in 1994 for the first time in 40 years. Age, ethnicity, and incumbency are other factors that determine the make-up of a particular legislative body. *(David Fish)*

CHAPTER SUMMARY

The chi-square distribution is an important distribution model with many applications. This chapter discussed three of those applications: testing for independence of qualitative variables, constructing confidence intervals for the population standard deviation, and testing goodness of fit. The symbol χ^2 is commonly used for the chi-square distribution as well as the chi-square test statistic.

When applying the chi-square distribution to hypothesis testing, use the same five-step method followed with the normal and t-distribution models.

Key Concepts, Terms, and Formulas

Independence of qualitative variables   *Page 441*

Formula for computing expected frequencies:

$$f_e = \frac{(RT) \cdot (CT)}{(GT)} \quad \textbf{(11-1)}$$ *Page 443*

Chi-square test statistic for independence of qualitative variables and goodness-of-fit testing:

$$\chi^2 = \sum \left(\frac{(f_o - f_e)^2}{f_e} \right) \quad \textbf{(11-2)}$$ *Page 444*

Chi-square distribution   *Page 444*

Guidelines for using the chi-square distribution model to test the independence of qualitative variables   *Page 448*

Value of chi square to construct a confidence interval for the standard deviation:

$$\chi^2 = \frac{(n - 1) \cdot s^2}{\sigma^2} \quad \textbf{(11-3)}$$ *Page 456*

Inequality to determine end points of a confidence interval for the standard deviation:

$$\sqrt{\frac{(n - 1) \cdot s^2}{\chi^2_{RIGHT}}} \leq \sigma \leq \sqrt{\frac{(n - 1) \cdot s^2}{\chi^2_{LEFT}}} \quad \textbf{(11-4)}$$ *Page 459*

Goodness-of-fit testing   *Page 461*

Videotape Suggestions

Program 11 in the series *Against All Odds: Inside Statistics* examines the association between variables displayed in a cross-tabs, or two-way, table and an interesting possibility called Simpson's paradox. *Program 24* in the same series provides more examples of cross-tabs tables and shows the chi-square distribution used in several practical applications.

(Atlanta, © 1994, Comstock, Inc.)

If no assumptions are
made about some traffic
patterns of this city, they
may be candidates for a
nonparametric test.

Chapter

12 Some Nonparametric Tests

John Arbuthnot. (*The Granger Collection, New York*)

A nonparametric, or **distribution-free, test** is a hypothesis test in which no assumptions are made about the shape of the distribution of the population.[1] Example 8 of Chapter 10 (page 423) used a nonparametric approach to construct a confidence interval for the difference of two populations' medians. In that example, no assumptions were made about the shape of the distribution of prices for rare coins and rare stamps.

Nonparametric tests are often very attractive because they tend to be easy to understand and apply. They are especially useful in situations where the assumption of a normally distributed population is unjustified. This chapter examines three of many nonparametric tests: the sign test, the number-of-runs test, and the Spearman rank-correlation test.

The **case study** at the end of this chapter focuses on the computer's roles in some new statistical methods.

12.1 THE SIGN TEST

The **sign test,** one of the easiest nonparametric tests to understand and apply, is a hypothesis test that determines whether statistical evidence exists that two populations differ with respect to the same characteristic.[2] For example, suppose that a particular advertising display is placed in various department stores for a week and then removed. In each store, the total sales (in hundreds of dollars) during a week with the display and during a week without the display are recorded, along with the sign (+ or −) of the difference in sales. The results of a random survey of 12 of the stores follow.

---

[1]The sampling distribution of the sample statistic, however, usually does have a specific shape. In this chapter, all nonparametric sample statistics have normal sampling distribution models.

[2]The sign test is just one of several nonparametric tests that analyze paired data. Another popular one is the Wilcoxon sign-rank test, which appears in most nonparametric statistics texts.

Store	Sales with Display, A	Sales without Display, B	Sign of A − B
1	22.5	20.7	+
2	18.3	20.1	−
3	16.5	16.5	0
4	20.0	20.0	0
5	18.7	18.9	−
6	23.4	20.9	+
7	21.8	16.4	+
8	27.9	27.3	+
9	18.5	18.3	+
10	21.8	23.7	−
11	22.4	21.9	+
12	18.7	16.3	+

If the presence of the display makes no difference in sales, then—based on the random selection of stores—there should be as many positive signs (+) as negative signs (−). That is, some stores should have sold more when the display was present, but just as many should have sold less when the display was present.

Let the sample statistic S stand for the number of plus signs. Then, if the display makes no difference, the expected value, or the average value of S, denoted by μ_S, can be found with the following formula.

Formula for the **Expected Number of Plus Signs, S**

$$\mu_S = \frac{n}{2}$$

(12-1)

In this formula, n is the number of nonzero differences.

In this instance there are two differences of zero. Therefore, if the display makes no difference, the expected number of plus signs is

$$\mu_S = \frac{12 - 2}{2}$$

$$= 5.$$

If the sample statistic S is treated as a binomial random variable, with probability of success (plus sign) $p = .5$, then the standard deviation of S can be found with the formula $\sigma = \sqrt{n \cdot p \cdot q}$ (presented in Chapter 6).

Formula for the **Standard Deviation of the Number of Plus Signs, S**

$$\sigma_S = \sqrt{n \cdot (.5) \cdot (.5)}$$

$$= .5 \cdot \sqrt{n}$$

(12-2)

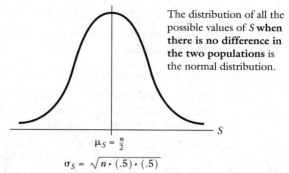

The distribution of all the possible values of S **when there is no difference in the two populations** is the normal distribution.

$$\mu_S = \frac{n}{2}$$

$$\sigma_S = \sqrt{n \cdot (.5) \cdot (.5)}$$

Figure 12-1 The sampling distribution of the number of plus signs.

In this case, if the display makes no difference in sales, the standard deviation of the number of plus signs should be

$$\sigma_S = \sqrt{10 \cdot (.5) \cdot (.5)}$$
$$= \sqrt{2.5}$$
$$\approx 1.58.$$

Also, for values of $n \geq 10$, the sampling distribution of the number of plus signs, S, may be considered to be normally distributed (Figure 12-1). This means that the normal distribution model may be used in a hypothesis test to determine whether the actual value of S is far enough from the expected value of S to warrant rejection of the null hypothesis. The next example illustrates this type of hypothesis test, called the sign test.

**E x a m p l e
1**

Perform a sign test at the 10% level of significance to determine whether there is statistical evidence that the advertising display just discussed makes a difference in sales.

Solution   Use the usual five-step hypothesis testing method.

Step 1   State the hypotheses.

The only possible outcomes of this hypothesis test are the conclusion that the display makes a difference in sales and reservation of judgment.

H_0:   The display makes no difference in sales.
H_A: The display makes a difference in sales.

In a sign test, the null hypothesis must always be that there is no difference in the two populations, because the formulas for the expected value (μ_S) and standard deviation (σ_S) of S are derived from a binomial random variable that assumes equal probabilities of success and failure.

Step 2   Decide on a model for the sampling distribution.

Use the normal distribution as the model for all the possible values of S under the null hypothesis.

Step 3   Determine the end points of the rejection region, and state the decision rule.

If the actual sales with and without the display are the same, then the obtained value of S should not be too far from its expected value, μ_S. The level of significance, $\alpha = .10$, determines how far from the expected value the obtained value of S must be to cause rejection of the null hypothesis.

Because this is a two-tail hypothesis test, the level of significance is split evenly between the outer left and outer right tails. Use Appendix A-2 and the following diagram to find the Z-scores that mark the end points of the rejection region. (As in earlier chapters, obtain the end points of the rejection region by searching the body of Appendix A-2 until the number closest to .4500 is found, which is .4495. The corresponding Z-scores that define this area are $z = \pm 1.64$.)

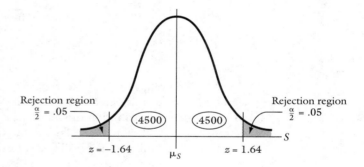

If $Z(S)$ represents the distance of S from its expected value, then the decision rule is:

Reject H_0 and accept H_A if $Z(S) > 1.64$ or if $Z(S) < -1.64$. Otherwise, reserve judgment.

Step 4   Compute the test statistic.

The value of S is 7, which is found by simply counting the number of plus signs. The test statistic is $Z(S)$, which is the number of standard deviations S is away from μ_S, is found with the familiar formula

$$Z(S) = \frac{S - \mu_S}{\sigma_S}.$$

The values of μ_S and σ_S were computed earlier as $\mu_S = 5$ and $\sigma_S \approx 1.58$. Therefore,

$$Z(S) = \frac{7 - 5}{1.58}$$

$$= 1.27.$$

Step 5   State the conclusion.

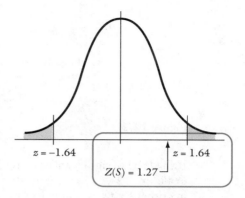

The value of $Z(S)$ is not in the rejection region. Therefore, **reserve judgment.** This result means that the expected number of plus signs and the actual number of plus signs do not differ enough to support the alternative hypothesis, that the display makes a difference in sales.   •

One-Tail Sign Tests

The sign test may also be conducted as a one-tail test. For example, suppose a researcher wants to compare the prices paid for breakfast cereals at supermarkets and at smaller stores. Thirteen different cereals are purchased and their prices recorded.

Brand of Cereal	Supermarket Price, *A*	Small Store Price, *B*	Sign of Difference, *A* − *B*
1	$2.43	$2.75	−
2	3.49	3.49	0
3	1.99	2.49	−
4	.99	1.49	−
5	1.99	1.95	+
6	4.59	4.99	−
7	1.49	.99	+
8	2.49	2.99	−
9	1.95	2.49	−
10	2.95	2.49	+
11	3.49	3.99	−
12	2.15	2.45	−
13	1.19	1.79	−

If the researcher seeks evidence that supermarket prices for breakfast cereals are less than the prices paid in smaller stores, the hypotheses are as follows:

H_0:   Cereal prices in supermarkets are greater than or equal to the prices paid in smaller stores.

H_A: Cereal prices in supermarkets are less than the prices paid in smaller stores.

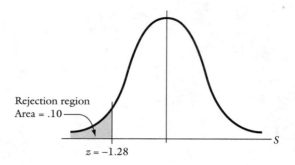

Figure 12-2

If the level of significance is, say, 10%, then the rejection region is located as shown in Figure 12-2. (Remember that the direction of the inequality of the alternative hypothesis always points toward the rejection region.) Exercise 10 completes this one-tail sign test.

Exercises 12.1

1. Although nonparametric tests are often easy to understand and apply, they also have certain disadvantages. Some statisticians would say that nonparametric tests waste information. Try to explain this criticism.

2. Explain why a sign test can provide only evidence of a difference in the characteristics of two populations, not evidence that the characteristics are the same.

3. The null hypothesis of a sign test claims that a certain characteristic is the same in two populations. As the level of significance increases, is this null hypothesis more likely or less likely to be rejected?

4. Explain why the number of plus signs, S, is treated as a binomial random variable with probability of success $p = .5$.

5. Researchers decide to use a sign test to determine whether there is evidence that a particular gasoline additive affects mileage. They select 15 different automobile models and record their gas mileages with and without the additive.

Automobile Model	With Additive (A)	Without Additive (B)	Sign of $A - B$
1	21.9	20.7	+
2	23.5	23.5	0
3	19.8	18.1	+
4	20.5	20.7	−
5	22.7	19.6	+
6	28.8	25.6	+
7	28.5	29.0	−
8	18.5	17.5	+
9	13.0	13.0	0
10	22.2	21.5	+

Automobile Model	With Additive (*A*)	Without Additive (*B*)	Sign of $A - B$
11	23.0	23.0	0
12	29.7	25.6	+
13	27.5	25.6	+
14	14.5	15.0	−
15	20.0	19.1	+

a. Perform the sign test at the 8% level of significance.

b. Estimate the attained level of significance of the test from Part a.

6. The following table shows the mean hourly compensation costs (in U.S. dollars)[3] for production workers in randomly selected countries for the years 1990 and 1993.

Country	1990	1993	Country	1990	1993
Australia	13.07	12.25	Austria	17.75	20.20
Belgium	19.22	21.38	Norway	21.47	20.20
Canada	15.94	16.36	Denmark	17.96	19.12
Finland	21.03	16.56	France	15.23	16.31
Germany	22.15	25.56	Hong Kong	3.20	4.31
Mexico	1.64	2.65	Italy	17.74	15.97
Japan	12.80	19.20	S. Korea	3.71	5.37
Netherlands	18.29	20.16	Portugal	3.77	4.60
Singapore	3.78	5.38	Spain	11.33	11.53
Sweden	20.93	17.91	Taiwan	3.95	5.23
U.K.	12.71	12.82	U.S.	14.91	16.79

Stat Fact

Although it is not listed in this table, the mean hourly compensation for production workers in Sri Lanka in 1990, 35¢, was one of the lowest in the world.

Perform a sign test at the 10% level of significance to determine whether there is evidence that hourly compensation costs for all production workers changed from 1990 to 1993.

7. The following table lists numbers of U.S. immigrants, in thousands, from ten randomly selected countries in the years 1991 and 1992.

Country of Birth	1991	1992	Country of Birth	1991	1992
Korea	26.5	19.4	Jamaica	23.8	18.9
Iran	19.6	13.2	Haiti	47.5	11.0
United Kingdom	13.9	20.0	Laos	10.0	8.7
Former Soviet Union	57.0	43.6	Taiwan	13.3	16.3
Mexico	946.2	213.8	El Salvador	47.4	26.2

Source: *Statistical Abstract of the United States, 1993* and *1994, p. 11.*

Perform a sign test at the 15% level of significance to determine whether levels of immigration from all countries changed from 1991 to 1992.

[3]Hourly compensation includes wages, premiums, bonuses, vacations, holidays and other leave, insurance, and benefit plans.

8. The following table presents the percent changes in total receipts for nonmanufacturing and service activities in some of the top counties of the United States in 1991–92.

County	Percent Change in 1991–92
Dade, FL	−4.8
Cook, IL	−10.5
San Francisco, CA	3.9
Harris, TX	−12.3
Orange, CA	−9.4
Philadelphia, PA	−8.8
Los Angeles, CA	−5.4
New York, NY	−8.9
Maricopa, AZ	−3.7
Fulton, GA	−4.9

Conduct a sign test at the 10% level of significance to determine whether a change occurred in receipts for the population during the years 1991 and 1992.

9. What is the difference between the type of claims investigated by one-tail sign tests and the type investigated by sign tests with two rejection regions?

10. Finish the one-tail sign test related to breakfast cereals (page 479) to determine whether evidence exists that cereal prices in supermarkets are less than those in small stores.

11. Refer to Exercise 5. Perform a one-tail sign test at the 5% level of significance to determine whether there is evidence that use of the gasoline additive results in a mileage greater than the mileage that would be obtained without the additive.

12. Refer to Exercise 8. Perform a one-tail sign test at the 0.5% level of significance to determine whether there is evidence that receipts for the population are less in 1992 than in 1991.

13. The following array shows the prevalences of some chronic health conditions among those under 45 years of age. All statistics are the numbers of such conditions per 1000 persons in 1992.

Chronic Condition	Male	Female
Arthritis	26.1	42.2
Dermatitus	32.6	48.6
Visual impairments	31.1	14.5
Cataracts	2.5	1.7
Hearing impairments	44.3	30.4
Diabetes	6.1	9.0
Migraine	26.0	60.8
Heart conditions	25.7	32.9
High blood pressure	37.5	30.2
Chronic bronchitis	41.0	58.0
Asthma	102.4	109.2
Chronic sinusitis	108.5	155.0

   a. Formulate a hypothesis that might be supported by the results of a one-tail sign test.

   b. Perform a one-tail sign test at the 10% level of significance to determine whether evidence exists to support the hypothesis from Part a.

14. Refer to Exercise 7 (in Exercises 10.6) on page 421.

   a. Perform a one-tail sign test at the 20% level of significance to determine whether evidence supports the claim that the mean salary of all 50 governors is less than the mean salary of all 50 chief justices.

   b. How does the conclusion from Part a compare with that from the original exercise?

   c. In what ways do the hypothesis tests from Part a and from the original exercise differ?

GROUP

15. The following table lists some of the top-rated syndicated television programs. Preference ratings from four different sample viewing groups are also indicated.

Program	Women	Men	Teens	Children
"Wheel of Fortune"	11.8	8.0	3.3	3.4
"Jeopardy"	9.9	6.9	3.0	2.3
"Star Trek"	7.1	9.2	6.4	5.1
"Entertainment Tonight"	6.4	5.0	2.1	1.9
"Magic II"	4.9	4.7	4.6	5.8
"Current Affair"	5.1	4.2	2.3	1.8
"Inside Edition"	5.6	4.0	1.8	1.6
"Kung Fu"	3.7	4.9	3.4	3.2
"National Geographic"	3.3	4.6	2.1	2.2
"World Wrestling Federation"	2.5	4.0	5.2	5.6

Source: *The World Almanac and Book of Facts, 1994,* p. 296.

   a. Perform a one-tail sign test at the 5% level of significance to determine whether there is evidence that women demonstrate a greater preference for these programs than men.

   b. Perform a one-tail sign test at the 5% level of significance to determine whether there is evidence that teens demonstrate a greater preference for these programs than children.

   c. For which two represented populations is $Z(S)$ closest to zero? For which two is $Z(S)$ greatest? Interpret the meaning of the relative values of $Z(S)$.

   Hint: There are six possible population pairs to consider.

12.2   THE NUMBER-OF-RUNS TEST

The **number-of-runs test** is a hypothesis test that determines whether there is statistical evidence that one of two possible outcomes in an experiment does not occur randomly—that is, that its occurrence can be predicted with reasonable

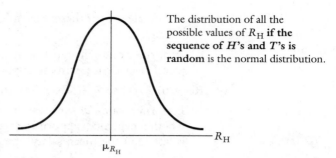

The distribution of all the possible values of R_H **if the sequence of *H*'s and *T*'s is random** is the normal distribution.

Figure 12-3 The sampling distribution of runs of heads.

certainty. The test is applicable when sample data can be separated into a sequence of outcomes arbitrarily defined as success and failure.

The number-of-runs test has many practical applications. For instance, suppose one sets out to determine whether a machine is producing defective parts randomly or is malfunctioning and needs adjustment. In other words, does the frequency of defective parts warrant suspicion of a machine performance problem, or is the frequency, relative to the number of parts being manufactured, attributable to randomness?

One approach is to sample many parts off the assembly line, in succession. For consistency of notation, H and T will denote the two possible outcomes.[4] That is, H will denote production of a defective part, and T will denote production of an acceptable part. If 23 parts are examined in succession, the results might be as follows:

T T H T T H H H H T T H T T T T H H H T T T.

The number-of-runs test always assumes under the null hypothesis that the variable under examination (defective parts, H) occurs randomly. One can determine the number of runs of H by using parentheses to separate blocks of different letters.

(T T)(H)(T T)(H H H H)(T T)(H)(T T T T)(H H H)(T T T).
    1   **2**   **3**   **4**

R_H denotes the number of runs of heads. From the preceding grouping, $R_H = 4$.

If the H's occur at random in the sequence, and if there are at least ten H's and ten T's, then the normal distribution model may be used for the sampling distribution of all the possible values of R_H (Figure 12-3). Furthermore, if the sequence of H's is random, the expected number (or mean number) of runs of heads, denoted by μ_{R_H}, can be found with the following formula.

---

[4]The letters H and T stand for heads and tails, the two possible outcomes of flipping a coin. This notation suggests the binomial nature of the number-of-runs test and is used throughout this chapter to represent success and failure.

Formula for the **Expected Number of Runs of Heads, R_H**

$$\mu_{RH} = \frac{n_H \cdot (n_T + 1)}{n_H + n_T} \qquad (12\text{-}3)$$

In this formula, n_H denotes the number of heads in the sequence and n_T the number of tails.

In this case, one can verify that $n_H = 10$ and $n_T = 13$, which means that in a random sequence of H's and T's 23 letters long, with 10 H's and 13 T's, one would expect the number of runs of heads to average to

$$\mu_{RH} = \frac{10 \cdot (13 + 1)}{10 + 13}$$

$$\approx 6.09.$$

The actual number of runs of heads is $R_H = 4$. The purpose of the number-of-runs hypothesis test is to determine whether this obtained value, 4, is so far from the expected value, 6.09, that the assumption of randomness must be rejected. To make that determination, the standard deviation of the number of runs, denoted by σ_{RH}, must be calculated.

Formula for the **Standard Deviation of the Number of Runs of Heads, R_H**

$$\sigma_{RH} = \sqrt{\frac{n_H \cdot (n_T + 1) \cdot (n_H - 1)}{(n_H + n_T)^2} \cdot \frac{n_T}{n_H + n_T - 1}} \qquad (12\text{-}4)$$

**E x a m p l e
2**

Referring to the preceding discussion on defective and acceptable machine parts, use the number-of-runs test to determine whether there is evidence that defective parts (H's) are not being produced at random. Perform this test at the 20% level of significance.

Solution   Use the five-step hypothesis test method.

Step 1   *State the hypotheses.*

H_0:   Defective parts are produced on a random basis (that is, the sequence of H's is random).

H_A: Defective parts are not produced at random.

The assumption of randomness must be assigned to the null hypothesis in order to apply the formulas for the expected number of runs and the standard deviation of the expected number of runs. This assumption of randomness is also needed to justify the use of the normal distribution model in the next step.

The only possible outcomes of this hypothesis test are the conclusion that defective parts are not produced at random and reservation of judgment.

Step 2 Decide on a model for the sampling distribution.

Use the normal distribution as the model for the number of runs of H in a random sequence.

Step 3 Determine the end points of the rejection region, and state the decision rule.

The rejection region is in the outer 20% of the normal distribution model. Because too few or too many runs of H could lead to rejection of the assumption of randomness, this is a two-tail test. Therefore, the rejection region is in the outer $\frac{20\%}{2} = 10\%$ of each tail.

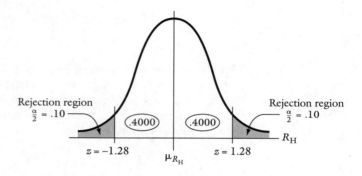

If $Z(R_H)$ represents how far R_H is from its expected value, μ_{R_H}, then, using Appendix A-2 and the above diagram, the decision rule is:

Reject H_0 and accept H_A if $Z(R_H) > 1.28$ or if $Z(R_H) < -1.28$. Otherwise, reserve judgment.

Step 4 Compute the test statistic.

The test statistic is $Z(R_H)$, the number of standard deviations R_H is from its expected value, μ_{R_H}. $Z(R_H)$ is found with the familiar formula

$$Z(R_H) = \frac{R_H - \mu_{R_H}}{\sigma_{R_H}}.$$

To compute this test statistic, first calculate the expected number of runs, μ_{R_H}, and the standard deviation, σ_{R_H}. It has already been shown that

$$\mu_{R_H} = \frac{n_H \cdot (n_T + 1)}{n_H + n_T}$$

$$= \frac{10 \cdot (13 + 1)}{10 + 13}$$

$$\approx 6.09.$$

Find the standard deviation by evaluating the following:

$$\sigma_{R_H} = \sqrt{\frac{n_H \cdot (n_T + 1) \cdot (n_H - 1)}{(n_H + n_T)^2} \cdot \frac{n_T}{n_H + n_T - 1}}$$

$$= \sqrt{\frac{10 \cdot (13 + 1) \cdot (10 - 1)}{(10 + 13)^2} \cdot \frac{13}{10 + 13 - 1}}$$

$$= \sqrt{\frac{10 \cdot 14 \cdot 9}{(23)^2} \cdot \frac{13}{22}}$$

$$= \sqrt{\frac{1260}{529} \cdot \frac{13}{22}}$$

$$= \sqrt{\frac{16,380}{11,638}}$$

$$\approx \sqrt{1.407}$$

$$\approx 1.19.$$

Therefore,

$$Z(R_H) = \frac{R_H - \mu_{R_H}}{\sigma_{R_H}}$$

$$= \frac{4 - 6.09}{1.19}$$

$$= \frac{-2.09}{1.19}$$

$$= -1.76.$$

Step 5 *State the conclusion.*

The value of the test statistic, $Z(R_H) = -1.76$, is in the rejection region.

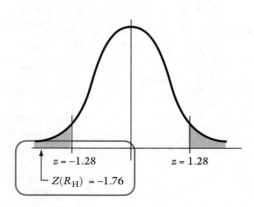

$z = -1.28$ $z = 1.28$

$Z(R_H) = -1.76$

Therefore, **reject** H_0. There is enough statistical evidence at the 20% level of significance to support the claim that defective parts are not being produced at random. The relatively large number of defective parts, 10 of 23, hinted at this result.   ●

Exercises 12.2

1. The number-of-runs test assumes that the sequence of H's occurs at random. The sequence is just as unlikely to have too many runs of H's as to have too few runs of H's. Explain why this assumption suggests a two-tail model for this hypothesis test.

2. In a number-of-runs test, does the expected value, μ_{R_H}, represent the mean number of heads or the mean number of runs of heads?

3. What is the maximum possible difference between R_H and R_T?

4. Refer to the conclusion of Example 2 on page 485. Is this necessarily a "good" result? Should the machine be adjusted? Explain.

Exercises 5–15 specify the variables for the number-of-runs tests so that answers will be consistent and may be checked more easily.

5. A coin is flipped 30 times, with these results:

   T T H H T H T H T H H H T H T T T T T T H T H T H T H T H H H T T H.

   Conduct a number-of-runs test at the 20% level of significance to determine whether there is evidence that heads are not occurring randomly and thus the coin is not fair.

6. A Little League baseball coach believes that his team's wins do not occur randomly. A partial record of his team's wins and losses is

   W W L L W L L W L W L L L W L W L L W W L W W.

   a. Perform a number-of-runs test at the 15% level of significance to determine whether there is evidence that this team's wins do not occur randomly.

   b. Determine the attained significance level of the test in Part a.

7. Is the following statement true or false? "The actual number of heads has a greater effect on the outcome of a number-of-runs test than does the actual number of runs of heads."

8. The exact amount of coffee dispensed into a 6-ounce cup by a vending machine is measured. Assume that the measurements are precise, so that each cup is either overfilled or underfilled. The owner of the machine thinks that overfilling does not occur at random. Use the following data from 40 consecutive overfills and underfills to test the owner's belief. Conduct a number-of-runs test at the 10% level of significance. U U O O O O U O O U U O O O O O U U O O U U U U O U O U O U U U U U O O O O O O O.

9. Foresters made a record of the diseased trees and the healthy trees along a certain highway:

    D D H H D H H H H H D D D H H D H H H H H D D D H.

    Are the diseased trees clustered (not occurring at random)? Perform a number-of-runs test at the 10% level of significance.

10. The gender sequence of the people waiting in line to see an exhibition of ballroom dancing is

    M M F F M M F M F M M F F F F F F F F F F M M F M.

    Is this evidence that men are not randomly identified? Perform a number-of-runs test at the 20% level of significance.

* 11. The number-of-runs test may be used to test the randomness of **runs above and below the median.** On 30 consecutive weekdays, on a route between downtown and the suburbs, a bus carried 34, 23, 19, 10, 44, 34, 45, 30, 29, 9, 28, 34, 40, 23, 45, 31, 27, 18, 20, 20, 33, 40, 37, 8, 27, 21, 19, 22, 33, and 19 passengers.

    a. Determine the median number of passengers carried.

    b. Let A represent a number of passengers carried above the median, and let B represent a number of passengers carried below the median. Form a sequence of A's and B's that shows the numbers of passengers carried above and below the median during these 30 consecutive weekdays. (Omit numbers equal to the median.)

    c. Use the results of Part b to determine whether there is evidence that the number of passengers carried above the median is not random. Conduct a hypothesis test at the 10% level of significance.

12. A student's quiz scores in a year-long introductory statistics class were 87, 90, 66, 88, 87, 90, 93, 56, 87, 90, 77, 75, 89, 92, 90, 78, 66, 67, 85, 75, 90, 45, 77, 67, 89, 67, 88, 55, 65, 83. Conduct a number-of-runs test at the 10% level of significance to determine whether these results are evidence that the student's performance above the median did not occur randomly.

    Hint: See Exercise 11.

13. Let E represent the digits 0, 2, 4, 6, and 8, and let O represent the digits 1, 3, 5, 7, and 9. Form a sequence of E's and O's 50 letters long by converting each of the digits in the first column of the random number table (Appendix A-3) to either an E or an O. Perform a number-of-runs test at the 10% level of significance to determine whether there is evidence that even numbers did not occur randomly.

GROUP * 14. Refer to the Monte Carlo method illustrated in Example 7 of Chapter 5A, on page 188. Write a computer program that simulates 50 tosses of a fair coin and records the results of the flips in a sequence of H's and T's. Perform a number-of-runs test at the 10% level of significance to determine whether there is evidence that the program is not simulating flips of heads on a random basis.

* 15. Many calculators can generate lists of random numbers. As shown in Chapter 7, (page 287), the following keystrokes activate this feature in the TI-82 graphics calculator.

- Press the [MATH] key.
- Press the [▶] key three times.
- Press the [1] key.

Now, everytime the [ENTER] key is pressed, a different random number is displayed.

a. In a manner similar to Exercise 13, form a sequence of 50 digits, converted to E's and O's, by using the second digit in each random number.

b. Perform a number-of-runs test at the 10% level of significance to determine whether there is evidence that the E's found in Part a did not occur randomly.

12.3 THE SPEARMAN RANK-CORRELATION TEST

The **Spearman rank-correlation test** is a hypothesis test that determines how well two sets of rankings are correlated. If both rankings tend to place the same items low and the same items high, they are **positively rank-correlated.** If items that are placed low in one ranking are placed high in the other, the two rankings are **negatively rank-correlated.** (Two sets of rankings are **rank-correlated** if they are either positively rank-correlated or negatively rank-correlated.)

The Spearman rank-correlation coefficient, denoted by r, can take any of the values between -1 and $+1$, inclusive. The following formula, which for now assumes that no ties occur within a set of rankings,[5] determines the value of r.

Formula for the **Spearman Rank-Correlation Coefficient, r**

$$r = 1 - \frac{6 \cdot \Sigma d_i^2}{n \cdot (n^2 - 1)} \qquad (12\text{-}5)$$

In this formula, n is the number of items to be ranked, d_i is the difference in ranks of item number i, and Σd_i^2 is the sum of all the squared differences.

For example, suppose two coaches rank four baseball players on their hitting, giving the best hitter rank number 1.

[5]Ties within a set of rankings are discussed in Section 12.5, on page 501.

Player	Coach A	Coach B
A	2	1
B	4	4
C	3	2
D	1	3

In this case, $n = 4$. The first difference in ranks (for player A) is

$$d_1 = 2 - 1$$
$$= 1,$$

and

$$d_1^2 = 1^2$$
$$= 1.$$

One can verify that $d_2^2 = 0$, $d_3^2 = 1$, and $d_4^2 = 4$. The sum of these squared differences is

$$\Sigma d_i^2 = (1 + 0 + 1 + 4)$$
$$= 6.$$

The value of r for these rankings can now be calculated:

$$r = 1 - \frac{6 \cdot \Sigma d_i^2}{n \cdot (n^2 - 1)},$$
$$r = 1 - \frac{6 \cdot 6}{4 \cdot (4^2 - 1)}$$
$$= 1 - \frac{36}{4 \cdot 15}$$
$$= .40.$$

A value of r close to $+1$ would have indicated that the two coaches ranked very similarly (ranked the same players high and the same players low). A value of r close to -1 would have indicated that the two coaches' rankings were opposite (players ranked high by one coach were ranked low by the other coach). A value of r close to either $+1$ or -1 would signify a strong rank correlation. In this case, $r = .40$, which is closer to zero than to one, indicates little or no correlation in ranking.

For less obvious values of r, one can decide whether a rank correlation exists by conducting a hypothesis test with the sampling distribution of r. When the two rankings of the variable of interest are not correlated and when $n \geq 10$, all the possible values of r may be considered normally distributed (Figure 12-4) with the following parameters.

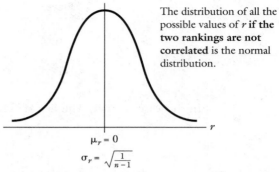

The distribution of all the possible values of r **if the two rankings are not correlated** is the normal distribution.

$$\mu_r = 0$$

$$\sigma_r = \sqrt{\frac{1}{n-1}}$$

Figure 12-4 The sampling distribution of the Spearman rank-correlation coefficient.

Formula for the **Mean and Standard Deviation of all Possible Values of** r

$$\mu_r = 0 \qquad\qquad\qquad (12\text{-}6)$$

$$\sigma_r = \sqrt{\frac{1}{n-1}} \qquad\qquad\qquad (12\text{-}7)$$

The following example illustrates how a Spearman rank-correlation hypothesis test is performed.

E x a m p l e 3

Two art critics ranked ten pieces of modern art, giving rank number 1 to the favorite piece and rank number 10 to the piece liked the least.

Work of Art	Rank of Critic A	Rank of Critic B
A	1	3
B	3	4
C	7	10
D	5	5
E	10	7
F	4	1
G	2	6
H	9	9
I	6	2
J	8	8

Determine whether there is evidence of a rank correlation between these two critics at the 5% level of significance.

Solution   Use the five-step method for hypothesis tests.

Step 1   State the hypotheses.

  H_0:   The rankings of the two critics are not correlated.
  H_A: The rankings of the two critics are correlated.

The only possible outcomes of this hypothesis test are the conclusion that the rankings of the two critics are correlated and reservation of judgment.

In a Spearman rank-correlation test, the null hypothesis is always that of no correlation.

Step 2   Decide on a sampling distribution model.

Use the normal distribution as the model for all the possible values of r under the null hypothesis.

Step 3   Determine the end points of the rejection region, and state the decision rule.

If the two rankings are not correlated, then the value of r should not be too far from its mean value, zero. That is, r should not be too far into either the left tail (strong negative correlation) or the right tail (strong positive correlation) of the normal distribution model. Therefore, the level of significance, α, is split evenly between the outer left and outer right tails. According to Appendix A-2 and the following diagram, the Z-scores that mark the rejection region are $z = \pm1.96$.

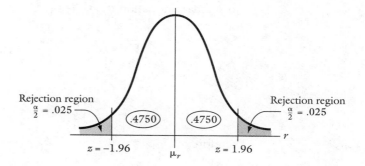

The decision rule is:

Reject H_0 and accept H_A if $Z(r) > 1.96$ or if $Z(r) < -1.96$. Otherwise, reserve judgment.

Step 4   Compute the test statistic.

One can compute the value of r more easily by adding two columns, d_i and $(d_i)^2$, to the rankings.

Work of Art	Critic A	Critic B	d_i	d_i^2
A	1	3	-2	4
B	3	4	-1	1
C	7	10	-3	9
D	5	5	0	0
E	10	7	3	9
F	4	1	3	9
G	2	6	-4	16
H	9	9	0	0
I	6	2	4	16
J	8	8	0	0
				$\Sigma d_i^2 = \overline{64}$

494 Chapter 12 Some Nonparametric Tests

$$r = 1 - \frac{6 \cdot \Sigma d_i^2}{n \cdot (n^2 - 1)}$$

$$= 1 - \frac{6 \cdot 64}{10 \cdot (10^2 - 1)}$$

$$= 1 - \frac{384}{990}$$

$$\approx .612.$$

The Z-score for this value of r is the number of standard deviations .612 is from the expected value of r, which is 0. Find $Z(r)$ with the usual formula:

$$Z(r) = \frac{r - \mu_r}{\sigma_r}.$$

Recall that the normal model for the sampling distribution or r has mean $\mu_r = 0$ and standard deviation $\sigma_r = \sqrt{\frac{1}{n-1}}$. Using the value of $n = 10$,

$$\sigma_r = \sqrt{\frac{1}{10 - 1}}$$

$$= \sqrt{\frac{1}{9}}$$

$$\approx .33.$$

Therefore,

$$Z(.612) = \frac{.612 - 0}{.33}$$

$$\approx 1.85.$$

Step 5 State the conclusion.

The value of $Z(r)$ is not quite in the right rejection region starting at 1.96. Thus, **reserve judgment.** At the 5% level of significance there is not enough statistical evidence to support the alternative hypothesis that the rankings are correlated. ●

Either Minitab or the TI-82 graphics calculator can be used to compute the Spearman rank-correlation coefficient. The next two examples illustrate the use of this software. In each application, the software calculates only the coefficient r; once that value is known, r is used in the calculation of $Z(r)$, and a hypothesis test may be completed.

E x a m p l e
4

Use Minitab to compute the Spearman rank-correlation coefficient, r, from Example 3.

Solution Enter the rankings into columns C1 and C2 of a Minitab worksheet.

	C1	C2	C3
↓	Critic A	Critic B	
1	1	3	
2	3	4	
3	7	10	
4	5	5	
5	10	7	
6	4	1	
7	2	6	
8	9	9	
9	6	2	
10	8	8	
11			

Next, obtain a Session window and type the macro command CORRELATION, followed by the locations of the data. Figure 12-5 shows the output once the ENTER key is pressed.

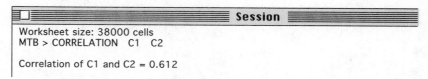

```
                                          Session
Worksheet size: 38000 cells
MTB > CORRELATION   C1   C2

Correlation of C1 and C2 = 0.612
```

Figure 12-5 Minitab's CORRELATION command. ●

E x a m p l e
5

Use a graphics calculator to compute the Spearman rank-correlation coefficient, r, from Example 3 on page 492.

Solution Use the TI-82 graphics calculator.[6]

A fast way to clear the first two lists of data is STAT, 4, 2nd, 1, 9, 2nd, 2, ENTER.

• Clear any previous data from lists L1 and L2.
• Enter the data from the two critics into lists L1 and L2, according to the following pattern.

[6]The TI-82 and most other calculators compute a correlation coefficient that is *not* the Spearman rank-correlation coefficient but the Pearson coefficient of correlation. However, when no ties occur within a set of rankings, these two correlation coefficients are exactly the same. When ties do occur within a set of rankings, the two coefficients may differ, but usually not by much. Chapter 14 discusses the Pearson coefficient of correlation.

	L1	L2
Work of art	Rank of critic A	Rank of critic B
A	1	3
B	3	4
C	7	10
·	·	·
·	·	·
·	·	·

When both lists of data are complete, the screen should appear as follows:

```
L1      L2      L3
1       3       ------
3       4
7       10
5       5
10      7
4       1
2       6
L3(1)=
```

- Press STAT, ▶ to obtain the following menu.

```
EDIT  CALC
1: 1-VarStats
2: 2-Var Stats
3: SetUp
4: Med-Med
5: LinReg(ax+b)
6: QuadReg
7↓CubicReg
```

- Type 5, ENTER to obtain the linear regression statistics shown in Figure 12-6.[7]

To be
explained
in Chapter 14 ──

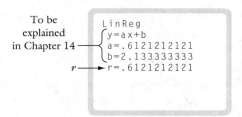

```
LinReg
⎧ y=ax+b
⎨ a=.6121212121
⎩ b=2.133333333
r ──▶ r=.6121212121
```

Figure 12-6 Linear regression statistics obtained from a TI-82 graphics calculator.

[7]Chapter 14 explains linear regression. For now, the only statistic of interest in this screen is the value of r.

Exercises 12.3 – 12.4

1. If the Spearman rank-correlation coefficient, r, is zero, then where on the model of the sampling distribution does the test statistic $Z(r)$ fall?

2. A strong rank correlation may be either positive or negative. If a 5-year-old child and a 30-year-old adult ranked ten television shows in order of their preferences, would you expect the value of r to be positive or negative?

GROUP

3. Intuitively decide which of the following pairs of rankings are most strongly rank-correlated and which are least strongly rank-correlated.

(A)	1	2	3	4	5	6	7	8	9	10
	10	9	8	6	7	5	4	2	3	1

(B)	1	2	3	4	5	6	7	8	9	10
	2	1	5	3	4	9	8	7	10	6

(C)	1	2	3	4	5	6	7	8	9	10
	8	5	1	9	10	2	8	4	6	3

4. Check your intuitive answers to Exercise 3 by computing the Spearman rank-correlation coefficient for each pair of rankings.

5. Following are rankings of the seven states that generate the greatest waste per year. The percentages of their waste that are recycled are also ranked.

	Waste Generated (tons/year)	Recycled (%)
CA	1	3
FL	3	2
IL	6	5
MI	7	1
NY	2	4
OH	5	7
TX	4	6

Based on data from *The 1993 Information Please Environmental Almanac*, pp. 54–55.

Determine the value of the Spearman rank-correlation coefficient for these data.

6. Mortality rates from sudden infant death syndrome (SIDS) in nine geographic regions of the United States have been computed and ranked for whites and blacks. A rank of 1 denotes the highest SIDS rate, and a rank of 9, the lowest.

Geographic Region	Rank for Whites	Rank for Blacks
New England	8	9
Middle Atlantic	9	8
East North Central	5	1
West North Central	3	2
South Atlantic	6	6
East South Central	2	5
West South Central	7	7
Mountain	1	4
Pacific	4	3

Source: *Statistical Bulletin*, January–March 1993, p. 12.

Compute the value of the Spearman rank-correlation coefficient for these data.

7. The 12 industrialized countries with the greatest life expectancies for males are ranked as follows, with rank number 1 assigned to the country with the greatest life expectancy. These 12 countries are then ranked with respect to female life expectancy.

Country	Male	Female	Country	Male	Female
Japan	1	1	Norway	7	8
Iceland	2	4	Canada	8	9
Sweden	3	3	France	9	5
Switzerland	4	2	Germany	10	12
Netherlands	5	6	United States	11	11
Australia	6	7	Finland	12	10

Based on data from *American Demographics*, October 1992, p. 45.

a. Compute the value of the Spearman rank-correlation coefficient for these data.

b. Determine whether there is evidence of a rank correlation between the life expectancies for males and females at the 10% level of significance.

S t a t   F a c t

The United States also conducted draft lotteries in 1917, 1918, and 1940. All encountered difficulties. Consult the case study in Chapter 7 (page 303) for further details.

8. In 1970 the U.S. Selective Service draft lottery "randomly" assigned a number between 1 and 366 to each day of the year. The mean of all the numbers assigned to the days of each month may be computed. The month with the smallest mean would have a priority rank of 1; in other words, the Selective Service would draw most heavily on that month for draftees (identified by their birth dates). The month with the greatest mean would have a priority rank of 12; the fewest draftees would be drawn from that month. The following table includes the priority ranking for each month.

Month	Calendar Ranking	Priority Ranking	Month	Calendar Ranking	Priority Ranking
January	1	5	July	7	8
February	2	4	August	8	9
March	3	1	September	9	10

Month	Calendar Ranking	Priority Ranking	Month	Calendar Ranking	Priority Ranking
April	4	3	October	10	7
May	5	2	November	11	11
June	6	6	December	12	12

a. If the lottery randomly chose draft priority days, should there be evidence of a rank correlation between monthly calendar rank and monthly draft priority rank?

b. Determine whether there is evidence of a rank correlation between monthly calendar rank and monthly draft priority rank at the 10% level of significance.

c. Estimate the attained significance level of the test from Part a.

9. Refer to Exercise 8. Reverse the order of the calendar rankings so that January has a ranking of 12 and December has a ranking of 1. Keep the months' priority rankings the same.

   a. What effect, if any, does the change in calendar rankings have on the value of the Spearman rank-correlation coefficient?

   b. Compute the Spearman rank-correlation coefficient for these transformed data, and compare it with the value from Exercise 8.

   c. Try to make a generalization on the basis of the results of Parts a and b.

10. The following table ranks the 15 restaurants that had the most franchises in 1990 in terms of numbers of domestic and foreign franchises (a rank of 1 signifying the most franchises).

Franchise	Rank by Number of Domestic Franchises	Rank by Number of Foreign Franchises
McDonald's	1	1
Kentucky Fried Chicken	2	2
Pizza Hut	3	3
Burger King	4	5
Domino's	5	7
Subway Sandwiches	6	10
Dairy Queen	7	6
Hardee's	8	15
Wendy's	9	9
Little Caesar's	10	12
Taco Bell	11	14
Baskin-Robbins	12	4
Arby's	13	11
Dunkin' Donuts	14	8
TCBY	15	13

Determine whether there is evidence of a rank correlation between the numbers of domestic and foreign franchises for this group. Test at the 15% level of significance.

Women smokers can be ranked according to occupation. (*American Cancer Society*)

S t a t   F a c t

Of those states reporting their tobacco production in 1992, North Carolina had the greatest (610 million pounds), and Massachusetts had the least (678,000 pounds.)

11. The following table ranks 14 occupations of women in terms of current smokers and former smokers. The occupation with the greatest percentage of (current or former) smokers received a rank of 1, and that with the smallest percentage received a rank of 14.

Women

Occupation	Ranking for Current Smokers	Ranking for Former Smokers
Executive/administrative	9	2
Professional/specialty	14	1
Technical and related	12	5
Sales	7	6
Administrative support	8	4
Private households	10	9
Protective service	1	14
Service	5	7
Farming, forestry, and fishing	11	13
Precision production	6	10
Machine operators	3	11
Transportation	4	3
Handlers/cleaners	2	12
Other	13	8

Based on data from *Statistical Bulletin* 73:4, October–December 1992, p. 17.

a. Compute the value of the Spearman rank-correlation coefficient for these data. Interpret the sign of r.

b. Perform a hypothesis test at the 5% level of significance to determine whether there is evidence of a rank correlation between current and former smokers in these occupations.

12. The following table contains the same information for men as Exercise 11 provided for women.

Men

Occupation	Ranking for Current Smokers	Ranking for Former Smokers
Executive/administrative	13	1
Professional/specialty	14	2
Technical and related	10	10
Sales	11	4
Administrative support	8	9
Private households	4	14
Protective service	7	11
Service	5	13

Occupation	Ranking for Current Smokers	Ranking for Former Smokers
Farming, forestry, and fishing	9	6
Precision production	6	3
Machine operators	3	7
Transportation	2	5
Handlers/cleaners	1	12
Other	12	8

Based on data from *Statistical Bulletin* 73:4, October–December 1992, p. 16.

Compute the value of the Spearman rank-correlation coefficient for these data, and compare this value of r with that in Exercise 11. Is there a stronger rank correlation for men than for women?

* 13. Refer to Exercises 11 and 12. Create a new table that compares rankings of current female smokers and current male smokers. Determine whether there is evidence of a rank correlation between these groups at the 5% level of significance.

GROUP 14. Refer to Exercises 11 and 12. Between which two of six possible pairs of rankings does the strongest rank correlation occur? The weakest rank correlation?

15. Does a positive correlation between two rankings necessarily imply that the rankings are valid? Does a negative rank correlation necessarily imply that the rankings are not valid? Explain.

12.5 TIES IN RANKINGS

It is possible and (in practical terms) common for two or more items to be ranked the same within a single set of rankings. For example, several students may have the same GPA, two stores in a group of chain stores may have the same sales in a given month, or two types of automobiles may achieve the same gasoline mileage.

On a more subjective basis, suppose each of 260 students is asked to vote for his or her favorite of five specified novels. The results of the voting follow.

Vote Tallies for Favorite Novels

Novel	Votes
A Tale of Two Cities	80
Gone With the Wind	65
Huckleberry Finn	65
To Kill a Mockingbird	40
Crime and Punishment	10

Gone With the Wind and *Huckleberry Finn* both occupy ranks 2 and 3. Therefore, they jointly occupy the rank that is the mean of these two ranks.

Rankings of Favorite Novels

Novel	Rank
A Tale of Two Cities	1
Gone With the Wind	2.5
Huckleberry Finn	2.5
To Kill a Mockingbird	4
Crime and Punishment	5

The same procedure may be used when ties of three or more ranks occur. These decimal, or fractional, rankings may then be used in the same manner as whole-number rankings to compute r and to perform hypothesis tests.

Exercises 12.5

1. Researchers polled one group of people composed of recent high-school graduates and another group composed of older adults, asking each participant which course in the following list was his or her favorite in high school.

Course	Recent High-School Graduate	Older Adult
U.S. History	45	42
Biology	20	20
Geometry	25	55
Algebra	20	3
English	27	10
Music	8	20
Art	45	35
Physical education	5	15
Foreign language	15	20
Technical	45	35

   a. Convert these votes to ranks.

   b. If $r < 0$ and a rank correlation exists, what may that mean? If $r > 0$ and a rank correlation exists, what may that mean?

   c. Determine whether there is evidence of a rank correlation at the 20% level of significance.

2. The family income levels of SAT test takers in 1992 are cross-tabulated by geographic location in the following array.

Income Level	Large Cities	Rural Areas
Over $70,000	18%	10%
$60,000 to under $70,000	6%	6%
$50,000 to under $60,000	9%	10%
$40,000 to under $50,000	11%	15%
$30,000 to under $40,000	16%	22%

Income Level	Large Cities	Rural Areas
$20,000 to under $30,000	16%	19%
$10,000 to under $20,000	15%	14%
Under $10,000	9%	5%

Rank these data, then compute their Spearman rank-correlation coefficient.

3. Refer to Exercise 2. In your own words, state the question that Part b attempts to answer.

GROUP * 4. The following table shows the number of motor-vehicle fatalities in each of the 50 states in 1992, along with the number of licensed drivers (in millions) in each state that year.

	Fatalities	Licensed Drivers		Fatalities	Licensed Drivers
AL	1031	2.98	AK	108	.41
AZ	809	2.51	AR	588	1.71
CA	4189	20.11	CO	522	2.39
CT	296	2.36	DE	140	.50
FL	2427	10.54	GA	1315	4.60
HI	129	.72	ID	243	.72
IL	1384	7.41	IN	901	3.80
IA	437	1.86	KS	387	1.69
KY	815	2.46	LA	883	2.62
ME	214	.92	MD	659	3.23
MA	485	4.17	MI	1298	6.48
MN	581	2.63	MS	766	1.63
MO	985	3.45	MT	192	.60
NE	269	1.13	NV	251	.96
NH	122	.85	NJ	764	5.29
NM	460	1.13	NY	1814	10.36
NC	1265	4.66	ND	88	.43
OH	1439	9.17	OK	613	2.29
OR	467	2.43	PA	1545	8.02
RI	79	.69	SC	807	2.40
SD	161	.50	TN	1153	3.49
TX	3059	11.44	UT	269	1.14
VT	96	.42	VA	839	4.70
WA	651	3.63	WV	420	1.31
WI	652	3.54	WY	118	.34

a. Estimate the Spearman rank-correlation coefficient for these data.

b. Compute the value of the Spearman rank-correlation coefficient for these data.

c. Determine whether there is evidence of a rank correlation between the number of fatalities and the number of licensed drivers. Test at the 15% level of significance.

5. Indicate which nonparametric test explained in this chapter would best apply to each of the following tasks.

a. Determine whether a machine is systematically overfilling (or underfilling) cereal boxes.

b. Determine whether a new advertising campaign has made a difference in sales.

c. Provide evidence that a coin is not fair.

d. Provide evidence that men and women rate several soft drinks the same.

e. Furnish "proof" that defective light bulbs are being manufactured consistently.

f. Determine whether prices for given items in one store are less than prices for the same items in another store.

g. Determine whether a company's spending priorities stayed the same from one year to the next.

Case Study   *Statistics in the Computer Age*

Today's powerful computers require fewer distribution assumptions than the more traditional methods of mathematical analysis. (*Space Science Telescope Institute/ NASA*).

Amazingly, today's statistician can afford to expend more computation on a single problem than a 1920s statistician expended in an entire year. The power of computers enables statisticians and scientists to explore data using algorithms that require fewer distribution assumptions than the more traditional methods of mathematical analysis.

The July 26, 1991, issue of *Science* contains an article titled "Statistical Data Analysis in the Computer Age" (pages 390–395). Obtain a copy of the article from your library, and use it to answer the following questions.

I. In what time period were most of the "classical" methods of confidence intervals and hypothesis testing developed? Name three scientists who were prominent in their development.

II. Although the power of computers is great and is changing many approaches to statistics, the article stresses that "mathematics has not disappeared from statistical theory." What role do the authors see for mathematics in future statistical theory?

III. In your own words, explain the "bootstrap."

IV. What are the steps necessary to compute the bootstrap estimate of the standard error?

V. Under what conditions is the trimmed mean of a sample a substantially more accurate estimator than the sample mean?

CHAPTER SUMMARY

This chapter examined three nonparametric, or distribution-free, hypothesis tests. Nonparametric tests make no assumptions about the shape of the distribution of the population from which the sample was taken. However, under certain reasonable conditions, the sampling distributions associated with these tests are assumed to be normal.

The sign test was developed first (and was the first test covered in this chapter). It is a hypothesis test, relatively simple to apply, that determines whether the populations from which two samples are taken are identical. The number-of-runs test determines the absence of randomness and may be extended to investigate the randomness of values above or below their median. The Spearman rank-correlation test determines how well two sets of rankings are correlated; examples included the special case of ties within a set of rankings.

Key Concepts, Terms, and Formulas

Nonparametric, or distribution-free, tests   *Page 475*

Expected number of plus signs:

$$\mu_S = \frac{n}{2} \quad \textbf{(12-1)}$$

Page 476

Standard deviation of the number of plus signs:

$$\sigma_S = .5 \cdot \sqrt{n} \quad \textbf{(12-2)}$$

Page 476

Expected number of runs of heads:

$$\mu_{R_H} = \frac{n_H \cdot (n_T + 1)}{n_H + n_T} \quad \textbf{(12-3)}$$

Page 485

Standard deviation of the number of runs of heads:

$$\sigma_{R_H} = \sqrt{\frac{n_H \cdot (n_T + 1) \cdot (n_H - 1)}{(n_H + n_T)^2} \cdot \frac{n_T}{n_H + n_T - 1}} \quad \textbf{(12-4)}$$

Page 485

Runs above and below the median   *Page 489*

Rank correlation   *Page 490*

Spearman rank-correlation coefficient:

$$r = 1 - \frac{6 \cdot \Sigma d_i^2}{n \cdot (n^2 - 1)} \quad \textbf{(12-5)}$$

Page 490

Mean value of the Spearman rank-correlation coefficient:

$$\mu_r = 0 \quad \textbf{(12-6)}$$

Page 492

Standard deviation of the Spearman rank-correlation coefficient:

$$\sigma_r = \sqrt{\frac{1}{n - 1}} \quad \textbf{(12-7)}$$

Page 492

Videotape Suggestions

Program 10 from the series *Against All Odds: Inside Statistics* shows the impact of computer technology on the study of statistics. Although nonparametric tests are not demonstrated, this program's graphic displays of multidimensional data and methods of organizing large data sets have many applications.

Analysis of variance
provides a procedure for
determining if there is a
difference in one of
several population
means.

13 Introduction to Analysis of Variance

T his chapter describes a statistical method for comparing the means of more than two samples in order to determine whether there is evidence that at least one of the population means is different from the others. The **case study** at the end of the chapter uses this method to examine the behaviors of victims of several types of crimes.

13.1 INTRODUCTION

Suppose one wishes to use the means of five samples to decide whether at least one of the five population means is different from the others. It may be tempting to conduct a series of pairwise hypothesis tests that involve five sample means compared in two ways. The number of necessary hypothesis tests would be $_5C_2 = 10$.

Test 1:	H_0: $\mu_1 = \mu_2$	vs.	H_A: $\mu_1 \neq \mu_2$.
Test 2:	H_0: $\mu_1 = \mu_3$	vs.	H_A: $\mu_1 \neq \mu_3$.
Test 3:	H_0: $\mu_1 = \mu_4$	vs.	H_A: $\mu_1 \neq \mu_4$.

$$\cdot$$
$$\cdot$$
$$\cdot$$

Test 10:	H_0: $\mu_4 = \mu_5$	vs.	H_A: $\mu_4 \neq \mu_5$.

Besides being tedious, this lengthy pairwise testing would be unwise for another reason. Recall that the level of significance of a hypothesis test, α, represents the probability of rejection of the null hypothesis when that hypothesis is, in fact, true.[1] If all five population means were indeed equal and if $\alpha = .05$, the probability that

---

[1] Chapter 9 referred to this as the probability of making a Type I error.

Consider the rejection of the null hypothesis as a binomial random variable with probability of success equal to α. The probability that none of the ten null hypotheses will be rejected is then $_{10}C_0 \cdot (\alpha)^0 \cdot (1 - \alpha)^{10-0}$.

the null hypothesis of equality would be rejected in at least one of the ten tests would be

$$1 - {_{10}C_0} \cdot (.05)^0 \cdot (1 - .05)^{10} \approx .4013.$$

Such a high probability of error is unacceptable. A special statistical procedure, the **analysis-of-variance method** (or the **analysis-of-variance test**), also referred to by the acronym **ANOVA,** tests more properly the claim of equal population means for more than two populations.

Unlike the distribution-free tests in Chapter 12, an analysis-of-variance test requires the following two assumptions about the populations.

Rule ▶

> Application of the ANOVA method requires that two assumptions be made about the populations from which the samples were taken.
>
> a. All of the populations are normally distributed, or very nearly so.
> b. All of the populations have approximately equal variances.

13.2 THE *F*-STATISTIC

With the just-mentioned assumptions in place, the aim of the analysis-of-variance method is to form a ratio called the *F*-statistic.

Definition ▶

The *F*-statistic,

$$F = \frac{\text{Variation among the samples}}{\text{Variation within the samples}},$$

is calculated with the formula

$$F = \frac{\text{Variance based on sample means}}{\text{Mean of sample variances}}. \qquad \textbf{(13-1)}$$

Stat Fact

The distribution of the *F*-statistic, or the *F*-distribution, is named after an important and prolific statistician, Sir Ronald A. Fisher (1890–1962). For more than half a century, Fisher published about one scholarly paper every 2 months, and most of his writings broke new ground. Throughout his life, Fisher received honors for his work in statistics. He was knighted in 1952.

The numerator (variation among the samples) and the denominator (variation within the samples) are estimates of a parent-population variance. The null hypothesis of equal population means will be rejected if these two estimates do not agree closely enough. Specifically, the calculated value of the *F*-statistic will be either inside or outside a rejection region used to determine whether at least one of the populations' means is different from the others.

Optional Excursion 10.7A was also concerned with an *F*-statistic, which was defined as the ratio of sample variances.

13.3 AN ILLUSTRATIVE EXAMPLE

The analysis-of-variance method is very effective for product comparisons, and the next step-by-step example explains it in detail. Because only one factor (the brand of insect repellent) is regarded as affecting the variable of interest (the number of insect bites), the analysis shown in the example is called a **one-factor** or **one-way ANOVA test**.[2]

E x a m p l e
1

Eighteen children are treated with three brands of insect repellent—six children being randomly assigned to each brand. All 18 are then sent out to play on a humid summer evening. The following cross-tabs table shows the number of mosquito bites each child received that evening.

Numbers of Mosquito Bites with Different Brands of Repellent

Child	Brand A	Brand B	Brand C
1	12	10	15
2	9	9	8
3	14	7	10
4	18	10	6
5	16	9	12
6	9	5	9

Obviously, the sample means for brands A, B, and C are not the same. This ANOVA test will determine if this difference in sample means provides evidence that the population means—the mean numbers of mosquito bites for all children who use these brands—are not all the same.

Determine whether there is statistical evidence that the mean numbers of mosquito bites of all children who use these brands are not the same. Perform a one-way ANOVA test at the 5% level of significance.

Solution   To perform an ANOVA test, it must be assumed that each of the three populations has a distribution at least close to normal and that all three have approximately equal variances. With these assumptions, perform the following four-step hypothesis test.

Step 1   State the hypotheses.

H_0:   The mean numbers of mosquito bites for all children who use each of the three brands are the same. That is, $\mu_A = \mu_B = \mu_C$.

If all three population means are not the same then either they are all different, or two are the same and one is different. This logic forms the alternative hypothesis statement.

   The only possible outcomes of this hypothesis test are the conclusion that the users of at least one brand of repellent have a mean number of mosquito bites different from the others, and reservation of judgment.

H_A: The users of at least one brand of repellent have a mean number of mosquito bites different from the others.

The assumption of equal population means is always the null hypothesis. This assumption is required for the following calculations.

Step 2   Determine the numerator and denominator of the F-statistic, and compute the F-statistic.

Recall that the F-statistic is defined as

$$F = \frac{\text{Variation among the samples}}{\text{Variation within the samples}}$$

[2]Consult other statistics texts for two-way and three-way ANOVA tests as well as tests involving unequal sample sizes.

and is actually calculated by evaluating the ratio

$$\frac{\text{Variance based on sample means}}{\text{Mean of sample variances}}.$$

The null hypothesis claims that $\mu_A = \mu_B = \mu_C$. Also, the underlying assumptions of this test are that the three populations (the numbers of mosquito bites for brands A, B, and C) are normally distributed with approximately equal variances. If the null hypothesis and these assumptions are true, these three samples must have come from the same normally distributed parent population. The variance of this parent population will be estimated in two different ways, forming the numerator and denominator of the *F*-statistic.

How to Compute the Numerator of the F-Statistic: Variation Among the Samples (This calculation is the variance based on sample means.) The first step is to compute the sample mean numbers of mosquito bites for each brand of repellent in the usual manner. For instance, the sample mean for brand A is

$$\overline{X}_A = \frac{12 + 9 + 14 + 18 + 16 + 9}{6}$$

$$= 13.$$

The following table shows the sample mean numbers of bites for all three brands.

Child	Brand A	Brand B	Brand C
1	12	10	15
2	9	9	8
3	14	7	10
4	18	10	6
5	16	9	12
6	9	5	9
Sample mean	$\overline{X}_A = 13$	$\overline{X}_B = 8.33$	$\overline{X}_C = 10$

Now use these three sample means (13, 8.33, and 10) to find the variation among the samples, either by hand or by utilizing the statistics capabilities of a TI-82 graphics calculator. Figure 13-1 shows these calculations after the three sample

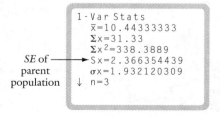

SE of parent population

```
1-Var Stats
x̄=10.44333333
Σx=31.33
Σx²=338.3889
Sx=2.366354439
σx=1.932120309
↓ n=3
```

Figure 13-1

means are entered. Since these calculations are based on sample means, the (rounded) value 2.37 is the estimated standard error of the parent population from which the three samples came. If σ represents the standard deviation of the parent population, then, as Chapter 8 noted,

$$SE = \frac{\sigma}{\sqrt{n}}.$$

Therefore,

$$2.37 \approx \frac{\sigma}{\sqrt{n}}.$$

Substituting the fact that $n = 6$,

$$2.37 \approx \frac{\sigma}{\sqrt{6}}.$$

Squaring both sides of this equation, an estimate for the variance of the parent population is

$$\sigma^2 \approx (2.37)^2 \cdot 6$$
$$\approx 33.70.$$

The value 33.70 is an estimate of the parent-population variance and is called the **variation among the samples.** It is the numerator of the *F*-statistic.

How to Compute the Denominator of the F-Statistic: Variation within the Samples (This calculation is the mean of the sample variances.) Another way to estimate the parent-population variance is to use the estimated population standard deviations based on a sample from each of the three brands. Enter the numbers of mosquito bites received by those using brands A, B, and C in three lists of a TI-82 graphics calculator and obtain the statistics shown in Figure 13-2. Thus, based on sample A, the estimated population variance of brand A is

$$(3.69)^2 \approx 13.62.$$

Brand A	Brand B	Brand C
1-Var Stats	1-Var Stats	1-Var Stats
$\bar{x}=13$	$\bar{x}=8.333333333$	$\bar{x}=10$
$\Sigma x=78$	$\Sigma x=50$	$\Sigma x=60$
$\Sigma x^2=1082$	$\Sigma x^2=436$	$\Sigma x^2=650$
$Sx=3.687817783$	$Sx=1.966384161$	$Sx=3.16227766$
$\sigma x=3.366501646$	$\sigma x=1.795054936$	$\sigma x=2.886751346$
$\downarrow$ n=6	$\downarrow$ n=6	$\downarrow$ n=6

Figure 13-2

The following table shows the estimates of the population variances of all three brands.

Child	Brand A	Brand B	Brand C
1	12	10	15
2	9	9	8
3	14	7	10
4	18	10	6
5	16	9	12
6	9	5	9
Mean	$\overline{X}_A = 13$	$\overline{X}_B = 8.33$	$\overline{X}_C = 10$
Estimated σ	$s_A \approx 3.69$	$s_B \approx 1.97$	$s_C \approx 3.16$
Estimated σ^2	$s_A^2 \approx \mathbf{13.62}$	$s_B^2 \approx \mathbf{3.88}$	$s_C^2 \approx \mathbf{9.99}$

Use the mean of these three estimated population variances to estimate the variance of the parent population.

$$\text{Estimated parent-population variance} = \frac{13.62 + 3.88 + 9.99}{3}$$

$$= \frac{27.49}{3}$$

$$\approx 9.16.$$

The value 9.16 is another estimate of the parent-population variance and is called the **variation within the samples.** It is the denominator of the *F*-statistic.

Computation of the F-Statistic   The *F*-statistic was defined earlier as

$$F = \frac{\text{Variation among the samples}}{\text{Variation within the samples}}.$$

Therefore,

$$F = \frac{33.70}{9.16}$$

$$\approx 3.68.$$

Now take the next step in the hypothesis test.

Step 3   Determine the end points of the rejection region and state the decision rule.
When the null hypothesis and assumptions about normality and equal population variances are true, the sampling distribution of all possible values of the *F*-statistic (the *F*-distribution) looks as shown in Figure 13-3.[3] Figure 13-3 shows that the most probable values of the *F*-statistic are those close to 1, indicating agreement between the two estimates of the parent-population variance. Therefore, when the calculated value of the *F*-statistic is close to 1, there is not enough statistical evidence

[3]As the degrees of freedom for this distribution increase, it will approach a normal distribution.

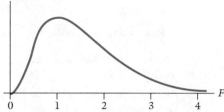

Figure 13-3 The sampling distribution of the *F*-statistic.

to warrant rejection of the null hypothesis of equal population means. The null hypothesis *is* rejected, however, for values of the *F*-statistic too far from $F = 1$. Since the *F*-statistic can never be negative, all of the rejection region is in the right tail.

Appendix A-6 lists values of the *F*-statistic that mark the rejection region for levels of significance $\alpha = .05, .025,$ and $.01$. To use this appendix, determine the degrees of freedom of the numerator and denominator of the *F*-statistic separately. The degrees of freedom of the numerator are

$$p - 1,$$

where *p* represents the number of different products being tested. In this case, three products (insect repellent brands A, B, and C) are being tested. Therefore, the numerator degrees of freedom for this example are

$$p - 1 = 3 - 1$$
$$= 2.$$

The degrees of freedom of the denominator are

$$p \cdot (n - 1),$$

where *n* represents the sample size (the number of each product tested). In this case, $n = 6$, and the degrees of freedom of the denominator are

$$p \cdot (n - 1) = 3 \cdot (6 - 1)$$
$$= 15.$$

Next, consult Appendix A-6 and find the intersection of the numerator degrees of freedom (column **2**) and the denominator degrees of freedom (row **15**) on the $F_{.05}$ page. This intersection indicates that an *F*-statistic value of 3.68 marks the rejection region.

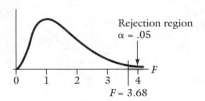

Chapter 13 *Introduction to Analysis of Variance*

Accordingly, the decision rule is:

Reject H_0 and accept H_A if $F > 3.68$. Otherwise, reserve judgment.

Step 4 State the decision.

The calculated value of $F = 3.68$ is not quite large enough to warrant rejection of H_0. Therefore, **reserve judgment.** There is not enough statistical evidence at the 5% level of significance to support the alternative hypothesis of at least one different population mean. The difference in sample means must be attributed to chance. (See Exercise 10 in Exercises 13.1–13.4 on page 516.) ●

13.4

Example 2

Use Minitab to find the value of the *F*-statistic in Example 1.

Solution For convenience, the data presented in Example 1 are repeated here.

Numbers of Mosquito Bites with Different Brands of Repellent

Child	Brand A	Brand B	Brand C
1	12	10	15
2	9	9	8
3	14	7	10
4	18	10	6
5	16	9	12
6	9	5	9

The first step is to enter these data into a Minitab worksheet in much the same format as just shown.

	C1	C2	C3	C4	C5
↓	Brand A	Brand B	Brand C		
1	12	10	15		
2	9	9	8		
3	14	7	10		
4	18	10	6		
5	16	9	12		
6	9	5	9		
7					

Untitled Worksheet

```
┌──────────────────────────────── Session ════════════════════┐
│ Worksheet size: 38000 cells                                   │
│ MTB > AOVONEWAY  C1  C2  C3                                    │
│                                                               │
│ ANALYSIS OF VARIANCE                                          │
│ SOURCE    DF       SS       MS      ┌─ F ─┐     P             │
│ FACTOR     2     67.11    33.56     │3.67 │   0.051           │
│ ERROR     15    137.33     9.16     └─────┘                   │
│ TOTAL     17    204.44                                        │
│                              INDIVIDUAL 95 PCT CI'S FOR MEAN   │
│                              BASED ON POOLED STDEV             │
│  LEVEL     N      MEAN     STDEV   --+---------+---------+---------+----
│  C1        6    13.000     3.688                 (-------*--------)
│  C2        6     8.333     1.966    (-------*--------)
│  C3        6    10.000     3.162          (-------*--------)
│                                    --+---------+---------+---------+----
│ POOLED STDEV =   3.026             6.0     9.0    12.0    15.0
└───────────────────────────────────────────────────────────────┘
```

Figure 13-4 Use of Minitab's AOVONEWAY command.

Next, obtain a Session window and type the macro command AOVONEWAY, followed by the locations of the data. Figure 13-4 shows the output once the RETURN key is pressed. (Notice that 95% confidence intervals for the population means are automatically included as part of the output of this command.)

Due to rounding differences, the F-statistic 3.67 computed by Minitab is just slightly less than the one computed by hand. However, it still leads to a reserved judgment. ●

Example 2 shows the power of statistical software. What would take many minutes to compute by hand can be achieved in just a couple of minutes with a computer and the right software. It is also possible to program many graphics calculators to compute the F-statistic.

Exercises 13.1–13.4

1. In the text, the probability that at least one null hypothesis will be falsely rejected in a series of ten pairwise hypothesis tests is given as

$$1 - {}_{10}C_0 \cdot (.05)^0 \cdot (1 - .05)^{10}.$$

   Explain where this formula came from.

* 2. Refer to Exercise 1. What is the probability that at least one null hypothesis will be falsely rejected if a series of eight pairwise hypothesis tests with a 10% level of significance is used instead of an ANOVA test?

3. What two assumptions must be made about the populations from which samples were taken in order for an analysis-of-variance test to be applied correctly?

4. Which seems to be more "work"—computing the numerator or the denominator of the F-statistic?

MTB   5. Refer to the Minitab output of Example 2 on page 514.

  a. Explain how the 95% confidence intervals shown are consistent with the results of the ANOVA test.

  b. What part of the ANOVA test would have to change to make the confidence intervals inconsistent with the results of the test?

6. An analysis-of-variance test lends itself well to product comparisons. Give five examples of possible applications of this test.

7. Although they are not treated in this text, two-factor and three-factor ANOVA tests are possible. Refer to Example 1 on page 509, and state another set of factors that could affect the numbers of mosquito bites received by each child.

8. Refer to Example 1 on page 509. Explain why the value of 2.37 is the estimated standard error of the parent population and not the estimated standard deviation of the parent population.

9. Explain why the rejection region for the F-statistic is only in the right tail.

10. Refer to the conclusion of Example 1 on page 509.

  a. Some might say that this test at the 5% level of significance is inconclusive. Why?

  b. Must the level of significance be raised slightly or lowered slightly to cause rejection of the null hypothesis?

11. The following F-statistics mark rejection regions for the given levels of significance.

  $F_{.05}$ with numerator d.f. = 10, denominator d.f. = 20

  $F_{.05}$ with numerator d.f. = 20, denominator d.f. = 10

  $F_{.025}$ with numerator d.f. = 10, denominator d.f. = 20

  $F_{.025}$ with numerator d.f. = 20, denominator d.f. = 10

  Referring to Appendix A-6, arrange these F-statistics from least to greatest.

12. Refer to Exercise 11. Which seems more likely to cause a rejection of the null hypothesis: a decrease in the degrees of freedom for the numerator or a decrease in the degrees of freedom for the denominator?

 A statistical software package or graphics calculator is recommended to work the following exercises, although a hand-held calculator may also be used.

Up to six different lists may be entered into a TI-82 graphics calculator.

13. The table in Example 1 on page 509 has been expanded to include one more brand of insect repellent.

Child	Brand A	Brand B	Brand C	Brand D
1	12	10	15	10
2	9	9	8	12
3	14	7	10	10
4	18	10	6	9
5	16	9	12	9
6	9	5	9	11

Perform a one-way ANOVA test at the 2.5% level of significance to determine whether there is statistical evidence that the mean numbers of mosquito bites for people who use these brands are not all the same.

14. Sixteen midsize cars (four groups of four) were randomly selected to test the mileage claims of four different brands of gasoline. The following table lists the miles per gallon obtained with each brand.

Mileage

Car	Brand A	Brand B	Brand C	Brand D
1	22.3	24.9	26.0	19.7
2	20.9	19.7	20.0	22.2
3	21.5	22.0	24.5	19.9
4	22.0	21.9	23.4	22.3

Use a one-way ANOVA test at the 1% level of significance to determine if there is evidence that the mean gas mileages obtained with these four brands of gasoline are not the same.

15. Can statistical evidence be found that total taxes paid, as percent of income, are different for income levels of $25,000, $50,000, and $75,000 for a family of four who own their own home? A random sample of tax data based on the mean taxes paid in six different cities reveals the following information.

City	$25,000	$50,000	$75,000
Des Moines, IA	7.4	8.2	9.0
Charleston, WV	6.8	7.1	8.5
Memphis, TN	6.9	5.8	6.0
Sioux Falls, SD	7.3	6.4	7.1
Philadelphia, PA	15.8	15.2	15.1
Chicago, IL	11.8	11.9	12.4

Source: *Statistical Abstract of the United States, 1994*, p. 310.

Perform an ANOVA test at the 5% level of significance.

GROUP   16. Three brands of toothpaste were tested on 24 people over a 5-year period. Eight people of the 24 were randomly assigned to each of the three brands. The following table shows the number of cavities each person had during this 5-year period.

Number of Cavities

Patient	Brand X	Brand Y	Brand Z
1	10	8	12
2	10	13	10
3	10	5	15
4	7	11	10
5	12	15	9
6	4	16	15
7	11	17	14
8	15	10	12

Perform a one-way ANOVA test at the 2.5% level of significance to determine whether there is statistical evidence that the mean numbers of cavities for people who use these brands are not the same.

17. The following array shows the numbers of full-time-equivalent employees for every 100 average daily patients in the community hospitals of eight states, selected at random, during 3 years a decade apart.

Employees per 100 Average Daily Patients

	1970	1980	1990
OK	296	404	585
WI	277	367	521
MT	247	302	397
HI	278	401	508
MD	354	403	566
PA	287	390	567
AZ	327	455	590
VT	318	348	574

Source: *Health United States, 1993*, U.S. Department of Health and Human Services, p. 217.

Perform a one-way ANOVA test at the 2.5% level of significance to show that the mean number of employees has not stayed the same.

18. An economist believes that U.S. incomes a century ago were not stable. To support her argument, she presents the following table, which shows daily pay for some standard jobs in the years 1895 through 1899.

Job	1895	1896	1897	1898	1899
Bricklayer	$3.34	3.87	3.45	3.41	3.60
Stationary engineer	2.52	2.67	3.00	3.17	2.63
Hod carrier	2.00	2.15	2.00	1.97	2.10
Painter	2.32	2.96	2.45	2.47	2.57
Plumber	3.74	3.49	3.73	3.74	3.19
Stonemason	3.94	3.92	3.38	3.67	3.06

Source: *The Value of a Dollar: Prices and Incomes in the United States, 1860–1989*, Gale Research, Inc., p. 16.

Perform a one-way ANOVA test at the 2.5% level of significance to determine whether there is evidence to support the economist's claim.

19. A researcher wants to determine whether at least one of three different brands of light bulbs has a different mean lifetime from the others. Six bulbs of each kind, all with the same wattage, are randomly sampled, with these results:

Lifetimes of Light Bulbs, in Hours

Bulb	Brand X	Brand Y	Brand Z
1	1200	1176	1405
2	1205	1100	1375
3	1140	1190	1275
4	1090	1150	1400
5	1185	1150	1380
6	1080	1110	1410

Perform a one-way ANOVA test at the 2.5% level of significance to determine whether there is evidence that at least one brand has a mean lifetime different from the others.

20. The numbers of new automobiles sold by four salesmen in 8 consecutive weeks are as follows:

    Salesman A: 16, 20, 22, 8, 12, 16, 22, 20

    Salesman B: 20, 20, 12, 20, 19, 11, 27, 17

    Salesman C: 11, 20, 23, 12, 22, 10, 10, 24

    Salesman D: 17, 19, 14, 18, 18, 19, 15, 17

    Perform a one-way ANOVA test at the 5% level of significance to determine whether there is statistical evidence that the expected mean yearly sales of at least one of these salesmen is different from the others.

21. A researcher tests four different brands of pea seeds to determine whether there is any difference in their mean plant yields per square yard. She randomly chooses three packets of each kind, plants the seeds, and records the following results.

    Number of Plants per Square Yard

Packet	Brand A	Brand B	Brand C	Brand D
1	72	68	55	70
2	70	70	70	66
3	70	71	70	68

   a. On the basis of only a quick mental inspection of these data, does it appear that there are any differences in the population mean yields per square yard for these four brands?

   b. Perform a one-way ANOVA test at the 5% level of significance to determine whether there is evidence of a difference in the population mean yield per square yard.

* 22. Refer to Exercise 21. By trial and error, and using a statistical software program, reverse the result of the ANOVA test by making minimal changes in the data for only brands C and D.

23. The monthly rents for randomly selected two-bedroom apartments in three fashionable neighborhoods in the Chicago area follow.

Apartment	Neighborhood A	Neighborhood B	Neighborhood C
1	$1175	$1400	$ 950
2	975	1800	1050
3	1000	995	1000
4	1025	1100	975
5	1250	1275	1100
6	980	1025	1000
7	975	1775	1050

Perform a one-way ANOVA test at the 2.5% level of significance to determine whether there is evidence that the mean rent for two-bedroom apartments in any of these three neighborhoods differs from the others.

GROUP   24. A consumer group wants to compare the on-time performances of four railroads that move commuters the same number of miles. The following table shows the numbers of minutes late 40 randomly selected commuters (ten per railroad) were on a randomly selected day.

Commuter	Railroad W	Railroad X	Railroad Y	Railroad Z
1	2	0	1	1
2	0	5	0	0
3	0	1	9	1
4	1	0	1	0
5	3	1	7	0
6	0	1	0	2
7	1	0	8	5
8	2	1	1	3
9	2	0	3	2
10	1	3	6	0

Perform a one-way ANOVA test at the 5% level of significance to determine whether there is evidence that the mean number of minutes a commuter can expect to be late on these four railroads are not the same.

* 25. Refer to Exercise 24. If railroad Y were removed from the study, would the numerator or the denominator of the F-statistic be affected more? Perform the necessary calculations to check your answer.

Case Study   *Crime Statistics*

An interesting line chart appears on page 566 in the Fall 1993 issue of *The Journal of Criminal Law and Criminology*. It compares the percentages of rape, robbery, and assault victims who reported their incidents to the police. Use a copy of this issue to answer the following questions.

I. Use your best estimates of the data presented to construct a cross-tabs table showing the percentages of rape, robbery, and assault victims who reported their incidents in each of the years 1973 through 1990. Round each estimated percentage to the nearest whole percent.

II. Based on the line graph in the journal, do you believe there is evidence of a difference in the percentages of victims who report these crimes? Which seems the greatest? Which seems the least?

III. Perform a one-way ANOVA test at the 5% level of significance to determine whether there is evidence that the mean percentages of victims who report their incidents differ according to the type of crime.

IV. For a proper ANOVA test, all population variances must be approximately equal. On the basis of the line graph, which crime seems most likely to have a variance different from the other two?

V. Construct 90% confidence intervals for each of the three population variances. What do these confidence intervals imply about your answer to Part IV?

CHAPTER SUMMARY

This chapter introduced the analysis-of-variance method, or ANOVA, a hypothesis test designed to determine whether there is evidence that several populations do not have the same mean. ANOVA requires two assumptions about the populations, and the test statistic—the ratio of the variation among the samples to the variation within the samples—can be tedious to compute by hand. The use of statistical software is recommended for the application of this test.

Key Concepts, Terms, and Formulas

Analysis-of-variance method	*Page 508*

Formula for the F-statistic:

$$F = \frac{\text{Variance based on sample means}}{\text{Mean of sample variances}} \quad (13\text{-}1)$$

	Page 508
One-way ANOVA test	*Page 509*
Variation among the samples	*Page 511*
Variation within the samples	*Page 512*
F-distribution	*Pages 512–513*

Videotape Suggestions

Program 26 in the series *Against All Odds: Inside Statistics* presents a timely case study on statistical testing of drugs. Although it does not illustrate ANOVA testing, the program provides a good overview of much of the material in this text.

(Science VU/Visuals Unlimited)

A satellite view of the
United States at night
piques our interest in
variables that are
correlated.

Chapter

14 Correlation and the Line of Best Fit

S ometimes it is important to know how closely two random variables are related. If increases in one correspond closely to increases (or decreases) in the other, then it may be possible to use values of one variable to predict values of the other.

The **case study** at the end of this chapter examines how the power of statistical software affects the study of correlation.

14.1 CORRELATION

Definition ▶

> Two random variables are **correlated** if there is a straight-line relationship between them such that increases (or decreases) in one variable correspond to increases (or decreases) in the other variable.

Two variables are **positively correlated** if increases in one match increases in the other, whereas two variables are **negatively correlated** if increases in one match decreases in the other. A few examples follow.

- A person's weight and height are positively correlated.
- In general, the number of years a person has been in school and that person's annual income are positively correlated.
- The number of passengers in an automobile and the number of miles per gallon traveled are negatively correlated.
- The number of hours spent watching television and the number of hours spent reading books are negatively correlated.

Although it may be possible to use the value of one variable to predict the value of a correlated variable, a correlation does not necessarily imply a cause-and-effect relationship. Some common sense is called for. There may, for instance, be a strong correlation between the price of a first-class postage stamp and the number of students enrolled in community colleges over the last 15 years. Yet the value of one of these variables certainly should not be used to predict the value of the other. A correlation of this kind is called a **nonsense correlation** or a **spurious correlation.**

This chapter first introduces two sample statistics that can be used to measure the correlation between two random variables. The second part of the chapter presents a mathematical model for relating two variables to each other, called the line of best fit.

14.2 SCATTER PLOTS

Consider the following table, which indicates some sample values of two random variables, X and Y: price per barrel of crude oil and price per gallon of gasoline.

Price per Barrel of Crude Oil, X	Price per Gallon of Gasoline, Y
$ 3.35	$ .38
3.75	.40
4.29	.43
5.80	.57
18.00	.80
21.30	.85
30.00	.98
35.75	1.20
37.05	1.15
42.50	1.35
45.75	1.40
45.75	1.25

These data can take the form of ordered pairs in which the first coordinate is a value from the random variable X (price per barrel of crude oil) and the second coordinate is the corresponding value from the random variable Y (price per gallon of gasoline). A graph that plots such ordered pairs is called a **scatter plot.**

If a straight line can be drawn that contains all the points in the scatter plot, then the variables X and Y have perfect correlation.[1] If no straight line can be drawn that contains all the points, then the correlation is a matter of degree; it can range from weak to strong, depending on how well a straight line "fits" the points. If the points do not seem to come close to any straight-line pattern (no placement of a line comes near many of them), then the two variables, X and Y, have either no correlation or very weak correlation.

Consider scatter plots (a) and (b) in Figure 14-1. They show a perfect correlation between X and Y; one is positive and the other negative. In actual practice,

---

[1] If the slope of such a line is positive, then X and Y have perfect positive correlation: as X increases, Y increases. If the slope of the line is negative, then X and Y have perfect negative correlation: as X increases, Y decreases.

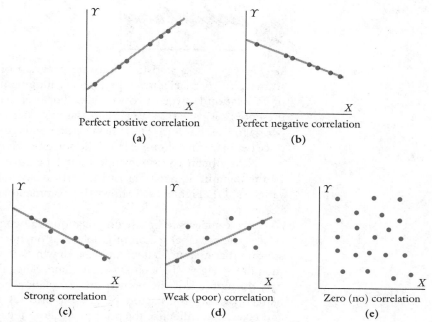

Figure 14-1 Scatter plots showing different degrees of correlation.

however, due to the random nature of samples and to possible measurement errors, perfect correlation is seldom achieved. Scatter plot (c) shows a strong correlation between the variables X and Y; a straight line comes very close to passing through all of the points. Scatter plot (d) shows some correlation: as X increases, generally so does Y, and the straight line in this diagram is "centered" among the points. However, very few points are actually on the line, and a few are far away from it. Even so, the correlation in this scatter plot is better than that shown in scatter plot (e), in which no straight line fit seems possible and there is no correlation whatsoever.

Of course, situations like the scatter plot shown in Figure 14-2 also occur. Other statistics texts explore curvilinear relationships between X and Y. In this text, however, the word "correlation" always implies a straight-line relationship.[2]

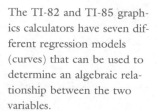

The TI-82 and TI-85 graphics calculators have seven different regression models (curves) that can be used to determine an algebraic relationship between the two variables.

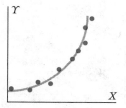

Figure 14-2

[2]The case study at the end of the chapter examines how the computer can be used to make some of these more complicated fits.

It is easy to use a statistical software program or graphics calculator to make a scatter plot that facilitates a preliminary judgment about the amount, or degree, of correlation between two variables. For example, consider once again the data presented on page 524, regarding the price per barrel of crude oil and the price per gallon of gasoline. These data can be entered into a Minitab worksheet—the price per barrel in column C1 and the corresponding price per gallon in column C2.

Next, obtain a Session window and type the macro command PLOT, followed by the column designations of the vertical axis variables, C2, and horizontal axis variables, C1. Figure 14-3 shows this command and the result once the [RETURN] key is pressed.[3]

There appears to be a strong relationship between the price per barrel of crude oil and the price per gallon of gasoline; as one increases, so does the other. It also appears that a straight line could be drawn that would be very close to most of the points. Figure 14-4 shows such a line. Since the line in Figure 14-4 does not pass through all of the points, the correlation between the two variables is not perfect. (Many other factors besides the price of crude oil, such as transportation and taxes, may influence the price of a gallon of gasoline.) The line does, however, pass very closely to most of the points, indicating a high degree of correlation. Example 1 measures the correlation and verifies this assumption.

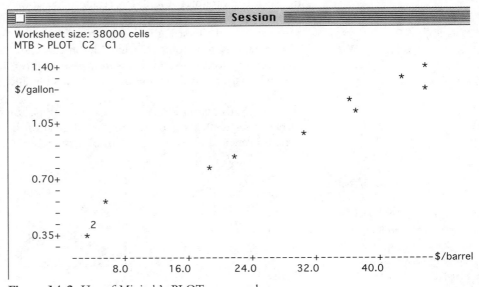

Figure 14-3 Use of Minitab's PLOT command.

---

[3]The small 2 in the lower left-hand corner of this scatter plot indicates two points graphed close together.

The scatter plot in Figure 14-4 is a high resolution graph of the data shown in Figure 14-3. This change occurred automatically when the line was added. The procedure to accomplish this will be explained later in the chapter.

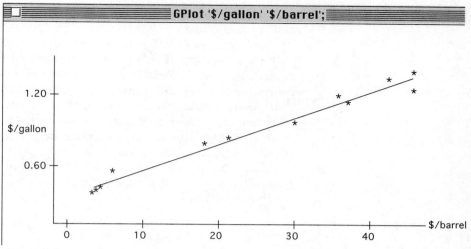

Figure 14-4

A later section in this chapter explains a procedure for finding the equation of the straight line that comes closest to the points in this scatter plot,[4] called the **line of best fit.** For now, however, the goal is simply to compute a statistic that measures the amount of correlation between the two variables.

14.4 MEASURES OF CORRELATION

Two statistics will be used to measure the correlation between two variables. One is the square of the other.

> Measures of correlation:
>
> a. The Pearson coefficient of correlation, p $(-1 \le p \le 1)$
> b. The coefficient of determination, p^2 $(0 \le p^2 \le 1)$

The sign of the Pearson coefficient of correlation, p, indicates either a negative correlation $(p < 0)$ or a positive correlation $(p > 0)$, and a value of $p = \pm 1$ shows a perfect correlation between two variables.[5] Also, the sign of p (positive or negative)

[4]A line, *L, comes closest* to passing through all the points in a scatter plot if the sum of the squares of the vertical distances from all the points to L is less than it would be for any other line. This topic is discussed in greater detail later in the chapter.

[5]In this chapter all final values of correlation are shown to three decimal places, with intermediate calculations performed to four decimal places.

indicates the slope of the line of best fit and can be used as an additional check on later calculations.

The square of p, the coefficient of determination, can be shown mathematically to be the percentage of variation in the variable Y that can be explained by a linear relationship with the variable X. Because of this interpretation, some statisticians prefer to use p^2 as a measure of correlation. Regardless of whether p or p^2 is used, remember that these sample statistics are only point estimates of the actual coefficient of correlation (or determination) of the two populations. Like the values of other sample statistics, the value of p may be used to compute a confidence interval for the true population parameter.[6]

The value of the coefficient of determination, p^2, is found by computing p and then squaring that value. The formula for p, the Pearson coefficient of correlation,[7] follows.

Formula for the **Pearson Coefficient of Correlation**

$$p = \frac{\Sigma(x \cdot y) - n \cdot \overline{X} \cdot \overline{Y}}{\sqrt{(\Sigma x^2 - n \cdot \overline{X^2}) \cdot (\Sigma y^2 - n \cdot \overline{Y^2})}} \qquad (14\text{-}1)$$

In Formula 14-1, $\overline{X}$ and $\overline{Y}$ represent the means of variables X and Y, respectively. $\Sigma(x \cdot y)$ is the sum of the product of all ordered pairs, and n is the number of ordered pairs. Σx^2 is the sum of the squares of all values of the variable X, and Σy^2 is the sum of the squares of all values of the variable Y.

This formula is tedious to apply by hand, but Example 1 shows all the steps in detail. Example 2 shows how to use a graphics calculator to find the Pearson coefficient of correlation and draw a scatter plot.

E x a m p l e   1

Determine the Pearson coefficient of correlation for the price per barrel of crude oil, X, and the price per gallon of gasoline, Y. Use the sample of 12 ordered pairs presented earlier.

Solution   For convenience, the prices given earlier are repeated here.

---

[6]The construction of this type of confidence interval is not part of this text.

[7]The Pearson coefficient of correlation and the Spearman rank-correlation coefficient from Chapter 12 are related. If no ties occur within a set of rankings, and if the rankings are considered as ordered pairs from two variables, then the Pearson coefficient of correlation and the Spearman rank-correlation coefficient are equal.

Price per Barrel of Crude Oil, X	Price per Gallon of Gasoline, Y
$ 3.35	$ .38
3.75	.40
4.29	.43
5.80	.57
18.00	.80
21.30	.85
30.00	.98
35.75	1.20
37.05	1.15
42.50	1.35
45.75	1.40
45.75	1.25

The Pearson coefficient of correlation is found with the formula given earlier:

$$p = \frac{\Sigma(x \cdot y) - n \cdot \overline{X} \cdot \overline{Y}}{\sqrt{(\Sigma x^2 - n \cdot \overline{X}^2) \cdot (\Sigma y^2 - n \cdot \overline{Y}^2)}}.$$

There are 12 ordered pairs—(3.35, .38), (3.75, .40), ..., (45.75, 1.25)—and many calculations must be made to compute the value of p. Taking these calculations in the order in which they appeared in the preceding formula,

$$\Sigma x \cdot y = (3.35 \cdot .38) + (3.75 \cdot .40) + \cdots + (45.75 \cdot 1.25)$$
$$= 1.273 + 1.500 + \cdots + 50.000$$
$$= 333.9487,$$

n = number of ordered pairs

$$= 12,$$

$n \cdot \overline{X} \cdot \overline{Y}$ = product of n, mean value of X, and mean value of Y.

First calculate the mean values $\overline{X}$ and $\overline{Y}$:

$$\overline{X} = \frac{3.35 + 3.75 + \cdots + 45.75}{12}$$
$$= 24.4408,$$
$$\overline{Y} = \frac{.38 + .40 + \cdots + 1.25}{12}$$
$$= .8967.$$

Therefore,

$$n \cdot \overline{X} \cdot \overline{Y} = 12 \cdot (24.4408) \cdot (.8967)$$
$$= 262.9928,$$
$$\Sigma x^2 = (3.35)^2 + (3.75)^2 + \cdots + (45.75)^2$$
$$= 11.2225 + 14.0625 + \cdots + 1600$$
$$= 10398.1591,$$

$$n \cdot \overline{X}^2 = 12 \cdot (24.4408)^2$$
$$= 12 \cdot 597.3527$$
$$= 7168.2325,$$
$$\Sigma y^2 = (.38)^2 + (.40)^2 + \cdots + (1.25)^2$$
$$= .1444 + .1600 + \cdots + 1.5625$$
$$= 11.2446,$$
$$n \cdot \overline{Y}^2 = 12 \cdot (.8967)^2$$
$$= 12 \cdot .8041$$
$$= 9.6489.$$

Now, substituting these calculations into the formula for the Pearson coefficient of correlation,

$$p = \frac{\Sigma(x \cdot y) - n \cdot \overline{X} \cdot \overline{Y}}{\sqrt{(\Sigma x^2 - n \cdot \overline{X}^2) \cdot (\Sigma y^2 - n \cdot \overline{Y}^2)}}$$

$$= \frac{333.9487 - 262.9928}{\sqrt{(10398.1591 - 7168.2325) \cdot (11.2446 - 9.6489)}}$$

$$= \frac{70.9559}{\sqrt{(3229.9266) \cdot (1.5957)}}$$

$$= \frac{70.9559}{\sqrt{5153.9939}}$$

$$\approx \frac{70.9559}{71.7913}$$

$$\approx .988.$$

Recall that a Pearson coefficient of 1 indicates a perfect positive correlation. This value of p, being very close to 1, implies a strong correlation between these two variables. •

The coefficient of determination for these data is

$$p^2 = (.988)^2$$
$$\approx .976.$$

This means that approximately 97.6% of the variation in gasoline prices can be correlated (in a straight-line relationship) to variation in crude oil prices.

**E x a m p l e
2**

Refer to the data in Example 1, and use a graphics calculator to:

a. Find the Pearson coefficient of correlation.

b. Construct a scatter plot.

Solution   Use a TI-82 graphics calculator.

Example 5 of Chapter 12 actually computed the Pearson coefficient of correlation, but—since there were no ties within either set of rankings—the Pearson coefficient of correlation and the Spearman rank-correlation coefficient were exactly the same. Thus, the first part of Example 2 is a review of the procedure described in Example 5 of Chapter 12.

A fast way to clear the first two lists of data is (STAT), (4), (2nd), (1), (,), (2nd), (2), (ENTER).

a. The first part of this example is nearly identical to Example 5 of Chapter 12 (page 495). The difference is that in this example variables represent oil and gasoline prices instead of ranks.

• Clear any previous data from lists L1 and L2.
• Enter the prices per barrel of crude oil in list L1 and the prices per gallon of gasoline in list L2.
• Press (STAT), (▶) to obtain the following menu.

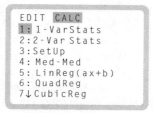

• Type (5), (2nd), (1), (,), (2nd), (2), (ENTER) to obtain the Pearson coefficient of correlation.

The screen in Figure 14-5 contains other important information as well. The value of $a \approx .022$ is the slope of the line of best fit, and $b \approx .360$ represents the Υ value at which the line of best fit intersects the Υ axis. Both of these statistics are discussed later in this chapter.

b. Using the data already entered in lists L1 and L2,

• Press (2nd), (Y=) to obtain the stat plot menu.
• Press (1).
• Use the arrow keys to turn the plot on; choose "scatter plot" as the display; specify L1 as the Xlist and L2 as the Ylist; arbitrarily choose the symbol "+" to mark points.

The shortcut (5), (ENTER) may also be used in the final step to obtain p.

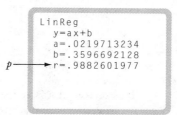

Figure 14-5

The completed screen should appear as follows:

- Press ⟨2nd⟩, ⟨Y=⟩ to return to the stat plot menu.
- Use the arrow keys to make sure all other plots are turned off.
- Press ⟨WINDOW⟩ to set the viewing rectangle.
- Use the arrow keys to set the following specifications.

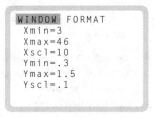

- Press the ⟨GRAPH⟩ button to obtain the scatter plot in Figure 14-6.

Instead of pressing ⟨WINDOW⟩ and using the arrow keys to set the viewing rectangle, the keystrokes ⟨ZOOM⟩, ⟨9⟩ automatically adjust the graph so that all statistical data points are displayed.

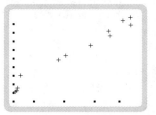

Figure 14-6 A scatter plot drawn with a TI-82 graphics calculator.

This scatter plot is very similar to the one obtained with Minitab, on page 526.   ●

Even when a scatter plot signifies no linear relationship, it may indicate a useful trend or pattern, as the following example illustrates.

E x a m p l e
3

The percentages of A's a mathematics instructor has given in her statistics classes in all the years she has taught that course follow.

Year	Years of Experience Teaching Statistics	Percentage of A's
1979	0	67.9
1980	1	70.5
1981	2	66.0
1982	3	50.4
1983	4	54.8
1984	5	42.7
1985	6	37.5
1986	7	32.6
1987	8	47.8
1988	9	31.0
1989	10	39.7
1990	11	47.9
1991	12	59.6
1992	13	58.0
1993	14	61.4

Draw a scatter plot for these data, and use it as a basis to suggest reasons for the instructor's pattern in grading.

Solution   Using a TI-82 graphics calculator, put the years of teaching experience on the horizontal axis (list L1) and the percentage of A's on the vertical axis (list L2). The scatter plot obtained in Figure 14-7 demonstrates that the percentage of A's was initially very high, declined, and is on the rise again. The coefficient of determination for these data is $p^2 \approx .077$. Such a small value of p^2 implies that very little grade variation (only 7.7%) can be explained by variation in years of teaching experience. Although the pattern of grades may be totally random, an accurately drawn scatter plot can suggest possibilities to consider:

a. The syllabus for the course has changed over the years, and students have had varying degrees of success coping with the changing content.
b. Textbooks were harder to understand 5 to 10 years ago.

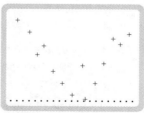

Figure 14-7 A scatter plot comparing years of teaching experience and the percentage of A's awarded.

c. Students were less prepared for the course 5 to 10 years ago.

d. Administrative pressure to raise or lower grades over the last 15 years has affected the distribution of statistics grades.

Without more information, it is impossible to know which, if any, of these possibilities explains the scatter plot.   ●

Exercises 14.1–14.4

In the following exercises Minitab may be used to find the Pearson coefficient of correlation by using the macro command CORRELATION C1 C2.

1. Describe the relationship between two negatively correlated random variables.

2. Give examples of pairs of random variables with near perfect correlation, with weak correlation, and with practically zero correlation.

3. If the line of best fit passes very close to all the points in a scatter plot, must a cause-and-effect relationship exist between the two variables? Explain.

4. Two students rank their personal preferences for ten different instructors, with a rank of 1 indicating lowest preference and a rank of 10 indicating highest preference. Each instructor is represented as a point on a scatter plot, with coordinates

    (rank from student A, rank from student B).

   a. How many points are represented on the scatter plot?

   b. If the line of best fit for these data has a negative slope, what does that imply about the preferences of the two students?

   * c. Under what condition(s) would the Pearson coefficient of correlation be equal to the Spearman rank-correlation coefficient (studied in Chapter 12) for these data?

5. Are there any advantages to using the coefficient of determination instead of the Pearson coefficient of correlation? Explain.

6. Is it possible for the sample coefficient of determination to be very close to 1 and the population coefficient of determination to be nearly zero? Explain.

7. The value of the Pearson coefficient of correlation is very tedious to compute by hand. Without a statistical software program or a graphics calculator, about how many separate computations must be made to find p?

GROUP

8. Use your best judgment to match the value of p in column A with the scatter plots in column B.

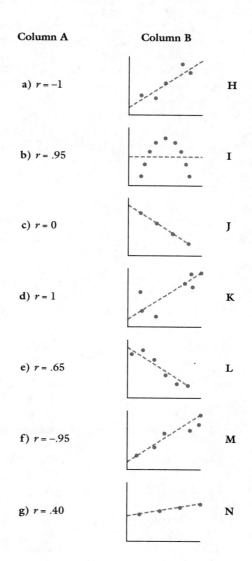

Column A

a) $r = -1$

b) $r = .95$

c) $r = 0$

d) $r = 1$

e) $r = .65$

f) $r = -.95$

g) $r = .40$

Column B

H

I

J

K

L

M

N

9. Refer to Example 3 on p. 533. Verify that $p^2 \approx .077$.

10. The following table shows the amount of fire damage to a residence in relation to its distance from a fire station. Draw a scatter plot for these sample data, then sketch a possible line of best fit on your drawing.

Distance to Fire Station (miles), X	Fire Damage (thousands of dollars), Y	Distance to Fire Station (miles), X	Fire Damage (thousands of dollars), Y
3.25	12.56	6.75	33.78
2.25	5.75	6.50	45.00
4.05	27.50	3.50	5.50
0.75	1.50	10.50	65.00
3.25	15.80	4.50	33.33
1.00	2.50	2.75	10.00

11. The following scatter plot shows years of experience and yearly numbers of days absent for a sample of white-collar workers in a certain industry. Use the scatter plot to suggest some possible reasons for this pattern of worker absences.

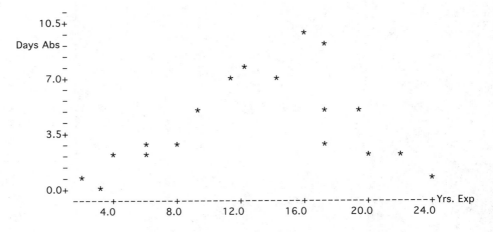

12. The following scatter plot shows the relative prices and the percentages of market share for a sample of $n = 18$ cities that carry a certain product.

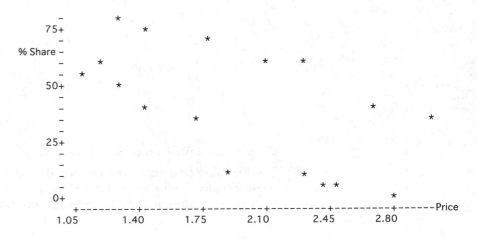

a. Use the scatter plot to suggest some sales patterns for this product.

b. Name some factors that could produce these sales patterns.

13. The following scatter plot shows the relationship between SAT scores and first-year GPAs for a sample of $n = 14$ students.

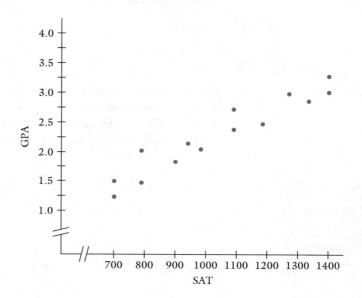

a. Sketch the line of best fit on this scatter plot.
b. Use the scatter plot to estimate, as well as possible, a set of ordered pairs: (SAT score, first-year GPA).
c. Use the ordered pairs from Part b to compute the Pearson coefficient of correlation between those two variables.
d. What percentage of variation in GPAs can be explained by a linear relationship with SAT scores?

GROUP

SAT scores may be cross-tabulated with many other qualitative variables besides those shown here. For example, data are available on SAT scores in comparison with natural sciences course work, mathematics course work, computer course work, citizenship status, and ethnic group membership.

14. Following are profiles of students who took the SAT test in 1993. One group is cross-tabulated by income level, and the other is cross-tabulated by foreign and classical language study.

Income	SAT Verbal Mean	SAT Math Mean
Less than $10,000	352	416
$10,000 to $20,000	379	434
$20,000 to $30,000	404	453
$30,000 to $40,000	418	469
$40,000 to $50,000	431	483
$50,000 to $60,000	440	493
$60,000 to $70,000	449	504
$70,000 or more	472	533

Foreign and Classical Language Study	SAT Verbal Mean	SAT Math Mean
French	445	494
German	462	518
Greek	457	516
Modern Hebrew	474	535
Italian	407	455
Russian	462	532
Spanish	418	470

Source: *1993 Profile of SAT and Achievement Test Takers,* The College Board, pp. 4, 7.

Use the Pearson coefficient of correlation to determine which cross-tabulation shows a stronger relationship between verbal and math scores.

15. The following scatter plots show men's and women's average scores on the mathematical section of the SAT test in the years 1969–1992.

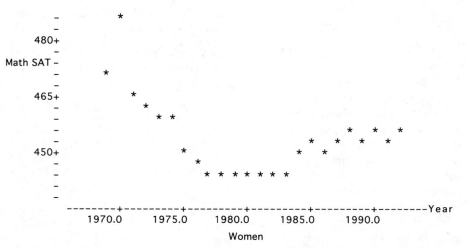

(Averages for 1969 through 1971 are estimates. Reports on college bound seniors were not prepared in those years.)

a. Use the scatter plots to estimate the mean SAT math score for men and the mean SAT math score for women in each of the years from 1969 through 1992.

b. Using the ordered pairs from Part a, estimate the values of the Pearson coefficient of correlation for men and for women.

c. Using the ordered pairs from Part a, compute the values of the Pearson coefficient of correlation for men and for women.

14.5 SIGNIFICANCE TESTING OF THE PEARSON COEFFICIENT OF CORRELATION

Consider the following statistics released by the College Board in its national report *1993 Profile of SAT and Achievement Test Takers.*

Student Profiles

Size of Senior Class	SAT Verbal Mean	SAT Math Mean
More than 1000	436	502
750–1000	415	473
500–749	421	485
250–499	424	482
100–249	425	475
Fewer than 100	434	479

The Pearson coefficient of correlation for these verbal and math scores is $p \approx .646$. Does this value of p show a significant correlation between verbal and math scores when they are cross-tabulated by the size of the senior class? The next example answers this question, but one must first understand that the value of $p = .646$ is a sample statistic, since there can be no guarantee that every school in the country was included in the survey or that the schools reported their sizes accurately. The value of p is an estimate of the Pearson coefficient of correlation computed from the entire population, denoted by $\mathbf{P}$. What p measures for samples, $\mathbf{P}$ measures for populations.

E x a m p l e 4

The only possible outcomes from this hypothesis test are the conclusion that a significant correlation exists between verbal and math SAT scores, and reservation of judgment.

Refer to the data just presented, relating SAT verbal and math scores to the size of the senior class. Perform a hypothesis test at the 10% level of significance to determine whether the calculated value of p is evidence of a significant correlation, $\mathbf{P}$, between verbal and math scores in the population.

Solution   Use the usual five-step method.

Step 1   State the hypotheses.

$$H_0: \quad \mathbf{P} = 0.$$
$$H_A: \quad \mathbf{P} \neq 0.$$

The null hypothesis always claims that there is no correlation in the population. The sample statistic provides evidence that there is some correlation in the population. If the hypothesis test rejects the null hypothesis, then evidence exists to support the alternative hypothesis and the claim of a significant correlation in the population.

Step 2   Decide on a model for the sampling distribution.

The sampling distribution is the model of all possible values of p under the null hypothesis. Use the t-distribution with $n - 2$ degrees of freedom.

Step 3   Determine the end points of the rejection region, and state the decision rule.

Consult Appendix A-7, "Critical Values of p When $\mathbf{P} = 0$." This is a two-tail hypothesis test with $\alpha = .10$. The intersection of the .10 column and the $6 - 2 = 4$ degrees of freedom row indicates the following rejection regions.

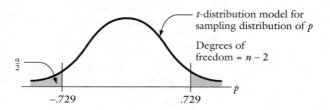

The decision rule is:

   Reject H_0 and accept H_A if $p > .729$ or if $p < -.729$. Otherwise, reserve judgment.

Step 4   Compute the test statistic.

The test statistic is p. On page 539 it was stated to be approximately .646.

Step 5   State the conclusion.

The value of $p = .646$ is less than .729; the test statistic is not in the rejection region. Therefore, **reserve judgment.** At the 10% level of significance, there is not enough statistical evidence to support the alternative hypothesis of a significant correlation between verbal and math scores in the population when cross-tabulated by the size of the senior class.   •

Exercises 14.5

1. In your own words, explain the difference between p and $\mathbf{P}$.

2. Is it possible for $\mathbf{P}$ to equal .10 and p to equal .90? Explain why or why not.

3. Examine Appendix A-7, and explain why critical values of p decrease for increasing degrees of freedom.

4. In what way is the five-step hypothesis test used in this section slightly different from other hypothesis tests in this text?

5. Refer to Example 1 on page 528 and Appendix A-7. Show that the sample statistic provides evidence of a significant correlation between crude oil prices and gasoline prices at nearly any level of significance.

6. Refer to Exercise 13 (in Exercises 14.1–14.4) on page 537. Determine whether the sample statistic calculated in Part c provides evidence of a significant correlation between SAT scores and GPAs at the 2% level of significance.

GROUP

7. The following data relating to births are cross-tabulated by nine geographic divisions.

Division	Percent of Births with Low Birth Weight (X)	Percent of Births to Teenage Mothers (Y)
Northeast	5.9	8.4
Mid-Atlantic	7.3	9.5
East North Central	7.1	13.2
West North Central	5.9	11.1
South Atlantic	7.9	14.4
East South Central	8.2	18.4
West South Central	7.3	16.3
Mountain	6.8	12.8
Pacific	5.7	11.5

Source: *Statistical Abstract of the United States, 1993*, p. 78.

a. Draw a scatter plot for these data.

b. Does the scatter plot indicate any correlation between the two variables?

c. Compute the value of the Pearson coefficient of correlation, and perform a hypothesis test at the 10% level of significance to determine whether the value of p is evidence of a significant correlation between low birth weights and births to teenage mothers.

d. How could the relationship of X to Y be better studied than by cross-tabulation with geographic divisions?

8. The following array shows the birth and death rates (occurrences per 1000 of the population) in the United States every ten years from 1910 to 1990.

Year	Birth Rate	Death Rate
1910	30.1	14.7
1920	27.7	13.0
1930	21.3	11.3
1940	19.4	10.8
1950	24.1	9.6
1960	23.7	9.5
1970	18.4	9.5
1980	15.9	8.8
1990	16.7	8.6

Source: *The Universal Almanac 1994*, p. 292.

a. Estimate the value of the Pearson coefficient of correlation.

b. Test at the 10% level of significance to determine whether there is evidence of a significant correlation between the death rate and the birth rate.

GROUP

9. Following is a list of wars that involved the United States, along with the total numbers of Americans who served and the total numbers of battle deaths. All numbers are in thousands.

War	Number Serving (X)	Battle Deaths (Y)
Revolutionary War	200.00	6.82
War of 1812	286.73	2.26
Mexican War	78.72	1.73
Civil War*	2213.36	140.41
Spanish-American War	306.76	.39
World War I	4743.83	53.51
World War II	16,353.66	292.13
Korean War	5764.14	33.63
Vietnam War	8744.00	47.32

*Includes Union forces only.

a. Draw a scatter plot for these data.

b. Determine the percentage of variation in the number of battle deaths that can be correlated with the variation in the number of soldiers serving in the conflict.

10. A check of supermarket shelves in May 1994 revealed the following prices and minimum percentages of crude fat for the specified brands of cat food.

Brand Name	Price per Pound (X)	Minimum % of Crude Fat (Y)
Friskies Senior	$0.97	3.5
Fancy Feast	2.29	3.0
Purina Tuna & Chicken	0.73	2.0
Kal Kan Optimum	0.58	5.0
Jewel Beef & Liver	0.84	6.5
President's Choice Gourmet	1.58	4.0
9-Lives Tuna & Whitefish	0.86	4.5
Amore Natural	1.97	2.0
Whiska's	0.86	5.0
Alpo Robust Stew	0.86	4.0

a. Construct a scatter plot for these sample data. Compute the Pearson coefficient of correlation to measure the strength of the relationship between price and crude fat content.

b. What conclusion seems obvious on the basis of Part a?

11. The following array shows average hourly compensations, in U.S. dollars, for production workers in selected countries in 1980 and 1990.

Country	1980	1990
Greece	3.73	6.71
Japan	5.52	12.74
United States	9.87	14.91
Austria	8.57	17.01
Mexico	2.18	1.64
Ireland	5.95	11.76
Taiwan	1.00	3.95
United Kingdom	7.56	12.71
Israel	3.79	8.55
Brazil	1.38	2.64
Sri Lanka	0.22	0.35
France	8.94	15.23

Source: *The Universal Almanac 1994*, p. 350.

a. What might the absence of a significant correlation between compensations in 1980 and 1990 imply?

b. Test at the 5% level of significance to determine whether there is evidence of a significant correlation between 1980 and 1990 compensations.

12. In May 1994, the following prices were recorded for popular brands of paper towels.

Brand Name	Cost per Roll, X	Sq. Ft. per Roll, Y
Viva Cuisine	$1.45	87.3
Bounty Microwave Big Roll	1.45	80.6
Scottowel Mega Roll	1.29	119.0
Jewel Jumbo Roll	0.99	92.8
Scottowels Recycled	0.84	79.0
Mardi Gras	0.79	58.0
Viva Border Design	0.89	59.5
Jewel Double Ply	0.68	61.8
Gala	0.79	59.0
Sparkle	0.69	61.8
Brawny Prestige Prints	0.89	63.0
Job Squad Super Towel	0.99	39.7

a. Construct a scatter plot for these data.

b. Use the coefficient of determination to find the percentage of variation in the number of square feet per roll that can be attributed to variation in price. What does this percentage imply?

13. Refer to Exercise 12. Perform a hypothesis test at the 2% level of significance to determine whether the value of p is evidence of a significant correlation between prices and the number of square feet per roll.

GROUP 14. The following table lists the 15 all-time top-ranked quarterbacks in the National Football League as of the start of the 1994 season. Also included are the number of pass attempts by each quarterback, X; the number of touchdowns thrown, Y_1; and the number of interceptions thrown, Y_2.

Quarterback	No. of pass attempts, X	No. of TD's, Y_1	No. of Int's., Y_2
Joe Montana	4898	257	130
Steve Young	1968	105	58
Dan Marino	5434	298	168
Jim Kelly	3494	179	126
Roger Staubach	2958	153	109
Neil Lomax	3153	136	90
Sonny Jurgensen	4262	255	189
Len Dawson	3741	239	183
Boomer Esiason	3851	190	140
Dave Krieg	4178	217	163
Ken Anderson	4475	197	160
Bernie Kosar	3213	119	81
Danny White	2950	155	132
Troy Aikman	1920	69	66
Bart Starr	3149	152	138

a. Do you think there is a greater correlation between the numbers of pass attempts and touchdowns or between the numbers of pass attempts and interceptions?

b. Determine the Pearson coefficient of correlation, p, between the number of pass attempts, X, and the number of touchdowns, Y_1.

c. Determine the Pearson coefficient of correlation, p, between the number of pass attempts, X, and the number of touchdowns, Y_2.

15. Refer to Exercise 14. Perform hypothesis tests at the 2% level of significance to determine whether there is evidence of a significant correlation between pass attempts and touchdowns or between pass attempts and interceptions.

14.6 THE LINE OF BEST FIT

A correlation of $r \geq .90$ or $r \leq -.90$ was arbitrarily chosen as an indication of a "strong" correlation.

If a strong correlation exists between two random variables (such as $r \geq .90$ or $r \leq -.90$), then a linear model can closely relate one variable to the other, and the observed values of one variable can then be used to predict values of the other variable. This procedure is an important application of the line of best fit.

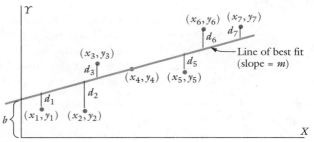

Figure 14-8 The line of best fit.

Definition ▶ | The **line of best fit,** or **simple**[8] **regression line,** is a straight line drawn in such a way that the sum of the squared vertical distances from all points in the scatter plot to the line is as small as possible.

Figure 14-8 shows the line of best fit. The line of best fit is not necessarily the line that passes through the most points; in fact, it does not need to pass through any of them. In this case, it passes through only one point, (x_4, y_4). Loosely speaking, the line of best fit is the line that comes the closest to the greatest number of points, and this condition is ensured when

$$(d_1)^2 + (d_2)^2 + (d_3)^2 + \cdots + (d_n)^2$$

is a minimum. Remember from algebra that, because the line of best fit is a line, its equation can be written in the form

$$y = m \cdot x + b.$$

If a line is written in the form $y = m \cdot x + b$, it is said to be written in slope-intercept form.

In this equation x and y are values of the two variables, m is the slope of the line, and b is the point at which the line intersects the Y axis.

Use the following formulas to determine the slope and Y-intercept of the line so that the sum of the squares of the vertical distances from all points to the line is a minimum.

Formula for the **Slope, m, of the Line of Best Fit**

$$m = \frac{\Sigma x \cdot y - n \cdot \overline{X} \cdot \overline{Y}}{\Sigma x^2 - n \cdot \overline{X}^2} \qquad \textbf{(14-2)}$$

Formula for the Y-**Intercept, b, of the Line of Best Fit**

$$b = \overline{Y} - m \cdot \overline{X} \qquad \textbf{(14-3)}$$

[8]The term "simple" implies a first-degree polynomial, or line. Many statistical software programs can be used to construct multiple regression models—that is, polynomials of higher degree that pass through the points of a scatter plot.

Once *m* and *b* are calculated, the line of best fit can be determined and a given value of one variable, *X*, can be used to find, or predict, the value of the other variable, *Y*. Accordingly, the line of best fit is also called the **prediction line.**[9]

The next three examples show how to find the equation of the line of best fit by hand, with Minitab, and with a graphics calculator.

The data from Example 1, which relate the price per gallon of gasoline to the price per barrel of crude oil, are repeated here.

Price per Barrel of Crude Oil, *X*	Price per Gallon of Gasoline, *Y*
$ 3.35	$ .38
3.75	.40
4.29	.43
5.80	.57
18.00	.80
21.30	.85
30.00	.98
35.75	1.20
37.05	1.15
42.50	1.35
45.75	1.40
45.75	1.25

Determine the equation of the line of best fit.

Solution First find the values of the line's slope and *Y*-intercept: *m* and *b*, respectively. By the formulas presented earlier,

$$m = \frac{\Sigma x \cdot y - n \cdot \overline{X} \cdot \overline{Y}}{\Sigma x^2 - n \cdot \overline{X}^2},$$

and

$$b = \overline{Y} - m \cdot \overline{X}.$$

The solution to Example 1 included all of the separate calculations necessary to find *m* and *b*. For convenience, the results of these calculations are repeated here.

$$\Sigma x \cdot y = 333.9487 \qquad \overline{X} = 24.4408 \qquad \overline{Y} = .8967$$
$$n \cdot \overline{X} \cdot \overline{Y} = 262.9928 \qquad \Sigma x^2 = 10,398.1591 \qquad n \cdot \overline{X}^2 = 7168.2325$$

Find the slope, *m*, of the line of best fit by substituting these values into the slope formula.

[9]This is assuming, of course, that a strong correlation exists between the two variables and that it is not a spurious or nonsense correlation.

$$m = \frac{\Sigma x \cdot y - n \cdot \overline{X} \cdot \overline{Y}}{\Sigma x^2 - n \cdot \overline{X}^2}$$

$$= \frac{333.9487 - 262.9928}{10398.1591 - 7168.2325}$$

$$= \frac{70.9559}{3229.9266}$$

$$\approx .022.$$

Since the slope is positive $(.022 > 0)$, the line of best fit will rise as values of X increase.

Find the Y-intercept, b, of the line of best fit by substituting values into the following formula.

$$b = \overline{Y} - m \cdot \overline{X}$$

$$= .8967 - (.022) \cdot (24.4408)$$

$$\approx .359.$$

Accordingly, the equation of the line of best fit is

$$y = m \cdot x + b$$

$$= .022 \cdot x + .359. \quad \bullet$$

E x a m p l e
6

Use Minitab to find the equation of the line of best fit for the data from Example 5. (*Note:* Minitab calls the equation of the line of best fit the "regression equation.")

Once the REGRESS command has been executed, Minitab may be used to draw a scatter plot that shows the line of best fit. To do so, choose **Scatter Plot. . .** from the **Graph** menu and use the **Lines** option.

```
                                   Session
Worksheet size: 38000 cells
MTB > REGRESS   C2   1   C1

The regression equation is
 Y = 0.360 + 0.0220 X

Predictor        Coef      Stdev     t-ratio        P
Constant      0.35967    0.03162      11.38    0.000
X            0.021971   0.001074      20.46    0.000

s = 0.06104     R-sq = 97.7%     R-sq(adj) = 97.4%

Analysis of Variance

SOURCE        DF         SS         MS        F        P
Regression     1     1.5592     1.5592   418.42    0.000
Error         10     0.0373     0.0037
Total         11     1.5965

Unusual Observations
 Obs.       X        Y       Fit    Stdev.Fit   Residual   St.Resid
  12     45.8   1.2500    1.3649       0.0289    -0.1149     -2.14R

R denotes an obs. with a large st. resid.
```

Figure 14-9 Use of Minitab's REGRESS command.

Solution   First enter the data into a Minitab worksheet, with the price per barrel in column C1, named the *X* column, and the corresponding price per gallon in column C2, named the *Y* column.

Next, obtain a Session window and type the following macro command.

REGRESS C2 1 C1

The command REGRESS is for regression, or the line of best fit. The rest of the command signifies that data in C2 will be fitted to one predictor, the data in C1.

Figure 14-9 shows the completed command line and output. Note the very slight difference due to rounding and the unusual observations, in respect to the calculated regression line, that are printed. This feature provides an opportunity to return to the original data and check their accuracy.   •

**E x a m p l e
7**

Use a TI-82 graphics calculator to find the equation of the line of best fit for the data from Example 5.

Solution   Refer to Example 2 on page 530. Enter the data in the same way, and obtain the screen that contains the Pearson coefficient of correlation. For convenience, Figure 14-10 repeats the screen. These values of the slope and *Y*-intercept are the same as those obtained by hand and with Minitab.

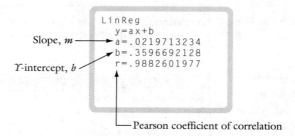

Slope, *m*
Y-intercept, *b*

```
LinReg
  y=ax+b
  a=.0219713234
  b=.3596692128
  r=.9882601977
```

Pearson coefficient of correlation

Figure 14-10   The slope and *Y*-intercept of the line of best fit.   •

Using the Line of Best Fit to Predict Values of a Variable

Figure 14-11 repeats the scatter plot relating the price of a barrel of crude oil to the price of a gallon of gasoline, along with the line of best fit. The figure illustrates that the line of best fit may not pass directly through many of the points. When the price of a barrel of crude oil is, for example, $x = 37.05$, the line of best fit can be used to predict the price of a gallon of gasoline. Using the slope and *Y*-intercept just obtained with a graphics calculator,

$$y = .022 \cdot x + .360$$
$$= .022 \cdot 37.05 + .360$$
$$\approx 1.18.$$

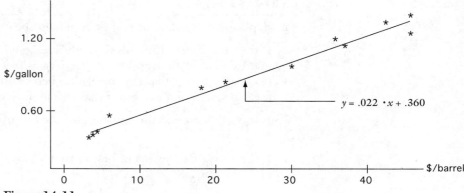

Figure 14-11

Notice that this **predicted value,** or **point estimate,** comes very close to the observed value, $y = 1.15$. However, like the coefficients of correlation and determination, the line of best fit is based on samples of two populations. Therefore, it may not be very close to the actual line of best fit for the two populations, known as the **true regression line.** As a result, point estimates, or predicted values, may have some error even if they are close to the observed values. Section 14.7 will attempt to quantify this error through the use of confidence intervals for predicted values.

The next example presents more instances of using the line of best fit to compute point estimates.

**E x a m p l e
8**

Use the equation of the line of best fit for the data in the preceding examples to predict the price per gallon of gasoline[10] if the price per barrel of crude oil is:

a. \$25.00 per barrel.
b. \$48.00 per barrel.

Solution   Nearly all of the necessary work was done in the preceding examples, which found the equation of the line of best fit. Recall that the line of best fit for these data is

$$y = .022 \cdot x + .360.$$

a. When $x = 25.00$,

$$y = .022 \cdot 25.00 + .360$$
$$= .91 \text{ per gallon.}$$

b. When $x = 48.00$,

$$y = .022 \cdot 48.00 + .360$$
$$\approx 1.42 \text{ per gallon.} \quad \bullet$$

---

[10]By solving the equation of the line of best fit for x, it is also possible to determine a value of the random variable X that "caused" a specified value of the random variable Y.

Exercises 14.6

1. Explain why the line of best fit could also be called the **least-squares line.**

2. If the line of best fit passes through three points in one scatter plot, and the line of best fit for another scatter plot passes through one point, does it necessarily follow that the correlation between the two variables described in the first scatter plot is stronger than that in the second? Explain why or why not.

3. Is it possible to find the equation for the line of best fit for data in any scatter plot? Explain.

4. Is it possible for the line of best fit to not pass through any points in a scatter plot? Explain.

5. What is the true regression line?

6. Refer to Exercise 10 (in Exercises 14.1–14.4) on page 535.

   a. Determine the equation of the line of best fit for those data.

   b. Use the line of best fit from Part a to find a point estimate for the amount of fire damage to a residence that is 8 miles from a fire station.

7. Refer to Exercise 13 (in Exercises 14.1–14.4) on page 537.

   a. Determine the equation of the line of best fit for those data.

   b. Use the line of best fit from Part a to find a point estimate for the first-year GPA of a student who had an SAT score of 1050.

   c. Solve the equation of the line of best fit from Part a for x, then use it to find a point estimate of the SAT score of someone who had a first-year GPA of 3.0.

8. Refer to Exercise 15 (in Exercises 14.1–14.4) on page 538. Estimate the average values of men's scores on the mathematics section for the years 1986 through 1992. Use these estimates to answer the following questions.

   a. Write the equation of the line of best fit for the data from 1986 through 1992.

   b. Use the line of best fit from Part a to predict the mean SAT score in the year 1999, supposing no recentering occurred.

9. Interpret the sign of the slope of the line of best fit from Exercise 8.

10. Refer to Exercise 14 (in Exercises 14.5) on page 544. Use a simple regression line to predict how many touchdowns a top-ranked NFL quarterback would throw in a career with 2000 pass attempts.

11. Some people believe that the crimes of robbery and burglary are related to each other. The numbers of reported robberies and burglaries in the United States in the years 1982 through 1992 in thousands, follow.

Year	Robberies (X)	Burglaries (Y)
1982	553	3447
1983	507	3130
1984	485	2984
1985	498	3073
1986	543	3241
1987	518	3236
1988	543	3218
1989	578	3168
1990	639	3074
1991	688	3157
1992	672	2980

Assuming that the number of robberies in a future year is 500,000, use a prediction line to find a point estimate for the number of burglaries that year.

12. Refer to Exercise 11.

    a. Perform a hypothesis test at the 2% level of significance to determine whether there is evidence of a significant correlation between the number of robberies and the number of burglaries.

    b. On the basis of the result from Part a, does the construction of a prediction line in Exercise 11 seem justified?

13. Following are the 1980 EPA combined mileage ratings and cylinder volumes for nine standard-transmission four-cylinder subcompacts.

Car	Cylinder Volume (cc), X	Miles per Gallon, Y
Volkswagen Rabbit	97	24
Datsun 210	85	29
Chevette	98	26
Dodge Omni	105	24
Mazda 626	120	24
Oldsmobile Starfire	151	22
Mercury Capri	140	23
Toyota Celica	134	23
Datsun 810	146	21

Use a simple regression line to find a point estimate for the miles-per-gallon rating of a four-cylinder car with a 130-cc cylinder volume in 1980.

14. The following table lists a random sample of $n = 10$ communities in the Chicago area, their populations (in thousands), and the amounts of money (in thousands of dollars) spent in their eating and drinking establishments.

City	Population, X	Money Spent Eating and Drinking, Y
Naperville	88.5	95.9
Hoffman Estates	46.6	28.1
Wheaton	53.4	23.0
Oak Park	53.7	37.1
Mount Prospect	53.2	39.7
Elmhurst	43.6	45.8
Des Plaines	53.2	100.7
Skokie	59.4	61.1
Oak Lawn	56.2	56.6
Downers Grove	48.6	73.1

Source: *Sales & Marketing Management,* August 24, 1992, p. C-59.

a. Perform a hypothesis test at the 10% level of significance to determine whether there is evidence of a significant correlation between population and the amount of money spent in eating and drinking establishments.

b. On the basis of the result of Part a, does it seem appropriate to use these data to derive the equation of a line of best fit, and to then use that line to estimate eating and drinking sales for a city of 50,000 people? Why or why not?

15. Following is a random sample of $n = 12$ states, showing the numbers of emergency room visits per 1000 of the population and the numbers of outpatient visits per 1000 of the population.

	Emergency Room Visits, X	Outpatient Visits, Y
GA	439.4	910.7
NY	385.9	1422.7
HI	322.0	2074.5
NE	250.0	1062.5
RI	473.1	815.3
MS	374.2	687.7
WA	342.9	1042.1
IL	349.0	1140.1
WV	503.3	1046.9
CT	397.0	1084.9
KS	313.0	1358.3
NM	335.2	1585.3

Source: *Statistical Bulletin* 73:2, April–June 1992, p. 26.

Draw a scatter plot for these data, and try to identify which point would be classified as an "unusual observation." If possible, use Minitab to confirm your identification.

16. Refer to Exercise 15. Explain why it does not seem appropriate to use the number of emergency room visits to predict the number of outpatient visits.

14.7 CONFIDENCE INTERVALS FOR PREDICTED VALUES

When the line of best fit is used to find a point estimate of a variable, it has been shown that some error may occur, since the estimate is based on random samples from two populations. In general, different samples result in different lines of best fit, which yield different point estimates. This uncertainty is made clear if a confidence interval is found instead of a point estimate.

The formulas for the construction of confidence intervals for predicted population values depend on two factors. The first is sample size, which is the number of ordered pairs. The second is whether the confidence interval is for an estimated population mean value or a random value of the population.

Construction of a Confidence Interval for the Mean Value of Y

Consider the following data from a random sample of $n = 10$ people who commute to work each morning on public transportation.

Distance of Commute (miles), X	Commuting Time (minutes), Y
20	60
18	62
30	85
10	27
8	22
20	54
10	32
5	10
4	12
22	59

A high correlation seems to exist between these two variables; in fact, one can verify that $p \approx .98$. It also seems reasonable to suppose that values of X (distance) affect the values of Y (time). Therefore, it is appropriate to use this sample of $n = 10$ ordered pairs to determine a 90% confidence interval for the mean morning commuting time of all people in the population who commute 15 miles.[11] The confidence interval is denoted $\mu_{Y,15}$, which should be read "the mean population value of Y for an x value of 15."

The first step in determining this confidence interval is to arrive at a point estimate of Y when $x = 15$ by obtaining the equation of the line of best fit. On a TI-82 graphics calculator, the equation of the line of best fit appears as shown in Figure 14-12.

---

[11] The next section of this chapter discusses the construction of a confidence interval for the commuting time of a person, chosen at random, who travels 15 miles.

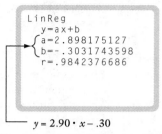

```
LinReg
  y=ax+b
 ╭ a=2.898175127
 ╰ b=-.3031743598
   r=.9842376686
```

$y = 2.90 \cdot x - .30$

Figure 14-12 Use of a graphics calculator to obtain the equation of the line of best fit.

Now, substituting $x = 15$ into this line-of-best-fit equation,

$$y = 2.90{\cdot}15 - .30$$
$$= 43.2 \text{ minutes.}$$

This point estimate, 43.2, will be the center of the desired confidence interval. Find the end points of the confidence interval by substituting the indicated quantities into this formidable-looking formula:

Formula for the **End Points of a Confidence Interval for the Mean Value of** Y**,** $\mu_{Y,X}$

$$\hat{Y} \pm t \cdot \sqrt{\frac{\Sigma y^2 - b\Sigma y - a\Sigma x \cdot y}{n - 2}} \cdot \sqrt{\frac{1}{n} + \frac{(x - \overline{X})^2}{\Sigma x^2 - n \cdot \overline{X}^2}} \qquad \textbf{(14-4)}$$

In this formula,

$\hat{Y} =$ center of the confidence interval, which in this case is 43.2;

$n =$ number of ordered pairs;

$t =$ t-score that encloses the middle 90% of the t-distribution model, with degrees of freedom $= n - 2$;

$a =$ slope of the equation of the line of best fit;

$b =$ Y-intercept of the equation of the line of best fit;

$\overline{X} =$ sample mean of the X-variable values.

Σy^2, Σx^2, $\Sigma x \cdot y$, Σy, and $n \cdot \overline{X}^2$ are calculated as described in Example 1 of this chapter.

One can verify that $\hat{Y} = 43.2$, $n = 10$, $t = 1.860$, $b = -.30$, $a = 2.90$, $x = 15$, $\overline{X} = 14.7$, $\Sigma x^2 = 2813$, $\Sigma x \cdot y = 8108$, $\Sigma y = 423$, and $\Sigma y^2 = 23{,}547$.

Find the end points of the confidence interval by evaluating

$$\hat{Y} \pm t \cdot \sqrt{\frac{\Sigma y^2 - b\Sigma y - a\Sigma x \cdot y}{n - 2}} \cdot \sqrt{\frac{1}{n} + \frac{(x - \overline{X})^2}{\Sigma x^2 - n \cdot \overline{X}^2}}.$$

Thus, the end points are at

$$43.2 \pm 1.860 \cdot \sqrt{\frac{23547 - (-.30) \cdot 423 - (2.90) \cdot 8108}{10 - 2}} \cdot \sqrt{\frac{1}{10} + \frac{(15 - 14.7)^2}{2813 - 10 \cdot 14.7^2}}$$

$$= 43.2 \pm 1.860 \cdot \sqrt{\frac{160.7}{8}} \cdot \sqrt{\frac{1}{10} + \frac{.09}{652.1}}$$

$$\approx 43.2 \pm 1.860 \cdot 4.48 \cdot .32$$

$$\approx 43.2 \pm 2.67.$$

Therefore, a 90% confidence interval for the mean morning commuting time for a distance of 15 miles, in the population from which these samples were taken, is approximately

$$43.2 \pm 2.67 = [40.53, 45.87] \text{ minutes.}$$

If the sample size had been large ($n \geq 30$), the calculation to find the end points of the confidence interval would have simplified a bit, to

$$\hat{Y} \pm Z \cdot \sqrt{\frac{\Sigma y^2 - b\Sigma y - a\Sigma x \cdot y}{n - 2}} \cdot \sqrt{\frac{1}{n}} \qquad \textbf{(14-4a)}$$

In this formula, Z represents the Z-score that encloses the middle 90% of the normal distribution model.

Construction of a Confidence Interval for a Random Value of Y

The confidence interval just found is for the *mean* morning commuting time of all people in the population who commute 15 miles. An alternative is a confidence interval for a single individual, chosen at random, who commutes 15 miles, denoted by $R_{Y,15}$. It can also be thought of as a confidence interval for the commuting time of the next person sampled. The formula for the end points of this confidence interval follows.

Formula for the **End Points of a Confidence Interval for a Random Value of Y, $R_{Y,X}$**

$$\hat{Y} \pm t \cdot \sqrt{\frac{\Sigma y^2 - b\Sigma y - a\Sigma x \cdot y}{n - 2}} \cdot \sqrt{\frac{1}{n} + \frac{(x - \overline{X})^2}{\Sigma x^2 - n \cdot \overline{X}^2} + 1} \qquad \textbf{(14-5)}$$

Construct a 90% confidence interval for the morning commuting time of a 15-mile commuter chosen at random and carefully compare it with the confidence interval just found. The new interval uses most of the same numbers and is very similar. Find the end points by calculating

$$\hat{Y} \pm t \cdot \sqrt{\frac{\Sigma y^2 - b\Sigma y - a\Sigma x \cdot y}{n - 2}} \cdot \sqrt{\frac{1}{n} + \frac{(x - \overline{X})^2}{\Sigma x^2 - n \cdot \overline{X}^2} + 1}.$$

Thus, the end points are at

$$43.2 \pm 1.860 \cdot \sqrt{\frac{23547 - (-.30) \cdot 423 - (2.90) \cdot 8108}{10 - 2}} \sqrt{\frac{1}{10} + \frac{(15 - 14.7)^2}{2813 - 10 \cdot 14.7^2} + 1}$$

$$= 43.2 \pm 1.860 \cdot \sqrt{\frac{160.7}{8}} \cdot \sqrt{\frac{1}{10} + \frac{.09}{652.1} + 1}$$

$$\approx 43.2 \pm 1.860 \cdot 4.48 \cdot 1.05$$

$$\approx 43.2 \pm 8.75$$

$$= [34.45, 51.95].$$

If the sample is large ($n \geq 30$), one may find the end points of a confidence interval for a random value of Y by evaluating the simpler expression

$$\hat{Y} \pm Z \cdot \sqrt{\frac{\Sigma y^2 - b\Sigma y - a\Sigma x \cdot y}{n - 2}}. \qquad \text{(14-5a)}$$

As before, Z represents the Z-score that encloses the middle 90% of the normal distribution model.

Exercises 14.7

1. Refer to the first confidence interval constructed in this section, on page 555. Verify that the values used for the quantities in its construction are correct.

* 2. Which will always be larger, $\mu_{Y,X}$ or $R_{Y,X}$? Explain why, both mathematically and logically.

Exercises 3, 4, and 5 are based on the data on morning commuting times presented on page 553.

3. Form an 80% confidence interval for the mean commuting time of those who commute 10 miles in the morning.

4. Form a 95% confidence interval for a random commuting time of those who commute 25 miles in the morning.

GROUP   5. Suppose that the following 25 additional morning commuting times are available.

Distance of Commute (miles), X	Commuting Time (minutes), Y
20	55
40	105
25	63
18	58
30	90
35	91
10	33
15	40

Distance of Commute (miles), X	Commuting Time (minutes), Y
12	30
17	48
22	55
30	95
10	30
5	10
2	12
6	8
18	65
60	158
30	86
10	25
10	32
8	34
12	28
30	80
14	45

Form 90% confidence intervals for both the mean commuting time and a random commuting time of a 25-mile trip. Base these confidence intervals on a combined sample of size $n = 35$.

6. Refer to Exercise 13 (in Exercises 14.1–14.4) on page 537. Estimate, as well as possible, a set of ordered pairs: (SAT score, first-year GPA). Use these estimates to construct an 80% confidence interval for the mean first-year GPA of all students who had SAT scores of 1060.

Case Study   *Nonparametric Regression*

   In many applications of statistics, the straight-line model for the line of best fit is replaced by a curved line, which may arise from a polynomial function or some other kind of function. "Statistical Data Analysis in the Computer Age" (an article in *Science*, July 26, 1991, addressed by the case study in Chapter 12) explains how computers are used to pass complicated curves near data points. Obtain a copy of this article from your library, and use it to answer the following questions.

I. Which mathematicians developed the least-squares estimation method, and when did they do this work?
II. In your own words, explain how the computer program Loess works.
III. Why is the curve that Loess fits to the data points called nonparametric?
IV. What type of data were used to illustrate the "generalized additive model?"
V. What surprising conclusion in the analysis of data with the generalized additive model might have been missed with a straight-line model?

CHAPTER SUMMARY

The strength of association, or correlation, between sample data from two random variables can be measured in several ways. Two of the most common measures are the Pearson coefficient of correlation and the coefficient of determination. The coefficient of determination can be interpreted as the percentage of variation in one variable that can be explained, through a straight-line relationship, by variation in the other variable.

Because measures of correlation do not take the nature of the data into account, a mathematically high degree of correlation is possible between two variables that have nothing to do with each other. Thus, a high degree of correlation between two variables does not necessarily imply that one variable affects, or causes, the other.

Two variables whose correlation is under consideration can be graphed as a scatter plot on the coordinate axes. A line can then be determined such that the sum of the squares of the vertical distances from all points in the scatter plot to that line is as small as possible. Called the line of best fit or regression line, it can be used to predict point estimates of one of the variables.

It is also possible to form confidence intervals instead of point estimates for the predicted value of a variable. Several formulas exist for this purpose, depending on the sample size and whether a confidence interval is to be found for the population mean value or for an individual value.

Although all of the necessary formulas are applied by hand in the examples, the material in this chapter lends itself well to the use of statistical software programs and graphics calculators.

Key Concepts, Terms, and Formulas

$$\hat{Y} \pm Z \cdot \sqrt{\frac{\Sigma y^2 - b\Sigma y - a\Sigma x \cdot y}{n - 2}} \cdot \sqrt{\frac{1}{n}} \quad \textbf{(14-4a)}$$

Page 555

Location of end points for a confidence interval for $R_{Y,X}$:

$$\hat{Y} \pm t \cdot \sqrt{\frac{\Sigma y^2 - b\Sigma y - a\Sigma x \cdot y}{n - 2}} \cdot \sqrt{\frac{1}{n} + \frac{(x - \overline{X})^2}{\Sigma x^2 - n \cdot \overline{X^2}} + 1} \quad \textbf{(14-5)}$$

Page 555

$$\hat{Y} \pm Z \cdot \sqrt{\frac{\Sigma y^2 - b\Sigma y - a\Sigma x \cdot y}{n - 2}} \quad \textbf{(14-5a)}$$

Page 556

Videotape Suggestions

Program 7 from the series *Against All Odds: Inside Statistics* reviews the geometry of straight lines and introduces the least-squares idea. *Program 8* describes scatter plots, and *Program 9* explores the properties of correlation. *Program 25* from the same series discusses linear prediction of mean response and individual response.

CUMULATIVE
REVIEW
CHAPTERS 11–14

▶ **O v e r v i e w**

Chapters 11 through 14 demonstrated the power and versatility of inferential statistics. Most of the work in these four chapters concerned hypothesis testing in a variety of applications, ranging from investigations of randomness to questions of significant correlation between two variables.

Chapter 11 addressed hypothesis testing with a model for the sampling distribution other than the normal or the *t*-distribution: the chi-square distribution. A variety of hypothesis tests examined data for independence of qualitative variables and goodness of fit. The chi-square distribution model was also used to find confidence intervals for the population standard deviation.

Chapters 12, 13, and 14 returned to the normal model for the sampling distribution and closely followed the five-step hypothesis testing procedure established earlier in the text. The nonparametric hypothesis tests in Chapter 12 provided a means to investigate such topics as randomness and rank correlation. Chapter 13 introduced the analysis-of-variance method, a technique for determining whether at least one of several population means is different from others. Finally, Chapter 14 returned to the topic of correlation (introduced through ranking in Chapter 12) and extended it to include confidence intervals.

P r o c e d u r e   I n d e x

EXERCISES ▶

1. Name three applications of the chi-square distribution.

2. Under what conditions should cells be combined in the computation of a chi-square test statistic?

3. The following cross-tabs table shows the numbers of women in two age groups who remarried after a divorce, cross-tabulated against the years between divorce and remarriage.

No. of Years from Divorce to Remarriage	Age at Time of Survey	
	40 to 49	*50 to 59*
3 or more	70	66
5 or more	73	68
10 or more	80	73
15 or more	86	77

Based on data from U.S. Bureau of the Census, *Current Population Reports*, P23-180, *Marriage, Divorce, and Remarriage in the 1990s,* U.S. Government Printing Office, Washington, D.C., 1992.

 a. What two qualitative variables appear in this table?

 b. Perform a chi-square hypothesis test at the 10% level of significance to determine whether this table presents statistical evidence that the two variables are not independent.

 c. In your own words, summarize the findings of Part b of this exercise.

4. Name three practical situations in which a confidence interval for a population's standard deviation would be important.

5. A computer chip manufacturer knows that the mean life of his chip is 4.65 years. However, to supply a guarantee to customers, he also needs some idea of the standard deviation. A random sample of $n = 10$ chips provided these lifetimes, in years.

4.67	4.33	3.57	5.10	4.39
4.00	3.99	5.87	6.00	4.45

 a. Form a 98% confidence interval for the population standard deviation.

 b. Use the greatest value in the confidence interval found in Part a to form an interval in which every time is within one standard deviation of the mean.

6. Refer to Exercise 5. Is the interval found in Part b a 95% confidence interval for the mean? Explain why or why not.

7. The following table summarizes the selling prices of a highly volatile stock.

Interval of Selling Price	Number of Days During Which Stock Sold in Price Interval
[0, 3.50)	2
[3.50, 7.00)	6
[7.00, 10.50)	14
[10.50, 14.00)	29
[14.00, 17.50)	11
[17.50, 21.00)	9
[21.00, 24.50)	1

Perform a goodness-of-fit test at the 10% level of significance to examine the claim that this sample comes from a normal population with parameters $\mu = 14.00$ and $\sigma = 4.00$.

8. Briefly summarize the types of situations in which each nonparametric test presented in Chapter 12 may be applied.

9. Following are the 1992 average SAT math scores of test takers in large cities and in rural areas, cross-tabulated by ethnic groups.

Ethnic Group	Large Cities	Rural Areas
American Indian	445	424
Asian American	518	481
Black	385	373
Mexican American	419	413
Puerto Rican	400	425
Other Hispanic	423	438
White	500	470
Other	460	445

Source: Press release from The College Board, August 27, 1992, p. 15.

a. Perform a sign test at the 8% level of significance to determine whether there is evidence (for these ethnic groups) of a difference in the SAT math averages in large cities and in rural areas.

b. Perform a sign test at the 5% level of significance to determine whether there is evidence (for these ethnic groups) that SAT math averages are greater in large cities than in rural areas.

10. Defective (D) and normal (N) parts are sampled off an assembly line. Test at the .02 level of significance to determine whether there is evidence that defective parts are not being produced at random.

N N D D N N N N N D N N N N N N N D D N D N N

11. Two people, chosen at random, ranked ten new gourmet coffees, giving a rank of 1 to the coffee liked best and a rank of 10 to the coffee liked least. The results follow.

Coffee	Rank from Person 1	Rank from Person 2
A	3	1
B	1	3
C	6	6
D	2	2
E	8	8
F	10	10
G	7	9
H	9	4
I	4	5
J	5	7

Test to see if there is a rank correlation between these two people at the 5% level of significance.

12. Refer to Exercise 11. Perform the same rank correlation test under the condition that person 1 cannot decide which to rank number 1 among coffees A, B, and D, and person 2 cannot decide which to rank number 1 between coffees A and D. (All other rankings remain the same.)

13. A college is searching for a new mathematics department chair. Four committees, made up of students, administrators, alumni, and present members of the math department, interview each candidate and rate him or her on a scale of 1 (unqualified) to 10 (highly qualified). The following cross-tabs table shows the mean scores for the first five candidates.

Candidate	Students	Administrators	Alumni	Math
A	5.1	6.5	4.4	5.4
B	8.6	6.4	8.4	7.0
C	3.8	9.7	3.2	1.9
D	5.2	5.1	5.6	5.4
E	9.1	5.3	7.1	7.2

The search coordinator wants to determine whether any of the committees tend to rate significantly different from the others. Someone suggests that a one-way ANOVA test be used for this purpose.

a. What assumptions are necessary for the ANOVA test?

b. Perform the ANOVA test at the 1% level of significance.

c. Interpret the meaning of the results from Part b.

14. The following table lists the numbers of federal and nonfederal hazardous waste sites for a sample of $n = 12$ states, chosen at random.

	No. of Federal Sites, X	No. of Nonfederal Sites, Y
TX	26	3
CO	13	3
SD	2	1
IL	34	4
MO	21	3
AL	10	2
MA	22	3
OK	11	1
FL	47	4
LA	10	1
WY	2	1
AK	2	4

Source: *The World Almanac and Book of Facts, 1991,* p. 250.

a. Draw a scatter plot for these data, and sketch the line of best fit.

b. Compute the Pearson coefficient of correlation for these data.

15. Refer to Exercise 14.

   a. Explain the difference between the calculated value of p in Part b and the actual value of **P**.

   b. Conduct a hypothesis test at the 10% level of significance to determine whether there is evidence of a significant correlation between the number of federal waste sites and the number of nonfederal waste sites.

16. Following, for a random sample of $n = 12$ people, are the amounts of time and money each person spent in a fashionable department store.

Time in Store (X minutes)	Money Spent (Y dollars)
30	27.75
10	12.75
65	52.10
5	0.00
35	30.59
115	110.95
45	45.25
60	49.95
25	40.10
58	50.00
14	13.50
8	12.00

Calculate the equation of the line of best fit for this data.

17. Refer to Exercise 16. Form a 95% confidence interval for the mean amount of money spent by a person who shopped 40 minutes at that store.

18. Refer to Exercise 16. Form a 95% confidence interval for the amount of money spent by a person, chosen at random, who shopped for 1 hour at that store.

A-1 Elementary Set Theory

A-1.1 INTRODUCTION AND BASIC DEFINITIONS

Set is one of the basic undefined terms in mathematics. It may be thought of as a collection of objects, numbers, names, or symbols grouped in such a way that it is clear which objects belong to the collection and which do not.

There are two ways of specifying, or naming, the **elements,** or **members,** of most sets. In the first, called the **roster method,** the elements in the set are listed between brackets—for example,

{Clinton, Bush, Reagan, Carter, Ford}.

The roster method can be employed for sets containing very many or even an infinite number of members if a series of three dots is used. For example, {2, 4, 6, . . . , 576} indicates the set containing the positive even integers up to and including 576, and {2, 4, 6, . . .} indicates the set containing all the positive even integers.

The second method, the **rule method,** states the rule by which elements are either included in or excluded from the set. {Presidents of the United States after Richard Nixon} is how the rule method would define the presidential set given a little earlier.

The order in which elements are listed in a set is not important. Two sets with exactly the same elements in them, in any order, are called **equal** sets.

The number of members in a set is the set's **cardinality.** For instance, the set {Clinton, Bush, Reagan, Carter, Ford}, has a cardinality of 5. When the cardinality of a set is large, the rule method is often the easier way of specifying it; when the cardinality is small, then the roster method is often easier.

For ease of reference, a set is identified by a capital letter. One could write

$$A = \{\text{Clinton, Bush, Reagan, Carter, Ford}\}$$

and refer to the set as set A or simply A.

Sets are often represented pictorially, as well. Figure A-1-1 is a **Venn diagram**—a picture that illustrates a set or, more commonly, a picture that shows the relationships among two or more sets. The U in the upper right corner of the rectangle stands for **universal set,** an arbitrarily chosen large set from which all

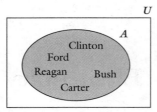

Figure A-1-1 A Venn diagram.

other sets under consideration can be formed. In this case, *U* could be {All U.S. presidents}.

The following special sets of numbers are used often.

$$K = \{\text{Complex numbers}\}$$
$$R = \{\text{All real numbers}\}$$
$$Q = \{\text{Rational numbers}\}$$
$$J = \{\text{Integers}\} = \{\ldots, -2, -1, 0, 1, 2, \ldots\}$$
$$W = \{\text{Whole numbers}\} = \{0, 1, 2, 3, \ldots\}$$
$$C = \{\text{Counting numbers}\} = \{1, 2, 3, 4, \ldots\}$$

Figure A-1-2 shows the Venn diagram for these sets, along with some of their members.

Recall that $i = \sqrt{-1}$, and real numbers do not involve even roots of negative numbers. Rational numbers are numbers that can be written as fractions—that is, either terminating or repeating decimals.

The use of a universal set allows for easy illustration of a set's complement. The **complement** of a given set is the set of all members of the universal set that are <u>not</u> in the given set. For example, the complement of the presidential set *A* is {All presidents of the United States except Clinton, Bush, Reagan, Carter, Ford} and is denoted by A^c. In a Venn diagram such as Figure A-1-1, the names of all presidents in the complement would be placed inside the rectangle but outside the oval.

The symbols $\in$ and $\notin$ provide a quick way of noting whether or not an element belongs to a set. The statement "Carter is a member of set *A*" is symbolized: Carter $\in A$. The statement "Lincoln is not in set *A*" is symbolized: Lincoln $\notin A$. Note that for any set *S* and any element *x*, if $x \notin S$, then it must follow that $x \in S^c$.

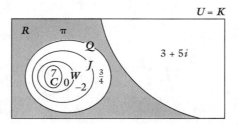

Figure A-1-2

A-1.2 *Set Operations*   **A.3**

One special set has no members and is called the **empty set** or **null set,** symbolized by either { } or ∅. This set occurs very frequently; there are many examples of it. If the set B = {Women presidents of the United States}, then B = ∅. It is important to realize that ∅ is not the same thing as {0}. The cardinality of {0} is one; the cardinality of ∅ is zero.

A-1.2 SET OPERATIONS

Just as there are operations for use with numbers, such as addition and subtraction, there are operations for use with sets, namely union and intersection. The **union** of two sets A and B, symbolized by $A \cup B$, is the set of all members that are either in A, or in B, or in both. The **intersection** of two sets A and B, symbolized by $A \cap B$, is the set of all members that are in both sets.

**E x a m p l e
1**

Consider the following presidential sets. To simplify matters, the universal set is now a list of 13 presidents.

U = {Lincoln, Hoover, FDR, Truman, Ike, JFK, LBJ, Nixon,
     Ford, Carter, Reagan, Bush, Clinton}
A = {Clinton, Bush, Reagan, Carter, Ford}
B = {Bush, Reagan}
C = {Bush, Nixon, Ike}
D = {Lincoln, FDR, JFK, LBJ}

Draw a Venn diagram, and determine all possible intersections and unions.

Solution Figure A-1-3 helps to find the following intersections and unions.

$A \cap B$ = {Bush, Reagan}
$A \cap C$ = {Bush}
$A \cap D$ = ∅
$B \cap C$ = {Bush}
$B \cap D$ = ∅
$C \cap D$ = ∅
$A \cup B$ = {Clinton, Bush, Reagan, Carter, Ford}
$A \cup C$ = {Clinton, Bush, Reagan, Carter, Ford, Nixon, Ike}
$A \cup D$ = {Clinton, Bush, Reagan, Carter, Ford, Lincoln, FDR, JFK, LBJ}
$B \cup C$ = {Bush, Reagan, Nixon, Ike}
$B \cup D$ = {Bush, Reagan, Lincoln, FDR, JFK, LBJ}
$C \cup D$ = {Bush, Nixon, Ike, Lincoln, FDR, JFK, LBJ}

Notice that each element is listed only once in the union and that the order in which members are listed in a set is not important.

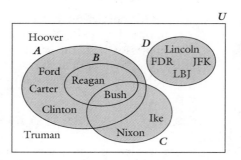

Figure A-1-3 ●

A-1.3 SUBSETS

The operations of intersection and union naturally give rise to the important concept of subset. A set X is a **subset** of set Y $(X \subset Y)$ if every member of X is also in Y. If not every element of X is in Y, then X is not a subset of Y $(X \not\subset Y)$.

By definition, any set is a subset of itself, and is often referred to as an **improper subset** of itself. The null set $(\varnothing)$ is also a subset of every set, including itself.

The intersection of any two sets is a subset of either set. In the example of the presidents,

$$(B \cap C) \subset B \quad \text{and also} \quad (B \cap C) \subset C.$$

If the intersection of two sets is the empty set, then their intersection, $\varnothing$, is still a subset of either set. By similar reasoning, any set is a subset of the union of itself and any other set. Note that

$$B \subset (B \cup C) \quad \text{and} \quad B \subset (B \cup D).$$

The notions of intersection and union may be repeated any number of times. For example,

$$(((A \cap B) \cap C) \cup D) = \{\text{Bush, Lincoln, FDR, JFK, LBJ}\}.$$

Such repetitions can be understood by working from inside out, just as with numbers. First compute $A \cap B$, then intersect that set with C. Finally, take the answer so far, $((A \cap B) \cap C)$, and determine the union of that set with D.

DeMorgan's Laws, to be stated shortly, are often the easiest way to approach repetitions of the operations of intersection and union with complements of sets.

Again consider the presidential sets in Example 1 and shown in Figure A-1-3. For ease of reference, the sets are repeated here.

$$U = \{\text{Lincoln, Hoover, FDR, Truman, Ike, JFK, LBJ,}$$
$$\text{Nixon, Carter, Reagan, Bush, Clinton}\}$$
$$A = \{\text{Clinton, Bush, Reagan, Carter, Ford}\}$$
$$B = \{\text{Bush, Reagan}\}$$
$$C = \{\text{Bush, Nixon, Ike}\}$$
$$D = \{\text{Lincoln, FDR, JFK, LBJ}\}$$

To determine $(A^c \cap B^c \cap C^c)$, one would need to find the set of all presidents that are *not* in A, intersect that set with the set of all presidents that are *not* in B, and finally intersect that resulting set with the set of all presidents *not* in C. Although this process is clearly possible, it would be tedious. DeMorgan's Laws could be applied in this type of situation.

A-1.4 DeMORGAN'S LAWS

DeMorgan's Laws say that (a) the intersection of the complements of any number of sets is equal to the complement of their unions and (b) the union of the complements of any number of sets is equal to the complement of their intersections. That is,

$$\text{(a)} \qquad \cap X_i^c = (\cup X_i)^c.$$
$$\text{(b)} \qquad \cup X_i^c = (\cap X_i)^c.$$

Computation of the presidential sets $(A^c \cap B^c \cap C^c)$ could occur using Part (a) of DeMorgan's Law:

$$(A^c \cap B^c \cap C^c) = (A \cup B \cup C)^c$$
$$= \{\text{Clinton, Bush, Reagan, Carter, Ford, Nixon, Ike}\}^c$$
$$= \{\text{Hoover, Truman, Lincoln, FDR, JFK, LBJ}\}.$$

E x e r c i s e s   A - 1 . 1 – A - 1 . 4

Make up your own sets to determine whether the statements in Exercises 1–14 are true or false. Remember that, in order to be marked as true, a statement must always be true for any set. In all problems, $U =$ universal set.

1. $(X \cap Y) \subset (X \cup Y)$.
2. $X \subset (X \cup B)$, and $X \subset (X \cap B)$.
3. $X \cap \varnothing = X$.
4. $G^c \cap G = \varnothing$.
5. $U^c \cap G = \varnothing$.
6. $(R \cup T)^c \subset (R^c \cap T^c)$.
7. $\varnothing^c \subset U$, and $U \subset \varnothing^c$.
8. $\varnothing \subset H \cap K$.
9. $0 \in \varnothing$.
10. $A \in (A \cup X)$.
11. $(X \cap Y) \cup (X \cup Y) = (X \cup Y)$.
12. $(X \cup Y) \cap (X \cap Y) = (X \cap Y)$.
13. $((X \cap Y) \cup Z) = (X \cap Z) \cup (Y \cap Z)$.

14. For the special number sets described on page A.2, $C \subset W \subset J \subset Q \subset R \subset K$.

15. Think of a set that has one member. List all of its subsets, including the set itself and Ø. How many subsets are there? Repeat this exercise for a set that has two members, then a set of cardinality 3, then one of cardinality 4. Keep increasing the cardinality until you think you have a formula for the number of subsets that can be formed from a set of cardinality n.

16. Here, once again, is the set of presidents used earlier.

$$U = \{\text{Lincoln, Hoover, FDR, Truman, Ike, JFK, LBJ,}$$
$$\text{Nixon, Ford, Carter, Reagan, Bush, Clinton}\}$$
$$A = \{\text{Clinton, Bush, Reagan, Carter, Ford}\}$$
$$B = \{\text{Bush, Reagan}\}$$
$$C = \{\text{Bush, Nixon, Ike}\}$$
$$D = \{\text{Lincoln, FDR, JFK, LBJ}\}$$

Illustrate DeMorgan's Laws using only sets A, B, and D. That is, show that

$$A^c \cup B^c \cup D^c = (A \cap B \cap D)^c$$

and

$$A^c \cap B^c \cap D^c = (A \cup B \cup D)^c$$

by working out each side separately, then comparing your answers.

For Exercises 17–21, specify the set operations necessary to obtain the shaded regions.

17.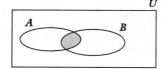

18.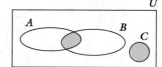

19.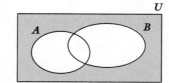

20.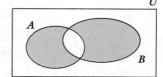

21.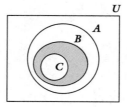

Use the following to determine the sets indicated in Exercises 22–30.

$$U = \{\text{Integers}\}$$
$$A = \{1, 2, 3, 4, \ldots, 600\}$$
$$B = \{2, 4, 6, 8, 10\}$$
$$C = \{8, 10, 12, 20\}$$
$$D = \{5, 10, 15, 20\}$$
$$E = \{5, 10, 15, 20, \ldots\}$$
$$F = \{0, -10, -20, -30, -40, \ldots\}$$

22. $(B \cup C)$.

23. $(D \cap E) \cap F$, and $D \cap (E \cap F)$.

24. $(C \cup D) \cap (A \cap B)$.

25. $(C \cap D)^c$, and $C^c \cap D^c$.

26. $B^c \cap D$.

27. $(F \cap \varnothing) \cup F$.

28. $(((D \cap E) \cap C) \cap B) \cap A$.

29. $C^c \cup D^c \cup E^c$.

30. $A^c \cap B^c \cap C^c \cap D^c$.

A-2 Standard Normal Curve Areas

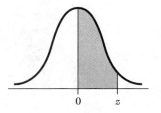

z	.00	.01	.02	.03	.04	.05	.06	.07	.08	.09
.0	.0000	.0040	.0080	.0120	.0160	.0199	.0239	.0279	.0319	.0359
.1	.0398	.0438	.0478	.0517	.0557	.0596	.0636	.0675	.0714	.0753
.2	.0793	.0832	.0871	.0910	.0948	.0987	.1026	.1064	.1103	.1141
.3	.1179	.1217	.1255	.1293	.1331	.1368	.1406	.1443	.1480	.1517
.4	.1554	.1591	.1628	.1664	.1700	.1736	.1772	.1808	.1844	.1879
.5	.1915	.1950	.1985	.2019	.2054	.2088	.2123	.2157	.2190	.2224
.6	.2257	.2291	.2324	.2357	.2389	.2422	.2454	.2486	.2518	.2549
.7	.2580	.2612	.2642	.2673	.2704	.2734	.2764	.2794	.2823	.2852
.8	.2881	.2910	.2939	.2967	.2995	.3023	.3051	.3078	.3106	.3133
.9	.3159	.3186	.3212	.3238	.3264	.3289	.3315	.3340	.3365	.3389
1.0	.3413	.3438	.3461	.3485	.3508	.3531	.3554	.3577	.3599	.3621
1.1	.3643	.3665	.3686	.3708	.3729	.3749	.3770	.3790	.3810	.3830
1.2	.3849	.3869	.3888	.3907	.3925	.3944	.3962	.3980	.3997	.4015
1.3	.4032	.4049	.4066	.4082	.4099	.4115	.4131	.4147	.4162	.4177
1.4	.4192	.4207	.4222	.4236	.4251	.4265	.4279	.4292	.4306	.4319
1.5	.4332	.4345	.4357	.4370	.4382	.4394	.4406	.4418	.4429	.4441
1.6	.4452	.4463	.4474	.4484	.4495	.4505	.4515	.4525	.4535	.4545
1.7	.4554	.4564	.4573	.4582	.4591	.4599	.4608	.4616	.4625	.4633
1.8	.4641	.4649	.4656	.4664	.4671	.4678	.4686	.4693	.4699	.4706
1.9	.4713	.4719	.4726	.4732	.4738	.4744	.4750	.4756	.4761	.4767
2.0	.4772	.4778	.4783	.4788	.4793	.4798	.4803	.4808	.4812	.4817
2.1	.4821	.4826	.4830	.4834	.4838	.4842	.4846	.4850	.4854	.4857
2.2	.4861	.4864	.4868	.4871	.4875	.4878	.4881	.4884	.4887	.4890
2.3	.4893	.4896	.4898	.4901	.4904	.4906	.4909	.4911	.4913	.4916
2.4	.4918	.4920	.4922	.4925	.4927	.4929	.4931	.4932	.4734	.4936
2.5	.4938	.4940	.4941	.4943	.4945	.4946	.4948	.4949	.4951	.4952
2.6	.4953	.4955	.4956	.4957	.4959	.4960	.4961	.4962	.4963	.4964
2.7	.4965	.4966	.4967	.4968	.4969	.4970	.4971	.4972	.4973	.4974
2.8	.4974	.4975	.4976	.4977	.4977	.4978	.4979	.4979	.4980	.4981
2.9	.4981	.4982	.4982	.4983	.4984	.4984	.4985	.4985	.4986	.4986
3.0	.4986	.4987	.4987	.4988	.4988	.4989	.4989	.4989	.4990	.4990
3.1	.4990	.4991	.4991	.4991	.4992	.4992	.4992	.4992	.4993	.4993
3.2	.4993	.4993	.4994	.4994	.4994	.4994	.4994	.4995	.4995	.4995
.3.3	.4995	.4995	.4995	.4996	.4996	.4996	.4996	.4996	.4996	.4997
3.4	.4997	.4997	.4997	.4997	.4997	.4997	.4997	.4997	.4997	.4998
3.6	.4998	.4998	.4999	.4999	.4999	.4999	.4999	.4999	.4999	.4999
3.9	.5000									

Source: *Standard Mathematical Tables,* 25th edition, p. 524, CRC Press, Inc.

A-3 Random Number Table

Random Numbers

71274	84346	75444	85690	35384	87841	97411	78698	46796	33552
64017	01373	14665	31891	80997	14321	47741	59980	87739	38174
43747	17686	11045	15549	52779	65135	00275	95434	36337	24041
59688	48689	41591	47042	83615	93034	25077	64835	67798	30547
95016	73467	11447	59500	94921	15166	69217	26267	11316	22651
65207	30591	65947	58339	00952	32111	45459	14986	57395	34492
34510	78657	08883	49489	85619	52912	01662	49854	78354	30631
56299	60624	91572	31734	18159	18927	31314	59682	41320	88602
02113	12579	86172	03819	69968	02616	72687	42699	04792	16510
00884	87979	45184	61572	20086	14498	29640	94263	90964	29278
89367	53577	97412	19603	57234	63055	49059	35761	72007	22751
99781	56740	42659	46617	21828	99831	45987	63450	66919	78252
92024	12100	76013	12587	86340	74880	79979	35906	38122	64917
82861	09215	87342	72789	76132	24468	93065	78968	03321	48081
11286	13011	67982	74101	44961	25468	14247	95934	50711	24492
68674	24686	14460	61242	92310	86810	87702	69811	53996	99517
24882	20749	94139	28785	74402	18561	79069	56838	30020	99707
21740	51134	39298	92203	66230	30636	58169	78982	50057	39908
46901	34825	28673	10404	97777	30782	04680	15319	84125	40937
92686	81702	74149	76326	01101	96278	90855	55145	89705	51199
98743	59366	94797	96803	69876	87533	19675	59246	65348	06606
11463	25619	38107	91053	58416	02720	86563	27443	99598	04074
76975	18636	54975	67422	57101	68857	35389	35641	34505	71552
67359	50379	81053	97357	00717	59504	34480	77127	23243	48682
34084	48031	27227	48912	10797	88917	93126	52945	79457	66528
94553	44441	83166	40056	73935	52103	13972	56781	31900	95037
20545	43211	34500	92233	53497	39401	78535	82360	57410	32060
05646	01152	13235	39168	69214	83852	00144	08105	94247	37189
54022	58295	96122	87620	74774	06884	81689	68392	25776	08748
32291	30700	32561	83579	94582	77930	06826	96855	97751	42664

Random Numbers (*continued*)

66212	75061	65891	96896	15107	12985	97616	64241	08592	81036
98059	44951	23078	92793	80756	52799	20340	62969	81775	31065
23157	52179	24394	39833	89427	58771	26992	00649	25475	50888
28911	81929	91368	49372	43335	44465	43257	66893	34761	60423
85959	90369	02100	28727	83001	84166	20473	35305	38088	54795
34459	31400	58760	17157	73816	55527	54133	24605	56153	35354
04073	57781	36894	93000	57834	29343	98195	58425	97275	71392
22126	91330	95667	75737	36869	55209	41663	04943	59401	17039
10288	61685	25302	84097	13088	86840	04020	43046	01043	43157
75431	32853	72907	99432	65482	23011	70466	87386	67471	77629
90800	17425	28042	53770	98924	31863	84115	82488	23239	82185
19083	89475	05207	41284	83405	55825	31117	59821	96455	63796
10405	67911	77238	46262	42766	07215	02391	47316	78724	41170
34711	77325	99768	63455	44335	91028	27740	86163	81474	08159
73334	61941	16883	05012	63191	35763	60157	09617	25501	44989
79452	68381	71937	23274	60273	47091	82876	24641	03825	50894
13864	28746	32434	88325	99996	96130	39471	74020	56077	22133
73082	50271	83240	80065	09328	02940	41686	32758	89467	73553
43060	88221	35010	79829	71520	80453	95049	66352	77495	83256
15172	42061	33264	63832	48528	23258	13520	83222	45659	39074

Source: RAND, *A Million Random Digits with 100,000 Normal Deviates*, p. 377, The Free Press, 1955.

A-4 Student

t-Distribution

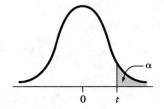

Degrees of Freedom	α									
	.4	.25	.1	.05	.025	.01	.005	.0025	.001	.0005
1	0.325	1.000	3.078	6.314	12.706	31.821	63.657	127.32	318.31	636.62
2	.289	.816	1.886	2.920	4.303	6.965	9.925	14.089	22.327	31.598
3	.277	.765	1.638	2.353	3.182	4.541	5.841	7.453	10.214	12.924
4	.271	.741	1.533	2.132	2.776	3.747	4.604	5.598	7.173	8.610
5	0.267	0.727	1.476	2.015	2.571	3.365	4.032	4.773	5.893	6.869
6	.265	.718	1.440	1.943	2.447	3.143	3.707	4.317	5.208	5.959
7	.263	.711	1.415	1.895	2.365	2.998	3.499	4.029	4.785	5.408
8	.262	.706	1.397	1.860	2.306	2.896	3.355	3.833	4.501	5.041
9	.261	.703	1.383	1.833	2.262	2.821	3.250	3.690	4.297	4.781
10	0.260	0.700	1.372	1.812	2.228	2.764	3.169	3.581	4.144	4.587
11	.260	.697	1.363	1.796	2.201	2.718	3.106	3.497	4.025	4.437
12	.259	.695	1.356	1.782	2.179	2.681	3.055	3.428	3.930	4.318
13	.259	.694	1.350	1.771	2.160	2.650	3.012	3.372	3.852	4.221
14	.258	.692	1.345	1.761	2.145	2.624	2.977	3.326	3.787	4.140
15	0.258	0.691	1.341	1.753	2.131	2.602	2.947	3.286	3.733	4.073
16	.258	.690	1.337	1.746	2.120	2.583	2.921	3.252	3.686	4.015
17	.257	.689	1.333	1.740	2.110	2.567	2.898	3.222	3.646	3.965
18	.257	.688	1.330	1.734	2.101	2.552	2.878	3.197	3.610	3.922
19	.257	.688	1.328	1.729	2.093	2.539	2.861	3.174	3.579	3.883
20	0.257	0.687	1.325	1.725	2.086	2.528	2.845	3.153	3.552	3.850
21	.257	.686	1.323	1.721	2.080	2.518	2.831	3.135	3.527	3.819
22	.256	.686	1.321	1.717	2.074	2.508	2.819	3.119	3.505	3.792
23	.256	.685	1.319	1.714	2.069	2.500	2.807	3.104	3.485	3.767
24	.256	.685	1.318	1.711	2.064	2.492	2.797	3.091	3.467	3.745
25	0.256	0.684	1.316	1.708	2.060	2.485	2.787	3.078	3.450	3.725
26	.256	.684	1.315	1.706	2.056	2.479	2.779	3.067	3.435	3.707
27	.256	.684	1.314	1.703	2.052	2.473	2.771	3.057	3.421	3.690
28	.256	.683	1.313	1.701	2.048	2.467	2.763	3.047	3.408	3.674
29	.256	.683	1.311	1.699	2.045	2.462	2.756	3.038	3.396	3.659
30	0.256	0.683	1.310	1.697	2.042	2.457	2.750	3.030	3.385	3.646
40	.255	.681	1.303	1.684	2.021	2.423	2.704	2.971	3.307	3.551
60	.254	.679	1.296	1.671	2.000	2.390	2.660	2.915	3.232	3.460
120	.254	.677	1.289	1.658	1.980	2.358	2.617	2.860	3.160	3.373
∞	.253	.674	1.282	1.645	1.960	2.326	2.576	2.807	3.090	3.291

Source: E. S. Pearson and H. O. Hartley, *Biometrika Tables for Statisticians*, Vol. I. (Cambridge: Cambridge University Press, 1966), p. 146.

A-5 The Chi-Square Distribution

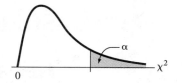

Critical Values of χ^2

Degrees of Freedom	α													
	.99	.98	.95	.90	.80	.70	.50	.30	.20	.10	.05	.02	.01	.001
1	.0³157	.0³628	.00393	.0158	.0642	.148	.455	1.074	1.642	2.706	3.841	5.412	6.635	10.827
2	.0201	.0404	.103	.211	.446	.713	1.386	2.408	3.219	4.605	5.991	7.824	9.210	13.815
3	.115	.185	.352	.584	1.005	1.424	2.366	3.665	4.642	6.251	7.815	9.837	11.345	16.266
4	.297	.429	.711	1.064	1.649	2.195	3.357	4.878	5.989	7.779	9.488	11.668	13.277	18.467
5	.554	.752	1.145	1.610	2.343	3.000	4.351	6.064	7.289	9.236	11.070	13.388	15.086	20.515
6	.872	1.134	1.635	2.204	3.070	3.828	5.348	7.231	8.558	10.645	12.592	15.033	16.812	22.457
7	1.239	1.564	2.167	2.833	3.822	4.671	6.346	8.383	9.803	12.017	14.067	16.622	18.475	24.322
8	1.646	2.032	2.733	3.490	4.594	5.527	7.344	9.524	11.030	13.362	15.507	18.168	20.090	26.125
9	2.088	2.532	3.325	4.168	5.380	6.393	8.343	10.656	12.242	14.684	16.919	19.679	21.666	27.877
10	2.588	3.059	3.940	4.865	6.179	7.267	9.342	11.781	13.442	15.987	18.307	21.161	23.209	29.588
11	3.053	3.609	4.575	5.578	6.989	8.148	10.341	12.899	14.631	17.275	19.675	22.618	24.725	31.264
12	3.571	4.178	5.226	6.304	7.807	9.034	11.340	14.011	15.812	18.549	21.026	24.054	26.217	32.909
13	4.107	4.765	5.892	7.042	8.634	9.926	12.340	15.119	16.985	19.812	22.362	25.472	27.688	34.528
14	4.660	5.368	6.571	7.790	9.467	10.821	13.339	16.222	18.151	21.064	23.685	26.873	29.141	36.123
15	5.229	5.985	7.261	8.547	10.307	11.721	14.339	17.322	19.311	22.307	24.996	28.259	30.578	37.697
16	5.812	6.614	7.962	9.312	11.152	12.624	15.338	18.418	20.465	23.542	26.296	29.633	32.000	39.252
17	6.408	7.255	8.672	10.085	12.002	13.531	16.338	19.511	21.615	24.769	27.587	30.995	33.409	40.790
18	7.015	7.906	9.390	10.865	12.857	14.440	17.338	20.601	22.760	25.989	28.869	32.346	34.805	42.312
19	7.633	8.567	10.117	11.651	13.716	15.352	18.338	21.689	23.900	27.204	30.144	33.687	36.191	43.820
20	8.260	9.237	10.851	12.443	14.578	16.266	19.337	22.775	25.038	28.412	31.410	35.020	37.566	45.315
21	8.897	9.915	11.591	13.240	15.445	17.182	20.337	23.858	26.171	29.615	32.671	36.343	38.932	46.797
22	9.542	10.600	12.338	14.041	16.314	18.101	21.337	24.939	27.301	30.813	33.924	37.659	40.289	48.268
23	10.196	11.293	13.091	14.848	17.187	19.021	22.337	26.018	28.429	32.007	35.172	38.968	41.638	49.728
24	10.856	11.992	13.848	15.659	18.062	19.943	23.337	27.096	29.553	33.196	36.415	40.270	42.980	51.179
25	11.524	12.697	14.611	16.473	18.940	20.867	24.337	28.172	30.675	34.382	37.652	41.566	44.314	52.620
26	12.198	13.409	15.379	17.292	19.820	21.792	25.336	29.246	31.795	35.563	38.885	42.856	45.642	54.052
27	12.879	14.125	16.151	18.114	20.703	22.719	26.336	30.319	32.912	36.741	40.113	44.140	46.963	55.476
28	13.565	14.847	16.928	18.939	21.588	23.647	27.336	31.391	34.027	37.916	41.337	45.419	48.278	56.893
29	14.256	15.574	17.708	19.768	22.475	24.577	28.336	32.461	35.139	39.087	42.557	46.693	49.588	58.302
30	14.953	16.306	18.493	20.599	23.364	25.508	29.336	33.530	36.250	40.256	43.773	47.962	50.892	59.703

Source: Chi-square distribution table of Fisher and Yates; *Statistical Tables for Biological Agricultural and Medical Research*, Published by Longman Group UK Ltd., 1974.

A-6 Distribution of the *F*-Statistic

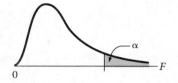

Denominator Degrees of Freedom	Values of F for α = .05																			
	Numerator Degrees of Freedom																			
	1	2	3	4	5	6	7	8	9	10	12	15	20	24	30	40	60	120	∞
1	161.4	199.5	215.7	224.6	230.2	234.0	236.8	238.9	240.5	241.9	243.9	245.9	248.0	249.1	250.1	251.1	252.2	253.3	254.3
2	18.51	19.00	19.16	19.25	19.30	19.33	19.35	19.37	19.38	19.40	19.41	19.43	19.45	19.45	19.46	19.47	19.48	19.49	19.50
3	10.13	9.55	9.28	9.12	9.01	8.94	8.89	8.85	8.81	8.79	8.74	8.70	8.66	8.64	8.62	8.59	8.57	8.55	8.53
4	7.71	6.94	6.59	6.39	6.26	6.16	6.09	6.04	6.00	5.96	5.91	5.86	5.80	5.77	5.75	5.72	5.69	5.66	5.63
5	6.61	5.79	5.41	5.19	5.05	4.95	4.88	4.82	4.77	4.74	4.68	4.62	4.56	4.53	4.50	4.46	4.43	4.40	4.36
6	5.99	5.14	4.76	4.53	4.39	4.28	4.21	4.15	4.10	4.06	4.00	3.94	3.87	3.84	3.81	3.77	3.74	3.70	3.67
7	5.59	4.74	4.35	4.12	3.97	3.87	3.79	3.73	3.68	3.64	3.57	3.51	3.44	3.41	3.38	3.34	3.30	3.27	3.23
8	5.32	4.46	4.07	3.84	3.69	3.58	3.50	3.44	3.39	3.35	3.28	3.22	3.15	3.12	3.08	3.04	3.01	2.97	2.93
9	5.12	4.26	3.86	3.63	3.48	3.37	3.29	3.23	3.18	3.14	3.07	3.01	2.94	2.90	2.86	2.83	2.79	2.75	2.71
10	4.96	4.10	3.71	3.48	3.33	3.22	3.14	3.07	3.02	2.98	2.91	2.85	2.77	2.74	2.70	2.66	2.62	2.58	2.54
11	4.84	3.98	3.59	3.36	3.20	3.09	3.01	2.95	2.90	2.85	2.79	2.72	2.65	2.61	2.57	2.53	2.49	2.45	2.40
12	4.75	3.89	3.49	3.26	3.11	3.00	2.91	2.85	2.80	2.75	2.69	2.62	2.54	2.51	2.47	2.43	2.38	2.34	2.30
13	4.67	3.81	3.41	3.18	3.03	2.92	2.83	2.77	2.71	2.67	2.60	2.53	2.46	2.42	2.38	2.34	2.30	2.25	2.21
14	4.60	3.74	3.34	3.11	2.96	2.85	2.76	2.70	2.65	2.60	2.53	2.46	2.39	2.35	2.31	2.27	2.22	2.18	2.13
15	4.54	3.68	3.29	3.06	2.90	2.79	2.71	2.64	2.59	2.54	2.48	2.40	2.33	2.29	2.25	2.20	2.16	2.11	2.07
16	4.49	3.63	3.24	3.01	2.85	2.74	2.66	2.59	2.54	2.49	2.42	2.35	2.28	2.24	2.19	2.15	2.11	2.06	2.01
17	4.45	3.59	3.20	2.96	2.81	2.70	2.61	2.55	2.49	2.45	2.38	2.31	2.23	2.19	2.15	2.10	2.06	2.01	1.96
18	4.41	3.55	3.16	2.93	2.77	2.66	2.58	2.51	2.46	2.41	2.34	2.27	2.19	2.15	2.11	2.06	2.02	1.97	1.92
19	4.38	3.52	3.13	2.90	2.74	2.63	2.54	2.48	2.42	2.38	2.31	2.23	2.16	2.11	2.07	2.03	1.98	1.93	1.88
20	4.35	3.49	3.10	2.87	2.71	2.60	2.51	2.45	2.39	2.35	2.28	2.20	2.12	2.08	2.04	1.99	1.95	1.90	1.84
21	4.32	3.47	3.07	2.84	2.68	2.57	2.49	2.42	2.37	2.32	2.25	2.18	2.10	2.05	2.01	1.96	1.92	1.87	1.81
22	4.30	3.44	3.05	2.82	2.66	2.55	2.46	2.40	2.34	2.30	2.23	2.15	2.07	2.03	1.98	1.94	1.89	1.84	1.78
23	4.28	3.42	3.03	2.80	2.64	2.53	2.44	2.37	2.32	2.27	2.20	2.13	2.05	2.01	1.96	1.91	1.86	1.81	1.76
24	4.26	3.40	3.01	2.78	2.62	2.51	2.42	2.36	2.30	2.25	2.18	2.11	2.03	1.98	1.94	1.89	1.84	1.79	1.73
25	4.24	3.39	2.99	2.76	2.60	2.49	2.40	2.34	2.28	2.24	2.16	2.09	2.01	1.96	1.92	1.87	1.82	1.77	1.71
26	4.23	3.37	2.98	2.74	2.59	2.47	2.39	2.32	2.27	2.22	2.15	2.07	1.99	1.95	1.90	1.85	1.80	1.75	1.69
27	4.21	3.35	2.96	2.73	2.57	2.46	2.37	2.31	2.25	2.20	2.13	2.06	1.97	1.93	1.88	1.84	1.79	1.73	1.67
28	4.20	3.34	2.95	2.71	2.56	2.45	2.36	2.29	2.24	2.19	2.12	2.04	1.96	1.91	1.87	1.82	1.77	1.71	1.65
29	4.18	3.33	2.93	2.70	2.55	2.43	2.35	2.28	2.22	2.18	2.10	2.03	1.94	1.90	1.85	1.81	1.75	1.70	1.64
30	4.17	3.32	2.92	2.69	2.53	2.42	2.33	2.27	2.21	2.16	2.09	2.01	1.93	1.89	1.84	1.79	1.74	1.68	1.62
40	4.08	3.23	2.84	2.61	2.45	2.34	2.25	2.18	2.12	2.08	2.00	1.92	1.84	1.79	1.74	1.69	1.64	1.58	1.51
60	4.00	3.15	2.76	2.53	2.37	2.25	2.17	2.10	2.04	1.99	1.92	1.84	1.75	1.70	1.65	1.59	1.53	1.47	1.39
120	3.92	3.07	2.68	2.45	2.29	2.17	2.09	2.02	1.96	1.91	1.83	1.75	1.66	1.61	1.55	1.50	1.43	1.35	1.25
∞	3.84	3.00	2.60	2.37	2.21	2.10	2.01	1.94	1.88	1.83	1.75	1.67	1.57	1.52	1.46	1.39	1.32	1.22	1.00

Source: E. S. Pearson and H. O. Hartley, *Biometrika Tables for Statisticians,* Vol. I (Cambridge: Cambridge University Press, 1966), pp. 171–73.

Denominator Degrees of Freedom	Values of F for α = .025 Numerator Degrees of Freedom																		
	1	2	3	4	5	6	7	8	9	10	12	15	20	24	30	40	60	120	∞
1	647.8	799.5	864.2	899.6	921.8	937.1	948.2	956.7	963.3	968.6	976.7	984.9	993.1	997.2	1001	1006	1010	1014	1018
2	38.51	39.00	39.17	39.25	39.30	39.33	39.36	39.37	39.39	39.40	39.41	39.43	39.45	39.46	39.46	39.47	39.48	39.49	39.50
3	17.44	16.04	15.44	15.10	14.88	14.73	14.62	14.54	14.47	14.42	14.34	14.25	14.17	14.12	14.08	14.04	13.99	13.95	13.90
4	12.22	10.65	9.98	9.60	9.26	9.20	9.07	8.98	8.90	8.84	8.75	8.66	8.56	8.51	8.46	8.41	8.36	8.31	8.26
5	10.01	8.43	7.76	7.39	7.15	6.98	6.85	6.76	6.68	6.62	6.52	6.43	6.33	6.28	6.23	6.18	6.12	6.07	6.02
6	8.81	7.26	6.60	6.23	5.99	5.82	5.70	5.60	5.52	5.46	5.37	5.27	5.17	5.12	5.07	5.01	4.96	4.90	4.85
7	8.07	6.54	5.89	5.52	5.29	5.12	4.99	4.90	4.82	4.76	4.67	4.57	4.47	4.42	4.36	4.31	4.25	4.20	4.14
8	7.57	6.06	5.42	5.05	4.82	4.65	4.53	4.43	4.36	4.30	4.20	4.10	4.00	3.95	3.89	3.84	3.78	3.73	3.67
9	7.21	5.71	5.08	4.72	4.48	4.22	4.20	4.10	4.03	3.96	3.87	3.77	3.67	3.61	3.56	3.51	3.45	3.39	3.33
10	6.94	5.46	4.83	4.47	4.24	4.07	3.95	3.85	3.78	3.72	3.62	3.52	3.42	3.37	3.31	3.26	3.20	3.14	3.08
11	6.72	5.26	4.63	4.28	4.04	3.88	3.76	3.66	3.59	3.53	3.43	3.33	3.23	3.17	3.12	3.06	3.00	2.94	2.88
12	6.55	5.10	4.47	4.12	3.89	3.73	3.61	3.51	3.44	3.37	3.28	3.18	3.07	3.02	2.96	2.91	2.85	2.79	2.72
13	6.41	4.97	4.35	4.00	3.77	3.60	3.48	3.39	3.31	3.25	3.15	3.05	2.95	2.89	2.84	2.78	2.72	2.66	2.60
14	6.30	4.86	4.24	3.89	3.66	3.50	3.38	3.29	3.21	3.15	3.05	2.95	2.84	2.79	2.73	2.67	2.61	2.55	2.49
15	6.20	4.77	4.15	3.80	3.58	3.41	3.29	3.20	3.12	3.06	2.96	2.86	2.76	2.70	2.64	2.59	2.52	2.46	2.40
16	6.12	4.69	4.08	3.73	3.50	3.34	3.22	3.12	3.05	2.99	2.89	2.79	2.68	2.63	2.57	2.51	2.45	2.38	2.32
17	6.04	4.62	4.01	3.66	3.44	3.28	3.16	3.06	2.98	2.92	2.82	2.72	2.62	2.56	2.50	2.44	2.38	2.32	2.25
18	5.98	4.56	3.95	3.61	3.38	3.22	3.10	3.01	2.93	2.87	2.77	2.67	2.56	2.50	2.44	2.38	2.32	2.26	2.19
19	5.92	4.51	3.90	3.56	3.33	3.17	3.05	2.96	2.88	2.82	2.72	2.62	2.51	2.45	2.39	2.33	2.27	2.20	2.13
20	5.87	4.46	3.86	3.51	3.29	3.13	3.01	2.91	2.84	2.77	2.68	2.57	2.46	2.41	2.35	2.29	2.22	2.16	2.09
21	5.83	4.42	3.82	3.48	3.25	3.09	2.97	2.87	2.80	2.73	2.64	2.53	2.42	2.37	2.31	2.25	2.18	2.11	2.04
22	5.79	4.38	3.78	3.44	3.22	3.05	2.93	2.84	2.76	2.70	2.60	2.50	2.39	2.33	2.27	2.21	2.14	2.08	2.00
23	5.75	4.35	3.75	3.41	3.18	3.02	2.90	2.81	2.73	2.67	2.57	2.47	2.36	2.30	2.24	2.18	2.11	2.04	1.97
24	5.72	4.32	3.72	3.38	3.15	2.99	2.87	2.78	2.70	2.64	2.54	2.44	2.33	2.27	2.21	2.15	2.08	2.01	1.94
25	5.69	4.29	3.69	3.35	3.13	2.97	2.85	2.75	2.68	2.61	2.51	2.41	2.30	2.24	2.18	2.12	2.05	1.98	1.91
26	5.66	4.27	3.67	3.33	3.10	2.94	2.82	2.73	2.65	2.59	2.49	2.39	2.28	2.22	2.16	2.09	2.03	1.95	1.88
27	5.63	4.24	3.65	3.31	3.08	2.92	2.80	2.71	2.63	2.57	2.47	2.36	2.25	2.19	2.13	2.07	2.00	1.93	1.85
28	5.61	4.22	3.63	3.29	3.06	2.90	2.78	2.69	2.61	2.55	2.45	2.34	2.23	2.17	2.11	2.05	1.98	1.91	1.83
29	5.59	4.20	3.61	3.27	3.04	2.88	2.76	2.67	2.59	2.53	2.43	2.32	2.21	2.15	2.09	2.03	1.96	1.89	1.81
30	5.57	4.18	3.59	3.25	3.03	2.87	2.75	2.65	2.57	2.51	2.41	2.31	2.20	2.14	2.07	2.01	1.94	1.87	1.79
40	5.42	4.05	3.46	3.13	2.90	2.74	2.62	2.53	2.45	2.39	2.29	2.18	2.07	2.01	1.94	1.88	1.80	1.72	1.64
60	5.29	3.93	3.34	3.01	2.79	2.63	2.51	2.41	2.33	2.27	2.17	2.06	1.94	1.88	1.82	1.74	1.67	1.58	1.48
120	5.15	3.80	3.23	2.89	2.67	2.52	2.39	2.30	2.22	2.16	2.05	1.94	1.82	1.76	1.69	1.61	1.53	1.43	1.31
∞	5.02	3.69	3.12	2.79	2.57	2.41	2.29	2.19	2.11	2.05	1.94	1.83	1.71	1.64	1.57	1.48	1.39	1.27	1.00

Values of F for $\alpha = .01$

Denominator Degrees of Freedom	Numerator Degrees of Freedom																		
	1	2	3	4	5	6	7	8	9	10	12	15	20	24	30	40	60	120	∞
1	4052	4999.5	5403	5625	5764	5859	5928	5981	6022	6056	6106	6157	6209	6235	6261	6287	6313	6339	6360
2	98.50	99.00	99.17	99.25	99.30	99.33	99.36	99.37	99.39	99.40	99.42	99.43	99.45	99.46	99.47	99.47	99.48	99.49	99.50
3	34.12	30.82	29.46	28.71	28.24	27.91	27.67	27.49	27.35	27.23	27.05	26.87	26.69	26.60	26.50	26.41	26.32	26.22	26.13
4	21.20	18.00	16.69	15.98	15.52	15.21	14.98	14.80	14.66	14.55	14.37	14.20	14.02	13.93	13.84	13.75	13.65	13.56	13.46
5	16.26	13.27	12.06	11.39	10.97	10.67	10.46	10.29	10.16	10.05	9.89	9.72	9.55	9.47	9.38	9.29	9.30	9.11	9.02
6	13.75	10.92	9.78	9.15	8.75	8.47	8.26	8.10	7.98	7.87	7.72	7.56	7.40	7.31	7.23	7.14	7.06	6.97	6.88
7	12.25	9.55	8.45	7.85	7.46	7.19	6.99	6.84	6.72	6.62	6.47	6.31	6.16	6.07	5.99	5.91	5.82	5.74	5.65
8	11.26	8.65	7.59	7.01	6.63	6.37	6.18	6.03	5.91	5.81	5.67	5.52	5.36	5.28	5.20	5.12	5.03	4.95	4.86
9	10.56	8.02	6.99	6.42	6.06	5.80	5.61	5.47	5.35	5.26	5.11	4.96	4.81	4.73	4.65	4.57	4.48	4.40	4.31
10	10.04	7.56	6.55	5.99	5.64	5.39	5.20	5.06	4.94	4.85	4.71	4.56	4.41	4.33	4.25	4.17	4.08	4.00	3.91
11	9.65	7.21	6.22	5.67	5.32	5.07	4.89	4.74	4.63	4.54	4.40	4.25	4.10	4.02	3.94	3.86	3.78	3.69	3.60
12	9.33	6.93	5.95	5.41	5.06	4.82	4.64	4.50	4.39	4.30	4.16	4.01	3.86	3.78	3.70	3.62	3.54	3.45	3.36
13	9.07	6.70	5.74	5.21	4.86	4.62	4.44	4.30	4.19	4.10	3.96	3.82	3.66	3.59	3.51	3.43	3.34	3.25	3.17
14	8.86	6.51	5.56	5.04	4.69	4.46	4.28	4.14	4.03	3.94	3.80	3.66	3.51	3.43	3.35	3.27	3.18	3.09	3.00
15	8.68	6.36	5.42	4.89	4.56	4.32	4.14	4.00	3.89	3.80	3.67	3.52	3.37	3.29	3.21	3.13	3.05	2.96	2.87
16	8.53	6.23	5.29	4.77	4.44	4.20	4.03	3.89	3.78	3.69	3.55	3.41	3.26	3.18	3.10	3.02	2.93	2.84	2.75
17	8.40	6.11	5.18	4.67	4.34	4.10	3.93	3.79	3.68	3.59	3.46	3.31	3.16	3.08	3.00	2.92	2.83	2.75	2.65
18	8.29	6.01	5.09	4.58	4.25	4.01	3.84	3.71	3.60	3.51	3.37	3.23	3.08	3.00	2.92	2.84	2.75	2.66	2.57
19	8.18	5.93	5.01	4.50	4.17	3.94	3.77	3.63	3.52	3.43	3.30	3.15	3.00	2.92	2.84	2.76	2.67	2.58	2.50
20	8.10	5.85	4.94	4.43	4.10	3.87	3.70	3.56	3.46	3.37	3.23	3.09	2.94	2.86	2.78	2.69	2.61	2.52	2.42
21	8.02	5.78	4.87	4.37	4.04	3.81	3.64	3.51	3.40	3.31	3.17	3.03	2.83	2.80	2.72	2.64	2.55	2.46	2.36
22	7.95	5.72	4.82	4.31	3.99	3.76	3.59	3.45	3.35	3.26	3.12	2.98	2.83	2.75	2.67	2.58	2.50	2.40	2.31
23	7.88	5.66	4.76	4.26	3.94	3.71	3.54	3.41	3.30	3.21	3.07	2.93	2.78	2.70	2.62	2.54	2.45	2.35	2.26
24	7.82	5.61	4.72	4.22	3.90	3.67	3.50	3.36	3.26	3.17	3.03	2.89	2.74	2.66	2.58	2.49	2.40	2.31	2.21
25	7.77	5.57	4.68	4.18	3.85	3.63	3.46	3.32	3.22	3.13	2.99	2.85	2.70	2.62	2.54	2.45	2.36	2.27	2.17
26	7.72	5.53	4.64	4.14	3.82	3.59	3.42	3.29	3.18	3.09	2.96	2.81	2.66	2.58	2.50	2.42	2.33	2.23	2.13
27	7.68	5.49	4.60	4.11	3.78	3.56	3.39	3.26	3.15	3.06	2.93	2.78	2.63	2.55	2.47	2.38	2.29	2.20	2.10
28	7.64	5.45	4.57	4.07	3.75	3.53	3.36	3.23	3.12	3.03	2.90	2.75	2.60	2.52	2.44	2.35	2.26	2.17	2.06
29	7.60	5.42	4.54	4.04	3.73	3.50	3.33	3.20	3.09	3.00	2.87	2.73	2.57	2.49	2.41	2.33	2.23	2.14	2.03
30	7.56	5.39	4.51	4.02	3.70	3.47	3.30	3.17	3.07	2.98	2.84	2.70	2.55	2.47	2.39	2.30	2.21	2.11	2.01
40	7.31	5.18	4.31	3.83	3.51	3.29	3.12	2.99	2.89	2.80	2.66	2.52	2.37	2.29	2.20	2.11	2.02	1.92	1.80
60	7.08	4.98	4.13	3.65	3.34	3.12	2.95	2.82	2.72	2.63	2.50	2.35	2.20	2.12	2.03	1.94	1.84	1.73	1.60
120	6.85	4.79	3.95	3.48	3.17	2.96	2.79	2.66	2.56	2.47	2.34	2.19	2.03	1.95	1.86	1.76	1.66	1.53	1.38
∞	6.63	4.61	3.78	3.32	3.02	2.80	2.64	2.51	2.41	2.32	2.18	2.04	1.88	1.79	1.70	1.59	1.47	1.32	1.00

A-7 Critical Values of p When P = 0

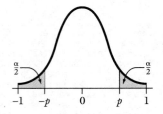

d.f. \ α	0.10	0.05	0.02	0.01
1	0.988	0.997	1.000	1.000
2	0.900	0.950	0.980	0.990
3	0.805	0.878	0.934	0.959
4	0.729	0.811	0.882	0.917
5	0.669	0.754	0.833	0.874
6	0.662	0.707	0.789	0.834
7	0.582	0.666	0.750	0.798
8	0.549	0.632	0.716	0.765
9	0.521	0.602	0.685	0.735
10	0.497	0.576	0.658	0.708
11	0.476	0.553	0.634	0.684
12	0.458	0.532	0.612	0.661
13	0.441	0.514	0.592	0.641
14	0.426	0.497	0.574	0.623
15	0.412	0.482	0.558	0.606
16	0.400	0.468	0.542	0.590
17	0.389	0.456	0.528	0.575
18	0.378	0.444	0.516	0.561
19	0.369	0.433	0.503	0.549
20	0.360	0.423	0.492	0.537
25	0.323	0.381	0.445	0.487
30	0.296	0.349	0.409	0.449
35	0.275	0.325	0.381	0.418
40	0.257	0.304	0.358	0.393
45	0.243	0.288	0.338	0.372
50	0.231	0.273	0.322	0.354
60	0.211	0.250	0.295	0.325
70	0.195	0.232	0.274	0.302
80	0.183	0.217	0.256	0.283
90	0.173	0.205	0.242	0.267
100	0.164	0.195	0.230	0.254

Source: E. S. Pearson and H. O. Hartley, *Biometrika Tables for Statisticians*, Vol. I. (Cambridge: Cambridge University Press, 1962), p. 138.

A-8 User's Guide to Minitab for Macintosh Computers[1]

Minitab is an extremely powerful software package with a wide range of capabilities for both data analysis and inferential statistics. The full capabilities of Minitab are beyond the scope of this introductory text. However, the following material can be a stepping stone to a fuller application of this software.

The first step in using Minitab is to enter the data onto a worksheet. Minitab then performs calculations on the data in one of two modes: interactive command mode or macro command–driven mode. It is easy to access either mode.

A-8.1 THE WORKSHEET

Figure A-8-1 shows what appears on the screen when Minitab is opened, or **launched.** The commands listed across the top

File Edit Calc Stat Window Graph Editor

compose the **menu bar. Select** any one of these commands by dragging the mouse on it. A command selected from the menu bar often leads to another menu, a **submenu.** For example, if the **Stat** command is selected from the menu bar, the submenu in Figure A-8-2 appears. A triangle (▶) after a command indicates that yet another menu will appear when that choice is selected.

The grid of boxes, or cells, shown in Figure A-8-1 is the **worksheet** on which the data are entered. It is very similar to a spreadsheet, and the first step in any application of Minitab (interactive mode or macro command–driven mode) is to enter the data onto the worksheet.

---

[1] Minitab version 8.2 is illustrated in this User's Guide. A new version scheduled for release, 10Xtra, has dual platform capability and operates very similarly for the applications in this text.

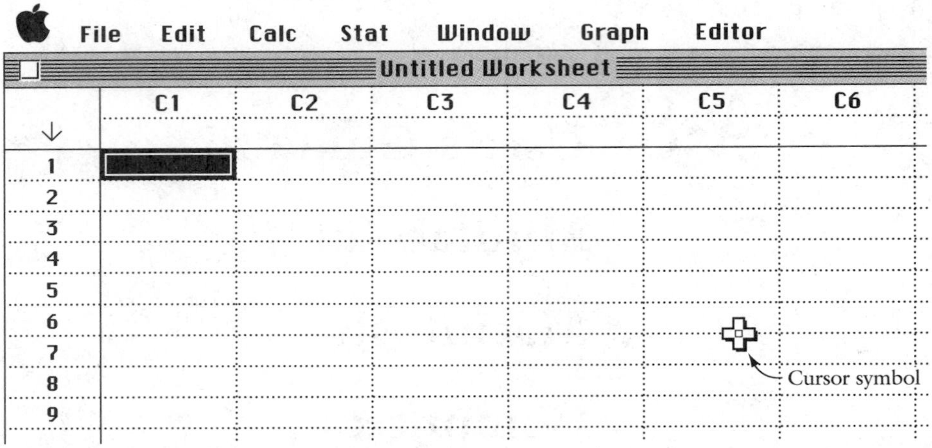

Figure A-8-1 A Minitab worksheet.

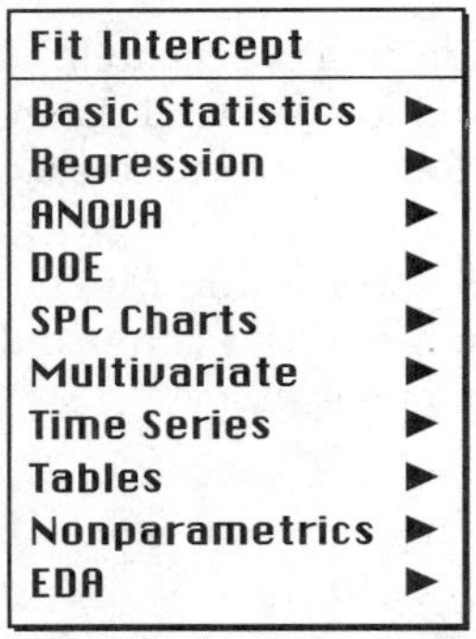

Figure A-8-2 Submenu for the `Stat` command.

The symbols `C1, C2, C3, . . .` across the top of the worksheet are column identifications. Each column contains data for a separate variable. A worksheet has 1000 available columns and as many rows as memory will allow. The **highlighted** cell (darkened in Figure A-8-1) shows where data typed on the keyboard will appear. Move the cursor symbol to the desired cell for data entry or data correction by simply clicking the mouse once on that cell. The downward direction of the arrow in the upper left corner of the worksheet indicates that use of the ENTER key will move the highlighted cell down to the next cell. Placing the cursor on the arrow and clicking the mouse once changes the arrow's direction so that use of the ENTER key will move the highlighted cell across the worksheet.

As an illustration, consider the data on cholesterol counts from Chapter 2. If these data were entered on a Minitab worksheet, the first few entries would appear as in Figure A-8-3. Notice that a space is provided above each column heading

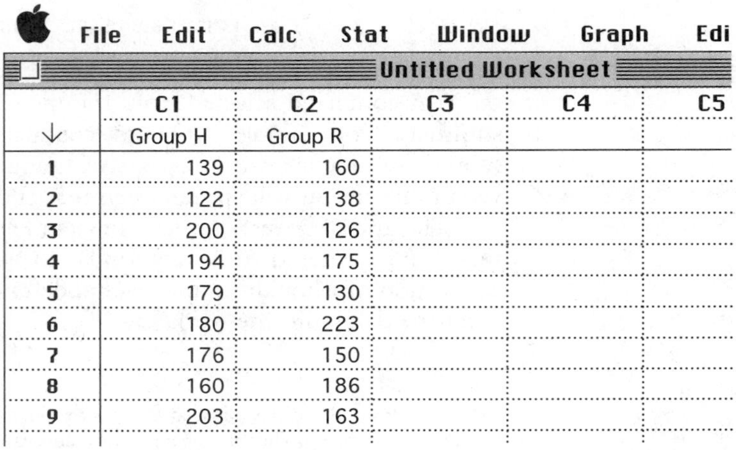

↓	C1 Group H	C2 Group R	C3	C4	C5
1	139	160			
2	122	138			
3	200	126			
4	194	175			
5	179	130			
6	180	223			
7	176	150			
8	160	186			
9	203	163			

Figure A-8-3

```
         🍎    File    Edit    Calc
        ┌───────────────────────────┐
        │▣══════════════════════════│
        │            │      C1       │
        │     ↓      │               │
        │────────────┼───────────────│
        │     1      │       23      │
        │     2      │       34      │
        │     3      │       23      │
        │     4      │       45      │
        │     5      │        *      │
        │     6      │               │
        │     7      │               │
        └───────────────────────────┘
```

Figure A-8-4

for a brief descriptive name of the data in that column. Data may be identified by either column number or descriptive name. For ease of reference, this guide identifies data by column number (such as C1).

Occasionally, typing errors occur and it is necessary to remove the entire contents of a cell. Highlighting the cell and then pressing the delete (backspace) key results in the placement of an asterisk within that cell when the cursor is moved, as shown in Figure A-8-4. The cell with the asterisk will now interfere with the proper execution of most Minitab commands. The *proper way* to completely remove the contents of a cell (or the troublesome asterisk) is as follows.

a. Highlight the cell.
b. Choose "Delete cell" from the Editor menu.

A-8.2   THE INTERACTIVE MODE

In interactive mode, the user responds to **dialog boxes** to direct calculations of data on the worksheet. The use of dialog boxes makes Minitab's interactive mode more user-friendly than the macro command–driven mode.

Once all the data have been entered on the worksheet, the user can access the interactive mode directly from the menu bar. For example, if the command **Graph** is selected (Figure A-8-5), the submenu in Figure A-8-6 appears. These subcommands appear on the screen as soon as the **Graph** command is selected. The three dots after each command in this new menu indicate that a dialog box

```
  🍎    File    Edit    Calc    Stat    Window   (Graph)   Editor
 ┌──────────────────────────────────────────────────────────────────┐
 │▣════════════════════════ H and R.MTW ════════════════════════════│
 │        │   C1      │   C2    │   C3   │   C4   │   C5   │   C6   │
 │   ↓    │  Group H  │ Group R │        │        │        │        │
 │────────┼───────────┼─────────┼────────┼────────┼────────┼────────│
 │   1    │    139    │   160   │        │        │        │        │
 └──────────────────────────────────────────────────────────────────┘
```

Figure A-8-5

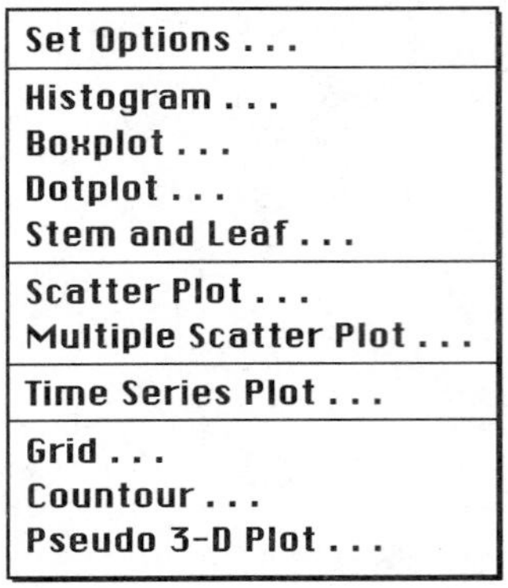

```
┌─────────────────────────────────┐
│ Set Options . . .               │
├─────────────────────────────────┤
│ Histogram . . .                 │
│ Boxplot . . .                   │
│ Dotplot . . .                   │
│ Stem and Leaf . . .             │
├─────────────────────────────────┤
│ Scatter Plot . . .              │
│ Multiple Scatter Plot . . .     │
├─────────────────────────────────┤
│ Time Series Plot . . .          │
├─────────────────────────────────┤
│ Grid . . .                      │
│ Countour . . .                  │
│ Pseudo 3-D Plot . . .           │
└─────────────────────────────────┘
```

Figure A-8-6

will request additional information when that subcommand is selected. Select a subcommand by dragging the **Graph** command to it and then releasing the mouse button.

For example, if the **Boxplot** . . . subcommand is selected, the screen in Figure A-8-7 appears. Figure A-8-7 is an example of a **dialog box,** or **dialog window.** Using the mouse and keyboard, the user can fill in the blanks of the dialog box to set the parameters for the calculations of the worksheet data. Default values are built into Minitab software, and not all the blanks need to be filled. Just typing C1 in the Variable box and then clicking on OK produces the high-resolution display shown in Figure A-8-8.

Since the **tick increment**—the number of units along the horizontal axis

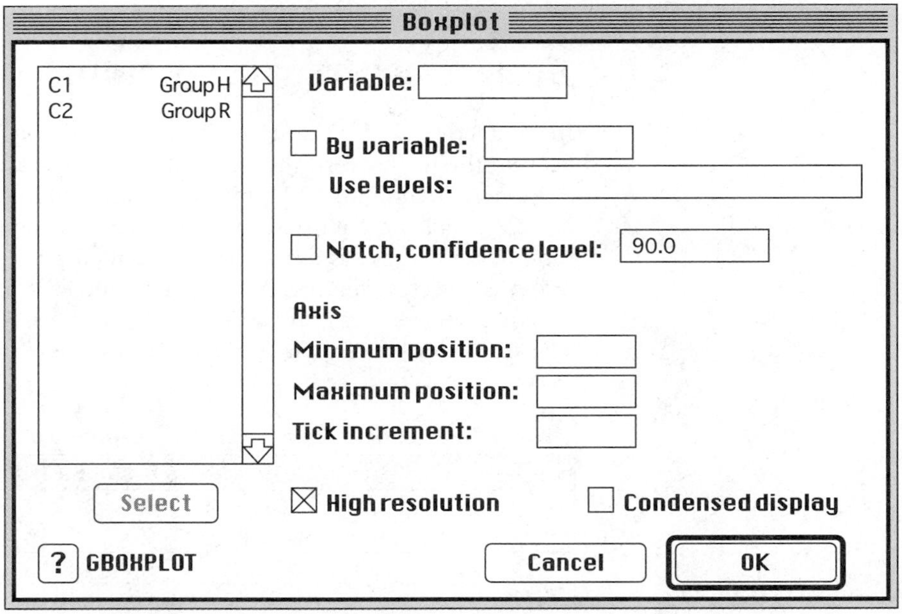

Figure A-8-7 Dialog box for the **Boxplot** command.

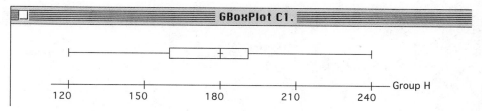

Figure A-8-8

between marks—was not supplied in the dialog window, it was set in default at 30. If the user specified a tick increment of, say, 50 in the dialog window, then the display shown in Figure A-8-9 would appear.

Not all of Minitab's commands result in high-resolution graphics. Low-resolution graphics are presented in a slightly different manner. For example, let **Dotplot** . . . be selected from the menu bar and the dialog box completed as in Figure A-8-10. For the sake of illustration, this time two variables, C1 and

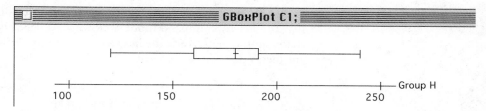

Figure A-8-9

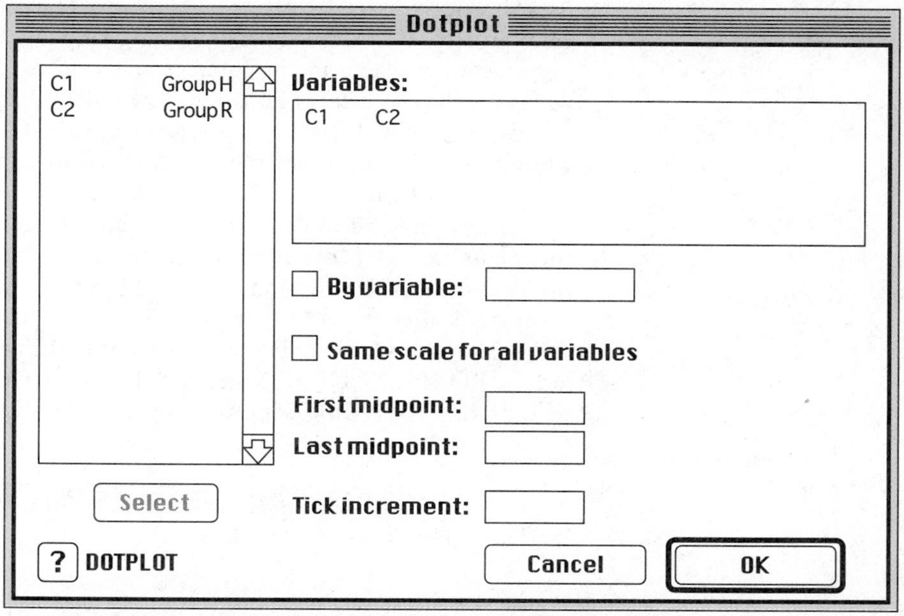

Figure A-8-10 A completed dialog box for the **Dotplot** command.

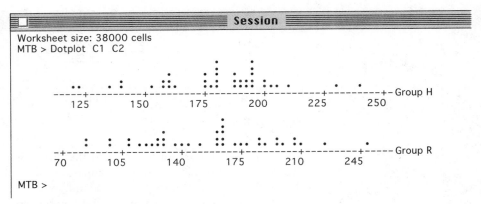

Figure A-8-11

C2, are specified. Figure A-8-11 shows what appears on the screen after OK is clicked. **Dotplot** is a low-resolution command, and low-resolution commands result in displays on windows titled **Session**. The presence of a Session window indicates that the macro command mode is now available, as well as the interactive mode. Before going on to study the macro command mode, independently explore different menu paths in the interactive mode.

A-8.3   MACRO COMMAND MODE

In addition to working in the interactive mode, Minitab can manipulate worksheet data through the use of macro commands. The set of macro commands to which Minitab responds forms a programming language that expands and enhances the possibilities available in the interactive mode. Unlike the interactive commands, which are executed through a pull-down menu bar, the user executes macro commands by typing them in a **Session** window. One way of obtaining a Session window—having it display the output of a low-resolution interactive command—was just explained.

A Session window can also be obtained at any time by selecting **Window** from the menu bar, then **Session** from the submenu. Figure A-8-12 shows what appears on the screen. The Session window is on top of the worksheet. Type macro commands after the **MTB**> prompt. To obtain a box plot for the data in column C1, type what's shown in Figure A-8-13 after the prompt, in uppercase or lowercase letters. BOXPLOT C1 in Figure A-8-13 is a macro command instructing Minitab to construct a box plot using the values of the variable in column C1. In general,

Figure A-8-12   Minitab Session window.

Figure A-8-13

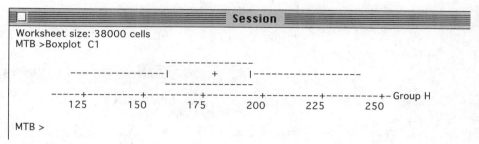

Figure A-8-14

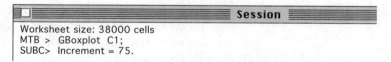

Figure A-8-15

a macro command (such as BOXPLOT) must be followed by a column identification (in this case, C1). Pressing the [RETURN] key has the result shown in Figure A-8-14, a low-resolution box plot. In general, prefixing the letter G to a macro command results in a high-resolution graph.

Many subcommands are available for each macro command. One of them, for the BOXPLOT macro command, is "Increment," which specifies the distance between tick marks. The macro commands shown in Figure A-8-15 would produce a high-resolution box plot for the values of the variable in column C1, with an increment of 75. Note that a semicolon (;) must be used after each macro command that is to be followed by a subcommand. Pressing the [RETURN] key automatically changes the next prompt to **SUBC**>. A period (.) must close the last subcommand. One can cancel the macro command session at any time with the command (or subcommand) ABORT.

Figure A-8-16 shows the display that would result from the command and subcommand in the last Session window.

A-8.4 WORKING MORE EFFICIENTLY

Minitab has many commands that can speed up your work, including STACK and "By."

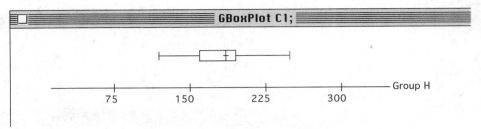

Figure A-8-16

Figure A-8-17

The STACK Command

When working with data from several different sources, it is often desirable to separate them into columns on the worksheet. For example, the results of a mathematics placement test might be stored in three columns: C1 for students under 20 years of age, C2 for students 20 to under 30, and C3 for students at least 30 years old. One could then execute the command for a box plot, for instance (in either interactive mode or macro command mode), three times, specifying a different column each time. This would produce a separate box plot for each of the three groups.

However, it might be useful to see a single box plot for the combined group. Instead of copying all the data in a new column, C4, it is more efficient to STACK the data from columns C1, C2, and C3 on top of one another and then place this new stack in a specified column (C4) on the worksheet. Figure A-8-17 shows the macro commands for stacking data and drawing a box plot for the combined group.

The "By" Subcommand

Like stacking, the "By" subcommand can reduce repetitive tasks. If data is entered into one column on the worksheet and one then wants to separate the data into several groups for analysis, the "By" subcommand can be used instead of re-entering the data. In the interactive mode, most dialog boxes offer the "By" subcommand feature.

Reconsider the example of the results of a mathematics placement test. Enter all the scores into one column, C1, then enter an identification code number for each student into column C2, next to his or her score. That is, 1 = a student under 20 years of age, 2 = a student from 20 to under 30, and 3 = a student at least 30 years of age. The macro command shown in Figure A-8-18 would then result in a box plot of all the scores. On the other hand, the macro command and subcommand shown in Figure A-8-19 would produce three box plots close together for the scores, grouped by age.

Figure A-8-18

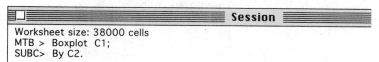

Figure A-8-19

A-8.5 ADDITIONAL MACRO COMMANDS

Some of the macro commands available on Minitab, along with their subcommands, follow in boldface type, listed in the order in which they were treated in this text. In this list, **K** denotes a constant, **C** denotes a column, **E** denotes either a constant or a column, and [ ] enclose an optional argument. Experiment with these commands independently. The *Minitab Mini-Manual* is a good source of further information on Minitab.

DOTPLOT C . . . C	Produces a dot plot for each column specified. (Use spaces, not commas, between column identifications.)
Increment = K	Specifies distance between tick marks.
Start = K [End = K]	Specifies first and last tick marks.
By C	Produces a dot plot for each column.
Same	Same scale used for all columns.
DESCRIBE C . . . C	Prints descriptive statistics for each column specified.
By C	
BOXPLOT C	Prints a box plot for the values of the variable in column C.
Increment = K	
Start = K	
By C	
Lines = K	Three lines are used to print each box plot, but this may be condensed by the subcommand "$K = 1$."
Notch [K%]	Produces a confidence interval.
Levels K . . . K	Used with the "By" subcommand; specifies which levels are to be used and in what order.
STEM AND LEAF C . . . C	Prints a stem-and-leaf display for each column specified.

Trim	Trims all values beyond the inner fences.
Increment = K	
By C	
HISTOGRAM C . . . C	Prints a separate histogram for each column specified.
Increment = K	
Start = K [End = K]	
By C	
Same	
TALLY C . . . C	Prints a one-way table for each column specified.
Counts	
Percents	
Cumcounts	Prints cumulative counts.
Cumpercents	Prints cumulative percents.
CDF	Cumulative distribution function.
Binomial $n = $ **K** $p = $ **K**	
Poisson $\mu = $ **K**	
Normal $[\mu = $ **K** $[\sigma = $ **K**$]]$	
INVCDF	Inverse function of CDF.
Binomial $n = $ **K** $p = $ **K**	
Poisson $\mu = $ **K**	
Normal $[\mu = $ **K** $[\sigma = $ **K**$]]$	
STDEV C	Calculates the sample standard deviation for the column specified.
ZINTERVAL [K% confidence] $\sigma = $ **K** for **C . . . C**	
	Calculates a K% confidence interval for the mean, separately for each column specified. (Use spaces, not commas, between the parameters.)
TINTERVAL [K% confidence] for **C . . . C**	
	Calculates a K% confidence interval for each column specified.
ZTEST $[\mu = $ **K**$]$ $\sigma = $ **K** for **C . . . C**	
	Performs a separate Z-test on the data in each column specified.
Alternative = K	Specifies a one-sided test. $K = -1$ gives $H_0: \mu < K$ and $K = +1$ gives $H_0: \mu > K$.
TTEST $[\mu = $ **K**$]$ on **C . . . C**	Performs a separate t-test on the data in each column specified.
TWOSAMPLE	Performs a two- (independent) sample t-test and confidence interval.
Alternative = K	
Pooled	Executes the TWOSAMPLE command with a pooled sample standard deviation.

A-9 User's Guide to the DOS Version of Minitab

Minitab is an extremely powerful software package with a wide range of capabilities for both data analysis and inferential statistics. The full capabilities of Minitab are beyond the scope of this introductory text. However, the following material can be a stepping stone to a fuller application of this software.

The illustrations of the DOS version of Minitab in this appendix will appear somewhat different from the text illustrations, which feature the Macintosh version. Nevertheless, on close comparison it can be seen that the two versions are remarkably similar; they use the same information and produce identical results.

The first step in using Minitab is to enter the data onto a worksheet or data screen. Minitab then performs calculations on the data in one of two modes: interactive command mode or macro command–driven mode. It is easy to access either mode.

A-9.1 OPENING MINITAB

This appendix is written for those using Minitab on IBM or IBM-compatible computers. Systems vary, but typically a DOS C prompt will appear on the start-up screen.

$$c:\>$$

After this prompt, type "cd Minitab" and press the RETURN key. After the next prompt, type "Minitab" and press the RETURN key. The program should then open with the start-up screen shown in Figure A-9-1, which is also called a Session window (its function will be discussed later). It is possible to enter data onto a worksheet or data screen for analysis from the Session window. However, for illustration purposes, obtain a data screen with the keystrokes ALT + D. Figure A-9-2 shows a data screen.

```
                                              Minitab
         File  Edit  Calc  Stat  Graph                          F1=Help
         Worksheet size: 16174 cells


         Press ALT + a highlighted letter to open a menu

         MTB >
```
Figure A-9-1 Minitab start-up screen, or Session window.

```
↓ C1    C2    C3    C4    C5    C6    C7    C8    C9    C10   C11   C12   C13   C14   C15
1
2
3
4
5
6
7
8
9
10
11
12
13
14
15
16
17
18
19
20
21
22
                                                        Enter Data   F10=Menu
```
Figure A-9-2 Minitab data screen.

A-9.2   THE DATA SCREEN

The data screen, an array of cells, is the worksheet on which the data are entered. It functions similarly to a spreadsheet. The first step in any application of Minitab (interactive mode or macro command–driven mode) is to enter the data onto the data screen. The symbols **C1, C2, C3, . . .** across the top of the data screen are column designations that may be used to stand for distinct variables. Each cell in a given column may receive a value of that variable. One thousand columns and as many rows as memory will allow are available for data entry. Keystrokes or a mouse may be used to move the cursor around the data screen.

   The downward direction of the small arrow in the upper left corner of the data screen indicates that use of the ENTER key will move the highlighted cell

Appendix A-2   Standard Normal Curve Areas

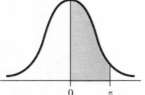

Z	.00	.01	.02	.03	.04	.05	.06	.07	.08	.09
.0	.0000	.0040	.0080	.0120	.0160	.0199	.0239	.0279	.0319	.0359
.1	.0398	.0438	.0478	.0517	.0557	.0596	.0636	.0675	.0714	.0753
.2	.0793	.0832	.0871	.0910	.0948	.0987	.1026	.1064	.1103	.1141
.3	.1179	.1217	.1255	.1293	.1331	.1368	.1406	.1443	.1480	.1517
.4	.1554	.1591	.1628	.1664	.1700	.1736	.1772	.1808	.1844	.1879
.5	.1915	.1950	.1985	.2019	.2054	.2088	.2123	.2157	.2190	.2224
.6	.2257	.2291	.2324	.2357	.2389	.2422	.2454	.2486	.2518	.2549
.7	.2580	.2612	.2642	.2673	.2704	.2734	.2764	.2794	.2823	.2852
.8	.2881	.2910	.2939	.2967	.2995	.3023	.3051	.3078	.3106	.3133
.9	.3159	.3186	.3212	.3238	.3264	.3289	.3315	.3340	.3365	.3389
1.0	.3413	.3438	.3461	.3485	.3508	.3531	.3554	.3577	.3599	.3621
1.1	.3643	.3665	.3686	.3708	.3729	.3749	.3770	.3790	.3810	.3830
1.2	.3849	.3869	.3888	.3907	.3925	.3944	.3962	.3980	.3997	.4015
1.3	.4032	.4049	.4066	.4082	.4099	.4115	.4131	.4147	.4162	.4177
1.4	.4192	.4207	.4222	.4236	.4251	.4265	.4279	.4292	.4306	.4319
1.5	.4332	.4345	.4357	.4370	.4382	.4394	.4406	.4418	.4429	.4441
1.6	.4452	.4463	.4474	.4484	.4495	.4505	.4515	.4525	.4535	.4545
1.7	.4554	.4564	.4573	.4582	.4591	.4599	.4608	.4616	.4625	.4633
1.8	.4641	.4649	.4656	.4664	.4671	.4678	.4686	.4693	.4699	.4706
1.9	.4713	.4719	.4726	.4732	.4738	.4744	.4750	.4756	.4761	.4767
2.0	.4772	.4778	.4783	.4788	.4793	.4798	.4803	.4808	.4812	.4817
2.1	.4821	.4826	.4830	.4834	.4838	.4842	.4846	.4850	.4854	.4857
2.2	.4861	.4864	.4868	.4871	.4875	.4878	.4881	.4884	.4887	.4890
2.3	.4893	.4896	.4898	.4901	.4904	.4906	.4909	.4911	.4913	.4916
2.4	.4918	.4920	.4922	.4925	.4927	.4929	.4931	.4932	.4934	.4936
2.5	.4938	.4940	.4941	.4943	.4945	.4946	.4948	.4949	.4951	.4952
2.6	.4953	.4955	.4956	.4957	.4959	.4960	.4961	.4962	.4963	.4964
2.7	.4965	.4966	.4967	.4968	.4969	.4970	.4971	.4972	.4973	.4974
2.8	.4974	.4975	.4976	.4977	.4977	.4978	.4979	.4979	.4980	.4981
2.9	.4981	.4982	.4982	.4983	.4984	.4984	.4985	.4985	.4986	.4986
3.0	.4986	.4987	.4987	.4988	.4988	.4989	.4989	.4989	.4990	.4990
3.1	.4990	.4991	.4991	.4991	.4992	.4992	.4992	.4992	.4993	.4993
3.2	.4993	.4993	.4994	.4994	.4994	.4994	.4994	.4995	.4995	.4995
3.3	.4995	.4995	.4995	.4996	.4996	.4996	.4996	.4996	.4996	.4997
3.4	.4997	.4997	.4997	.4997	.4997	.4997	.4997	.4997	.4997	.4998
3.6	.4998	.4998	.4999	.4999	.4999	.4999	.4999	.4999	.4999	.4999
3.9	.5000									

Source: *Standard Mathematical Tables*, 25th edition, p. 524, CRC Press, Inc.

(D)

Difference of population means

$$\mu_{\overline{X}_A - \overline{X}_B} = (\mu_A - \mu_B).$$

$$SE = \sigma_{\overline{X}_A - \overline{X}_B} = \sqrt{\frac{\sigma_A^2}{n_A} + \frac{\sigma_B^2}{n_B}} \approx \sqrt{\frac{(s_A)^2}{n_A} + \frac{(s_B)^2}{n_B}}$$

$$(s_p)^2 = \frac{(n_A - 1) \cdot (s_A)^2 + (n_B - 1) \cdot (s_B)^2}{n_A + n_B = 2} \quad \text{(pooled variance)}$$

Degrees of freedom $= n_A + n_B - 2$

Endpoints of confidence interval for difference of population means

$\overline{X}_A - \overline{X}_B \pm Z(\overline{X}_A - \overline{X}_B) \cdot SE$ (normal model)

$\overline{X}_A - \overline{X}_B \pm t(\overline{X}_A - \overline{X}_B) \cdot SE$ (t-distribution model)

$\overline{D} \pm Z(\overline{D}) \cdot SE$ or $\overline{D} \pm t(\overline{D}) \cdot SE$ (matched pair sampling)

Test statistic for difference of population means

$$Z(\overline{X}_A - \overline{X}_B) \text{ (or } t(\overline{X}_A - \overline{X}_B)) = \frac{(\overline{X}_A - \overline{X}_B) - (\mu_A - \mu_B)}{SE}$$

Difference of population proportions

$$SE = \sqrt{\frac{P_A \cdot (1 - P_A)}{n_A - 1} + \frac{P_B \cdot (1 - P_B)}{n_B - 1}}$$

$(P_A - P_B) \pm Z(P_A - P_B) \cdot SE$ (endpoints of confidence interval)

The chi-square distribution

$\chi^2 = \Sigma \frac{(f_o - f_e)^2}{f_e}$ (for independence and goodness-of-fit testing)

d.f. = Number of adjusted classes $-$ number of estimated population parameters $- 1$

$\chi^2 = \frac{(n - 1) \cdot s^2}{\sigma^2}$ (confidence interval for σ)

$$\sqrt{\frac{(n - 1) \cdot s^2}{\chi^2_{\text{RIGHT}}}} \leq \sigma \leq \sqrt{\frac{(n - 1) \cdot s^2}{\chi^2_{\text{LEFT}}}} \quad \text{(endpoints of confidence interval for } \sigma)$$

Appendix A-4 Student *t*-Distribution

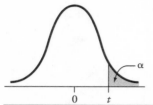

Degrees of Freedom	α									
	.4	.25	.1	.05	.025	.01	.005	.0025	.001	.0005
1	0.325	1.000	3.078	6.314	12.706	31.821	63.657	127.32	318.31	636.62
2	.289	.816	1.886	2.920	4.303	6.965	9.925	14.089	22.327	31.598
3	.277	.765	1.638	2.353	3.182	4.541	5.841	7.453	10.214	12.924
4	.271	.741	1.533	2.132	2.776	3.747	4.604	5.598	7.173	8.610
5	0.267	0.727	1.476	2.015	2.571	3.365	4.032	4.773	5.893	6.869
6	.265	.718	1.440	1.943	2.447	3.143	3.707	4.317	5.208	5.959
7	.263	.711	1.415	1.895	2.365	2.998	3.499	4.029	4.785	5.408
8	.262	.706	1.397	1.860	2.306	2.896	3.355	3.833	4.501	5.041
9	.261	.703	1.383	1.833	2.262	2.821	3.250	3.690	4.297	4.781
10	0.260	0.700	1.372	1.812	2.228	2.764	3.169	3.581	4.144	4.587
11	.260	.697	1.363	1.796	2.201	2.718	3.106	3.497	4.025	4.437
12	.259	.695	1.356	1.782	2.179	2.681	3.055	3.428	3.930	4.318
13	.259	.694	1.350	1.771	2.160	2.650	3.012	3.372	3.852	4.221
14	.258	.692	1.345	1.761	2.145	2.624	2.977	3.326	3.787	4.140
15	0.258	0.691	1.341	1.753	2.131	2.602	2.947	3.286	3.733	4.073
16	.258	.690	1.337	1.746	2.120	2.583	2.921	3.252	3.686	4.015
17	.257	.689	1.333	1.740	2.110	2.567	2.898	3.222	3.646	3.965
18	.257	.688	1.330	1.734	2.101	2.552	2.878	3.197	3.610	3.922
19	.257	.688	1.328	1.729	2.093	2.539	2.861	3.174	3.579	3.883
20	0.257	0.687	1.325	1.725	2.086	2.528	2.845	3.153	3.552	3.850
21	.257	.686	1.323	1.721	2.080	2.518	2.831	3.135	3.527	3.819
22	.256	.686	1.321	1.717	2.074	2.508	2.819	3.119	3.505	3.792
23	.256	.685	1.319	1.714	2.069	2.500	2.807	3.104	3.485	3.767
24	.256	.685	1.318	1.711	2.064	2.492	2.797	3.091	3.467	3.745
25	0.256	0.684	1.316	1.708	2.060	2.485	2.787	3.078	3.450	3.725
26	.256	.684	1.315	1.706	2.056	2.479	2.779	3.067	3.435	3.707
27	.256	.684	1.314	1.703	2.052	2.473	2.771	3.057	3.421	3.690
28	.256	.683	1.313	1.701	2.048	2.467	2.763	3.047	3.408	3.674
29	.256	.683	1.311	1.699	2.045	2.462	2.756	3.038	3.396	3.659
30	0.256	0.683	1.310	1.697	2.042	2.457	2.750	3.030	3.385	3.646
40	.255	.681	1.303	1.684	2.021	2.423	2.704	2.971	3.307	3.551
60	.254	.679	1.296	1.671	2.000	2.390	2.660	2.915	3.232	3.460
120	.254	.677	1.289	1.658	1.980	2.358	2.617	2.860	3.160	3.373
∞	.253	.674	1.282	1.645	1.960	2.326	2.576	2.807	3.090	3.291

Source: E. S. Pearson and H. O. Hartley, *Biometrika Tables for Statisticians,* Vol. I. (Cambridge: Cambridge University Press, 1966), p. 146.

Discrete random variables

$$\mu_X = \Sigma x \cdot P(X = x)$$

$$\sigma_X = \sqrt{\Sigma(x - \mu)^2 \cdot P(X = x)}$$

$$P(X = r) = {}_nC_r \cdot p^r \cdot q^{n-r} \quad \text{(binomial formula)}$$

$$\mu_B = n \cdot p$$

$$\sigma_B = \sqrt{n \cdot p \cdot q}$$

$$P(X = r) = \frac{e^{-\lambda} \cdot \lambda^r}{r!} \quad \text{(Poisson formula)}$$

Sampling

$$\overline{X} = \frac{\Sigma x}{n}$$

$$s = \sqrt{\frac{\Sigma(x - \overline{X})^2}{n - 1}} = \sqrt{\frac{\Sigma x^2 - \frac{(\Sigma x)^2}{n}}{n - 1}}$$

$$\mu_{\overline{X}} = \mu$$

$$\sigma_{\overline{X}} = SE = \frac{\sigma}{\sqrt{n}} \approx \frac{s}{\sqrt{n}}$$

$$\text{Finite correction factor} = \frac{\sigma}{\sqrt{n}} \cdot \sqrt{\frac{N - n}{N - 1}}$$

The normal probability distribution

$$Z(x) = \frac{x - \mu}{\sigma} \quad \text{(standardization formula)}$$

Flowchart for the construction of confidence intervals

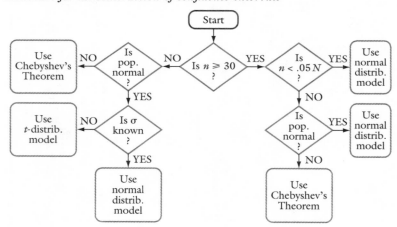

Endpoints of a confidence interval

$\overline{X} \pm Z(\overline{X}) \cdot SE$ (normal distribution model)
$\overline{X} \pm t(\overline{X}) \cdot SE$ (t-distribution model)
$\overline{X} \pm k(\overline{X}) \cdot SE$ (Chebyshev's theorem)
$P \pm Z(P) \cdot SE$ (population proportion)

Margin of error for a population proportion:
$$ME = Z(P) \cdot SE$$

$$n \geq \left(\frac{Z(P) \cdot .5}{ME}\right)^2 \quad \text{(minimum sample size)}$$

Test statistic

$$Z(\overline{X}) \, (\text{or } t(\overline{X})) = \frac{\overline{X} - \mu_{\overline{X}}}{SE}$$

If $P < \alpha$, then reject H_O and accept H_A.
Otherwise, reserve judgment.

Percentiles

To find the score that corresponds to a given percentile, P_k:

Data rank $R = \left(\dfrac{k}{100}\right) \cdot n$ (n = number of scores).

If R is a whole number, then P_k is half the sum of the scores with data rank R and $(R + 1)$.

If R is not a whole number, then round R up to the next whole number, and P_k is the score with data rank R.

To find the percentile corresponding to a given score, S:

$$\dfrac{\text{Number of scores less than } S}{\text{Total number of scores}}$$

1st quartile $= Q_1 = P_{25}$
2nd quartile $= Q_2 = P_{50} =$ median
3rd quartile $= Q_3 = P_{75}$

Interquartile range $= \text{IQR} = Q_3 - Q_1$

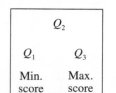 five-number summary

Box plots

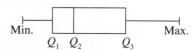

$\dfrac{1.57 \cdot \text{IQR}}{\sqrt{n}} =$ Notch width of confidence interval for median

Inner fences located at $Q_3 + 1.5 \cdot \text{IQR}$ and $Q_1 - 1.5 \cdot \text{IQR}$
Outer fences located at $Q_3 + 3 \cdot \text{IQR}$ and $Q_1 - 3 \cdot \text{IQR}$

Central location and dispersion

$\mu = \dfrac{\text{Sum of scores}}{\text{Total number of scores}}$

Mode $=$ Most often repeated score

5% trimmed mean $=$ Mean of the middle 90% of the scores

Midrange $= \dfrac{\text{Minimum score} + \text{Maximum score}}{2}$

Midquartile $= \dfrac{Q_1 + Q_2}{2}$

Standard deviation $\sigma = \sqrt{\dfrac{\Sigma(x - \mu)^2}{N}} = \sqrt{\dfrac{(\Sigma x^2) - \dfrac{(\Sigma x)^2}{N}}{N}}$

Coefficient of variation $v = \dfrac{\sigma}{\mu}$

Histograms

Sturgess' Rule:

Number of intervals $\approx 1 + 3.3 \cdot \log(n)$

Estimated $\mu = \dfrac{\Sigma m \cdot f}{N}$

Estimated $\sigma = \sqrt{\dfrac{\Sigma f \cdot (m - \mu)^2}{N}}$

Estimated mode $= L + \left(\dfrac{a}{a + b}\right) \cdot w$

Estimated median $= L + \left(\dfrac{\dfrac{N}{2} - F}{f}\right) \cdot w$

Probability

$P(A^c) = 1 - P(A)$
$P(A \text{ or } B) = P(A \cup B) = P(A) + P(B) - P(A \cap B)$
If A and B are independent:

$$P(A \text{ and } B) = P(A \cap B) = P(A) \cdot P(B)$$

If A and B are not independent:

$$P(A \text{ and } B) = P(A \cap B) = P(A) \cdot P(B \mid A)$$

$P(A \mid B) = \dfrac{P(A \cap B)}{P(B)}$

Bayes's Theorem:

$$P(A \mid B) = \dfrac{P(A \cap B)}{\Sigma P(A) \cdot P(B \mid A)}$$

$_nP_r = \dfrac{n!}{(n - r)!}$

$_nC_r = \dfrac{n!}{(n - r)! \cdot r!}$

$O(E) = \dfrac{P(E)}{1 - P(E)}$

Nonparametric tests

The sign test:

$$\mu_S = \frac{n}{2}, \quad \sigma_S = .5 \cdot \sqrt{n}$$

The number of runs test:

$$\mu_{RH} = \frac{n_H \cdot (n_T + 1)}{n_H + n_T}$$

$$\sigma_{RH} = \sqrt{\frac{n_H \cdot (n_T + 1) \cdot (n_H - 1)}{(n_H + n_T)^2} \cdot \frac{n_T}{n_H + n_T - 1}}$$

Linear regression

$$p = \frac{\Sigma(x \cdot y) - n \cdot \overline{X} \cdot \overline{Y}}{\sqrt{(\Sigma x^2 - n \cdot \overline{X}^2) \cdot (\Sigma y^2 - n \cdot \overline{Y}^2)}} \quad \text{(Pearson coefficient)}$$

Coefficient of determination $= p^2$

The line of best fit

$$m = \frac{\Sigma x \cdot y - n \cdot \overline{X} \cdot \overline{Y}}{\Sigma x^2 - n \cdot \overline{X}^2}, \quad b = \overline{Y} - m \cdot \overline{X}$$

Location of endpoints for a confidence interval for $\mu_{y,X}$

$$\hat{Y} \pm t \cdot \sqrt{\frac{\Sigma y^2 - b\Sigma y - a\Sigma x \cdot y}{n - 2}} \cdot \sqrt{\frac{1}{n} + \frac{(X - \overline{X})^2}{\Sigma X^2 - n \cdot \overline{X}^2}} \quad (n < 30)$$

$$\hat{Y} \pm Z \cdot \sqrt{\frac{\Sigma y^2 - b\Sigma y - a\Sigma x \cdot y}{n - 2}} \cdot \sqrt{\frac{1}{n}} \quad (n \geq 30)$$

Location of endpoints for a confidence interval for $R_{Y,X}$

$$\hat{Y} \pm t \cdot \sqrt{\frac{\Sigma y^2 - b\Sigma y - a\Sigma x \cdot y}{n - 2}} \cdot \sqrt{\frac{1}{n} + \frac{(x - \overline{X})^2}{\Sigma x^2 - n \cdot \overline{X}^2} + 1} \quad (n < 30)$$

$$\hat{Y} \pm Z \cdot \sqrt{\frac{\Sigma y^2 - b\Sigma y - a\Sigma x \cdot y}{n - 2}} \quad (n \geq 30)$$

The Spearman rank-correlation test:

$$r = 1 - \frac{6 \cdot \Sigma d_i^2}{n \cdot (n^2 - 1)}$$

$$\mu_r = 0$$

$$\sigma_r = \sqrt{\frac{1}{n - 1}}$$

ANOVA

$$F = \frac{\text{Variance based on sample means}}{\text{Mean of sample variances}}$$

TI-82 graphics calculator functions

To clear a list:

[STAT], [4], list designation

To sort a list:

[STAT], [2] or [3], list designation

To obtain one-variable statistics:

[STAT], [▶], [1], list designation

To obtain linear regression statistics:

[STAT], [▶], [5], [ENTER]

To draw a scatter plot, box plot, or histogram:

[2nd], [Y=]

Use arrow keys to:

- Turn plot on
- Choose type of graph
- Specify data list and frequency

[WINDOW] or [ZOOM], [9]

Use arrow keys to set limits of viewing window and scale.
To generate random numbers and compute $_nP_r$, $_nC_r$, or $n!$:

[MATH], [▶], [▶], [▶], designate function

(F)

↓ C1	C2	C3	C4	C5	C6	C7	C8	C9	C10	C11	C12	C13	C14	C15
1														
2														
3				Help					F1					
4														
5				Edit within a Cell					F2					
6				Change Entry Direction					F3					
7				Compress					F4					
8				Go to Cell...					F5					
9				Go to Next Row/Column					F6					
10														
11				Insert Cell					F7					
12				Insert Row				Shift+F7						
13				Delete Cell					F8					
14				Delete Row				Shift+F8						
15														
16				Format Columns					F9					
17				Name Columns				Alt+N						
18														
19				Go to Minitab Session				Alt+M						
20														
21														
22														

Enter Data   F10=Menu

Figure A-9-3

down to the next cell. Placing the cursor on the arrow and clicking the mouse once changes the arrow's direction so that use of the ENTER key will move the highlighted cell across the worksheet. One can also accomplish this directional change with keystrokes.

Data may be efficiently edited and formatted on the data screen. Pressing the F10 menu key produces the editing options shown in Figure A-9-3. As an illustration, consider entering the data on cholesterol counts from Chapter 2, page 42, onto the data screen. If these data were entered, the first 22 entries appear as in Figure A-9-4. One may type a brief descriptive name for each column's data above the column designation. Data may be identified by either column designation or descriptive name, such as C1 or "Group H." For ease of reference, this guide identifies data by column number (such as C1).

A-9.3   THE INTERACTIVE MODE

In interactive mode, the user responds to **dialog boxes** to direct calculations involving the scores on the data screen. In some instances, the use of dialog boxes makes Minitab's interactive mode more user-friendly than the macro command–driven mode.

Once all the data have been entered onto the data screen, the user can access the interactive mode from the Session window. Use the keystrokes ALT + M to

```
↓  C1   C2   C3   C4   C5   C6   C7   C8   C9   C10  C11  C12  C13  C14  C15
1  139  160
2  122  138
3  200  126
4  194  175
5  179  130
6  180  223
7  176  150
8  160  186
9  203  163
10 194  210
11 157  198
12 188   85
13 180  208
14 190  208
15 194  120
16 193   98
17 157  186
18 120  161
19 210  145
20 189  123
21 160  140
22 194   98
```

                                                              Enter Data   F10=Menu

Figure A-9-4

                                         Minitab
   File Edit Calc Stat Graph   F1=Help
Worksheet size: 16174 cells

Press ALT + a highlighted letter to open a menu

MTB >
Figure A-9-5 Minitab start-up screen.

obtain this window, which was the original start-up window (Figure A-9-5). The
commands listed at the upper left of the start-up screen

                    File Edit Calc Stat Graph

are the interactive commands. Selecting one of them with the mouse or appropriate
keystrokes leads to a menu of options. For example, Figure A-9-6 shows what
appears on the screen if the **Graph** command is selected.

    The three dots after each choice indicates that a submenu will be offered for
that choice. For instance, if one selects the command

                       Boxplot . . .

by using the keystrokes ALT + B , the screen shown in Figure A-9-7 appears.

```
                                      Minitab
     File  Edit  Calc  Stat  Graph                           F1=Help
     Worksheet size: 16174 ce
                                   Set Options...
     Press ALT + a highlighte
                                   Histogram...
                                   Boxplot...
                                   Dotplot...
                                   Stem-and-Leaf...

                                   Scatter Plot...
                                   Multiple Scatter Plot...

     MTB >                         Time Series Plot...

                                   Grid...
                                   Contour...
                                   Pseudo 3-D Plot...
```

Figure A-9-6

```
                                Minitab
  File  Edit  Calc  Stat  Graph                              F1=Help
  Worksheet size: 16174 cells
                              Boxplot

                      Variable: [        ]
        C1
        C2            _ [ ] By variable: [       ]
                      _     Use levels:  [                      ]
                      _
                      _ [ ] Notch, confidence level: [90.0    ]
                      _
                      _ Axis
                      _ Minimum position: [     ]
                      _ Maximum position: [     ]
                        Tick increment:   [     ]

                        [X] High resolution
       <  Select  >      [ ] Store in file      [ ] Condensed display

  <?> GBOXPLOT                          µ   OK    .  < Cancel  >
```
Figure A-9-7

Figure A-9-7 is an example of a **dialog box,** or **dialog screen.** Using a mouse or keystrokes, the user can fill in the blanks of the dialog box to set the parameters for the calculation of scores on the data screen. Default values are built into Minitab software, and not all the blanks need to be filled. Typing C1 anywhere in the brackets after the "By variable:" option and then choosing OK results in the high-resolution display shown in Figure A-9-8.

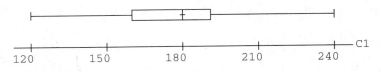

Type P to Print, N for Next, or Q to Quit

Figure A-9-8

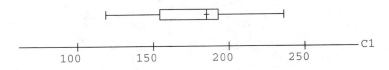

Type P to Print, N for Next, or Q to Quit

Figure A-9-9

```
                              Minitab
    File  Edit  Calc  Stat  Graph                              F1=Help

MTB >                              Dotplot

                        Variables:

        C1
        C2                   C1 C2


                        [ ] By variable: [       ]
                        [ ] Same scale for all variables
                         First midpoint: [    ]
                        Last midpoint:  [    ]

     < Select  >     Tick increment: [    ]
    <?> DOTPLOT                           μ  OK   .  < Cancel >
```

Figure A-9-10

The **tick increment**—the number of units along the horizontal axis between marks was not supplied by the user in the dialog box, it was set in default at **30**. If one specified a tick increment of, say, **50** in the dialog box, then Figure A-9-9 would appear.

Not all of Minitab's commands result in high-resolution graphics. Low-resolution graphics are presented in a Session window. For example, select

Dotplot . . .

from the menu and complete the dialog box as shown in Figure A-9-10.

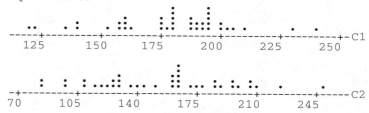

                                    Minitab
       File   Edit   Calc   Stat   Graph   F1=Help

MTB > Dotplot C1 C2

Figure A-9-11

For the sake of illustration, this time two variables, C1 and C2, are specified. Figure A-9-11 shows the dot plots that appear after OK is selected. DOTPLOT is a low-resolution command, and low-resolution commands result in Session window displays that have the **MTB**> prompt, which indicates that the macro command mode is available. Before going on to study the macro command mode, independently explore different menu paths in the interactive mode.

A-9.4 MACRO COMMAND MODE

In addition to working in the interactive mode, Minitab can be programmed to perform statistical calculations on the scores in a data screen. The commands to which Minitab responds, called macro commands, make up a programming language that expands and enhances the possibilities available in the interactive mode. Unlike the interactive commands, which are executed through choices in a series of dialog boxes, the user executes macro commands by typing them in a Session window. A Session window is always the first thing that appears when Minitab is started, and it displays the outputs of many interactive commands. At any time, the keystrokes ALT + M can be used to obtain a Session window.

In the Session window, type macro commands after the following prompt.

MTB>

Instead of using the interactive mode, one can obtain a box plot for the data in column C1 by typing the macro command BOXPLOT C1 after the **MTB**> prompt in a Session window. Once the RETURN key is pressed, the screen appears as shown in Figure A-9-12. Notice that the **MTB**> prompt is at the bottom of the screen, indicating that Minitab is ready to execute the next macro command.

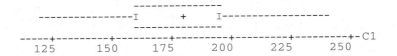

```
                                    Minitab
        File   Edit   Calc   Stat   Graph                        F1=Help




MTB > Boxplot C1

                              ---------------
          ---------------I       +      I------------------
                              ---------------
        -----+---------+---------+---------+---------+---------+-C1
            125       150       175       200       225       250


MTB >
```

Figure A-9-12

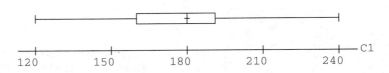

```
        Type P to Print, N for Next, or Q to Quit
```
Figure A-9-13 A high-resolution box plot.

In general, most macro commands (such as BOXPLOT) must be followed by a column identification (in this case, C1). Also, BOXPLOT is a low-resolution graph. Prefixing the letter G to many macro commands results in a high-resolution graph.

MTB> GBOXPLOT C1

When the (RETURN) key is pressed, the high-resolution graph in Figure A-9-13 appears.

Many subcommands are available for each macro command. One of them, for the BOXPLOT macro command, is "Increment," which specifies the distance between tick marks on a box plot or dot plot. The macro commands shown in Figure A-9-14 would produce a high-resolution box plot for the values of the variable in column C1, with an increment of 75. A semicolon (;) must be used after each macro command that is to be followed by a subcommand. Pressing the (RETURN) key automatically changes the next prompt to a subcommand prompt.

SUBC>

A period (.) must close the last subcommand. One can cancel the macro command session at any time with the command (or subcommand) ABORT.

Figure A-9-15 shows the display that would result from the command and subcommand in Figure A-9-14.

Minitab

File   Edit   Calc   Stat   Graph   F1=Help

```
MTB > GBoxplot C1;
SUBC> Increment=75.
```
Figure A-9-14

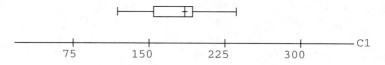

```
        Type P to Print, N for Next, or Q to Quit
```
Figure A-9-15

A-9.5   WORKING MORE EFFICIENTLY

Minitab has many commands that can speed up your work, including STACK and "By."

The STACK Command

When your data are from several different sources, it is often desirable to separate them into columns, which can represent distinct variables, on the data screen. For example, the results of a mathematics placement test might be stored in three columns: C1 for students under 20 years of age, C2 for students 20 to under 30, and column C3 for students at least 30 years old. One could then execute the command for a box plot, for instance (in either interactive mode or macro command mode), three times, specifying a different column each time. This would produce a separate box plot for each of the three groups.

However, it might be useful to see a single box plot for the combined group. Instead of copying all the data in a new column, C4, it is more efficient to STACK the data from columns C1, C2, and C3 on top of one another and then place this new stack in a specified column (C4) on the data screen.

As an illustration, Figure A-9-16 shows a data screen where three columns, C1, C2, and C3, are used. To stack these data, obtain a Session window with the keystrokes ALT + M. Figure A-9-17 shows what must be typed in the Session window to stack all the data into column C4, then draw a box plot for the combined data. The output is also shown. If the data screen is brought into view with the keystrokes ALT + D, it shows the data of the first three columns stacked into column C4 (Figure A-9-18)[1].

---

[1] Because of printing limitations, only the first 22 scores are shown in column C4.

```
↓   C1   C2   C3   C4   C5   C6   C7   C8   C9   C10   C11   C12   C13   C14   C15
1   80   90   50
2   70   80   60
3   75   85   30
4   70   85   30
5   80   85   40
6   50   70   20
7   60  100   40
8   75  100   35
9   75   95   30
10  80   90   40
11
12
13
14
15
16
17
18
19
20
21
22
                                                       Enter Data   F10=Menu
```

Figure A-9-16

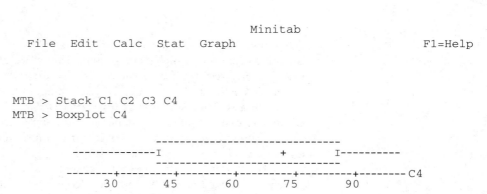

Figure A-9-17

The "By" Subcommand

Like stacking, the "By" subcommand can reduce repetitive tasks. If data is entered into one column on the data screen then the "By" subcommand can be used to separate the data into several groups for analysis. Use the "By" subcommand instead of re-entering the data into different columns on the data screen. In the interactive mode, most dialog boxes offer the "By" subcommand feature.

↓	C1	C2	C3	C4	C5	C6	C7	C8	C9	C10	C11	C12	C13	C14	C15
1	80	90	50	80											
2	70	80	60	70											
3	75	85	30	75											
4	70	85	30	70											
5	80	85	40	80											
6	50	70	20	50											
7	60	100	40	60											
8	75	100	35	75											
9	75	95	30	75											
10	80	90	40	80											
11				90											
12				80											
13				85											
14				85											
15				85											
16				70											
17				100											
18				100											
19				95											
20				90											
21				50											
22				60											

                                                    Enter Data   F10=Menu

Figure A-9-18 Data from C1, C2, and C3 stacked on C4.

Returning to the stacking example, all the scores can at first be entered into one column, C1. Next enter an identification code number for each student into column C2, next to his or her score. That is,

> 1 = Student under 20 years of age,
>
> 2 = Student 20 to under 30,
>
> 3 = Student at least 30 years of age.

The first 15 lines appear as shown in Figure A-9-19. The macro command

MTB > BOXPLOT C1

would then result in a single box plot of all the scores. On the other hand, the macro command and subcommand

MTB > BOXPLOT C1;

SUBC > By C2.

would produce three box plots. The values in column C2 define the three groups. Figure A-9-20 shows the output of this command.

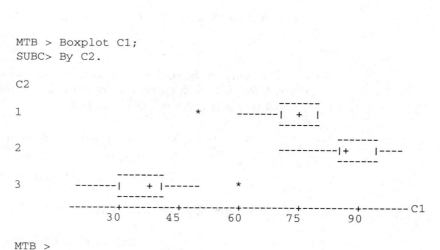

```
↓  C1  C2  C3  C4  C5  C6  C7  C8  C9  C10  C11  C12  C13  C14  C15
 1  70   1
 2  75   1
 3  68   1
 4  77   1
 5  79   1
 6  79   1
 7  80   1
 8  64   1
 9  55   1
10  70   1
11  98   2
12  90   2
13 100   2
14 100   2
15  92   2
```

```
                                                   Enter Data   F10=Menu
```

Figure A-9-19

```
                                          Minitab
      File   Edit   Calc   Stat   Graph                         F1=Help



MTB > Boxplot C1;
SUBC> By C2.

C2
                                            -------
1                                   *     -------| + |
                                            -------

                                                -------
2                                     ----------|+    |----
                                                -------

                 --------
3        -------|   +  |------      *
                 --------
        --------+---------+---------+---------+---------+---------+-------- C1
               30        45        60        75        90

MTB >
```
Figure A-9-20

A-9.6   ADDITIONAL MACRO COMMANDS

Some of the macro commands available on Minitab, along with their subcommands, follow in boldface type, listed in the order in which they were treated in this text. In this list, **K** denotes a constant, **C** denotes a column, **E** denotes either a constant or a column, and **[ ]** enclose an optional argument. Experiment with these commands independently. The *Minitab Mini-Manual* is a good source of further information on Minitab.

DOTPLOT C . . . C	Produces a dot plot for each column specified. (Use spaces, not commas, between column identifications.)
Increment = K	Specifies distance between tick marks.
Start = K [End = **K**]	Specifies first and last tick marks.
By C	Produces a dot plot for each column.
Same	Same scale used for all columns.
DESCRIBE C . . . C	Prints descriptive statistics for each column specified.
By C	
BOXPLOT C	Prints a box plot for the values of the variable in column C.
Increment = K	
Start = K	
By C	
Lines = K	Three lines are used to print each box plot, but this may be condensed by the subcommand "$K = 1$."
Notch [K%]	Produces a confidence interval.
Levels K . . . K	Used with the "By" subcommand; specifies which levels are to be used and in what order.
STEM-AND-LEAF C . . . C	Prints a stem-and-leaf display for each column specified.
Trim	Trims all values beyond the inner fences.
Increment = K	
By C	
HISTOGRAM C . . . C	Prints a separate histogram for each column specified.
Increment = K	
Start = K [End = **K**]	
By C	
Same	
TALLY C . . . C	Prints a one-way table for each column specified.
Counts	
Percents	
Cumcounts	Prints cumulative counts.
Cumpercents	Prints cumulative percents.
CDF	Cumulative distribution function.
Binomial $n = $ **K** $p = $ **K**	
Poisson $\mu = $ **K**	
Normal [$\mu = $ **K** [$\sigma = $ **K**]]	

INVCDF	Inverse function of CDF.
Binomial $n = K$ $p = K$	
Poisson $\mu = K$	
Normal $[\mu = K$ $[\sigma = K]]$	
STDEV C	Calculates the sample standard deviation for the column specified.
ZINTERVAL [K% confidence] $\sigma = K$ for **C . . . C**	Calculates a K% confidence interval for the mean, separately for each column specified. (Use spaces, not commas, between the parameters.)
TINTERVAL [K% confidence] for **C . . . C**	Calculates a K% confidence interval for each column specified.
ZTEST $[\mu = K]$ $\sigma = K$ for **C . . . C**	Performs a separate Z-test on the data in each column specified.
Alternative = K	Specifies a one-sided test. $K = -1$ gives H_0: $\mu < K$ and $K = +1$ gives H_0: $\mu > K$.
TTEST $[\mu = K]$ on **C . . . C**	Performs a separate t-test on the data in each column specified.
TWOSAMPLE	Performs a two- (independent) sample t-test and confidence interval.
Alternative = K	
Pooled	Executes the TWOSAMPLE command with a pooled sample standard deviation.

A p p e n d i x

A-10 User's Guide to the TI-85 Graphics Calculator

The TI-85 has many characteristics that are similar to those of the TI-82, which is illustrated throughout the text. There are also some differences between the two models, and this appendix is meant to bridge the gap, making users of the TI-85 comfortable with the calculator instructions in the text.

The TI-85 keyboard (Figure A-10-1) is similar to the TI-82 keyboard except that the STAT key is in a different place. Also, the F keys of the TI-85 are often used to make menu selections.

A-10.1 CLEARING THE STATISTICS MEMORY AND ENTERING DATA

Consider the set of scores {53, 18, 32, 17, 70, 41, 116, 12, 788, 70, 100, 68, 3, 27, 69}, which were part of Example 8 in Chapter 1 (page 28). To enter these scores into the TI-85 for statistical calculations, use the following steps.

• Press the STAT key.
• Activate the editing option by pressing the F2 key.

After these two keystrokes, the window shown in Figure A-10-2 appears.

Options for naming data lists are now available. "xStat" refers to each score in the data, and "yStat" to the frequency of each score. The default yStat frequency value is 1.

In this case the data lists will not be named, so press the ENTER key twice. Then, to clear the statistics memory of any previous data that may have been entered, press the F5 key. The screen now looks like Figure A-10-3.

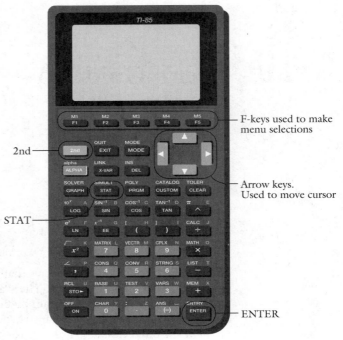

Figure A-10-1 The TI-85 keyboard.

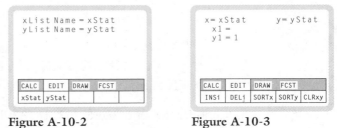

Figure A-10-2   **Figure A-10-3**

One may now enter the data

- Type in the first score, 53.
- Type [ENTER], [ENTER]. (The [ENTER] key is pressed twice, once to enter the score of 53 and once to enter the default frequency for that score, which is 1.)
- Enter the remaining 14 scores in the same manner.

A-10.2 SORTING SCORES

Once all the scores in a data set have been entered, execute the sorting option by pressing the [F3] key. For the data set just given, the result is as shown in Figure A-10-4. On this screen, x1 is the score with data rank 1, and it is equal to 3; x2

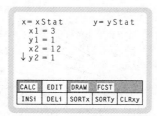

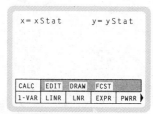

Figure A-10-4 Sorted data.   **Figure A-10-5**

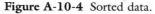

Figure A-10-6 Basic calculations.

is the score with data rank 2, and it is equal to 12. Scroll up and down this list by using the arrow keys in the upper right corner of the keypad.

A-10.3 BASIC CALCULATIONS

Consider the scores {1, 2, 3, 40, 40, 40, 40, 40, 53, 54, 101}, which were part of Example 3 in Chapter 3 (page 89). First clear the statistics memory and enter all the scores, using the method described in Section A-10.1. Then activate the calculation menu with the following keystrokes.

- Press the key labeled 2nd.
- Press the F1 key.
- Press the ENTER key twice.

The screen now looks like Figure A-10-5.

Now press the F1 key to display the calculations shown in Figure A-10-6.

A-10.4 DRAWING HISTOGRAMS

Consider the construction of a histogram for the data in Example 6 of Chapter 3 (page 103). For convenience, the data are repeated here.

720	610	700	520	760	440	380	600	720	600
610	800	320	460	580	610	780	770	760	570
640	520	600	480	500	520	680	780	780	680
660	580	490	620	520	780	700	680	760	520

- Clear the statistics memory and enter the data.

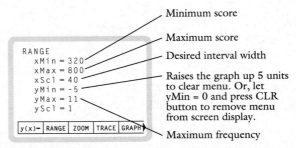

Figure A-10-7 Range specifications for drawing a histogram.

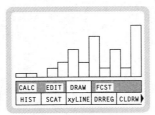

Figure A-10-8

When all the data have been entered make the following keystrokes.

- Press ⟨2nd⟩, ⟨EXIT⟩ to quit the statistics menu.
- Press the ⟨GRAPH⟩ key.
- Press the ⟨F2⟩ key.
- Use the arrow keys to complete the viewing range of the histogram, which should appear as shown in Figure A-10-7.

Finally,

- Press the ⟨STAT⟩ key.
- Press the ⟨F3⟩ key.
- Press the ⟨F1⟩ key.

The resulting histogram looks like Figure A-10-8.

A-10.5 CORRELATION AND LINEAR REGRESSION

One can easily use the TI-85 graphics calculator to find the equation of the line of best fit and the Pearson coefficient of correlation for the data in Example 1 of Chapter 14 (page 528). The data are repeated here.

Price per Barrel of Crude Oil, X	Price per Gallon of Gasoline, Y
3.35	.38
3.75	.40
4.29	.43
5.80	.57
18.00	.80
21.30	.85
30.00	.98
35.75	1.20
37.05	1.15
42.50	1.35
45.75	1.40
45.75	1.25

• Clear the statistics memory and enter the data.

The first four entries should appear as in Figure A-10-9. When the last score has been entered make the following keystrokes.

• Press the 2nd key.
• Press the F1 key.
• Press the ENTER key twice.
• Press the F2 key.

The calculations shown in Figure A-10-10 will then be displayed.

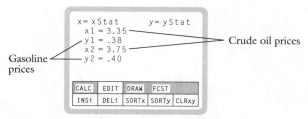

Figure A-10-9 Ordered pairs entered on a TI-85 graphics calculator.

An unfortunate inconsistency in notation exists between the TI-85 and the TI-82. Take care to remember that the statistic labeled b on the TI-85 calculator display (Figure A-10-10) is not the Y-intercept but the slope.

It is possible to calculate Spearman's rank-correlation coefficient by letting the X values represent one ranking and the Y values the other.

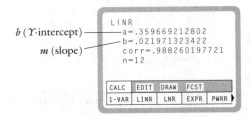

Figure A-10-10

Answers to Odd-Numbered Exercises

Chapter 1

Exercises 1.1–1.3

1. Yes. If the same score appears multiple times, it holds more than one data rank.
3. An array is simply a rectangular block of scores in no particular order. A stem-and-leaf display is a list that shows the scores in order from least to greatest.
5. Yes. Consider the following set of scores.

   30
   29
   25
    5
    2

   The median is 25, which is much closer to the maximum score, 30, than it is to the minimum score, 2.
7. (a) 11.08 seconds.   (b) The winning times are usually decreasing from one Olympiad to the next.
9. The median is .338. Puckett and Oliva (in 1971) came closest to batting the median. Players whose batting averages are in the top 20% are Brett, Carew (twice), Boggs (three times), and Olerud.
11. (a) The median yearly average is 7 h, 4 m; the median average for February is 7 h, 35 m; the median average for July is 6 h, 31 m.   (b) Answers may vary.
13. (a) Median = 35. About 42%.   (b) Oldest 10%: J. Tandy, K. Hepburn (three times), Geraldine Page and S. MacLaine. Youngest 10%: Marlee Matlin; Julie Christie and A. Hepburn (tied for second youngest); G. Kelly, B. Streisand, L. Minelli, and J. Foster (tied for third youngest); J. Woodward and S. Loren (tied for fourth youngest); and J. Holliday and E. Taylor (tied for fifth youngest).
15. (a) NC and NY. Answers will vary.   (b) NY and AR.

Exercises 1.4–1.5

1. A percentage is a number that stands by itself, whereas a percentile is a number that divides a data set into two parts.
3. Second.
5. It is not precise because quartiles are technically percentiles, which are single numbers. The probable meaning is that the score is among the lowest 25% of all scores.
7. As the number of scores increases, the number of possible values to correspond to different percentiles also increases. When scores repeat.
9. Answers will vary. Some possibilities follow.
   (a) A: 10, 20, 30, 40, 50, 60, 70, 80, 90, 100
       B: 1, 2, 3, 4, 5, 6, 7, 8, 9, 10
   (b) A: 10, 15, 20, 25, 30, 35, 40, 45, 50, 55
       B: 9, 15, 22, 25, 30, 35, 40, 45, 50, 55
   (c) A: 10, 15, 20, 21, 22, 23, 24, 40, 109, 110
       B: 0, 1, 5, 6, 7, 8, 9, 45, 46, 50
   (d) A: 0, 10, 20, 21, 22, 23, 24, 30, 40, 50
       B: 0, 10, 20, 21, 22, 23, 24, 60, 70, 100
11. (a) Singles (73,800), LPs/EPs (89,700), CDs (125,900), cassettes (363,250).   (b) 90th percentile; 70th percentile; 0 percentile.   (c) about 900,000.
13. Victoria, George III, Henry III, Edward III, Elizabeth I, Elizabeth II, Henry VI, Henry VIII, Henry I, Henry II, Henry V, Edward VII, William IV, Edward VI, Mary II, Mary I, James II, Richard III, Edward VIII, Edward V.
15. (a) Median = 3438, IQR = 5583 − 2188 = 3395, range = 36,555.   (b) More than ten times greater. Yes, and indicates the presence of extremely large and/or small scores.   (c) Yes, a few scores that are very large stretch this set of data in the direction of the higher values.

Exercises 1.6

1. Stanines are like deciles in the sense that they separate the data into groups. They are different from deciles in that they are sets of scores, whereas deciles are numbers.
3. Approximately 56%.
5. One possibility is to use a cutoff score of 71 or more for the higher-ability classes and a cutoff score of 19 or less for the lower-ability class.

7. (a) Top five: elementary school teacher, dental assistant, registered nurse, and data entry worker. Bottom five: carpenter, airline pilot, dentist, mail carrier, and physician. (b) Most consistent: dental assistant, registered nurse, social worker, bartender, college-level teacher, computer programmer, and dentist. Most variable: librarian.

9. (a) 51%. (b) No, because nearly as many states fail to rank similarly as rank similarly.

Chapter 2

Note: *All five-number summaries and box plots show percentiles that were computed by hand. If Minitab or another statistical software program is used to arrive at these percentiles, the answers may vary slightly from those shown here.*

Exercises 2.1–2.4

1. A few possibilities are the type of automobile a person drives, a person's profession, and the location of residence.

3. If the number of scores is divisible by 4 and if there are no repeated scores in key positions.

5.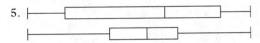

7. Dot plots would appear as follows:

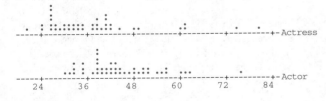

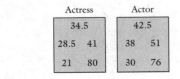

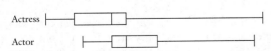

(a) 7 years. (b) 11.1%. (c) 88.9%. (d) E. Burstyn and K. Bates. (e) S. Hayward, L. Fletcher, J. Fonda, I. Bergman, Cher, S. Poitier, B. Kingsley, G. Hackman, J. Voight, L. Marvin, and Y. Brynner.

9. (a)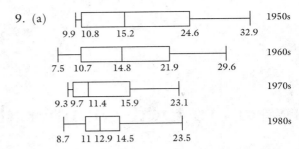

(b) No, they are decreasing. This may indicate keener competition for the top ratings. Also, the median rating is generally decreasing.

Exercises 2.4A

1. (a)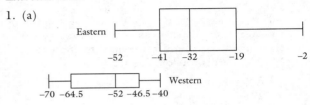

(b) AZ.

3.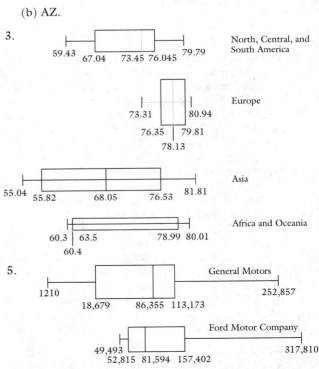

Exercises 2.5

1. Because the data are based on several samples that may or may not be typical of their populations.
3. Decreases.
5. The results might be different if students scored similarly on both sections.
7. (a)

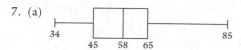

(b) [53.95, 62.05].

9. (a)

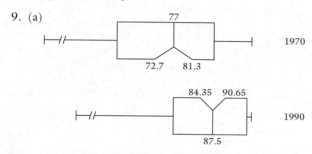

The difference in medians is statistically significant. (b) Kuwait. (c) There is less deviation than expected.

11.

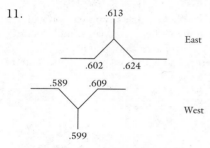

The difference in median winning percentages is not statistically significant and must be attributed to chance.
13. Not necessarily. A division with several very weak teams could produce a division champion with a winning percentage higher than that held by a division winner from an evenly balanced division.
15. (a) Panama. (b) West Germany. (c) Generally high.

Exercises 2.6

1. It is the largest (or smallest) score that is not questionable.
3. No. The placement of the fences depends upon the IQR.
5. Answers will vary.
7. Monthly sales revenues, the number of reservations made to an airline and so forth.

9. There are no possible outliers, but fried beef liver and a large cooked egg are probable outliers. The adjacent scores are 0 and 96.
11. None of these rates might be outliers if they are included in the group from Example 5.
13. (a) Geraldine Page and K. Hepburn won the Oscar at unusual ages, and Jessica Tandy won it at a rare age. (b) Henry Fonda won the Oscar at an unusual age; no one won at a rare age. (c) K. Hepburn (74), Henry Fonda (76), and J. Tandy (80) would be considered unusual for this combined group. There are no rare ages. These people were also in Parts (a) and (b).

15. (a)

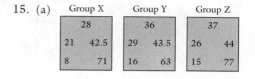

(b)

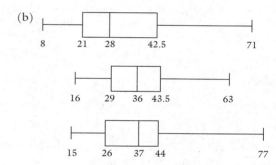

(c)

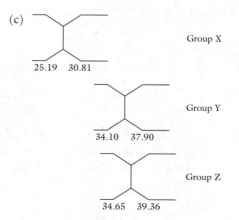

There is a statistically significant difference in medians between Group X and Group Y and between Group X and Group Z.

(d)

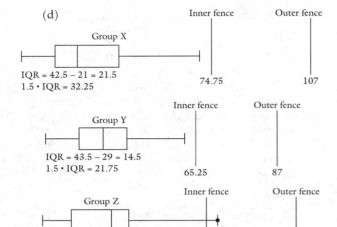

IQR = 42.5 − 21 = 21.5
1.5 · IQR = 32.25

IQR = 43.5 − 29 = 14.5
1.5 · IQR = 21.75

IQR = 44 − 26 = 18
1.5 · IQR = 27

Group Z's typing speed, 77 words per minute, is the only possible outlier. There are no probable outliers.

Chapter 3

All answers in this chapter have been rounded to the nearest hundredth.

Exercises 3.1–3.3

1. Because the value of the midrange would be greatly shifted in the direction of the outlier.
3. Answers may vary, but it should be clear that A has the smallest midrange and C has the largest midrange and largest mean.
5. Probably the mode.
7. There are many possibilities. One set of examples follows. (a) 3, 5, 10, 50, 60, 60, 60. (b) 48, 49, 49, 50, 60, 60, 60. (c) 40, 40, 40, 70, 71, 72, 73. (d) 40, 40, 40, 70, 190, 191, 192. (e) 1, 65, 65, 70, 71, 72, 73. (f) 1, 2, 3, 4, 5, 5, 90.
9. Length ≈ 168.92. Wheelbase ≈ 97.42. Width ≈ 65.96. The Nissan Sentra four-door, Geo Prizm sedan, or Volkswagen Jetta four-door.
11. (a) $\sigma = \sqrt{\dfrac{187.550}{20}} \approx 3.06$

(b) $\sigma = \sqrt{\dfrac{1813.566 - \dfrac{180.334^2}{20}}{20}} \approx 3.06$.

13. (a) True. (b) False. The value of the IQR depends upon the values of two quartiles. The remaining scores in the middle half of the data may or may not be clustered closely about the mean. (c) False. For example, the IQR was used earlier to determine outliers. (d) True. (e) False. The range does not take every single score into account.
15. Mean = 45.93, standard deviation = 7.83, range = 28. Yes.

Exercises 3.4

1. Years experience: $v = .80$. Salaries: $v = .30$. Years of experience are more variable.
3. If you are computing by hand, the variance is easier to compute than the standard deviation because it eliminates one step: finding the square root. No.
5. House spending was more variable.
7. Winning times for men are more variable.
9. (a) 1980, 1970, 1960, 1990. (b) 1960, 1970, 1980, 1990. (c) No.

Exercises 3.5

Due to differences in the rounding of interval widths, the histograms obtained for some of the exercises in this section may appear slightly different from those shown here.

1. Some examples are physical data from men and women, weather data from two different climates, and economic data from two very different populations.
3. It will make changes appear more gradual and smooth them out.
5.

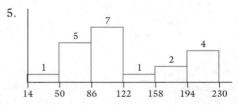

7.

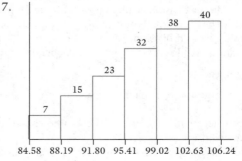

9.

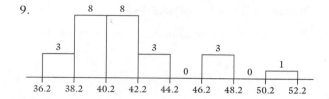

(b)

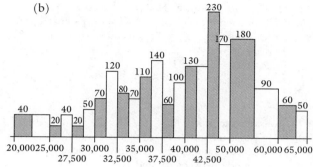

(a) Airplane pilot, structural metal worker, and operating engineer are in the youngest interval; barber is in the oldest interval. (b) The second interval, [38.2, 40.2), and the third interval, [40.2, 42.2) are the modal intervals. (c) The median and mean appear to be nearly the same—about 40 or 41.

11. Go to the "1-Var Stats" menu and use $minX = Xmin$ and $maxX = Xmax$.
13. Let $Xscl = 35$.
15. Let $Xmin = 35$, $Xmax = 52$, $Xscl = 2$, $Ymin = 0$, $Ymax = 30$, and $Yscl = 1$.

Exercises 3.6

1. (1) Do not omit data, and (2) avoid truncation if possible.
3. Space limitations or two axes with greatly different scales sometimes make truncated graphs necessary.
5. The area of the dollar bill in the second graph is four times the area of the dollar bill in the first graph because both the length and width were doubled. One way to correct this inaccuracy would be to double only one of the dimensions, not both. The new dollar bill would then look out of scale, however, so perhaps two dollar bills should be pictured for "This Year."

Chapter 4

Exercises 4.1–4.3

1. No, because only the midpoints of the intervals are used.
3. Estimated mean $= \Sigma m_i \cdot (\frac{f_i}{N}) = 4.5 \cdot (.25) + 7.5 \cdot (.025) + 10.5 \cdot (.175) + 13.5 \cdot (.175) + 16.5 \cdot (.125) + 19.5 \cdot (.10) + 22.5(.025) + 25.5 \cdot (.10) + 28.5 \cdot (.025) = 13.35$. Similarly, the estimated standard deviation is $\sqrt{\Sigma \frac{f_i}{N} \cdot m_i^2 - \mu^2} =$
$$\sqrt{.25 \cdot (4.5)^2 + .025 \cdot (7.5)^2 + .175 \cdot (10.5)^2 + \cdots}$$
$$\sqrt{+ (.025) \cdot (28.5)^2 - (13.35)^2}$$
$$\approx 7.03.$$
5. (a) \$39,750.

7. Estimated mean ≈ 50.44. Estimated standard deviation ≈ 27.70.
9. (a) Estimated mean ≈ 20.66. Estimated standard deviation ≈ 3.08. (b) The mean pay for the nation takes into account the relative size of each state and would be a weighted mean of the data presented in the histogram. The number of working people in each state.
11. (a) Estimated mean ≈ 17.33. Estimated standard deviation ≈ 12.68. (b) Approximately 93%.
    (c)

Accumulated Mileage	Tally
[40, 60)	24
[15, 40)	353
[5, 15)	150
[.5, 5)	52
[0, .5)	18

Increased.

13. (a) Declined: $\approx 24.59\%$. Increased: $\approx 71.29\%$.
    (b) Approximately 62.68%.
    (c) $-24.59\% \cdot \frac{408}{4542} + 71.29\% \cdot \frac{4134}{4542} \approx 62.68\%$.
15. (a) .72%. (b) Approximately .81%.

Exercises 4.4

1. Only if the histogram or frequency curve is mound-shaped; since it is not, use of the word "must" is not justified.
3. Chebyshev's Theorem. Consider that $(1 - \frac{1}{k^2}) = \frac{3}{4}$ if $k = 2$. Therefore, $\frac{3}{4}$ of the distribution will be within two standard deviations of the mean, which is less than 95% for the Empirical Rule. Thus, to enclose 95% of the distribution using Chebyshev's Theorem, more than two standard deviations will be needed, forming a wider interval than if the Empirical Rule was used.

5. (a) Estimated mean ≈ 12.38%. (b) [6.52, 18.23].
7. (a) Estimated mean ≈ 114.78 seconds. Estimated standard deviation ≈ 4.57 seconds. (b) [110.21, 119.35] seconds.
9. [$32,818.67, $52,881.33].

Exercises 4.5

1. Consider the scores [1, 7, 7, 11, 12, 13, 14, 16, 17, 18}. The actual mode is 7, but a histogram that shows these scores could appear as follows, indicating an estimated mode between 10 and 15.

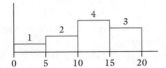

3. Estimated median ≈ 9.91%. Estimated mean ≈ 11.22%.
5. Yes; the mode.
7. Estimated mean ≈ $217.86 billion. Estimated median = $175 billion. Estimated mode ≈ $61.11 billion.
9. Estimated mean ≈ 6897.32. Estimated median = 175. Estimated mode ≈ 103.23.

Exercises 4.6

1. About 32%. About 5%.

3. (a)

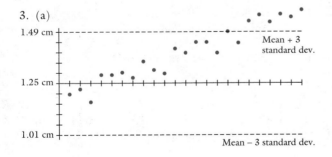

(b) The product is being manufactured by a machine that needs adjustment. The diameters tend to be too great.
5. P-charts would be most appropriate for manufactured items that either work or don't work, such as light bulbs, fuses, and various switches.

Cumulative Review, Chapters 1–4

1. No, not necessarily. Consider the following set of scores.

Stem	Leaf
19	0, 2, 2, 5
18	5
17	6, 7, 7, 9

The median is 185, which is in the shortest row.

3.

Stem	Leaf
2.8	1
2.7	1, 6
2.5	2
2.4	0, 3, 8
2.3	9, 4
2.2	1, 1, 2, 4, 5, 8, 8
2.1	8
2.0	7, 8
1.9	8
1.7	6
1.6	9
1.5	3

The median ERA is 2.25.
5. $P_{20} = 480$, $P_{40} = 580$, $P_{60} = 665$, $P_{80} = 735$. The range is 460, and the IQR is $710 - 500 = 210$.
7. The box plot would be greatly distorted. A dot plot would best show the distribution of these scores.
9. Hinges are medians of the upper and lower halves of the scores. They are slightly more difficult to determine than quartiles, because the half must first be identified and then the median of that group found. Hinges are often the same as quartiles, but, depending upon the nature of the data, may differ by one data rank.
11. The mode and the median.
13. Annual vehicle miles are more variable.
15. (a)

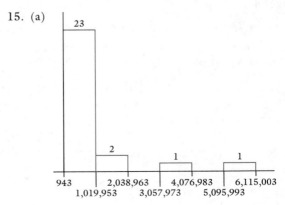

(b)

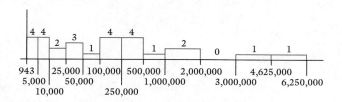

17. The estimated mean would change the most.
19. (a) [19.64, 45.66].   (b) 6.63 hours.

Chapter 5A

Exercises 5A.1–5A.2

1. $\dfrac{\text{Total number of “yes” answers minus } .25 \cdot P\%(1000)}{.50 \cdot P\%(1000)}.$
3. No.
5. Lower, because the denominator of the probability fraction is less.
7. (a) 80.   (b) 2 and 18, each with a probability equal to $\frac{1}{80}$.   (c) $P(6) = \frac{1}{16}$.   (d) $P(9) = \frac{1}{10}$.
9. $P(\text{Black}) = \frac{26}{51}$, $P(\text{Queen or jack}) = \frac{8}{51}$, $P(\text{Ace or red numbered card}) = \frac{21}{51}$. $P(\text{Ace or red numbered card})$ decreased.

Exercises 5A.3

1.

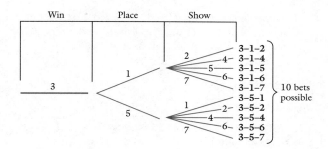

3. 72.
5. Four times.
7. $\frac{3}{50}$.
9. (a) $\frac{1}{16}$.   (b) $\frac{5}{8}$.   (c) $\frac{13}{14}$.   (d) $\frac{1}{3}$.   (e) $\frac{3}{4}$.   (f) $\frac{1}{3}$.
11. Type C is most likely to pay; type A is least likely to pay.
13. Answers will vary.

15.

```
10 Randomize timer
20 For n = 1 to 5000
30 Let X = Rnd
40 Let Y = Rnd
50 If X^2 + Y^2 < 1 then Shaded = Shaded + 1
60 Next n
70 Print "Probability of hitting the shaded region =" Shaded/5000
80 End
```

$P(\text{hitting shaded region}) = \frac{\pi}{4}.$

Exercises 5A.4–5A.5

1. $P(X \cup Y)$ could be greater than 1.
3. (a) $P(X \cap Y) = \frac{17}{50}$.   (b) $P(X \cup Y) = \frac{7}{9}$.   (c) $P(X) = .28$.   (d) The probabilities given in this part of the exercise cannot be related in the way shown.
5. .75.
7. .80.
9. Triplets are about 87 times less likely than twins; quadruplets are about 88 times less likely than triplets; and quintuplets are about 125 times less likely than quadruplets.
11. .0113673075
13. Genetic factors and the quality of prenatal care.
15. The set composed of $A \cup B \cup C$ is $X = \{2, 3, 4, 5, 6, 7, 8, 9, 10, 12, 13\}$. Let the set R be determined by first listing all of the elements in A or B or C, with their repetitions: $R = \{2, 3, 4, 8, 12, 2, 4, 5, 6, 7, 13, 3, 4, 7, 9, 10\}$. Then remove from R those elements in $A \cap B$, $A \cap C$, or $B \cap C$, which are $\{2, 4, 4, 7, 3, 4\}$. Finally, put back in R the element in $A \cap B \cap C$, which is $\{4\}$. The resulting set R is now the same as set X.

Exercises 5A.6

1. All three are always true. In Part c the numerator of both conditional probability fractions is $P(A \cap B) = 0$, and since $P(A) \neq 0$, and $P(B) \neq 0$, then both $P(A \mid B) = P(B \mid A) = 0$.
3. Not necessarily. There is no way that the ball can “remember” what has happened on the previous spins.

5. (a) The probabilities indicated along each branch would change:

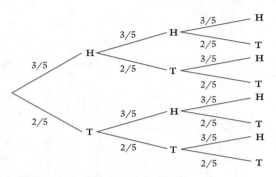

(b) $\frac{27}{125}$.   (c) $\frac{54}{125}$.   (d) $\frac{117}{125}$.

7. (a)

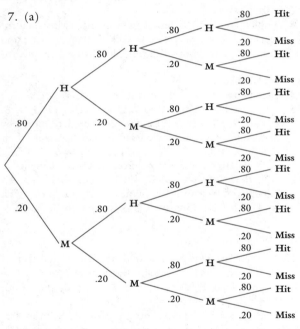

(b) .1536.   (c) Three hits or 4 hits are most likely.

9. (a) $P(B) = \frac{7}{20}$. $P(A \mid B) = \frac{4}{7}$. $P(B \mid A) = \frac{4}{15}$.   (b) $P(B) = \frac{3}{4}$. $P(A) = \frac{5}{6}$. $P(A \cup B) = \frac{11}{12}$.   (c) $P(A \cap B) = \frac{1}{100}$. $P(A \mid B) = \frac{1}{50}$. $P(A \cup B) = \frac{69}{100}$.   (d) $P(A \cup B) = 7k$. $P(A \mid B) = \frac{1}{3}$. $P(B \mid A) = \frac{1}{5}$.

11. (a) $\frac{2}{15}$.   (b) $\frac{1}{3}$.   (c) No, because together they do not include all possible outcomes.   (d) $\frac{2}{3}$.   (e) The probability for choosing a woman for the second is $\frac{4}{10} \cdot \frac{6}{9} + \frac{6}{10} \cdot \frac{5}{9} = \frac{54}{90} = \frac{3}{5}$, which is not the same as the answer to Part d.

Exercises 5A.7

1. $P(G) = \frac{5}{8}$, $P(G \cap H) = \frac{3}{5}$.
3. $P(L \cup P) = \frac{100}{400} + \frac{150}{400} - \frac{90}{400} = \frac{2}{5}$.
5. (a) $\frac{1}{4}$.   (b) $\frac{3}{25}$.   (c) $\frac{1}{5}$.   (d) $\frac{12}{35}$.   (e) $\frac{8}{15}$.

Exercises 5A.8

1. If a woman is chosen at random from a group of people, that occurrence does not in any way affect the probability that the person chosen is a parent. Thus, the two events—being a woman and being a parent—are independent. On the other hand, if a mother is chosen from the group, that does affect the probability that the person is a parent. Thus, the events of choosing a mother and choosing a parent are not independent.

3.

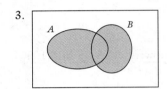

This is a necessary, but not a sufficient, condition for independence.

5.

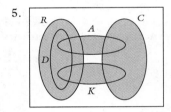

Possibly independent: D and A, D and K, C and A, C and K, A and R, K and R. *Definitely not independent:* D and C, D and R, C and R, A and K.

Chapter 5B

Exercises 5B.1

1. (a) No.   (b) 800.   (c) 8,000,000.
3. 6,392,000,000.
5. (a) True.   (b) False.   (c) False.   (d) True.
7. At least 18 people but no more than 28.
9. Although the probability of any single ticket's winning is practically zero, if enough tickets (25,827,165) with different number combinations are purchased, someone must win it.
11. Three-of-a-kind $(xxxyz) = {}_{13}C_1 \cdot {}_4C_3 \cdot {}_{12}C_2 \cdot {}_4C_1 \cdot {}_4C_1$. Flush (all the same suit) $= {}_4C_1 \cdot {}_{13}C_5$. Two pairs

($xxyyz$) = $_{13}C_2 \cdot {}_4C_2 \cdot {}_4C_2 \cdot {}_{11}C_1 \cdot {}_4C_1$. Two aces, two kings, one queen = $_4C_2 \cdot {}_4C_2 \cdot {}_4C_1$. One pair ($xxwyz$) = $_{13}C_1 \cdot {}_4C_2 \cdot {}_{12}C_3 \cdot ({}_4C_1)^3$. Three aces and two kings = $_4C_3 \cdot {}_4C_2$. All red cards = $_{26}C_5$. All cards of one color = $_2C_1 \cdot {}_{26}C_5$. Straight = $_{10}C_1 \cdot ({}_4C_1)^5$.

13. (a) 258,336,000. (b) 411,278,400. (c) 175,032,900.

15. Suppose that a given odds ratio is $\frac{a}{b}$, or $a{:}b$. Then ($a + b$) is the total number of possible outcomes, and the probability of the event happening is $\frac{a}{a + b}$.

17. To "even the odds" means to place a bet that has a probability of winning equal to $\frac{1}{2}$. To "beat the odds" means to win a bet on something that was unlikely to occur.

19. (a) 56.   (b) 1:55.

Exercises 5B.2

1. (a) 3.   (b) 30.   (c) 50,400.

3. 840.

5. (a) $\frac{23!}{14! \cdot 9!}$ = 817,190.   (b) $\frac{228690}{817190}$ ≈ .28.
   (c) .28 · .28 = .0784.

Exercises 5B.3–5B.4

1. The grand total. A row or column total.

3. (a) .24. (b) .34. (c) .65. (d) .67. (e) .61.
   (f) .29. (g) .71. (h) .42.

5.

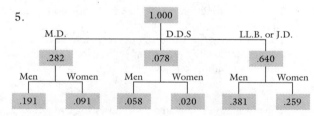

7. The given tree diagram might be more useful in situations that require an analysis of gender preferences for each brand. The diagram in Part d is more helpful for examination of the brand preferences for each gender.

9.

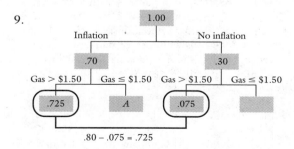

The probability estimates are not coherent.

Exercises 5B.5

1. Using the weighting factor
$$\left(\frac{\% \text{ of copies made by machine}}{\text{total } \% \text{ of copies made}}\right), .10 \cdot \left(\frac{.60}{1.00}\right)$$
$$+ .20 \cdot \left(\frac{.30}{1.00}\right) + .40 \cdot \left(\frac{.10}{1.00}\right) = .16.$$

3. If people are selected totally at random to be tested for the disease, then the group of participants is so large that even a highly accurate test may result in more people being mistakenly identified as having the disease than actually have it. This is the case when a relatively small part of the population being tested has the disease.

5. $\frac{11}{18}$.

7. (a) .921. (b) .875. (c) He is more likely to guess correctly. Because more items will be answered correctly due to guessing the overall probability of his actually knowing an answer that was marked correct diminishes.

9. (a) $\frac{1}{4}$. (b) $\frac{1}{15}$.

11. (a) .67. (b) .95.

Chapter 6

A hand-held calculator was used to solve most of the exercises in this chapter. If software programs such as Minitab are used instead, some answers may differ slightly due to rounding.

Exercises 6.1

1. No, it indicates that, of the traffic citations received, about 28% will be for parking.

3. $12,362.50.

5. (a)

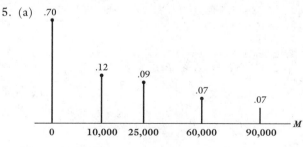

(b) μ_M = $9450.

Exercises 6.2–6.3

1. A Bernoulli process is related to a binomial random variable in the sense that a binomial random variable keeps track of the number of successes in a Bernoulli process.

3. (a) This is not a binomial random variable, because there are more than two possible outcomes (there are six) each time a die is thrown. (b) This is binomial. (c) This is not binomial, because the probability of receiving a ticket is most likely not the same for all trips. (d) This is binomial. (e) This is not binomial, because there are more than two possible outcomes (means). (f) This is binomial if the thousand parts are all assumed to have the same probability of being produced defectively.

5. There cannot be more successes than trials.

7. 45.

9. If the chances of the drug's helping all patients are not the same, then it is not valid to use the binomial random variable model.

11. (a) .2607. (b) .1815. (c) 1.

13. $\frac{5}{9}$.

15. Nine tanks should be sent.

Exercises 6.4

1. Using Part a as an example, $\mu = n \cdot p = 4 \cdot (.5) = 2$. $\mu = 0 \cdot (.0625) + 1 \cdot (.2500) + 2 \cdot (.3750) + 3 \cdot (.2500) + 4 \cdot (.0625) = 2$.

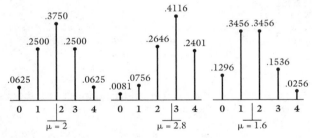

3. (a) .3487. (b) .0001. (c) .9999. (d) 9.

5. (a) 14.25. (b) 1.0. (c) .9638.

7. .2816.

9. $\Sigma_{s=0}^{4} P(B = s) = P(B = 0) + P(B = 1) + P(B = 2) + P(B = 3) + P(B = 4) = {}_4C_0 \cdot p^0 \cdot q^{4-0} + {}_4C_1 \cdot p^1 \cdot q^{4-1} + {}_4C_2 \cdot p^2 \cdot q^{4-2} + {}_4C_3 \cdot p_3 \cdot q^{4-3} + {}_4C_4 \cdot p^4 \cdot q^{4-4} = {}_4C_0 \cdot p^0 \cdot q^4 + {}_4C_1 \cdot p^1 \cdot q^3 + {}_4C_2 \cdot p^2 \cdot q^2 + {}_4C_3 \cdot p^3 \cdot q^1 + {}_4C_4 \cdot p^4 \cdot q^0 = \Sigma_{s=0}^{4} {}_4C_s \cdot p^s \cdot q^{4-s} = (p + q)^4 = (1)^4 = 1$.

Exercises 6.4A

1. A Poisson random variable is applied in situations that cannot be structured as repeated trials but in which the total number of successes for a given time or space can be determined. The total number of successes can then be divided by the appropriate time span or space restrictions to arrive at the average or mean value.

3. (a) 48 ties. (b) ≈ .0310. (c) .3513.

5. (a) .9554. (b) .2157. (c) .6010.

7. .9994.

9.

Number of Wrong Telephone Numbers, w	$P(W = w)$
8	.0655
9	.0874
10	.1048
11	.1144
12	.1144
13	.1056
14	.0905
15	.0724

Exercises 6.5

1. This is the same result that was obtained earlier.

3. (a)

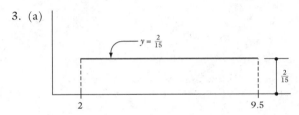

The density function is $y = \frac{2}{15}$. (b) $\frac{7}{15}$ for both probabilities. (c) $\frac{2}{15}$; 0. (d) 0; 0.

5. There are many possible examples. A few of them are postage stamp rates, fines for speeding, income tax rates, and sales commission rates.

Exercises 6.6

1. There are infinitely many possibilities for the mean, μ, and the standard deviation, σ. No table could contain all the possibilities.

3. (a) .7022. (b) .0811. (c) .7486.

5. For the normal probability distribution, $2 \cdot (.3413) = .6826 = 68.26\%$ of the area is within one standard deviation of the mean, $2 \cdot (.4772) = .9544 = 95.44\%$ is within two standard deviations, and $3 \cdot (.4986) = .9972 = 99.72\%$ is within three standard deviations.

7. *Between fences:* .74%. *Beyond outer fence:* 0%.

9. (a) .1308. (b) .5009. (c) .0934; .1587. (d) .7191. (e) .3047; .6915.

11. 1746 drives.

13. 96th percentile.

15. (a) $S ≈ 80.81$; $S ≈ 85.98$. (b) $S = 66.305$.

17. (a) Company A. (b) Company B. (c) Company B.

19. 9.248 ounces.

Chapter 7

Exercises 7.1–7.2

1. Answers will vary.
3. 3,190,187,286.
5. $\overline{X} = \$33,250$.
7. $\overline{X} \approx 1{,}119.83$ thousands.
9. There are many ways to do this. One way is to use a five-digit random number block and let the first two digits signify the number of the case to be tested. Then use the next two digits in that block to choose the number of a box in that case. Finally, use the first three digits of the next five-digit random number block to choose the number of the bulb to be tested from that box.

Exercises 7.3–7.4

1. (a) 10. (b) The calculated population parameters are $\mu = 10.00$ and $\sigma \approx 6.36$. A table of all the possible values of $\overline{X}$ and s for samples of size $n = 3$ follows.

Sample No.	Scores in Sample	$\overline{X}$	s
1	{2, 6, 10}	6.00	4.00
2	{2, 6, 11}	6.33	4.51
3	{2, 6, 21}	9.67	10.02
4	{2, 10, 11}	7.67	4.93
5	{2, 10, 21}	11.00	9.54
6	{2, 11, 21}	11.33	9.50
7	{6, 10, 11}	9.00	2.65
8	{6, 10, 21}	12.33	6.34
9	{6, 11, 21}	12.67	7.64
10	{10, 11, 21}	14.00	6.08

(c) No. (d) No.
3. True.
5. The estimated coefficient of variation for females is $\dfrac{s}{\overline{X}} \approx .01$. The estimated coefficient of variation for males is $\dfrac{s}{\overline{X}} \approx .03$. Based on these point estimates of the coefficient of variation, life expectancy is more variable for males in the population that these six countries represent.
7. It seems reasonable to assume that tire life is at least bell-shaped, because tires are mass-produced of identical materials on an assembly line. Thus, the factors that would affect tire life are the driving habits of the people who use the tires and the condition of their cars. Some drivers are very conservative and careful and would obtain greater mileage from a set of tires, while others are less careful and would obtain less mileage. It also seems reasonable to assume that the tires would be installed on some cars with excellent wheel alignment and on other cars with poor alignment.
9. $\overline{X} \approx 47.56$ cents, and $s \approx 26.08$ cents.

Cumulative Review, Chapters 5A–7

1. (a)

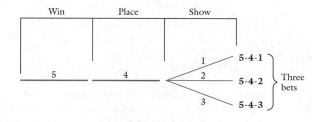

(b) $\frac{1}{20}$.
3. (a) $P(C) = P(D) = P(H) = P(S) = \frac{13}{52} = \frac{1}{4}$. $P(W) = \frac{8}{52} = \frac{2}{13}$. $P(X) = P(Y) = P(Q) = P(T) = \frac{2}{52} = \frac{1}{26}$. (b) $P(H \cup W) = \frac{19}{52}$. (c) $P(Z \cup W) = \frac{15}{26}$, $P(Z \cup C) = \frac{39}{52}$. (d) $P(K \cap W) = \frac{3}{26}$, $P(K \cup W) = \frac{41}{52}$, $P(K \cup H) = \frac{39}{52}$, $P(K \cup S) = 1$.
5. .5177.
7. 32.
9. (a) 360. (b) 105. (c) 69,300.
11. .39.
13. (a) 300. (b) .1209. (c) .2318. (d) .5941.
15. (a)

Probability density function: $y = .06$

$h = \frac{1}{15.7}$

2.3   18

(b) .4578; .3439. (c) $P(C = 4) = 0$, and the density at $C = 4$ is equal to $\frac{1}{15.7} \approx .06$.
17. The 1970 draft lottery could have been conducted using a random number table by simply assigning to each possible birth date one of the numbers from 1 through 366: January 1 = 001, January 2 = 002, January 3 = 003, . . . , December 31 = 366. Pages of random numbers could have been chosen and the first three digits used in an arbitrary pattern to determine the draft order. The reason this procedure was not followed could have been that the public would not have understood how a random number table is used, or that the procedure was not dramatic enough.

Chapter 8

A hand-held calculator was used to solve most of the exercises in this chapter. If a software program such as Minitab is used instead, some answers may differ slightly due to rounding.

Exercises 8.1–8.2

1. For a given sample size, say $n = 30$, with $N = 100$, then $\sqrt{\frac{N-n}{N-1}} \approx .84$; if $N = 500$, then $\sqrt{\frac{N-n}{N-1}} \approx .97$; if $N = 750$, then $\sqrt{\frac{N-n}{N-1}} \approx .98$. From this it can be seen that as N increases the finite correction factor becomes closer to 1.

3. The notations SE and $\sigma_{\overline{X}}$ stand for the standard error of a sampling distribution and are equal to $\frac{\sigma}{\sqrt{n}}$. When σ, the standard deviation of the population, is unknown, then $S_{\overline{X}}$ represents the estimated value of the standard error, found with the formula $S_{\overline{X}} = \frac{s}{\sqrt{n}}$.

5.

Sample No.	Scores in Sample	$\overline{X}$	s
1	{30, 40, 60}	43.33	15.28
2	{30, 40, 80}	50.00	26.46
3	{30, 40, 85}	51.67	29.30
4	{30, 40, 100}	56.67	37.86
5	{30, 60, 80}	56.67	25.17
6	{30, 60, 85}	58.33	27.54
7	{30, 60, 100}	63.33	35.12
8	{30, 80, 85}	65.00	30.41
9	{30, 80, 100}	70.00	36.06
10	{30, 85, 100}	71.67	36.86
11	{40, 60, 80}	60.00	20.00
12	{40, 60, 85}	61.67	22.55
13	{40, 60, 100}	66.67	30.55
14	{40, 80, 85}	68.33	24.66
15	{40, 80, 100}	73.33	30.55
16	{40, 85, 100}	75.00	31.22
17	{60, 80, 85}	75.00	13.23
18	{60, 80, 100}	80.00	20.00
19	{60, 85, 100}	81.67	20.21
20	{80, 85, 100}	88.33	10.41

(a) *Characteristic 1.* The mean of this population is
$$\mu = \frac{30 + 40 + 60 + 80 + 85 + 100}{6}$$
$$= \frac{43.33 + 50.00 + 51.67 + \cdots + 88.33}{10}.$$
Characteristic 2. $\sigma \approx 24.90$. $SE = 11.14 \approx \frac{\sigma}{\sqrt{n}} \cdot \sqrt{\frac{N-n}{N-1}} = \frac{24.90}{\sqrt{3}} \cdot \sqrt{\frac{6-3}{6-1}}$. (b) The sample that yields the value of s closest to σ is the sample {40, 80, 85}. No.

Exercises 8.3–8.5

1. A large sample is one in which $n \geq 30$. A relatively large population satisfies the inequality $n < .05\ N$.

3. If the cars were not of the same size, there could be several quite different modes, resulting in a distribution that was not normal or not even roughly mound-shaped.

5. (a) [103.65, 115.35]   (b) [104.42, 114.58]   (c) [106.28, 112.72].

7. (a) True.   (b) True.   (c) True.   (d) False.   (e) False.

9. It would be greater.

11. [628.11, 690.55].

13. (a) [7.59, 9.91].   (b) [7.93, 9.57].   (c) [7.24, 10.26]. (d) [8.00, 9.50].   (e) [8.27, 9.23].   (f) [6.95, 10.55].

15. (a) The population consists of the resident abortion rates of all 50 states; μ represents the mean resident abortion rate of all 50 states.   (b) [19.80, 25.70]. (c) Yes. There appear to be a few states in the sample that have very low rates (WY and SD) and a few states with very high rates (NY and CA), with the rest fairly evenly spaced out between them.

17. 80%: [3:33.08, 3:33.44].   90%: [3:33.03, 3:33.49]. 98%: [3:32.93, 3:33.59].

19. (a) [24,934.25, 28,918.01]. (b) [24,829.72, 28,520.35].

Exercises 8.6

1. Answers will vary.

3. (a) 92.68% wider.   (b) Yes.

5. [17.46, 28.04].

Exercises 8.7

1. The population is usually much too large to survey completely.

3. This exercise may be started as follows:

	①	②	3	4	⑤	6	⑦	⑧	P
1	×	×	×	×	×				3/5 = .60
2	×	×	×	×			×		2/5 = .40
3	×	×	×	×				×	3/5 = .60
4	×	×	×	×				×	3/5 = .60
55		×	×	×	×			×	2/5 = .40
56			×	×	×	×		×	3/5 = .60

The numbers 1, 2, 5, 7, and 8 were randomly chosen. The pattern of X's represents different choices of five of the eight numbers in the population. P is determined as the percentage of $\{1, 2, 5, 7, 8\}$ in each of the 56 different choices of five of eight numbers. One can verify that there are ten choices (patterns) that lead to $P = \frac{2}{5}$, 30 choices that lead to $P = \frac{3}{5}$, 15 choices that lead to $P = \frac{4}{5}$, and one choice that leads to $P = \frac{5}{5}$.

5. (a) π would represent the percentage of all customers buying this new detergent who prefer it to their old ones.   (b) [.513, .601].

7. (a) *ME* for Kennedy = 2.5%; *ME* for Johnson = 2.7%; *ME* for Humphrey = 2.5%; *ME* for McGovern = 2.5%.   (b) Percentage of confidence, P, and sample size.

9. If fewer people were surveyed, *ME* would have to increase. $Z(P)$, which determines the percentage of confidence, would not have to change.

Exercises 8.8

1. Decrease.

3. As the sample size, n, increases, the value of *SE* decreases because *SE* is found by dividing s by $\sqrt{n}$. Since the end points of the confidence interval are found by evaluating $\overline{X} \pm Z(\overline{X}) \cdot SE$, the resulting confidence interval becomes narrower.

5. Using Minitab, the confidence interval is (585.7, 659.3). This confidence interval is slightly different because the TINTERVAL command uses a *t*-distribution model, and Example 3 used a normal distribution model. However, for $40 - 1 = 39$ degrees of freedom these two distributions are almost identical which is why the two confidence intervals are only slightly different.

7. [82.30, 90.06].

9. (a) The population is the U.S. birth rates for each of the years since 1910.   (b) Since the sample size is small, to use the *t*-distribution the population must be normally distributed with unknown standard deviation.   (c) [18.05, 21.25].

Exercises 8.9

1. (a) [818.27, 1770.07].   (b) [945.57, ∞).   (c) Assumption 1 assures that the sample is fairly random, and assumption 2 justifies the use of the *t*-distribution model. If these two assumptions were not made, Chebyshev's Theorem would have to be used to construct the confidence interval in Part a.

3. [7.40, ∞) minutes.

5. [6.39, ∞) minutes.

Chapter 9

Exercises 9.1–9.3

1. The level of significance, or size, of the hypothesis test.

3. The sample size.

5. Statistical evidence can support only the alternative hypothesis. Therefore, each of the following statements, which corresponds to H_A in each part of Exercise 4, could be supported by statistical evidence.

   a. The mean height of NBA players is not 6 feet 7 inches.

   b. The mean salary of major league baseball players is not $875,000.

   c. The mean salary for professional football players is different from the mean salary for professional baseball players.

   d. The mean price for minivans in the Chicago area is not $16,775.

   e. The gasoline additive does change mileage.

   f. Women do have different SAT scores from men.

Exercises 9.4–9.6

1. Luck plays a part in the results of a hypothesis test because the chance selection of scores in the sample determines the sample mean, and the nearness of the sample mean to the claimed population mean ultimately determines whether or not the null hypothesis is rejected.

3. The outer tail or tails, which are so far from the claimed value of μ that any test statistic that reaches that region results in the rejection of the null hypothesis.

5. To "nullify" a statement or claim is to disprove it. The null hypothesis is chosen as the opposite of what one wants to prove. It is hoped that the null hypothesis will be rejected (or nullified), establishing the alternative hypothesis.

7. One specifying a 10% level of significance.

9. A relatively small P value.

11. $\alpha = 1.5\%$.

13. Minitab's ZTEST requires the user to input the population standard deviation; TTEST does not.

15. $H_0: \mu = 25$; $H_A: \mu \neq 25$. Reject H_0 and accept H_A if $Z(\overline{X}) > 1.64$ or if $Z(\overline{X}) < -1.64$; otherwise, reserve judgment. $Z(\overline{X}) \approx 1.70$. Therefore, reject H_0 and accept H_A. There is statistical evidence that the process is not running correctly.

17. (a) Because the population is normally distributed with known standard deviation.   (b) $H_0: \mu = \$19,750$; H_A:

$\mu \neq \$19{,}750$. Reject H_0 and accept H_A if $Z(\overline{X}) > 1.96$ or if $Z(\overline{X}) < -1.96$; otherwise, reserve judgment. $Z(\overline{X}) \approx -1.09$. Reserve judgment.

19. H_0: $\pi = .44$; H_A: $\pi \neq .44$. Reject H_0 and accept H_A if $Z(P) > 1.64$ or if $Z(P) < -1.64$; otherwise, reserve judgment. $Z(P) \approx 1.14$. Reserve judgment. The veterinarian's sample does not provide enough statistical evidence to challenge the journal's statement.

21. H_0: $\pi = .50$; H_A: $\pi \neq .50$. Reject H_0 and accept H_A if $Z(P) > 1.96$ or if $Z(P) < -1.96$; otherwise, reserve judgment. $Z(P) = 8$. Reject H_0 and accept H_A. There is statistical evidence to support the claim that gender is a factor.

23. H_0: $\mu = .25$; H_A: $\mu \neq .25$. Reject H_0 and accept H_A if $t(\overline{X}) > 1.711$ or if $t(\overline{X}) < -1.711$. $t(\overline{X}) \approx 1.70$. Therefore, reserve judgment. There is not enough statistical evidence to support the alternative hypothesis that the process is not running correctly.

25. (a) The number of people injured in automobile accidents in cities with populations of 25,000 to 50,000 in large states. (b) H_0: $\mu = 500$; H_A: $\mu \neq 500$. Reject H_0 and accept H_A if $t(\overline{X}) > 2.160$ or if $t(\overline{X}) < -2.160$; otherwise, reserve judgment. $t(\overline{X}) \approx -2.39$. Reject H_0 and accept H_A. There is statistical evidence that the mean is not 500. (c) $p = .032$.

27. H_0: $\mu = 6.0$; H_A: $\mu \neq 6.0$. Reject H_0 and accept H_A if $t(\overline{X}) > A$ or if $t(\overline{X}) < -A$. A is determined as shown.

α	.1%	.5%	1%	5%	10%
A	3.883	3.174	2.861	2.093	1.729

$t(\overline{X}) \approx -1.80$. Therefore, reject H_0 and accept H_A at the 10% level of significance, but do not reject H_0 at the .1%, .5%, 1%, and 5% levels of significance.

29. H_0: $\mu = 12$; H_A: $\mu \neq 12$. Reject H_0 and accept H_A if $t(\overline{X}) > 2.145$ or if $t(\overline{X}) < -2.145$; otherwise, reserve judgment. $t(\overline{X}) \approx -1.31$. Therefore, reserve judgment.

Exercises 9.7

1. In a one-tail hypothesis test, the rejection region is concentrated in one end of the distribution model. This configuration moves the end point of the rejection region on that side closer to the claimed value of μ.

3. H_0: $\mu \geq \$19{,}750$; H_A: $\mu < \$19{,}750$. Reject H_0 and accept H_A if $Z(\overline{X}) < -1.88$; otherwise, reserve judgment. $Z(\overline{X}) \approx -1.09$. Reserve judgment. There is not enough statistical evidence to support the claim that the mean was less than $19,750.00.

5. H_0: $\mu \geq \$170$; H_A: $\mu < \$170$. Reject H_0 and accept H_A if $t(\overline{X}) < -1.383$; otherwise, reserve judgment. $t(\overline{X}) \approx -1.60$. Reject H_0 and accept H_A. The claim

that the average state benefit is less than $170 can be supported at the 10% level of significance.

7. H_0: $\mu \geq 3.75$; H_A: $\mu < 3.75$. Reject H_0 and accept H_A if $t(\overline{X}) < -1.729$; otherwise, reserve judgment. $t(\overline{X}) \approx -.08$. Reserve judgment. There is not enough statistical evidence to support the alternative hypothesis that the population mean is less than 3.75 days.

9. H_0: $\mu \geq 75$; H_A: $\mu < 75$. *For $\alpha = 10\%$:* Reject H_0 and accept H_A if $Z(\overline{X}) \leq -1.64$; otherwise, reserve judgment. $Z(\overline{X}) \approx -2.26$. Reject H_0 and accept H_A. There is statistical evidence to support the drug's claimed reaction time. *For $\alpha = 5\%$:* Reject H_0 and accept H_A if $Z(\overline{X}) < -1.96$. Since $Z(\overline{X}) \approx -2.26$, reject H_0 and accept H_A. There is statistical evidence to support the drug's claimed reaction time.

11. Let A represent the maximum amount of postage that could be spent by Representative A. Then $A < \$94{,}606.70$.

13. At the 5% and 10% levels of significance, statistical evidence supports the claim that the mean price for that minivan is less than $16,500. At the 1% level of significance, judgment would be reserved.

15. 33.

17. (a) The sales of America's 100 fastest growing companies in the four quarters dated from August 9, 1993. (b) H_0: $\mu \geq \$600$ million, H_A: $\mu < \$600$ million. At the 2% level of significance there is not enough statistical evidence to support the alternative hypothesis.

19. (a) H_0: $\mu \leq \$50$, H_A: $\mu > \$50$. At the 10% level of significance there is not enough statistical evidence to support the alternative hypothesis. (b) Under this null hypothesis the rejection region would be in the left tail. However, the test statistic is positive, which makes it impossible for the test statistic to fall in the rejection region.

Chapter 10

Exercises 10.1–10.4

1. That the two populations are normally distributed and that they have approximately equal standard deviations.

3. Both samples are small, with unknown (but approximately equal) standard deviations.

5. No.

7. True.

9. A significance level between $\alpha = .05$ and $\alpha = .025$—say, $\alpha = .0375$.

11. 78%.

13. 50%:   [6.47, 6.73].   75%:   [6.37, 6.83].   95%: [6.21, 6.99].

15. H_0: $\mu_A - \mu_B = 0$; H_A: $\mu_A - \mu_B \neq 0$. Reject H_0 and accept H_A if $Z(\overline{X}_A - \overline{X}_B) < -1.64$ or if $Z(\overline{X}_A - \overline{X}_B) > 1.64$; otherwise, reserve judgment. $Z(\overline{X}_A - \overline{X}_B) \approx 5.88$. Reject H_0. There is enough statistical evidence to support the manager's belief.

17. (a) [.32, 2.68].   (b) H_0: $\mu_A - \mu_B = 0$; H_A: $\mu_A - \mu_B \neq 0$. Reject H_0 and accept H_A if $Z(\overline{X}_A - \overline{X}_B) < -1.96$ or if $Z(\overline{X}_A - \overline{X}_B) > 1.96$; otherwise, reserve judgment. $Z(\overline{X}_A - \overline{X}_B) \approx 2.09$. Reject H_0 and accept H_A. At the 5% level of significance there is statistical evidence that the mean numbers of knots per board on the two freight cars are different.   (c) Reject H_0 and accept H_A if $Z(\overline{X}_A - \overline{X}_B) > 2.33$; otherwise, reserve judgment. $Z(\overline{X}_A - \overline{X}_B) = 2.09$. Reserve judgment. There is not enough statistical evidence at the 1% level of significance that the mean number of knots per board in car A is more than that in car B.

19. [$2.29, $2.71].

21. [$2.78, $3.22].

23. [$0.45, $0.75].

25. It would make the confidence intervals narrower.

27. H_0: $\mu_N - \mu_O = 0$; H_A: $\mu_N - \mu_O \neq 0$. Reject H_0 and accept H_A if $t(\overline{X}_N - \overline{X}_O) > 1.697$ or if $t(\overline{X}_N - \overline{X}_O) < -1.697$; otherwise, reserve judgment. $t(\overline{X}_N - \overline{X}_O) \approx 4.19$. Reject H_0 and accept H_A. At the 10% level of significance there is statistical evidence that the new process changes the expected tire life.

29. Using Minitab, 50%: [20, 60]. 75%: [6, 74]. 80%: [2, 78]. 90%: [−9, 89]. 95%: [−19, 98]. 99%: [−38, 118]. Confidence intervals with both end points positive imply a difference in mean SAT scores for the classes of 1990 and 1993. Confidence intervals with one end point negative and the other positive imply that there is no difference in the mean SAT scores for the classes of 1990 and 1993.

Exercises 10.5

1. (a) It is possible that there is no difference between the percentages of employees who favor the new package and who favor the old one.   (b) Lowered.   (c) Yes, if the sample size were increased enough. If so, then *SE* would become much smaller, and since the middle of the confidence interval, $P_{New} - P_{Old}$, is positive, a smaller amount, $Z \cdot SE$, would be subtracted from it, so the left end point would be positive.

3. 60%:   [.007, .061].   80%:   [−.007, .075].   95%: [−.029, .097]. Since 0 is in the last two confidence intervals, the two population proportions do not differ at these levels of confidence.

5. Any percentage confidence interval less than 19.74% has both end points positive and does not contain 0. However, a confidence interval with such a low percentage of confidence has little or no practical value.

Exercises 10.6

1. Answers will vary.

3. Because matched-pair sampling eventually simplifies to a simple random sample taken from a single population and is then treated with the ZINTERVAL, TINTERVAL, ZTEST, or TTEST command.

5. 80%: [$3083.48, $3743.68].   90%: [$2983.20, $3843.95].

7. H_0: $\mu_D = 0$; H_A: $\mu_D \neq 0$. Reject H_0 and accept H_A if $t(\overline{D}) > 2.201$ or if $t(\overline{D}) < -2.201$; otherwise, reserve judgment. $t(\overline{D}) \approx -1.13$. Reserve judgment. There is not enough statistical evidence at the 5% level of significance to support the claim that the mean salary for all 50 governors is different from the mean salary for all 50 chief justices.

9. H_0: $\mu_D = 0$; H_A: $\mu_D \neq 0$. Reject H_0 and accept H_A if $t(\overline{D}) > 1.372$ or if $t(\overline{D}) < -1.372$; otherwise, reserve judgment. $t(\overline{D}) \approx 3.51$. Reject H_0 and accept H_A. Statistical evidence supports the claim that a difference exists between approval ratings and performance ratings.

Exercises 10.7

1. When the populations are not necessarily normally distributed or do not have approximately equal standard deviations.

3. [$1,695,040, $6,216,730].

5. [2, 7]. There is a difference in the population medians, and one may be 95% confident that it is between 2 and 7 minutes.

Exercises 10.7A

1. The special case of small-sample confidence intervals requires that the two populations be normally distributed, but the Mann-Whitney technique does not.

3. Wider, because the formula for degrees of freedom will result in a lower degree of freedom than if the population standard deviations are assumed to be approximately equal.

5. [1.08, 8.92].

7. [14.31, 56.89].

9. H_0: $\sigma_1 = \sigma_2$; H_A: $\sigma_1 \neq \sigma_2$. Reject H_0 and accept H_A if $F > 3.18$; otherwise, reserve judgment. $F \approx 55.253$. Reject H_0 and accept H_A. There is statistical evidence

at the 10% level of significance that the two population standard deviations are not approximately the same.

11. The rejection region seems to have only one tail (an upper one) because the calculated value of the F-statistic, $F = \left(\frac{s_1}{s_2}\right)^2$, with $s_1 > s_2$, is always greater than 1. Thus, very small values of F, which would fall in the lower tail, cannot occur.

Cumulative Review, Chapters 8–10

1. (a) A population mean is the mean of a population, such as the mean income in a certain city or the mean gas mileage of a particular type of automobile. A population proportion is the percentage of the population that has a certain characteristic or characteristics, such as the percentage of the voting-age population of the city of Chicago who are registered Democrats. A population proportion is always between 0 and 1, whereas a population mean has no such restriction.
(b) The standard deviation is a measure of dispersion of individual scores in a set of data. The standard error is a measure of the dispersion of a set of means. The standard error is a function of the standard deviation.
(c) Both the normal distribution and the t-distribution are mound-shaped distribution models. The t-distribution nears the normal distribution as the sample size increases, and for sample sizes of 30 or more the two distributions are practically the same.

3. (a) [$47,995.96, $59,098.04].   (b) [$49,283.39, ∞).

5. A hypothesis test provides statistical evidence that supports a claim (the alternative hypothesis) through the rejection of the opposite claim (the null hypothesis). Thus, to support the statement S, the null hypothesis must be the opposite of S and be rejected.

7. No.

9. H_0: $\mu \leq 78$ inches; H_A: $\mu > 78$ inches. Reject H_0 and accept H_A if $t(\overline{X}) > 1.796$; otherwise, reserve judgment. $t(\overline{X}) \approx 1.90$. Reject H_0 and accept H_A. At the 5% level of significance, these cards provide statistical evidence that the mean height is more than 78 inches.

11. H_0: $\pi \leq .500$; H_A: $\pi > .500$, where π represents the percentage of women voters who favor the candidate. Reject H_0 and accept H_A if $Z(P) > 1.64$; otherwise, reserve judgment. $Z(P = 2.8$. Reject H_0 and accept H_A. At the 5% level of significance there is statistical evidence that supports the candidate's claim.

13. The two populations are assumed to be normally distributed and to have approximately equal standard deviations. The Mann-Whitney nonparametric approach for

the difference of two population medians is used when one or both of these assumptions do not seem valid.

15. [−2,651.04, 26,360.32]. Since 0 is in this confidence interval, the difference in mean incomes is not significant and may be attributed to the chance selection of incomes.

17. 50%:   [.027, .061].   75%:   [.015, .073].   85%: [.008, .080]. 90%: [.003, .085]. 95%: [−.005, .093]. 99%: [−.020, .108]. Since 0 is in the last two intervals, there is no difference between the numbers of cavities associated with the two brands at those levels of confidence.

Chapter 11

Exercises 11.1–11.2

1. The χ^2 statistic is computed by summing squared differences, and any number squared is either zero or positive.

3. It reduces the value of the χ^2 statistic.

5. If gender and type of traffic violation are not statistically independent, then (for example) males are more likely to obtain speeding tickets than women. On the other hand, perhaps police are simply biased toward men (or women) in relation to certain types of offenses.

7. Cells (C3, Row 1), (C3, Row 3), and (C4, Row 4). Yes, cell (C4, Row 4).

9. H_0: gender and tea preference are independent; H_A: gender and tea preference are not independent. Reject H_0 and accept H_A if $\chi^2 > 6.064$; otherwise, reserve judgment. $\chi^2 = 22.271$. Reject H_0 and accept H_A. Evidence exists to support the alternative hypothesis.

11. (a) H_0: type of injury and state are independent; H_A: type of injury and state are not independent. Reject H_0 and accept H_A if $\chi^2 > 8.558$; otherwise, reserve judgment. $\chi^2 = 165.246$. Therefore, reject H_0 and accept H_A. Evidence exists to support the alternative hypothesis.   (b) The numbers are so large that the value of the chi-square statistic is almost certain to exceed that stated in the decision rule.

13. H_0: mode of travel is independent of the decade; H_A: mode of travel is not independent of the decade. Reject H_0 and accept H_A if $\chi^2 > 13.362$; otherwise, reserve judgment. $\chi^2 = 252.158$. Reject H_0 and accept H_A. Evidence exists to support the alternative hypothesis.

15. H_0: level of career commitment is independent of educational level; H_A: level of career commitment is not independent of educational level. Reject H_0 and accept H_A if $\chi^2 > 5.991$; otherwise, reserve judgment. $\chi^2 = 33.319$.

Reject H_0 and accept H_A. Evidence exists to support alternative hypothesis.

Exercises 11.3

1. Answers will vary.
3. After a certain point, higher values of s^2 are less likely to be obtained.
5. 50%: area to the right of $A = .75$; area to the right of $B = .25$. 70%: area to the right of $A = .85$; area to the right of $B = .15$.
7. (a) The value of s is computed from a simple random sample, and the population is normally distributed.   (b) [1.01, 3.26].   (c) She might conclude that there is less variation than expected in the mean time a person must wait.
9. (a) [.004498, .01126] mg.   (b) No. The value of .007 should have been the upper limit of the confidence interval in order for the machine to be considered safe to use.

Exercises 11.4

1. Some intervals had to be combined.
3. H_0: the GPAs for the school in question are normally distributed with a mean of 2.85 and a standard deviation of 1; H_A: the GPAs are not normally distributed with a mean of 2.85 and a standard deviation of 1. Reject H_0 and accept H_A if $\chi^2 > 7.779$; otherwise, reserve judgment. $\chi^2 \approx 59.62$. Therefore, reject H_0 and accept H_A. Evidence exists to support the alternative hypothesis.
5. (a) Combining the first two rows and the last three rows produces the following:

Interval	f_o	f_e
[40, 60)	24	9.20
[20, 40)	108	$.3936 \cdot (597) = 234.98$
[15, 20)	245	$.1897 \cdot (597) = 113.25$
[10, 15)	120	$.1747 \cdot (597) = 104.30$
[5, 10)	30	$.1154 \cdot (597) = 68.89$
[1, 5)	30	$.0575 \cdot (597) = 34.33$
[0, 1)	40	32.05

H_0: the gasoline mileages for this fleet of cars are normally distributed with unknown (estimated) mean and standard deviation; H_A: the gasoline mileages are not normally distributed. Reject H_0 and accept H_A if $\chi^2 > 5.989$; otherwise, reserve judgment. $\chi^2 \approx 272.53$. Therefore, reject H_0 and accept H_A. Evidence exists to support the alternative hypothesis.   (b) The attained significance level must be extremely low.

7. (a) Moving down the first column and then down the second, these lengths (in years) may be computed: {9, 24, 17, 13, 21, 17, 22, 16, 23, 21, 17, 16, 21, 19, 17, 14, 12, 11, 21, 23, 14, 20, 17, 21, 22, 8, 19, 18, 15, 22, 20, 17, 23, 18, 13, 9, 17, 23, 21, 20, 19, 18, 13, 21}.

(b)

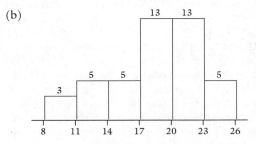

(c) Using the histogram from Part (b) and combining intervals where necessary, the following data are obtained.

Interval	f_o	f_e
[20, 26)	18	12.57
[17, 20)	13	$.2902 \cdot (44) = 12.77$
[14, 17)	5	$.2426 \cdot (44) = 10.67$
[8, 14)	8	6.71

H_0: the baseball careers are normally distributed with a mean of 18 years and a standard deviation of 4 years; H_A: the baseball careers are not normally distributed with a mean of 18 and a standard deviation of 4. Reject H_0 and accept H_A if $\chi^2 > 7.815$; otherwise, reserve judgment. $\chi^2 \approx 5.611$. Therefore, reserve judgment. There is not enough evidence to support the alternative hypothesis.

Exercises 11.4A

1. Each type of "success" is independent of any other type.
3. Yes, because the value of χ^2 in the first cell alone is so great.

Chapter 12

Exercises 12.1

1. Using the sign test as an example, information about the relative differences is unused, or wasted.
3. More likely.
5. (a) H_0: the gasoline additive makes no difference in mileage; H_A: the gasoline additive makes a difference in mileage. Reject H_0 and accept H_A if $Z(S) > 1.75$ or

if $Z(S) < -1.75$; otherwise, reserve judgment. $Z(S) \approx$ 1.732. Therefore, reserve judgment. There is not enough evidence to support the alternative hypothesis. (b) 8.36%.

7. H_0: there is no difference in the numbers of immigrants, by country, from 1991 to 1992; H_A: there is a difference in the numbers of immigrants, by country, from 1991 to 1992. Reject H_0 and accept H_A if $Z(S) > 1.44$ or if $Z(S) < -1.44$; otherwise, reserve judgment. $Z(S) \approx -1.898$. Therefore, reject H_0 and accept H_A. There is enough evidence to support the alternative hypothesis.

9. In a one-tail sign test, the statement that may be supported, H_A, specifies a relationship between the variables (such as "greater than"). In a two-tail sign test, H_A only claims that there *is* a difference between the variables; it does not specify the relationship.

11. H_0: the additive produces a gasoline mileage less than or equal to the mileage that would be obtained without the additive. H_A: the additive produces a gasoline mileage greater than the mileage that would be obtained without the additive. Reject H_0 and accept H_A if $Z(S) > 1.64$; otherwise, reserve judgment. $Z(S) \approx$ 1.732. Therefore, reject H_0 and accept H_A. There is enough evidence to support the alternative hypothesis.

13. (a) H_0: males have at least as many chronic health conditions as women; H_A: males have fewer chronic health conditions than women. (b) Let a plus sign be determined by the difference (No. of male conditions − No. of female conditions). Reject H_0 and accept H_A if $Z(S) < -1.28$; otherwise, reserve judgment. $Z(S) \approx -1.16$. Therefore, reserve judgment. There is not enough evidence to support the alternative hypothesis.

15. (a) H_0: women have a lesser or equal preference for these shows than men; H_A: women have a greater preference for these shows than men. Reject H_0 and accept H_A if $Z(S) > 1.64$; otherwise, reserve judgment. $Z(S) \approx .633$. Reserve judgment. There is not enough evidence to support the alternative hypothesis. (b) H_0: teens have a lesser or equal preference for these shows than children; H_A: teens have a greater preference for these shows than children. Reject H_0 and accept H_A if $Z(S) > 1.64$; otherwise, reserve judgment. $Z(S) = .633$. Reserve judgment. There is not enough evidence to support H_A. (c) For women and men, teens and children, $Z(S) = .633$. $Z(S)$ is greatest for women and teens. $Z(S)$ near 0 implies that the difference in preferences is due to chance, selection of, whereas a relatively large value of $Z(S)$ implies there is a real difference in preferences.

Exercises 12.2

1. If there are too many runs of H, then $Z(R_H)$ will be much greater than 0, and if there are too few runs of H, then $Z(R_H)$ will be much less than 0.

3. One.

5. H_0: the sequence of H's is random; H_A: the sequence of H's is not random. Reject H_0 and accept H_A if $Z(R_H) > 1.28$ or if $Z(R_H) < -1.28$; otherwise, reserve judgment. $Z(R_H) \approx .78$. Therefore, reserve judgment. There is not enough evidence to support the alternative hypothesis.

7. False. With either some very short runs or some very long runs, the actual number of runs might be very close (or equal) to the expected number of runs. The actual numbers of heads in the two cases would be quite different, but the result would be the same: reserve judgment.

9. H_0: the diseased trees occur at random (the sequence of D's is random); H_A: the diseased trees do not occur at random (the sequence of D's is clustered). Reject H_0 and accept H_A if $Z(R_D) > 1.64$ or if $Z(R_D) < -1.64$; otherwise, reserve judgment. $Z(R_D) \approx -1.17$. Therefore, reserve judgment. There is not evidence that the diseased trees are clustered.

11. (a) 27.5. (b) {A, B, B, B, A, A, A, A, A, B, A, A, A, B, A, A, B, B, B, B, A, A, A, B, B, B, B, B, A, B}. (c) H_0: the number of passengers carried above the median is random; H_A: the number of passengers carried above the median is not random. Reject H_0 and accept H_A if $Z(R_A) > 1.64$ or if $Z(R_A) < -1.64$; otherwise, reserve judgment. $Z(R_M) \approx -1.44$. Therefore, reserve judgment. There is not enough evidence to support the alternative hypothesis.

13. The sequence is {O, E, E, O, O, E, O, O, E, E, E, O, O, E, O, E, E, E, E, O, O, O, O, E, O, O, E, E, O, O, E, O, E, E, E, O, E, E, O, O, O, O, O, O, O, O, O, O, E, O}. H_0: E's occur randomly; H_A: E's do not occur randomly. Reject H_0 and accept H_A if $Z(R_E) > 1.64$ or if $Z(R_E) < -1.64$; otherwise, reserve judgment. $Z(R_E) \approx -.926$. Therefore, reserve judgment. There is not enough evidence to support the alternative hypothesis.

15. Proceed as in the solution to Exercise 13.

Exercises 12.3–12.4

1. At the center of the model of the sampling distribution.

3. (A) most strongly; (C) least strongly.

5. $r \approx .036$.

7. (a) $r \approx .874$. (b) H_0: the rankings of life expectancies

are not correlated; H_A: the rankings of life expectancies are correlated. Reject H_0 and accept H_A if $Z(r) > 1.64$ or if $Z(r) < -1.64$; otherwise, reserve judgment. $Z(r) \approx 2.89$. Therefore, reject H_0 and accept H_A. There is enough evidence to support the alternative hypothesis.

9. (a) It changes the sign of the computed value of r. (b) $r \approx -.839$. (c) If one set of rankings is reversed but the other is kept the same, only the sign of r will change.

11. (a) $r \approx -.468$. The fact that $r < 0$ implies that occupations that rank high for current smokers may rank low for former smokers. (b) H_0: the two rankings are not correlated; H_A: the two rankings are correlated. Reject H_0 and accept H_A if $Z(r) > 1.96$ or if $Z(r) < -1.96$; otherwise, reserve judgment. $Z(r) \approx -1.69$. Therefore, reserve judgment. There is not enough evidence to support the alternative hypothesis.

13.

Current Smokers

Occupation	Rank for Women	Rank for Men
Executive/administrative	9	13
Professional/specialty	14	14
Technical and related	12	10
Sales	7	11
Administrative support	8	8
Private households	10	4
Protective service	1	7
Service	5	5
Farmer, forestry, and fishing	11	9
Precision production	6	6
Machine operators	3	3
Transportation	4	2
Handlers/cleaners	2	1
Other	13	12

H_0: the two rankings are not correlated; H_A: the two rankings are correlated. Reject H_0 and accept H_A if $Z(r) > 1.96$ or if $Z(r) < -1.96$; otherwise, reserve judgment. $Z(r) \approx 2.68$. Therefore, reject H_0 and accept H_A. There is enough evidence to support the alternative hypothesis.

15. No, it only means that two separate rankings of the data are nearly the same. Similarly, if two rankings are negatively rank-correlated, it only means that the data are ranked two different ways when judged by two different persons, who could both have valid, though different, criteria.

Exercises 12.5

1. (a)

Course	Recent High-School Graduate	Older Adult
U.S. history	2	2
Biology	6.5	6
Geometry	5	1
Algebra	6.5	10
English	4	9
Music	9	6
Art	2	3.5
Physical education	10	8
Foreign language	8	6
Technical	2	3.5

(b) Possibly one's opinions about high-school course work change over time. Possibly one's opinions about high-school course work stay essentially the same over time. (c) H_0: the two rankings are not correlated; H_A: the two rankings are correlated. Reject H_0 and accept H_A if $Z(r) > 1.28$ or if $Z(r) < -1.28$; otherwise, reserve judgment. $Z(r) \approx 1.59$. Therefore, reject H_0 and accept H_A. There is enough evidence to support the alternative hypothesis.

3. Is there evidence that large-city and rural SAT test takers are distributed about the same among the income level?

5. (a) The number-of-runs test. (b) The sign test. (c) The number-of-runs test. (d) The Spearman rank-correlation test. (e) The number-of-runs test. (f) The sign test. (g) The sign test.

Chapter 13

Exercises 13.1–13.4

1. For each pairwise comparison, acceptance or rejection is a binomial random variable, with probability of success $p = .05$. The probability of no successes in 10 (trials) is $_{10}C_0 \cdot (.05)^0 \cdot (1 - .05)^{10}$. By the property of complementary events, the desired probability is $1 - _{10}C_0 \cdot (.05)^0 \cdot (1 - .05)^{10}$.

3. All of the populations are normally distributed, or very nearly so, and have approximately equal variances.

5. (a) The result of the ANOVA test was reservation of judgment, which implies that not enough statistical evidence was present to cause the rejection of the null

hypothesis, that the population means were the same. All three 95% confidence intervals overlapped (from approximately [10, 11]), suggesting the possibility that all three population means are the same.   (b) The level of significance.

7. Answers will vary.

9. The F-statistic is a function of variances, which are never negative, so the F-statistic must be nonnegative. This implies that the rejection region must be in the right tail.

11. The least is $F_{.05}$ with numerator d.f. = 10 and denominator d.f. = 20. It is followed by $F_{.05}$ with numerator d.f. = 20 and denominator d.f. = 10 and $F_{.025}$ with numerator d.f. = 10 and denominator d.f. = 20, which are equal, and finally $F_{.025}$ with numerator d.f. = 20 and denominator d.f. = 10.

13. H_0: the mean numbers of mosquito bites for each of the different brands of repellent are the same; H_A: for at least one brand of repellent, the mean number of mosquito bites received by children is different from the others. Reject H_0 and accept H_A if $F > 3.86$; otherwise, reserve judgment. $F = 3.12$. Therefore, reserve judgment. There is not enough evidence to support the alternative hypothesis.

15. H_0: the mean percentages of taxes paid for all three income levels are the same. H_A: for at least one of the three income levels the mean percentage of taxes paid is different from the others. Reject H_0 and accept H_A if $F > 3.68$. $F = .04$. Therefore, reserve judgment. There is not enough statistical evidence to support the alternative hypothesis.

17. H_0: the mean number of employees has stayed the same; H_A: for at least 1 year the mean number of employees has differed. Reject H_0 and accept H_A if $F > 4.42$; otherwise, reserve judgment. $F = 48.67$. Therefore, reject H_0 and accept H_A. There is enough evidence to support the alternative hypothesis.

19. H_0: the mean lifetimes of the three brands shown are the same; H_A: at least one brand has a mean lifetime different from the others. Reject H_0 and accept H_A if $F > 4.77$; otherwise, reserve judgment. $F = 44.62$. Therefore, reject H_0 and accept H_A. There is enough evidence to support the alternative hypothesis.

21. (a) It is difficult to tell, but it appears that at least one brand (brand C) may have a mean yield per square yard different from the others.   (b) H_0: the mean yields per square yard for the four brands shown are the same; H_A: at least one brand has a mean yield per square yard different from the others. Reject H_0 and accept H_A if $F > 4.07$; otherwise, reserve judgment. $F = .89$.

Therefore, reserve judgment. There is not enough evidence to support the alternative hypothesis.

23. H_0: the mean monthly rents in the neighborhoods shown are the same; H_A: at least one neighborhood has a mean monthly rent different from the others. Reject H_0 and accept H_A if $F > 4.56$; otherwise, reserve judgment. $F = 5.01$. Therefore, reject H_0 and accept H_A. There is enough evidence to support the alternative hypothesis.

25. The numerator would be affected more.

Chapter 14

Exercises 14.1–14.4

1. When values of one increase, values of the other decrease.

3. No. Both variables could be affected by a third, unknown variable.

5. The coefficient of determination gives a precise measure of the percentage of variation in one variable that can be explained by, or linked to, a linear relationship with the other variable.

7. At least 13.

9. Work will vary.

11. Answers will vary.

13. (a) The line of best fit may be sketched somewhat as follows:

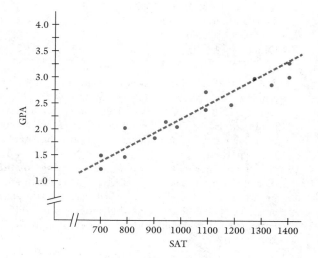

(b) Some good estimates might be:

SAT Score	GPA
700	1.25
700	1.50
800	1.45
800	2.10
900	1.85
960	2.40
1000	2.15
1100	2.45
1100	2.75
1200	2.50
1270	3.00
1330	2.80
1400	3.00
1400	3.35

(c) $p \approx .940$.   (d) .8836.

15. (a) Estimates close to the actual scores are as follows:

Year	Men	Women
1969	513	470
1970	509	485
1971	506	466
1972	505	461
1973	502	460
1974	501	459
1975	495	449
1976	497	446
1977	497	445
1978	494	444
1979	493	443
1980	491	443
1981	492	443
1982	493	443
1983	493	445
1984	495	449
1985	499	452
1986	501	451
1987	500	453
1988	498	455
1989	500	454
1990	499	455
1991	497	453
1992	499	456

(b) Estimates will vary, but good ones will be approximately .900.   (c) $p = .907$.

Exercises 14.5

1. The symbol p denotes the value of the Pearson coefficient of correlation for the sample data. The symbol **P** denotes the value of the Pearson coefficient of correlation for the two populations represented by the sample data.

3. The sample size n increases, and the sample becomes a truer picture of the population. Accordingly, if **P** $= 0$, then p should be closer to 0 for larger samples or more degrees of freedom.

5. From Appendix A-7 it can be seen that this value of p clearly falls in the rejection region for any of the given values of α.

7. (a)

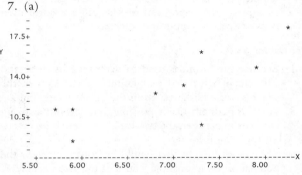

(b) Yes.   (c) H_0: **P** $= 0$; H_A: **P** $\neq 0$. Reject H_0 and accept H_A if $p > .582$ or if $p < -.582$. $p \approx .727$. Reject H_0 and accept H_A. There is enough evidence, at the 10% level of significance, to support the alternative hypothesis.   (d) One possibility is the quality of prenatal care.

9. (a)

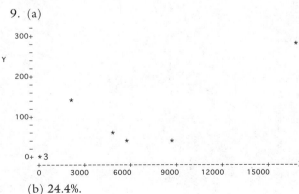

(b) 24.4%.

11. (a) It might imply a correlation among the factors that influence worker compensation throughout the world. (b) H_0: $\mathbf{P} = 0$; H_A: $\mathbf{P} \neq 0$. Reject H_0 and accept H_A if $p > .576$ or if $p < -.576$. $p \approx .957$. Reject H_0 and accept H_A. There is enough evidence, at the 5% level of significance, to support the alternative hypothesis.

13. H_0: $\mathbf{P} = 0$; H_A: $\mathbf{P} \neq 0$. Reject H_0 and accept H_A if $p > .658$ or if $p < -.658$. $p \approx .599$. Reserve judgment. There is not enough evidence to support the alternative hypothesis.

15. H_0: $\mathbf{P} = 0$; H_A: $\mathbf{P} \neq 0$. Reject H_0 and accept H_A if $p > .592$ or if $p < -.592$. $p \approx .929$ for pass attempts and touchdowns. Reject H_0. At the 2% level of significance there is enough evidence to support the alternative hypothesis of a significant correlation between pass attempts and touchdowns. Also, there is enough evidence to support the hypothesis of a significant correlation between pass attempts and interceptions.

Exercises 14.6

1. Because the formulas used to determine its slope and Y-intercept are based on minimizing the sum of the squares of the distances from the points to the line.

3. Yes, it is mathematically possible, but without a strong correlation between the two variables there is no practical application for it.

5. The line of best fit that would be constructed for the population of all possible values of the two variables.

7. (a) GPA $= .00242 \cdot$ SAT $- .212$. (b) GPA ≈ 2.33. (c) SAT ≈ 1300.

9. Average SAT scores are generally decreasing over time.

11. 3174.

13. Approximately 23.119 mpg.

15.

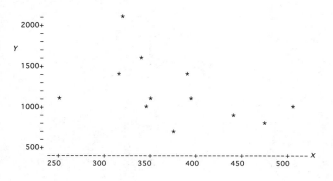

Unusual observation: (322, 2074.5).

Exercises 14.7

1. Work will vary.

3. [26.70, 30.70].

5. $\mu_{Y,25}$: [67.61, 70.93]. $R_{Y,25}$: [59.50, 79.04].

Cumulative Review, Chapters 11–14

1. Hypothesis testing for the independence of two qualitative variables, a model for constructing a confidence interval for σ, and testing assumptions about a population's distribution.

3. (a) Age at time of survey and number of years from divorce to remarriage. (b) H_0: age and number of years from divorce to remarriage are independent; H_A: age and number of years from divorce to remarriage are not independent. Reject H_0 and accept H_A if $\chi^2 > 6.251$; otherwise, reserve judgment. $\chi^2 = .058$. Therefore, reserve judgment. (c) There is not enough evidence to support the alternative hypothesis.

5. (a) [.515, 1.659] (b) [2.978, 6.296].

7. H_0: these selling prices come from a normal population with a mean of 14.00 and a standard deviation of 4.00; H_A: these selling prices do not come from a normal population with a mean of 14.00 and a standard deviation of 4.00. Reject H_0 and accept H_A if $\chi^2 > 7.779$; otherwise, reserve judgment. $\chi^2 = 12.9611$. (Adjust the first two and last two classes to ensure that no more than 20% of the intervals have an expected frequency less than 5.) Reject H_0 and accept H_A. There is enough evidence to support the alternative hypothesis.

9. (a) H_0: there is no difference in SAT math averages for these ethnic groups in large cities and rural areas; H_A: there is a difference in SAT math averages for these ethnic groups in large cities and rural areas. Reject H_0 and accept H_A if $Z(S) > 1.75$ or if $Z(S) < -1.75$; otherwise, reserve judgment. $Z(S) = 1$. Therefore, reserve judgment. There is not enough evidence to support the alternative hypothesis. (b) H_0: SAT math averages for these ethnic groups in large cities are less than or equal to what they would be in rural areas; H_A: SAT math averages for these ethnic groups are greater in large cities than in rural areas. Reject H_0 and accept H_A if $Z(S) > 1.64$; otherwise, reserve judgment. $Z(S) = 1$. Therefore, reserve judgment. There is not enough evidence to support the alternative hypothesis.

11. H_0: the rankings are not correlated; H_A: the rankings are correlated. Reject H_0 and accept H_A if $Z(r) > 1.96$ or if $Z(r) < -1.96$; otherwise, reserve judgment. $Z(r) \approx 2.24$. Therefore, reject H_0 and accept H_A. There is enough evidence to support the alternative hypothesis.

13. (a) The ratings are normally distributed for the four committees and have approximately equal standard deviations. (b) H_0: the mean ratings of the four committees are the same; H_A: at least one committee has a mean rating different than the others. Reject H_0 and accept H_A if $F > 5.29$; otherwise, reserve judgment. $F = 0.36$. Therefore, reserve judgment. (c) There is not enough evidence to support the alternative hypothesis.

15. (a) The value of p is the Pearson coefficient of correlation for the sample data. The value of **P** is the Pearson coefficient of correlation for the two populations being sampled. (b) H_0: **P** $= 0$; H_A: **P** $\neq 0$. Reject H_0 and accept H_A if $p > .497$ or if $p < -.497$. $p \approx .646$. Therefore, reject H_0 and accept H. There is enough evidence to support the alternative hypothesis.

17. $[33.42, 42.22]$.

Answers to Exercises from Appendix A-1

1. True.
2. False. (The second part of this statement is false, so the entire statement is false.)
3. False.
4. True.
5. True.
6. True. (By DeMorgan's Laws, these two sets are equal.)
7. True.
8. True.
9. False.
10. False. (The notation is incorrect, which makes the statement false.)
11. True.
12. True.
13. True.
14. True.
15. There are 2^n subsets.
16. Illustrating the first law, $A^c \cup B^c \cup D^c$ = {Lincoln, Hoover, FDR, Truman, Ike, JFK, LBJ, Nixon} ∪ {Lincoln, Hoover, FDR, Truman, Ike, JFK, LBJ, Nixon, Ford, Carter, Clinton} ∪ {Hoover, Truman, Ike, Nixon, Ford, Carter, Reagan, Bush, Clinton} = {Lincoln, Hoover, FDR, Truman, Ike, JFK, LBJ, Nixon, Ford, Carter, Reagan, Bush, Clinton} = U. $A \cap B \cap D$ =

$\emptyset$, so $(A \cap B \cap D)^c = \emptyset^c = U$. Illustrating the second law, $A^c \cap B^c \cap D^c$ = {Lincoln, Hoover, FDR, Truman, Ike, JFK, LBJ, Nixon} ∩ {Lincoln, Hoover, FDR, Truman, Ike, JFK, LBJ, Nixon, Ford, Carter, Clinton} ∩ {Hoover, Truman, Ike, Nixon, Ford, Carter, Reagan, Bush, Clinton} = {Hoover, Truman, Ike, Nixon}. This is also the same set as $(A \cup B \cup D)^c$ = (Lincoln, FDR, JFK, LBJ, Clinton, Bush, Reagan, Carter, Ford)^c = {Hoover, Truman, Ike, Nixon}.
17. $A \cap B$
18. $A \cap B \cup C$
19. $(A \cup B)^c$
20. $(A \cup B) \cap (A \cap B)^c$
21. $B \cap C^c$
22. {2, 4, 6, 8, 10, 12, 20}
23. $\emptyset$
24. {5, 8, 10, 12, 15, 20}
25. $(C \cap D)^c$ = {All integers except 10 and 20}, $C^c \cap D^c$ = {All integers except 5, 8, 10, 12, 15, 20}
26. {5, 15, 20}
27. {0, −10, −20, −30, −40, . . .}
28. {10}
29. {All integers except 10 and 20{
30. {All integers greater than 600}

General Index

The Index of Applications appears on page xiii.